Introduction to Engineering Materials

Introduction to Engineering Materials

Vernon John
BSc, MSc, CEng, MIMMM

Fourth Edition

First edition 1972
Reprinted 7 times
Second edition 1983
Reprinted 3 times
Third edition 1992
Reprinted 7 times
Fourth edition 2003
Published by
PALGRAVE MACMILLAN
Houndmills, Basingstoke, Hampshire RG21 6XS and
175 Fifth Avenue, New York, N.Y. 10010
Companies and representatives throughout the world

PALGRAVE MACMILLAN is the global academic imprint of the Palgrave Macmillan division of St. Martin's Press, LLC and of Palgrave Macmillan Ltd. Macmillan® is a registered trademark in the United States, United Kingdom and other countries. Palgrave is a registered trademark in the European Union and other countries.

ISBN 0–333–94917–X

This book is printed on paper suitable for recycling and made from fully managed and sustained forest sources.

A catalogue record for this book is available from the British Library.

10   9   8   7   6   5   4   3   2   1
12   11   10   09   08   07   06   05   04   03

Printed and bound in Great Britain by
Ashford Colour Press Ltd, Gosport

Contents

PART I   INTRODUCTION   1

PART II   MATERIALS SCIENCE   15

PART III   THE MATERIALS OF ENGINEERING   231

PART IV   FORMING AND FABRICATION OF
MATERIALS   361

Preface to the first edition

Not many years ago it was possible to obtain a first degree or other qualification in engineering with no knowledge whatsoever of metallurgy or the other materials of engineering. Today it is fully accepted that a sound knowledge of the science and technology of materials is very necessary to the engineer if he is to be able efficiently to translate a design into functional 'hardware'. The selection of materials and manufacturing route is an integral part of design procedure. Because of this, materials science and technology now feature prominently in the educational programme for all engineering technologists and technicians. Engineering, including materials science, is also beginning to appear in the curricula of some schools.

This book was conceived against the general background of the changing patterns in engineering education, and the aim was to produce a text which dealt with the basic principles of materials science and technology in a simple, yet meaningful manner.

It is my belief that no aspect of engineering should be studied in a vacuum, but that academic studies be related to our real cost-conscious world. It is for this reason that I have included some materials' costs and a short note on the selection of materials. Although actual costs will vary from year to year, this section should be of value as it indicates the principle that any work or processing performed on a material is reflected in an increase in the cost of the material.

I wish to acknowledge the help and advice received from various individuals and organisations, and in particular to thank Professor Bob Fergusson for providing the initial stimulus for this work. I am indebted to some of my colleagues at PCL, especially Mr C. J. Beesley, Mr G. E. Drabble, and Mr M. D. Munro Mackenzie, for their assistance in reading the manuscript and suggesting improvements. I also wish to acknowledge the assistance, in the way of photographs and information, which I received from the following firms and organisations: British Aluminium Company Ltd, British Metals Sinterings Association, Bound Brook Ltd, Copper Development Association, I.C.I. Ltd (Plastics Division), International Nickel Ltd, Raleigh Industries Ltd, Sintered Products Ltd, and the Zinc Development Association. Several colleges and polytechnics provided me with question papers, and I wish to record my appreciation of their help. Finally, I am very grateful to my wife for not merely putting up with me while I was struggling with the manuscript, but for helping considerably by sustaining me with refreshment and also typing some of the manuscript.

Vernon John

1972

Preface to the second edition

Ten years have elapsed since this book was first published and after such a period of time an updating is necessary for any technical text. In the preparation of this edition I have attempted to satisfy two objectives; firstly to keep a basic and straightforward approach so that the text can be readily understood by students beginning their studies, and secondly to broaden the scope of the volume and so increase its suitability for degree and diploma students of both mechanical and production engineering in the later years of their courses.

When preparing a new edition of a book an author is torn between the desire to tear everything up and start again with a ream of blank paper or merely to make small cosmetic changes. This book was generally well received when it was first published and students appeared to like the general style and content. I have, therefore, retained much of the original text and confined myself to meeting most of the criticisms (fortunately, these were not too numerous) levelled at the first edition and also to making changes that reflect the development in my own teaching at PCL.

Some of the major differences between this text and the first edition are: (1) the inclusion of more worked examples within the text, (2) an increase in the sections dealing with dislocations and plastic deformation within metals, (3) a substantial enlargement of the chapter on material-forming processes, (4) a major increase in the chapters devoted to plastics materials, (5) a more detailed treatment of oxidation and corrosion and (6) the development of a small section on welding, formerly placed as an Appendix, into a full chapter within the book. In addition to these changes, an introduction to the principles of fracture mechanics has been included as an Appendix to the main text. It is my hope that these changes will find favour with both students and teachers.

I wish to thank my colleagues in the Materials section at PCL for the helpful comments freely given during the preparation of this revision and to record my indebtedness to my wife for her great help and for the typing of the draft.

Vernon John   *1983*

Preface to the third edition

There is a process of continuous development for curricula for courses in engineering. One significant development of the last decade is that the subject of engineering materials has moved into a centre-stage position, especially as a subject field within mechanical and manufacturing engineering courses. There has also been debate on the subject of a 'core curriculum'. During the preparation of this revised and enlarged edition, I have attempted to cater for current curriculum developments and changes in emphasis within syllabi. One possible failing of the last edition was, perhaps, an overemphasis on metals. I have attempted to redress this with increased coverage of non-metallic materials. My aim has been to present that body of core knowledge in materials science and technology which, in my opinion, should be a necessary part of the educational programme for any engineer.

The earlier editions of this book were well received and feedback from students indicated that the style of writing and presentation was clear and understandable. In consequence, I have retained much of the original text and illustrations in this new edition, even though the general structure of the book has been changed significantly and the number of chapters increased. The principal changes from the second edition include an introductory chapter outlining the range of materials, their basic characteristics, costs and availability to set the scene. The materials science section has been enlarged with elasticity and plasticity separated into two chapters and with new chapters dealing with polymer formation, visco-elasticity, toughness and fracture, phase transformations, and optical and thermal properties. Worked examples have been included in the text where appropriate. There has been a major increase in the textual matter covering polymers, ceramics and glasses. The formation and fabrication of materials has been reorganised with separate chapters for each class of material and a new chapter for material removal processes. The importance of non-destructive testing has been recognised by including a new chapter devoted to this subject. Material which appeared as appendices in the last edition has now been incorporated into the main text. It is hoped that these changes will find general acceptance by both teachers and students.

I hope that those engineering students who read and use this book will be stimulated and encouraged in their studies. Also, that they will develop an understanding and a feel for materials so that, as practising engineers, they will be able to select and use materials effectively and efficiently.

Questions requiring numerical answers are included at the ends of chapters. For descriptive answer questions and further numerical problems, refer to *Work Out Engineering Materials*, also by Vernon John and published by Macmillan.

Vernon John

1991

Preface to the fourth edition

Little did I think when I wrote the first preface in 1972 that thirty years later, in the 21st century, I would be writing another to introduce a new edition. The aim of this edition is still to present that body of core knowledge in materials science and technology that I believe to be a necessary part of the educational programme for engineers, particularly that for mechanical and aeronautical engineers.

Engineering education in the United Kingdom has changed considerably in the last decade and the educational requirement for joining the register to progress to Chartered Engineer status is now either the four-year MEng degree or the three-year BEng degree plus a one-year additional approved course of study. In this new edition I have aimed to satisfy the requirements in materials education for BEng courses in mechanical engineering and for the first three-years of MEng courses. In recent years, much of the basic science dealing with atomic structure, bonding and crystallography has been deleted from the engineering curricula of many universities. I have retained these topics in this edition as I believe them to be necessary for a full understanding of materials technology. I hope that these sections will be read and appreciated by students.

Earlier editions of this book were well received with favourable comments about the clarity of style and presentation. Consequently, much of the text of the third edition has been retained. The main changes over earlier editions have been the introduction of reasons for study and learning objectives at the commencement of each chapter and the inclusion of summaries at the end of chapters, In addition, many new worked examples have been included at appropriate points in the text and the number of questions placed at the end of chapters has been increased. The three chapters on polymer materials in the third edition have been consolidated into one. There has been an expansion of the chapters on composite materials, fatigue and materials selection and a brief survey of surface engineering has been added to Chapter 20. It is hoped that these changes will be of benefit to both teachers and students.

I wish to formally acknowledge and thank the various individuals and companies who have helped me by providing both textual material and illustrations. Among these I wish to thank especially Margaret Bartlett and Peter Head of Maunsell Structural Plastics; Mark Basely of Rolls-Royce plc, Derby; Mark Challoner, Sue Cutts and Karen Mitchell of Airbus UK, Broughton; Julia Dunn of Timet UK; Nicholas Hayward and Peter Webster of The Copper Development Association; Jonathan Lee of BAE Systems, Warton; Alison Murray of Krautkramer UK Ltd; Peter Smith of Schunk UK Ltd; Jamie Worthy and Graham Wright of Ticona UK Ltd; David Wiggins and Jonathan Wroe of The European Powder Metallurgy Association; Roger Elton; and Bill Parrott. I also wish to thank my wife Gillian for her assistance and her general forbearance during the preparation of this manuscript.

I hope that all who read and use this book will find stimulation and that it will be of considerable help to them in their studies.

Vernon John   *2003*

Symbols used in text

Quantity	Symbol	
Atomic mass number (atomic weight)	M	
Atomic number	Z	
Avogadro's number	N_0	$N_0 = 6.023 \times 10^{23}$ molecules per mole
Bulk modulus of elasticity	K	
Density	ρ	
Direct strain	ε	
Direct stress	σ	
Force	F	
Fracture toughness	K_c	
Glass transition temperature	T_g	
Melting temperature	T_m	
Modulus of rigidity	G	
Poisson's ratio	ν (Greek nu)	
Shear strain	γ	
Shear stress	τ	
Temperature	T	
Time	t	
Universal constant	R_0	$R_0 = 8.314\,$kJ/kmol K
Viscosity	η	
Young's modulus of elasticity	E	

Units

The units used throughout this book conform to the SI system. The principal units that are quoted in the text are given below. Preferred SI units are printed in bold type.

Quantity	Unit	Symbol
mass	**kilogramme**	**kg** (1 kg = 2.205 lb)
	gramme	g
	tonne	t (Mg) (1 t = 1000 kg = 0.984 ton)
length	**metre**	**m** (1 m = 39.37 in)
	millimetre	mm
time	**second**	**s**
	minute	min
	hour	h
temperature	degree Kelvin	K
	degree Celsius	°C
amount of substance	mole	mol
	kilomole	**kmol**
area	**square metre**	**m^2**
	square millimetre	mm^2
volume	**cubic metre**	**m^3** (1 m^3 = 35.315 ft^3)
	cubic millimetre	mm^3
density	**kilogramme per cubic metre**	**kg/m^3** (1 kg/m^3 = 10^{-3} g/cm^3 = 0.062 lb/ft^3)
force	**newton**	**N** (1 N = 0.225 lbf)
	kilonewton	kN
	meganewton	MN
stress (pressure)	**newton per square metre**	**N/m^2** (1 N/m^2 = 0.000145 lbf/in^2)
	meganewton per square metre	MN/m^2 (1 MN/m^2 = 0.0648 tonf/in^2)
	giganewton per square metre	GN/m^2
	pascal	**Pa** (1 Pa = 1 N/m^2)
	bar	bar or b (1 bar = 10^5 N/m^2)

Quantity	Unit	Symbol
energy	joule	J (N m)
	electron volt	eV (1 eV $=1.602 \times 10^{-19}$ J)
electric current	ampere	A
voltage	volt	V
quantity of electricity	coulomb	C (A s)
electrical resistance	ohm	Ω (V/A)
electrical resistivity	ohm metre	Ω m
magnetic flux	weber	Wb (V s)
magnetic flux density	tesla	T (Wb/m^2) (1 T $= 10^4$ gauss)
frequency	hertz	Hz (s^{-1})

PART I

INTRODUCTION

CHAPTER 1

The materials of engineering

Why should an engineer study materials?

Every single thing that has ever been made has been constructed from some material or another. It is necessary, therefore, for engineers to have a sound understanding of materials so that they may select the most suitable material to fit a particular design requirement. To this end it is necessary not only to know the range of available materials and the major properties of each material, but also how the properties of a material are dependent on structure and how many properties can be modified by a variety of treatments and processes. Atomic structure, the type of bonding between atoms and the manner in which atoms and molecules are arranged in bulk materials all play a part in determining the final properties of the material. Also, the processing methods used to make the required shape and the type of heat treatment given will have a major influence in determining the properties. Knowledge of these factors will enable the engineer to utilise materials both effectively and efficiently.

In addition to understanding the nature and properties of materials it is very necessary that today's engineers are aware of the environmental and social factors associated with the production and utilisation of materials. The earth's mineral resources are non-renewable and thought must be given in the product design stage to the eventual decommissioning and recycling of the product.

LEARNING OBJECTIVES

1. Appreciate that there are complex interrelationships between a material, manufacturing process and design.

2. Know the major classifications of materials and the main characteristics of each class.

3. Have some knowledge of the basic costs of some materials and how cost and energy content are affected by processing.

1.1 ┃ Introduction

The development of civilisation has been governed by man's ability to invent or adapt and use materials. In prehistoric times our forebears could use only those materials which were available naturally – stones, mud and clay, fallen tree trunks, animal skins and animal and vegetable fibres. The development of simple tools during the Stone Ages permitted man to dress and shape timber and stone into the designs required. Early man also discovered the art of moulding clay for the production of bowls, beakers and other utilitarian articles. One early constructional material was brick, moulded from mud and dried in the sun. Soon it was realised that a brick with better properties could be obtained if straw was mixed with the mud or clay before shaping and drying. This same principle is used today in the manufacture of fibre reinforced plastics. Around 4500 BC, in the Middle East, the metallic alloy bronze was first made and this became a major material for the manufacture of tools, weapons and other artefacts for the next few millennia. At about 1100 BC man discovered and first began to use iron. Early civilised peoples discovered and used many metals including copper, gold, lead, silver and tin. They also discovered that the properties of a metal could be altered by alloying with other metals and, through empirical means, a number of useful alloys were developed. The first steels were produced about 1000 years ago by heating bars of iron in charcoal. This allowed carbon to diffuse into the iron to give greater hardness and strength.

The pace of development was slow and, in general, new designs and new forms of construction could not proceed in advance of the development of materials. As an example of this, let us consider the progress of bridge design. The major constructional materials of earlier centuries – stone and brick – are strong in compression but weak in tension. Some early stone bridges, known in Britain as 'clapper' bridges, merely consisted of large stone slabs laid as beams across the gaps between stone piers (Figure 1.1(a)). The maximum span between piers is up to about 4 m. The bridging of larger gaps by means of beam bridges had to wait until the late 19th century, when steels of consistent quality became available in quantity, and the 20th century, with the development of reinforced and prestressed concrete – materials with a good strength in tension. Spans of up to 300 m are possible in reinforced concrete beam bridges.

Meanwhile, in Roman times, the masonry arch was developed. The individual stones or bricks in a semi-circular arch are stressed in compression (see Figure 1.1(b)) and this design enables much larger gaps to be spanned. There is, however, a limit to the span possible with a masonry arch and multi-arch bridges were necessary for the crossing of wide gaps. Most major bridges from Roman times up to the end of the 18th century were of the masonry arch type. The largest masonry arch bridge in existence is the railway bridge across the Thames at Maidenhead (Figure 1.2). This was designed and built by

Figure 1.1 *(a) Old 'clapper' bridge; (b) masonry arch showing force distribution.*

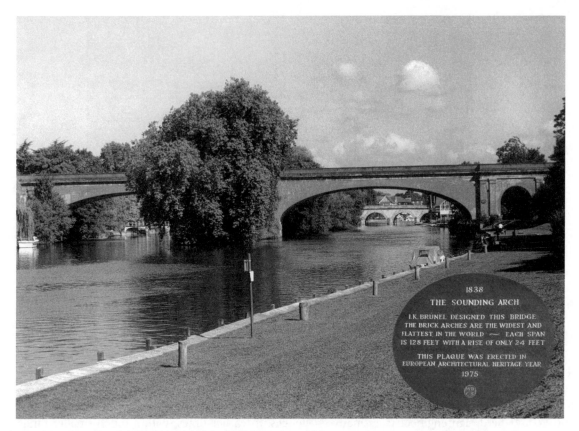

Figure 1.2 *Brunel's railway bridge of 1838 with brick arches of 39 m span.*

I. K. Brunel in 1838 and has two flat elliptical brick arches, each of 39 m span. It is a tribute to Brunel's design that when, at the beginning of the 20th century, the number of tracks was increased from two to four, the decision was taken to widen the existing bridge, adhering rigidly to Brunel's original design, rather than build a new bridge.

Towards the end of the 18th century wrought iron, as strong in tension as in compression, became available as a constructional material. This enabled engineers to design and construct suspension bridges using suspension chains fabricated from wrought iron links. The first large span bridge of this type (176 m) was the Menai bridge linking the island of Anglesey with the mainland (Figure 1.3). This was the first permanent link across the Menai Straits. It was constructed by Thomas Telford and, when opened in 1826, permitted much faster transit from London to Holyhead, the main ferry port for Ireland. This bridge is still in use today.

The cables for modern suspension bridges are fabricated from high tensile steel wire and great spans are possible. The Humber bridge, for example, has a span of 1410 m between the main towers.

Enormous developments have occurred in the materials field during the last 100 years or so. Metals which were merely curiosities at one time, such as aluminium, magnesium and nickel, are now commonplace; many new alloys have been devised; a complete industry – the plastics industry – has been created; many modern industrial ceramics have come into use; and a wide range of composite materials has been developed. While early materials development was based on empiricism there has been an intensive study of the science of materials during the 20th century and present-day engineering materials are based on sound scientific theory and practice. Running parallel to the invention of new and improved materials there have been equally important developments in materials processing, including vacuum melting and casting, new forming techniques and new joining technologies.

Figure 1.3 *Thomas Telford's suspension bridge of 1826 across the Menai Straits.*

1.2 | Design, materials and manufacture

Every manufactured item is made from some material or other. Equally, every item has to be designed and, today, the design engineer faces many challenges. In earlier times, with a much smaller range of materials available, engineers often produced their designs and manufactured the product by a process of trial and error, in many cases using far more material than was necessary. Today, with diminishing reserves of many economically important minerals and the need to be increasingly energy efficient, it is a major requirement that new designs utilise materials in the most effective and efficient manner possible and with the provision that the materials used can be recycled when the product has reached the end of its useful life. Also, the design engineer must consider the environmental and sociological impact of all aspects of the design from initial concept, through realisation and use, to scrapping and reclamation.

There are complex interrelationships between design, material and manufacture (Figure 1.4). The shape desired will influence the manufacturing processes which can be used. The properties required in the product will determine the range of possible materials and the choice of material will help determine the manufacturing process. The choice of manufacturing process will affect the properties, for example, a casting will possess different properties from a forging made of the same basic material, and so on. Many factors, therefore have to be taken into consideration when choosing possible materials to fit a design and manufacturing requirement, including:

- Does the material possess the necessary properties?
- Can the material be formed to the desired shape?
- Will the properties of the material alter with time during service?
- Will the material be adversely affected by environmental conditions and resist corrosion and other forms of attack?
- Will the material give sufficient reliability and quality?
- Will the material be acceptable on aesthetic grounds?
- Can the product be made at an acceptable cost?

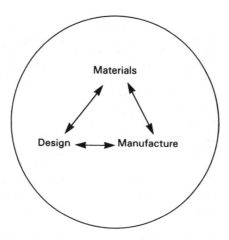

Figure 1.4 *Interrelationships between design, materials and manufacturing.*

1.3 | Properties of engineering materials

There are many thousands of different engineering materials available today but they can be placed into one or other of the following categories:

METALS, POLYMERS, CERAMICS & INORGANIC GLASSES and COMPOSITES.

Composites, as the term implies, are composed of more than one type of material, for example, glass fibres in a polyester matrix. Many polymers, such as polyesters, are of low density but are also of relatively low strength – the use of glass, carbon or other fibres as reinforcement gives increased strength and stiffness without adding an excessive weight penalty. Similarly, the low-friction characteristics of a polymer such as PTFE can be combined with the load-bearing qualities of a metal in metal particle/PTFE composites which have been developed as bearing materials. All materials exhibit many different characteristics and properties which can be classified as physical, mechanical, electrical and so on, and these are itemised in Table 1.1.

Each of the main categories of materials has certain characteristics, for example, metals are electrical conductors whereas ceramics and polymers possess extremely high electrical resistivities and many are

Table 1.1 *Material properties and qualities.*

Properties	Qualities
Physical properties	Density, melting point, hardness, elastic moduli, damping capacity.
Mechanical properties	Yield, tensile, compressive and torsional strengths, ductility, fatigue strength, creep strength, fracture toughness.
Manufacturing properties	Ability to be shaped by moulding, casting, plastic deformation, powder processing, machining. Ability to be joined by adhesives, welding and other processes.
Chemical properties	Resistance to oxidation, corrosion, solvents and environmental factors.
Other non-mechanical properties	Electrical, magnetic, optical and thermal properties.
Economic properties	Raw material and processing costs. Availability.
Aesthetic properties	Appearance, texture and ability to accept special finishes

Table 1.2 *Comparison of properties of metals, ceramics and polymers.*

Property	Metals	Ceramics	Polymers
Density (kg/m^3 × 10^{-3})	0.5–22 (average 8)	2–17 (average 5)	1–3
Melting points	Low to high, Sn 232°C, W 3400°C	High, up to 4000°C	Low
Hardness	Medium	High	Low
Machineability	Good	Poor	Low
Tensile strength (MPa)	Up to 2500	Up to 400	Up to 120
Compressive strength (MPa)	Up to 2500	Up to 5000	Up to 350
Young's modulus (E) (GPa)	40–400	50–450	0.001–3.5
Plasticity	Can be shaped by plastic deformation	Not plastically deformable	Shaped by plastic moulding
High-temperature creep resistance	Poor	Excellent	Nil
Thermal expansion	Medium to high	Low to medium	Very high
Thermal conductivity	Medium	Medium but often decreases rapidly with temperature	Very low
Thermal shock resistance	Good	Generally poor	—
Electrical properties	Conductors	Insulators	Insulators
Chemical resistance	Low to medium	Excellent	Generally good
Oxidation resistance at high temperatures	Poor, except for rare metals	Oxides excellent, SiC and Si$_3$N$_4$ good	—

used as insulating materials. A comparison of the properties metals, ceramics and polymers is given in Table 1.2.

Very many properties or qualities of materials have to be considered when choosing a material to meet a design requirement. These include a wide range of physical, chemical and mechanical properties, forming or manufacturing characteristics, availability, raw material and manufacturing cost data and, in addition, subjective assessments of aesthetic qualities such as appearance and texture. The ranges of some important properties, E, tensile yield strength, tensile strength, fracture toughness and density, for some common groups of materials are shown in Table 1.3.

Metals, when subject to a stress, strain in an elastic manner up to a certain level and then, with increasing stress, deform plastically. The *yield stress*, which indicates the change from elastic to plastic deformation, is an important parameter for this class of material. Thermoplastics possess the property of viscoelasticity and many will strain slowly with time at comparatively low levels of stress. Many thermosetting polymers, together with ceramics and glasses, show no plasticity and deform when stressed in an elastic manner until the level of stress is high enough to cause fracture. The concept of yield stress is not applicable to these groups of materials.

Let us now look at two examples. The first is an item of sports equipment familiar to readers and illustrates how progressive changes of material and design lead to an improved product. The second example, electrical transmission lines, shows that the material with the highest electrical conductivity cannot give an acceptable engineering solution.

Table 1.3 *Typical properties (at 25°C) of some groups of materials.*

Material	E (GPa)	Yield strength (MPa)	Tensile strength (MPa)	Fracture toughness (MPa m$^{0.5}$)	Density (kg m^{-3} × 10^{-3})
Steels	200–220	200–1800	350–2300	80–170	7.8–7.9
Cast irons	150–180	100–500	300–1000	6–20	7.2–7.6
Aluminium alloys	70	25–500	70–600	5–70	2.7–2.8
Copper alloys	90–130	70–1000	720–1400	30–120	8.4–8.9
Magnesium alloys	40–50	30–250	60–300		1.7–1.8
Nickel alloys	180–220	60–1200	200–1400	>100	7.9–8.9
Titanium alloys	100–120	180–1400	350–1500	50–100	4.4–4.5
Zinc alloys	70–90	50–300	150–350		6.7–7.1
Polyethylene (LDPE)	0.12–0.25		1–16	1–2	0.91–0.94
Polyethylene (HDPE)	0.45–1.4		20–38	2–5	0.95–0.97
Polypropylene (PP)	0.5–1.9		20–40	3.5	0.90–0.91
PTFE	0.35–0.6		17–28		2.1–2.25
Polystyrene (PS)	2.8–3.5		35–85	2	1.0–1.1
Rigid PVC	2.4–4.0		24–60	2.4	1.4–1.5
Acrylic (PMMA)	2.7–3.5		50–80	1.6	1.2
Nylon (PA)	2.0–3.5		60–100	3–5	1.05–1.15
PF resins	5–8		35–55		1.25
Polyester resins	1.3–4.5		45–85	0.5	1.1–1.4
Epoxy resins	2.1–5.5		40–85	0.3–0.5	1.2–1.4
GFRP	10–45		100–300	20–60	1.55–2.0
CFRP	70–200		70–650	30–45	1.40–1.75
Soda glass	74		50*	0.7	2.5
Alumina	380		300–400*	3–5	3.9
Silicon carbide	410		200–500*		3.2
Silicon nitride	310		300–850*	4	3.2
Concrete	30–50		7*	0.2	2.4–2.5

* Modulus of rupture value.

EXAMPLE 1.1

Materials selection for a tennis racquet
    Consider suitable materials from which to make a tennis racquet frame.

SOLUTION

A major requirement of the material for a tennis racquet is that it combines high strength and stiffness with low weight. Tennis racquet frames can and have been made from a variety of materials including laminated wood, steel, aluminium and fibre reinforced composites. Until the early 1970s almost all racquets were made from laminated wood. One of the drawbacks with wood is that it can absorb water which can lead to variations in performance and also can cause warping of the frame. In the 1970s frames made from steel and aluminium came into use. These materials, with high strength/weight ratios, permitted strong, rigid, yet light weight frames to be made. One disadvantage of these metals is that they possess low damping capacities and, consequently, the amount of vibration transmitted to the player's hand and arm is increased. By the early 1980s composite construction using continuous glass or carbon fibres in a polyester or epoxy resin matrix largely superseded the use of aluminium. These composite materials give a lightweight frame with high strength and stiffness and a lower level of vibration because they have higher damping capacities than metals.

EXAMPLE 1.2

Materials selection for overhead electrical transmission wires
 Consider suitable materials from which to make overhead electrical power transmission lines.

SOLUTION

The material must have a high electrical conductivity. The four best electrical conductors, in descending order, are silver, copper, gold and aluminium (see Chapter 13). Also, for the highest conductivities, these metals should be of high purity as the presence of other elements either as impurities or as alloying elements will cause a reduction in conductivity. Silver and gold are much too expensive for use in this application and copper is considerably more expensive than aluminium, especially when considered on a volume basis (see Table 1.4). The selected material, then, is pure aluminium. However, pure aluminium has a very low tensile strength. Consequently overhead transmission cables are made of aluminium wires braided around a steel wire core to give strength.

An exception to the above is the case of overhead power wires for electric railways and tramways. Here the wire needs to be hard to resist the abrasive effect of the rubbing contact with the pantograph arms of the vehicles. The solution is to use wire made from an alloy of copper containing 1 per cent of cadmium. Cadmium greatly increases the hardness of the copper with little adverse effect on conductivity.

1.4 | Cost and availability

Cost and availability are very important factors which affect the selection and use of materials and in many cases the purchase cost of materials accounts for about one-half of the total works cost of the finished product. From this it follows that the use of a cheaper raw material should have a significant effect on the final product cost but this is not always the case. In some instances the choice of a more expensive material may permit the use of simpler and lower cost processing techniques than those needed in conjunction with a cheaper material.

It is usual to see the cost of raw materials quoted per unit mass, for example £100 per tonne. This may give a misleading picture as often it is the volume of material used which is important rather than its mass. The relative position of a material in a league table of costs may change when the criterion is altered from £ ($) per tonne to £ ($) per unit volume. This effect can be seen in Table 1.4.

The costs of some basic engineering materials, both by weight and volume, are given in Table 1.4. However, it should be noted that the data in Table 1.4 only indicate an approximate comparison between raw materials. The figures are based on the European bulk commodity prices ruling in October/November 2002. The prices are quoted in pounds sterling (£), euros (€) and US dollars ($). In early November 2002 the sterling/dollar exchange rate was £1 = 1.55$ and there was almost parity between the euro and the US dollar. The costs of some materials are relatively unchanged over fairly long periods of time but others may be subject to great fluctuations, for example the costs of some polymer materials have fluctuated by more than £100 per tonne during 2001/2002. The cost data for metals in Table 1.4 are for refined metal in ingot form. In general the cost of alloys will be higher than the cost of unalloyed metals, for example, bronze – an alloy of copper and tin – will be more costly than pure copper. Every process and heat treatment will give added value and increase the cost of the material or product. From this it will be realised that the costs of processed metal products such as sheet, plate, sections and forgings will be higher than those of ingot metal. The way in which processing adds to the cost of one material – low carbon steel – is shown in Table 1.5 but the principle that processing adds to the value of a material holds equally for other metals, polymer materials and ceramics.

The choice of a material for a particular application can be influenced also by availability. In the major growth period of railways in the 19th century most railway bridges in Britain were constructed of masonry or wrought iron. In the same period when tracks were being pushed westwards in the United States of America and Canada, bridges and viaducts were usually of timber trestle construction owing to the ready availability of suitable timber close to the point of use. The same principle holds today and often a material or source of material supply will be chosen on the basis of proximity of availability.

Table 1.4   *Costs of some materials, by mass and by volume.**

Material	Cost		Material	Cost	
	(£/kg)	(€/kg – $/kg)		(£/100 cm³)	(€/100 cm³ – $/100 cm³)
Gold	10030	15550	Gold	19380	30040
Germanium	1065	1650	Germanium	622	963
Silver	93	145	Silver	97.7	152
Cobalt	9.1	14.1	Cobalt	7.92	12.28
Titanium	6.04	9.35	Nickel	3.78	5.86
Nickel	4.25	6.6	Titanium	2.72	4.22
Tin	2.48	3.85	Tin	1.81	2.81
PVC	1.95	3.02	Copper	0.83	1.29
Nylon 66 (PA 66)	1.75	2.71	Zinc	0.34	0.53
Polycarbonate (PC)	1.75	2.71	Lead	0.29	0.45
Acrylic (PMMA)	1.65	2.56	PVC	0.28	0.43
Acetal (POM)	1.50	2.33	Aluminium	0.22	0.3
Nylon 6 (PA 6)	1.45	2.25	Acetal	0.21	0.33
ABS	0.97	1.50	Polycarbonate	0.21	0.33
Copper (grade A ingot)	0.93	1.44	Silicon	0.204	0.32
Silicon	0.85	1.30	PA 66	0.20	0.31
U-F resin thermoset	0.85	1.30	Acrylic (PMMA)	0.19	0.295
Aluminium (ingot)	0.82	1.27	PA 6	0.165	0.26
P-F thermoset	0.75	1.16	U-F thermoset	0.13	0.20
Polystyrene (PS)	0.63	0.98	Mild steel ingot	0.105	0.16
Polypropylene (PP)	0.57	0.88	ABS	0.103	0.16
Zinc (ingot)	0.47	0.73	Cast iron	0.10	0.155
Polyethylene (LDPE)	0.43	0.67	P-F thermoset	0.09	0.14
Polyethylene (HDPE)	0.35	0.54	Polystyrene	0.07	0.108
Lead (ingot)	0.26	0.41	Polypropylene	0.05	0.078
Mild steel (ingot)	0.135	0.21	LDPE	0.041	0.064
Cast iron	0.08	0.12	HDPE	0.034	0.053
Portland cement	0.06	0.09	Portland cement	0.02	0.03
Common brick	0.04	0.06	Common brick	0.012	0.019
Concrete (ready mixed)	0.03	0.045	Concrete	0.009	0.014

* The figures are based on European bulk prices in October/November 2002. At that time the Sterling/US dollar conversion was 1£ = 1.55US$. There was almost parity between the euro (€) and the US$.

Table 1.5   *Cost build-up (steel products).**

Material	Cost	
	(£ per tonne)	($ per tonne or € per tonne)
Iron from blast furnace	76	118
Low carbon steel (ingot)	135	210
Low carbon steel (hot rolled strip coil)	184	285
Low carbon steel (hot rolled bar)	190	295
Low carbon steel (hot rolled sections)	220	341
Low carbon steel (cold rolled strip coil)	222	344
Low carbon steel (galvanised sheet)	245	380
Low carbon steel (cold drawn bright bar)	247	383

* Approximate costs in October 2002. £1 = 1.55US$. (There is almost parity between the euro (€) and the US$.)

1.5 | Some trends and difficulties

At the beginning of the 20th century major industrial development was taking place in many parts of the world and man was quite profligate in his use of the world's mineral resources. Today it is realised that our mineral resources are non-renewable and finite and the present known reserves of some metals of economic importance, including copper, lead, silver, tin and zinc, could be exhausted within our lifetime. Also, the smelting of ores to produce metals, the melting of metals, the production of polymers and the processing of all materials are activities which require large energy consumption. In recent years man has become increasingly aware of the damage that unfettered exploitation of the earth's resources can cause to the environment. Also, these activities, including the production of electrical energy from fossil fuels, generate large quantities of the greenhouse gases thought to contribute to global warming. This means that engineers and designers face major challenges as they have a responsibility to select and utilise materials wisely and efficiently. The process of design must also include consideration of eventual disposal and recycling of the materials used.

A considerable amount of energy is used in the production of materials. In the case of a metal, for example, energy is consumed in mining the ore deposit, concentrating and smelting the ore, and refining the metal. Further energy is consumed in reheating a metal ingot and forming it by hot and cold working processes into, say, metal sheets. The approximate energy content of some materials is given in Table 1.6. The additional energy content of metals derived from secondary sources, namely recycled scrap, is comparatively small as it only involves remelting with, perhaps, some refining.

Forecasts that known reserves of some metallic ores will be exhausted within a relatively short time are based on knowledge of the ore sources that are being worked at present. It does not necessarily mean that these metals will not be available at all. These elements are present in small quantities, lower than in current workable ores, in other rock formations. If it became necessary to extract metals from such low level deposits then the cost of extraction and, more importantly, the energy consumption for extraction, would rise considerably. Major emphasis will be placed on the use of secondary metal, that is metal derived from recycled scrap. Design engineers will also look for processing methods which will give a product with the required shape and dimensions without generating much scrap. As an example, in conventional sand casting approximately 50 per cent of the metal melted forms scrap

Table 1.6 *Approximate energy content of some materials (GJ/tonne).*

A Materials derived from primary sources			
Titanium bar	560	Zinc (castings)	70
Magnesium extruded bar	425	Mild steel bar	60
Aluminium (ingot)	280	Cast iron (castings)	50
Aluminium (sheet)	300	Glass	20
PVC (rod or tube)	180	Reinforced concrete	12
Nylon (PA66 rod)	180	Cement	8
Polyethylene sheet	110	Brick	4
Stainless steel (sheet)	110	Timber	2
Copper (tubing)	100	Gravel	0.1
B Additional energy content of metals from secondary sources			
Aluminium	45	Mild steel	20
Copper	30	Cast iron	17

which has to be recycled, whereas, making parts by a powder metallurgy process will result in virtually no scrap.

Some possible supply problems may be overcome by substitution of one material for another. As examples of this aluminium has replaced copper for many electrical conduction applications and has replaced much tin-plated steel sheet for can manufacture. Also polymers have replaced metals in many applications. The future supply position for polymers may be uncertain as, largely, they are derived from petroleum although research into the biological generation of polymer materials is now taking place. The future supply situation for ceramic materials is much different. The earth's crust is composed mainly of silicates and alumino-silicates and so the raw materials for ceramic manufacture are present in great abundance.

There has been much research and development in the fields of materials science and materials engineering in the last fifty years and the pace of development is steadily increasing. Formerly an engineering designer would select a suitable material from a list of common materials. Today, because of our greater knowledge of the relationships between the properties of a material and its microstructure, in many instances we can specifically tailor a material to suit a particular purpose. This trend will continue to be developed. Materials with memory that will return to their original shape after deformation already exist. Further development of 'smart' materials will occur. Developments into renewable sources of materials, based on vegetable starch and other matter, will continue and probably become of great importance. These and other developments will present major challenges to engineers and designers.

Now, having tried a little crystal gazing, it is time to get down to basics. The next few chapters present the background science necessary for a proper understanding of the properties and behaviour of the various classes of materials.

SUMMARY

There are complex interrelationships between materials, manufacture and design and many factors influence the final properties of a component.

There are three main classes of material, metals, ceramics and polymers. Composite materials have been developed to combine some attractive properties of two material classes.

Metals, generally, possess good strength and elasticity, can be plastically deformed, and are good thermal and electrical conductors.

Ceramics, generally, have high elasticity and high compressive strengths but are weak in tension. They show some thermal conductivity but are electrical insulators.

Polymer materials, generally, are of low strength and elasticity and are thermal and electrical insulators.

Material costs vary widely. Final costs include the basic cost plus the cost of processing. The energy content of materials is an important parameter to be considered.

Engineers and designers require a good knowledge of materials and processing in order that they may utilise a scarce resource both wisely and efficiently.

PART II

MATERIALS SCIENCE

CHAPTER 2

Atomic structure and bonding

Why study atomic structure and bonding?

The structure of atoms, and in particular the manner in which the electrons are arranged, determines the type of bonding which occurs between atoms. The nature of the bonding between atoms and molecules is one of the important factors that help determine the properties of bulk materials.

LEARNING OBJECTIVES

1. Understand the significance of atomic number and atomic mass number and the existence of isotopes.

2. Understand the concept of the *mole*, know the value of Avogadro's number, and be able to convert moles to mass and vice versa.

3. Know that four quantum numbers are needed to define the energy state of an electron and that no two electrons in an atom can possess the same set of quantum numbers.

4. Using quantum theory, describe how the electrons in atoms builds into an arrangement of shells and sub-shells.

5. Know that radioactive nuclei can emit α-particles, β-particles and γ-radiation, understand the terms *activity constant* and *half-life*, as applied to radioactive nuclei and be able to calculate changes in activity with time.

6. Understand the mechanisms of the formation of ionic, covalent, metallic and secondary bonds.

7. Know the nature of the attractive and repulsive forces existing between atoms, the Mie equation and the relative bond energies and strengths of the various bonding types.

2.1 Introduction

It is necessary for the proper understanding of materials technology to have some knowledge of the underlying science and the fundamental principles that control the properties of materials. There are many thousands of materials available for use by today's engineers and, clearly, it is impossible for one person to have a full and detailed knowledge of all of them. However, as we have seen in Chapter 1, materials can be classified within several well-defined groups. There is a further common feature – all substances are composed of atoms. There is a comparatively limited number of different atomic species or chemical elements (the total number of elements, including the artificially created transuranic elements, is 103) and all these elements are composed of the same basic sub-atomic particles. The manner in which these sub-atomic particles are assembled into atoms and the ways in which the various atoms are bonded to one another in the make-up of bulk materials play a major part in determining the final properties of any material. For example, it is the different nature of the interatomic bonding in metals, as compared with polymer materials, which allows for electrical conduction in the former while making the latter dielectrics or insulators.

2.2 | Elementary particles

A large number of sub-atomic particles have been identified by physicists but, in terms of our general understanding of atoms and atomic structure, it is convenient to consider three types of sub-atomic particles only. These are the *proton, neutron* and *electron*. The proton is a particle that has a positive electrical charge and a mass of 1.672×10^{-27} kg. The neutron has a mass which is almost identical with that of a proton but possesses no electrical charge. The mass of an electron is extremely small, being 1/1836 of that of a proton, and possesses an electrical charge which is equal in magnitude but opposite in sign to that of a proton. The characteristics of these sub-atomic particles are summarised in Table 2.1.

Table 2.1 *Sub-atomic particles.*

Particle	Mass (kg)	Charge (C)	Relative mass	Relative charge
Proton	1.672×10^{-27}	1.602×10^{-19}	1	+1
Neutron	1.675×10^{-27}	0	1	0
Electron	0.901×10^{-30}	1.602×10^{-19}	0	−1

Within an atom the protons and neutrons together form a small compact *nucleus* and the electrons are positioned around this nucleus in a series of energy shells (refer to section 2.6). The general term *nucleon* may be used to describe both nuclear protons and neutrons. Generally, atoms contain an equal number of protons and electrons and, thus, are electrically neutral. When, as a result of excitation or interaction with other atoms, there is an imbalance between the number of protons and electrons and the atom possesses a net positive or negative charge, this charged atom is termed an *ion*.

2.3 | Atomic number and atomic mass number

There are 103 different chemical elements and each is characterised by having a different number of protons in its nucleus. The number of protons in the nucleus is termed the *atomic number*, Z. In an electrically neutral atom, therefore, there must be Z electrons. The nucleus is composed of protons and neutrons, except in the case of the simplest atom, hydrogen, in which the nucleus comprises one proton only. It is the protons and neutrons which contribute to virtually all the mass of an atom and the *atomic mass number* (or *atomic weight*), M, indicates the total number of nucleons in the nucleus. For the simplest atom, hydrogen, $Z = 1$ and $M = 1$, so that the nucleus is simply one proton but for oxygen, $Z = 8$ and $M = 16$, the nucleus contains eight protons and eight neutrons. The quoted values of atomic mass number are based on a unit termed the *atomic mass unit (a.m.u.)* and this is one-twelfth of the mass of an atom of the *isotope* of carbon which has a nucleus containing six protons and six neutrons. The a.m.u. has a measured mass of 1.6598×10^{-27} kg.

The quoted values for the atomic mass numbers of chemical elements are not all integers. There are two reasons for this, *mass defect* and the existence of *isotopes*. The measured mass of an atom does not equate with the sum of the masses of the protons, neutrons and electrons present. For example, the mass of an oxygen atom containing eight protons, eight neutrons and eight electrons is 26.556×10^{-27} kg but the sum of the masses of eight protons, eight neutrons and eight electrons is 26.784×10^{-27} kg. This discrepancy is termed the *mass defect* and, generally, is expressed as mass defect per nucleon. From the above figures the mass defect for oxygen is 14.25×10^{-30} kg. According to Einstein's theory of relativity, mass and energy are related according to the expression $E = mc^2$ where m is mass (kg), E is energy (J) and c is the velocity of light (3×10^8 m s^{-1}). In the case of oxygen the mass defect per nucleon is equivalent to an energy of 1.28×10^{-12} J, or about 8 MeV. (An electron volt (eV) is a unit of energy and is the energy change when an electron moves through a potential difference of one volt.) The energy equivalent of the mass defect per nucleon is referred to as the binding energy per nucleon. The value of mass defect per nucleon and, hence, nuclear binding energy varies with the atomic mass number as shown in Figure 2.1. The elements with the greatest mass defect values are those with atomic mass numbers in the range from 40 to 100.

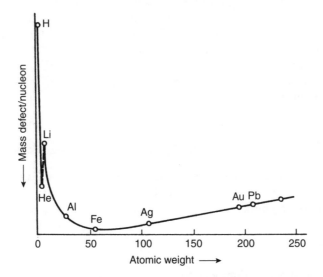

Figure 2.1   *Relationship between the mass defect and atomic mass number.*

It can be seen from Figure 2.1 that when elements of high atomic mass number disintegrate into nuclei of lower mass number, that is with higher mass defect values, mass will be converted into a large energy emission. This is the principle involved in nuclear fission reactions and is harnessed in nuclear reactors for power generation. It will be seen also that a nuclear fusion of hydrogen nuclei to form helium would result in a very large emission of energy. This is the principle involved in the 'hydrogen' bomb but as yet it has not proved possible to use nuclear fusion reactions for the controlled release of energy for power generation.

In chemical reactions, as opposed to nuclear reactions, the changes which occur are confined to outer electron shells. Energy changes in chemical reactions are of an extremely small order in comparison with the energy changes accompanying nuclear reactions, so small that any corresponding mass change is immeasurable.

2.4 | Isotopes and isotones

Isotopes are atoms of the same element which possess different mass numbers, that is there is a variation in the number of neutrons in the nuclei. Isotopes exist in the majority of chemical elements and the relative abundance of the various isotopes in any sample of an element is virtually constant, irrespective of the source. Some examples are given. Two isotopes of hydrogen occur naturally, normal hydrogen ($Z = 1$, $M = 1$) and deuterium ($Z = 1$, $M = 2$) while a third isotope, tritium ($Z = 1$, $M = 3$) which is radioactive, has been formed by the bombardment of lithium and boron by neutrons. The proportion of deuterium in a sample of hydrogen is about 0.2 per cent.

Isotopes may be symbolised in one of two ways, as for example $^{12}_{6}C$, where the superscript denotes atomic mass number and the subscript denotes atomic number, or as carbon-12 (C-12). Any sample of oxygen contains three isotopes with mass numbers of 16, 17 and 18 while any sample of carbon has three isotopes with mass numbers of 12, 13 and 14, the last one, present only in very small quantities, being radioactive. It is the existence of this last radioactive isotope which permits the use of carbon dating technique for determining the age of ancient remains. Living matter absorbs carbon throughout its life and the proportion of C-14 atoms within the total carbon content is virtually constant. When the plant or animal dies no further carbon is absorbed. Thereafter, as the C-14 atoms slowly decay radioactively, the proportion of C-14 atoms and, hence the total radioactivity, in the substance diminishes with time (see section 2.9). In addition to the isotopes of elements which occur naturally, other isotopes may be created by means of nuclear reactions and these are almost invariably radioactive, disintegrating spontaneously with the emission of radiation. A number of these artificially made isotopes are useful in medical and industrial applications (see section 2.10).

Isotones, or *isobars*, are nuclei of different elements which possess the same atomic mass number. Examples of this are argon-40 (a nucleus with 18 protons and 22 neutrons) and calcium-40 (a nucleus with 20 protons and 20 neutrons).

EXAMPLE 2.1

Magnesium contains three isotopes with atomic mass numbers of 24, 25 and 26 in the following proportions: 78.7 per cent of Mg-24, 10.13 per cent of Mg-25 and 11.17 per cent of Mg-26. Calculate the atomic mass number of magnesium.

SOLUTION

The atomic mass number of magnesium will be

$(24 \times 0.787) + (25 \times 0.1013) + (26 \times 0.1117) = 24.325$

2.5 | The mole and Avogadro's number

Some substances are termed atomic while others are termed molecular, a *molecule* being a number of atoms in combination as, for example in gaseous oxygen, O_2 (two atoms per molecule, or CO_2 (one carbon atom and two oxygen atoms per molecule). For a molecular substance the *molecular weight*, or molecular mass number, is the sum of the atomic mass numbers of the atoms in the molecule. This brings us to the concept of *molar mass*. The SI units of molar mass are the *mole* (abbreviation mol) and the *kilomole* (abbreviation kmol). A mole is the amount of a substance with a mass in grammes numerically equal to its atomic mass number, for a monatomic element, or to its molecular weight, for a molecular substance. The larger unit, the kilomole, is the amount of substance with a mass in kilogrammes numerically equal to the atomic mass number or molecular weight.

Equal molar masses of any substance will always contain the same number of molecules (the same number of atoms in the case of monatomic substances). The number of molecules in a mole or kilomole is given by the universal constant Avogadro's number, N_0. The numerical value of Avogadro's number is 6.023×10^{23}/mol or 6.023×10^{26}/kmol.

EXAMPLE 2.2

How many atoms are present in: (a) 10 kg of aluminium, and (b) 15 kg of copper?
The atomic mass numbers of aluminium and copper are 26.98 and 63.54 respectively.

SOLUTION

The number of kilomoles is given by $\dfrac{\text{mass}}{\text{atomic mass number}}$

(a) 10 kg of aluminium contains $\dfrac{10}{26.98} = 0.3706$ kmol $= 0.3706 \times 6.023 \times 10^{26}$ atoms

$$= 2.232 \times 10^{26} \text{ atoms}$$

(b) 15 kg of copper contains $\dfrac{15}{63.54} = 0.2361$ kmol $\quad = 0.2361 \times 6.023 \times 10^{26}$ atoms

$$= 1.422 \times 10^{26} \text{ atoms}$$

2.6 Atomic structure and quantum numbers

From experiments conducted in 1911 Rutherford deduced that an atomic nucleus is extremely small and that much of an atom is void space. For example, the effective diameter of an atom of gold is approximately 3×10^{-10} m but the diameter of the nucleus is about 1/10,000 of this size.

One of the first postulations about electrons was that they are in orbit around the central nucleus in much the same way as planets are in orbit around the Sun. On the basis of classical mechanics this picture cannot be valid. The electron is a charged particle and in following a circular or elliptical path will be subject to angular acceleration. But from electromagnetic theory an electrical charge will radiate energy when it is accelerated. (It is the acceleration of charges in an aerial that is responsible for the transmission of energy as radio waves.) This would not allow stability in atoms. If orbiting electrons were continually losing energy, the orbits would not remain constant. The electrons would follow spiral paths and eventually collapse into the nucleus. Electrons, therefore, cannot simply be regarded as corpuscular particles. They exhibit some of the characteristics of waves – an electron beam may be diffracted – and the motion of electrons can be described not by classical mechanics, but by quantum mechanics and wave mechanics.

In 1913, Bohr suggested a quantum model for an atom, making the following assumptions:

(a) the electrons exist in stable circular orbits of fixed energy, with the angular momentum of an electron in orbit being an integral multiple of $h/2\pi$, where h is Planck's constant,

(b) an electron can emit or absorb energy only when making a transition from one possible orbit to another.

According to Bohr, when an atom is excited by the input of energy an electron moves to an orbit of greater radius. Subsequently, when the electron moves back to the 'ground' state of energy, energy will be released at some specific frequency giving rise to a spectral line. Planck postulated that energy is not absorbed or emitted in a continuous manner, but in discrete packets termed *quanta*, with one quantum of energy being hf, where h is Planck's universal constant ($h = 6.62 \times 10^{-34}$ J s) and f is the frequency of the radiation.

The modern concept is that the electrons exist in a series of energy levels, or shells, surrounding the nucleus. The quantum theory of atoms requires that four quantum numbers be defined to give a complete description of the energy state of an electron. The first, or principal, quantum number, n, which may have positive integral values, 1, 2, 3, 4, etc., is a description of the general energy level of the electron in terms of distance from the nucleus, that is the layer or shell containing the electron. Often, these principal shells are denoted by letters: $K = 1$, $L = 2$, $M = 3$, $N = 4$, $O = 5$, $P = 6$, $Q = 7$.

The secondary quantum number, l, is a measure of the angular momentum of the electron. l may have integral values from 0 to $n - 1$ and the values of l are denoted by the letter symbols s, p, d and f, where s refers to $l = 0$, p for $l = 1$, d for $l = 2$ and f for $l = 3$. A value of $l = 0$ does not mean that the electron is stationary, but rather that there is an equal probability of finding an s state electron at any point on the surface of a spherical orbital. The probability envelope for a p state electron, that is when $l = 1$, has a dumb-bell shape (Figure 2.2(b)).

The third quantum number, m_l, describes the magnetic moment of an electron. A moving electron will produce a magnetic field and this, in turn, will be affected by an external magnetic field. The quantum number m_l may have any integral value, including 0, between $-l$ and $+l$. Consider the p state ($l = 1$) where m_l may have three possible values, -1, 0 and $+1$. Three mutually perpendicular dumb-bell shaped orbitals, corresponding to the three values of m_l are possible (Figure 2.2(c)). The fourth quantum number considers motion within the electron. It is assumed that the electron is spinning about an axis as well as being in orbit around a nucleus. This spin may be deemed either clockwise or counter-clockwise. The spin quantum number, m_s, may have the values of either $+1/2$ or $-1/2$.

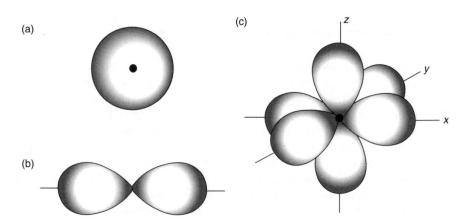

Figure 2.2 *(a) Spherical probability envelope for the s state. (b) Probability envelope for the p state. (c) Three mutually perpendicular p state envelopes for m_l values of -1, 0 and $+1$.*

2.7 | The Pauli exclusion principle

In the 'ground state' of energy the electrons in an atom occupy the lowest possible energy levels. Pauli deduced that no two electrons in the same atom can have the same set of quantum numbers. This would be equivalent to two bodies occupying exactly the same space. We are now in a position to see how the electronic structure of atoms can be built up.

For a principal quantum number of $n = 1$ (that is the first or K shell), l can only have the value 0 (s). Therefore, m_l can only be 0. The spin quantum number can be $+1/2$ or $-1/2$ meaning that only two electrons can be held in the K shell, these being in the 1s state. In the L shell, $n = 2$, therefore l may be 0 or 1 (s or p). When l is 0, m_l is 0. As m_s can have two values then two s state electrons can be placed in the L shell. But when $n = 2$, l can also have the value of 1, allowing three values of m_l, -1, 0 or $+1$. With two values of m_s possible in each case a total number of six electrons can exist in this, the p state of the L shell, giving the L shell a total of eight electrons. Similarly we can show that the M shell can contain eighteen electrons, two in the $3s$, six in the $3p$ and ten in the $3d$ state. Also that the N, or fourth, shell can hold thirty-two electrons: two $4s$, six $4p$, ten $4d$ and fourteen in the $4f$ state. Table 2.2 gives the 'ground state' electron configuration for the first 38 elements in ascending order of atomic number.

As the atomic number increases, at least up to argon ($Z = 18$) each additional electron takes up a place in the appropriate shell, filling first the $1s$, then $2s$ and $2p$, followed by $3s$ and $3p$. When the $3p$

Table 2.2 *Electronic configuration of the elements.*

Element	Symbol	Atomic number Z	Atomic weight M	Number of electrons in each shell and sub-shell K (1s)	L (2s)	(2p)	M (3s)	(3p)	(3d)	N (4s)	(4p)	(4d)	(4f)	O (5s)	(5p)	(5d)	(5f)	P (6s)	(6p)	(6d)	(6f)	Q (7s)
Hydrogen	H	1	1.008	1																		
Helium	He	2	4.003	2																		
Lithium	Li	3	6.940	2	1																	
Beryllium	Be	4	9.013	2	2																	
Boron	B	5	10.82	2	2	1																
Carbon	C	6	12.011	2	2	2																
Nitrogen	N	7	14.008	2	2	3																
Oxygen	O	8	16.000	2	2	4																
Fluorine	F	9	19.000	2	2	5																
Neon	Ne	10	20.183	2	2	6																
Sodium	Na	11	22.991	2	2	6	1															
Magnesium	Mg	12	24.32	2	2	6	2															
Aluminium	Al	13	26.98	2	2	6	2	1														
Silicon	Si	14	28.09	2	2	6	2	2														
Phosphorus	P	15	30.975	2	2	6	2	3														
Sulphur	S	16	32.066	2	2	6	2	4														
Chlorine	Cl	17	35.457	2	2	6	2	5														
Argon	Ar	18	39.944	2	2	6	2	6														
Potassium	K	19	39.10	2	2	6	2	6		1												
Calcium	Ca	20	40.08	2	2	6	2	6		2												
Scandium	Sc	21	44.96	2	2	6	2	6	1	2												
Titanium	Ti	22	47.90	2	2	6	2	6	2	2												
Vanadium	V	23	50.95	2	2	6	2	6	3	2												
Chromium	Cr	24	52.01	2	2	6	2	6	5	1												
Manganese	Mn	25	54.94	2	2	6	2	6	5	2												
Iron	Fe	26	55.85	2	2	6	2	6	6	2												
Cobalt	Co	27	58.94	2	2	6	2	6	7	2												
Nickel	Ni	28	58.69	2	2	6	2	6	8	2												
Copper	Cu	29	63.54	2	2	6	2	6	10	1												
Zinc	Zn	30	65.38	2	2	6	2	6	10	2												
Gallium	Ga	31	69.72	All K, L and M shells are completely filled						2	1											
Germanium	Ge	32	72.60							2	2											
Arsenic	As	33	74.91							2	3											
Selenium	Se	34	78.96							2	4											
Bromine	Br	35	79.916							2	5											
Krypton	Kr	36	83.80							2	6											
Rubidium	Rb	37	85.48							2	6			1								
Strontium	Sr	38	87.63							2	6			2								

state is complete the nineteenth electron (for potassium) goes not into the $3d$ state but into the $4s$ state, as this state has a slightly lower energy level than $3d$. The energy level of the $3d$ state lies between those of $4s$ and $4p$. This means that after element 20 (calcium), when the $4s$ state is full, additional electrons go into the $3d$ state. In passing from element 21 (scandium) to 30 (zinc) the $3d$ sub-shell is filled completely and filling of the $4p$ sub-shell is commenced with element 31 (gallium). This can be seen in Table 2.2.

Energy levels are fairly close together for all states above the $4p$ state, as shown in Figure 2.3, and the general order in which the shells fill up is $4s$, $3d$, $4p$, $5s$, $4d$, $5p$, $6s$, $4f$, $5d$, $6p$, $7s$, $6d$ and $5f$. This can be seen in the periodic table (Table 2.3).

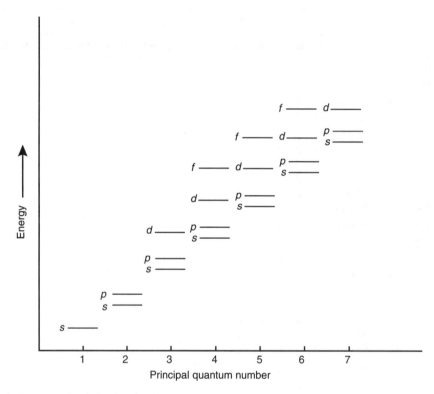

Figure 2.3 *Relative energy levels for the electron states in atoms.*

2.8 | The periodic table

The number of electrons in the outermost shell of any atom has an important bearing on the properties of the element. Some elements, those with atomic numbers of 2, 10, 18, 36, 54 and 86, are extremely stable and are inert chemically. These, the inert or noble gases, are characterised by having all their energy levels completely filled with electrons. Apart from helium, which possesses only two electrons, these elements all have a complement of eight electrons in their outermost shell. All other elements, which have at least one incompletely filled shell or sub-shell, will take part in chemical reactions. It is the outermost shell electrons, the valence electrons, that are responsible for the chemical reactivity of elements. Elements in combination attempt to achieve an outer shell configuration of eight electrons (see sections 2.12 and 2.13).

Table 2.3 *Periodic table of the elements.*

EXAMPLE 2.3

Indium, $Z = 49$, has all of its electron states, in sequential order, filled, except for the 4*f* state, which is empty. What is the expected valence of indium?

SOLUTION

The filled electron states of indium are 1*s* (2), 2*s* (2), 2*p* (6), 3*s* (2), 3*p* (6), 3*d* (10), 4*s* (2), 4*p* (6) and 4*d* (10), giving a total of 46 electrons. The three remaining electrons must be in the 5th or outer shell (5*s* (2), 5*p* (1)), thus giving indium a valence of 3.

It was noticed more than a hundred and thirty years ago, long before deductions had been made about the inner structure of the atom, that there existed certain groups of elements showing some similarity in properties. The alkali metals, lithium, sodium and potassium, formed one such group, and the halogens, fluorine, chlorine, bromine and iodine, formed another. When the various elements known at that time were placed in order of ascending atomic weight, a certain periodicity of elements with specific properties, such as the alkali metals and the halogens, was observed. In 1870, Mendeleev arranged the elements in a table in a manner that highlighted this effect. This was the forerunner of the modern periodic table of elements. With the discovery of the electron and the build-up of electrons in shells and sub-shells, as outlined above, came a fuller understanding of the properties of the elements.

Table 2.3 gives the full periodic table of the elements. The figures in front of each row of elements, for example 3*p* in front of elements 13 to 18, indicate the particular sub-shell being filled with electrons. The electrons in the outermost shell are the valence electrons. For the vertical columns IA, IIA, IIIB, IVB, VB, VIB, VIIB and 0, within each column the elements possess the same number of outer shell electrons and, hence, possess similar properties. But, considerable differences occur between the elements of one column and those of another.

In the transition series, the first of which is the horizontal row containing the elements 21 to 30, the sub-shell being filled is not in the outer shell. This means that all the elements in the row possess somewhat similar properties. However, because the 4*s*, 4*p* and 3*d* energy levels are close to one another, there is a tendency for electrons to move from one sub-shell to another. This causes these elements to possess variable valency characteristics. For example, iron (element 26) has a valency of three in the ferric (or iron III) state, and a valency of two in the ferrous (or iron II) state. In the case of the two long series, the lanthanides (elements 58 to 71) and the actinides (elements 90 to 103), the sub-shell being filled is well in towards the core of the atom and so the chemical differences between each member of these two groups is very small.

2.9 | The nucleus and radioactivity

It is known that the nucleus is extremely small in size, and that its diameter is only about 1/10,000 of the effective diameter of the atom. Particles of like electrical charge should repel one another strongly, and yet the nuclei of most elements are extremely stable. The presence of neutrons seems to play a major role in ensuring the stability of an atomic nucleus, but the nature of the nuclear binding is imperfectly understood. The lighter elements have roughly equal numbers of protons and neutrons in their nuclei. As the atomic number increases, so the ratio of neutrons to protons increases from 1:1 to about 1.5:1 in stable nuclei. Elements with atomic numbers greater than 83 do not possess stable nuclei.

These large nuclei disintegrate spontaneously, emitting energy and particles, these emissions being termed *radioactivity*.

Natural radioactivity was first discovered in 1896 by Bequerel. Subsequent research workers found that all elements of high atomic number were radioactive, emitting small particles and very high frequency electromagnetic radiation, similar to X-radiation. Two types of particle emission were identified, these being termed α-particles and β-particles. The α-particle had a mass number of 4 and possessed two unit positive charges (or, in other words, a helium nucleus), while the β-particle was identified as an electron. The very high frequency radiation emitted was termed γ-radiation.

The activity of a radioactive substance decreases exponentially with time. It is impossible to predict when any individual atom will disintegrate but, statistically, the number of atoms that will disintegrate in a given time can be forecast accurately. The rate of radioactive decay at some particular time is proportional to the activity at that instant. If I is the activity at some time t then

$$\frac{dI}{dt} = \lambda I \text{ where } \lambda \text{ is a constant.}$$

Integration gives $I = I_0 \exp(-\lambda t)$ where I_0 is the initial activity at zero time. The constant λ is known as the radioactive constant. An important characteristic of any radioactive substance is its *half life*. This is the time required for half of the radioactive nuclei originally present to disintegrate, or in other words, the time needed for the activity to reduce to half of its original value. If the half life is T then

$$I_T = \frac{I_0}{2} = I_0 \exp(-\lambda T)$$

Therefore $\qquad \lambda T = \ln 2$

$$\lambda = \frac{0.693}{T}$$

EXAMPLE 2.4

The radioactive isotope of cobalt, cobalt-60, which has a half life period of 5.3 years, is used as a source of γ-radiation in industrial radiography. What is the radioactive constant for cobalt-60? What is the useful life of a cobalt-60 source if it can be used until its activity is reduced to 80 per cent of its original value?

SOLUTION

T for cobalt-60 = 5.3 years = 1.67×10^8 s

$$\lambda = \frac{0.693}{T} = 4.14 \times 10^{-9} \text{ s}^{-1}$$

To find t when I_t is 80 per cent of I_0 $\qquad I_t = 0.8 I_0 = I_0 \exp(-\lambda t)$

$\ln 0.8 = -\lambda t$

$$t = \frac{-\ln 0.8}{\lambda}$$

$= 5.39 \times 10^7 \text{s} = 1.71 \text{ years}$

The useful life of the radioactive source is 1.71 years.

The half-life periods for radioactive elements vary considerably, and range from about 10^{10} years for uranium and thorium to less than a second for some isotopes. The eventual product of natural radioactive disintegration will be some stable nucleus. In fact, in all cases this end product is an isotope of lead. As an example, the disintegration sequence for thorium is given in Table 2.4. The loss of an α-particle from a nucleus means the loss of two protons and two neutrons and, hence, a decrease of two in the atomic number and a decrease of four in the atomic mass number. Similarly, the loss of a β-particle from a nucleus means a gain of one in the number of unit positive charges remaining in the nucleus, or an increase in one in the atomic number.

Table 2.4 *Radiation disintegration sequence for thorium.*

Nucleus		Isotope symbol	Particle emitted	Half life
Thorium		$^{232}_{90}\text{Th}$	α	1.39×10^{10} years
Mesothorium I	(radium)	$^{228}_{88}\text{Ra}$	β	6.7 years
Mesothorium II	(actinium)	$^{228}_{89}\text{Ac}$	β	6.13 hours
Radiothorium I	(thorium)	$^{228}_{90}\text{Th}$	α	1.90 years
Thorium X	(radium)	$^{224}_{88}\text{Ra}$	α	3.64 days
Thorium emanation	(radon)	$^{220}_{86}\text{Rn}$	α	54.5 seconds
Thorium A	(polonium)	$^{216}_{84}\text{Po}$	α	0.16 second
Thorium B	(lead)	$^{212}_{82}\text{Pb}$	β	10.6 hours
Thorium C	(bismuth)	$^{212}_{83}\text{Bi}$	(a) β, (b) α	47 minutes
(a) Thorium C'	(polonium)	$^{212}_{84}\text{Po}$	α	3×10^{-7} second
Thorium D	(lead)	$^{208}_{82}\text{Pb}$	stable	
(b) Thorium C''	(thallium)	$^{208}_{81}\text{Tl}$	β	2.1 minutes
Thorium D	(lead)	$^{208}_{82}\text{Pb}$	stable	

2.10 | Artificial radioactive materials

The naturally occurring elements with atomic numbers of less than 84 possess stable nuclei, but unstable nuclei of low atomic number elements may be made artificially. The natural form of phosphorus has an atomic mass number of 31, but in 1933 Joliot-Curie produced a radioactive isotope of phosphorus, with an atomic mass number of 30, by bombarding aluminium with α-particles. The nuclear reaction can be written as follows:

$$^{27}_{13}\text{Al} + {}^{4}_{2}\text{He} \,(\alpha\text{-particle}) = {}^{30}_{15}\text{P} + {}^{1}_{0}\text{n}$$

Subscripts denote atomic number, Z, and superscripts denote atomic mass number, M. Thus $^{1}_{0}\text{n}$ represents the neutron formed in the reaction.

α-particle bombardment is not a very efficient means of producing radioactive isotopes, as the positively charged atomic nuclei repel the bombarding particles of equal sign, so greatly limiting the number of collisions. The use of neutrons for the bombardment of nuclei is a much better technique. A typical nuclear reaction involving neutron bombardment is the formation of radioactive sodium from aluminium, with the emission of an α-particle, as shown below:

$$^{27}_{13}\text{Al} + {}^{1}_{0}\text{n} = {}^{24}_{11}\text{Na} + {}^{4}_{2}\text{He} \,(\alpha\text{-particle})$$

The $^{24}_{11}$Na nucleus produced in this reaction is an unstable nucleus and this decays radioactively into a stable isotope of magnesium with the emission of a β-particle (electron):

$$^{24}_{11}\text{Na} = ^{24}_{12}\text{Mg} + ^{0}_{-1}\text{e (electron)}$$

Radioisotopes for industrial and medical use are produced commercially by subjecting materials to neutron bombardment in nuclear reactors.

EXAMPLE 2.5

Give three possible applications in engineering for radioactive isotopes.

SOLUTION

1. Radioactive isotopes which emit high-intensity γ-radiation are used widely in radiography for the detection of flaws in metal components or as a means of checking for the correct assembly of a complex device. (For more information refer to section 28.7.)
2. Radioactive isotopes may be used in research programmes as, for example, in the investigation of surface wear rates. The cylinder linings of an experimental internal combustion engine can be made with a small proportion of radioactive iron isotope. Rates of wear can be determined through monitoring the radioactivity in the lubricating oil.
3. If small amounts of radioactive argon are added to the working gas the mixture can be used to check for leakage in pressurised pipe-work systems by monitoring the outside of the pipes for radiation.

Some of the radioactive isotopes that are used commercially are given in Table 2.5.

Table 2.5 *Some radioactive isotopes used in engineering.*

Element	Symbol	Radiation emitted	Half life	Application
Argon	$^{41}_{18}$Ar	β, γ	109 minutes	Leak detection
Caesium	$^{137}_{55}$Cs	β, γ	30 years	Radiography
Cobalt	$^{60}_{27}$Co	β, γ	5.3 years	Radiography
Iridium	$^{192}_{77}$Ir	β, γ	74.4 days	Radiography
Iron	$^{59}_{26}$Fe	β, γ	45 days	Engine wear
Strontium	$^{90}_{38}$Sr	β	28 years	Thickness gauges
Tungsten	$^{187}_{74}$W	β, γ	24.1 hours	Tool wear

2.11 | Interatomic and intermolecular bonding

The fluid and solid substances that we are familiar with are composed of very large aggregates of atoms and the properties of these materials derive, in part, from the manner in which the individual atoms are bonded together and the strengths of these bonds. There are several types of interatomic bonding possible and so the bonding which exists between atoms is not the same for all materials. Generally, the bonding involves some degree of interaction between the outer shell, or *valence*, electrons and is,

Figure 2.4 *(a) Variation of force with atomic separation distance. (b) Variation of energy with atomic separation distance. (The resultant curves in (a) and (b) are shown as continuous lines.)*

therefore, dependent on the number and distribution of electrons within the atom. It has been stated already (section 2.8) that a completely filled outer electron shell confers a very high degree of stability to an atom, this being the electronic structure of the inert or noble gases. In the formation of inter-atomic bonds, atoms with incomplete outer electron shells attempt, in combination with other atoms, to achieve filled outer shells, thus satisfying this condition for stability.

Interatomic bonds are termed *primary* bonds and are relatively strong. Other types of bonding, weaker than primary bonds, can occur also within substances. Generally, these weaker *secondary* bonds exist between molecules. The principal types of interatomic bond are the *ionic* bond, the *covalent* bond and the *metallic* bond. The *hydrogen* bond and *van der Waal's* bonds are the main secondary bond types.

Interactive forces exist between atoms, the magnitude of which is a function of the separation distance between the atoms. There is a gravitational attractive force between the nuclear masses of two atoms and this force increases as the distance between two nuclei decreases. There may also be an electrostatic attractive force acting between atoms (see section 2.12). As two atoms get close to one another there will be a strong force of repulsion generated between the negatively charged electron shells of both atoms. There will be one particular separation distance, r_0, at which both the forces of attraction and repulsion will be equal and the net force on the atoms will be zero (Figure 2.4(a)). At this distance the potential energy of the atoms will be a minimum (Figure 2.4(b)). This is the stable separation distance between the atoms and input of energy would be necessary to move the two atoms either closer to one another or further apart. This minimum energy value is termed the *binding energy*. Materials with high values of binding energy tend to have high strengths and melting points.

The potential energy of atoms, in relation to the separation distance, r, is given by the Mie equation, the general form of which is

$$\text{P.E.} = -\frac{A}{r^m} + \frac{B}{r^n}$$

where A, B, m and n are constants whose values depend on the particular system and type of bond. The negative first term is the potential energy of attraction while the second term is that due to repulsion.

The ranges of binding energies for the various types of bonding are given in Table 2.6.

Table 2.6 *Range of binding energies.*

Bond type	Binding energy	
	eV	MJ/kmol
Ionic	6.5–16	625–1550
Covalent	5.4–13	520–1250
Metallic	1.0–8.3	100–800
Secondary	0.002–0.3	0.2–30

2.12 The ionic bond

Elements in the lower groups of the periodic table would expose a completely filled electron shell if they could lose the electrons present in their incomplete outer shell. Group IA elements, the alkali metals, would need to shed only one electron per atom, while Group IIA elements would need to lose two electrons from each atom. Conversely, atoms of the elements in Groups VIB and VIIB can complete their unfilled outer shell by capturing electrons. An element of Group IA, therefore, could react chemically with an element of Group VIIB, resulting in the transfer of one electron from one atom to the other. For example, sodium ($Z = 11$) would transfer its one electron from the third principal shell, so exposing a completely filled second shell, to chlorine ($Z = 17$), so giving the chlorine atom the complement of eight electrons in its outer layer (Figure 2.5). By losing an electron, the sodium atom has become out

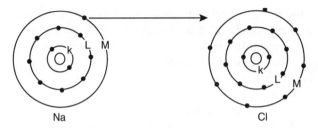

Figure 2.5 *Electron transfer from sodium to chlorine.*

of balance electrically, and is said to be in an ionised state:

$$Na = Na^+ + e^-$$

Similarly, by gaining an additional electron, the chlorine atom has become ionised

$$Cl + e^- = Cl^-$$

An ion is a charged particle and the net charge may be positive or negative, depending on whether electrons are lost or gained. The magnitude of the net charge may be one, two, three, or in a few cases, even four units, a unit being the charge of a single electron. An ion is symbolised by using the symbol for the chemical element with a superscript indicating both the number and sign of the electrical charge, as with Na^+, Al^{3+}, Cl^-, O^{2-} etc.

The sodium and chlorine ions, being of opposing charge, will be strongly attracted to one another, the attractive force, F, being given by

$$F = q^2/4\pi\varepsilon_0 r^2.$$

where q is the charge on each ion, ε_0 is the permittivity of a vacuum and r is the separation distance between the ions. Also, as stated in section 2.11 above, there is a gravitational attractive force between the nuclear masses of the two ions and an electrostatic repulsive force when the two ions are close to one another and the outer electron shells begin to overlap. When the opposing forces are in balance the potential energy, as given by the Mie equation, will be at a minimum.

In the interaction of Na^+ and Cl^- ions to form sodium chloride the transfer of only one electron is involved in the formation of one ionic molecule. In the case of a Group IIA element reacting with a Group VIB element, two electrons would be transferred to make one ionic molecule. For example, for calcium and sulphur

$$Ca = Ca^{2+} + 2e^-$$

$$S + 2e^- = S^{2-}$$

or we may simply write

$$Ca + S = CaS.$$

EXAMPLE 2.6

What type of bonding will occur between atoms of magnesium and chlorine?

SOLUTION

Magnesium has two electrons in its outermost shell (from Table 2.2) while chlorine has seven electrons in its outer shell. Magnesium can ionise by losing its two outer shell electrons to give Mg^{2+}. Chlorine atoms ionise by gaining an electron to form Cl⁻. So the ionisation of one magnesium atom provides sufficient electrons to ionise two chlorine atoms. Magnesium and chlorine bond ionically, the resulting compound having the formula $MgCl_2$ (Figure 2.6).

Figure 2.6 *Electron transfer from magnesium to two chlorine atoms.*

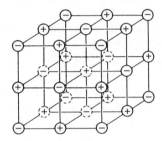

Figure 2.7 *Symmetrical arrangement of ions in NaCl structure.*

The ionic bond is a strong bond and, in the solid state, individual ionic molecules do not exist. Instead, the ions pack into symmetrical crystalline arrays. In the case of a sodium chloride crystal, for example, the ions pack into a cubic array with each Na⁺ ion surrounded by six Cl⁻ ions, and vice versa. The cubic crystalline pattern for sodium chloride is shown in Figure 2.7. The charged ions occupy fixed sites within the crystal lattice and little, if any, movement of ions is possible, even in a strong electrical field, rendering the ionically bonded crystal an electrical insulator.

Most ionic crystals, however, are soluble in a polar solvent such as water and in solution the charge carrying ions have mobility. Thus, in an electrical field they will move preferentially constituting an electrical current. Ionic conductive solutions are termed *electrolytes*. Similarly, when an ionic crystalline substance melts the ions have mobility and molten salts are electrically conductive.

2.13 The covalent bond

In covalent bonding the stable arrangement of eight electrons in an outer shell is achieved by a process of electron sharing rather than electron transfer. In the case of chlorine, for example, individual atoms combine to form diatomic molecules. The reaction can be written

$$2Cl = Cl_2$$

The bond is achieved by the sharing of a pair of electrons. One electron from each atom enters into joint orbit around both nuclei, so giving each nucleus an effective complement of eight outer shell electrons (Figure 2.8(a)). This may be represented symbolically as

Cl : Cl or Cl–Cl

(a) Chlorine molecule Cl_2. One pair
of electrons shared

(b) Oxygen molecule O_2. Two pairs
of electrons shared

Figure 2.8 *Covalent bonds in (a) chlorine and (b) oxygen.*

where a pair of dots, or a hyphen, represents a pair of electrons shared between adjacent atoms, namely then, one covalent bond. An oxygen atom has six outer shell electrons and in an oxygen molecule, O_2, two pairs of electrons are shared between two adjacent atoms – a double covalent bond – thus giving each atom a complement of eight outer shell electrons (Figure 2.8(b)).

In a manner similar to that described for the ionic bond, covalently bonded atoms are subject to both attractive and repulsive forces and the energy of the bond can be represented by the Mie equation with $m < n$.

Covalent bonds tend to be highly directional. The shape of the water molecule, H_2O, is shown in Figure 2.9. There are two covalent bonding orbitals, each containing two electrons, linking the hydrogen nuclei to the oxygen. The four remaining outer shell electrons of oxygen occupy two non-bonding orbitals, with two electrons in each, and these are much smaller than the two bonding orbitals. The electron sharing is not equal within the two covalent bonds and the shared electrons spend a greater portion of the time close to the oxygen nucleus than they do in the vicinity of the hydrogen nuclei. This means that the water molecule is polarised, with the two hydrogen ends being slightly positive relative to the two non-bonding orbitals of the oxygen. Each H_2O is, thus, a small dipole and so secondary bonds, *hydrogen bonds*, based on electrostatic attraction, occur between these polar molecules (refer also to section 2.16).

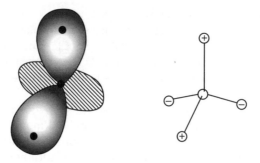

Figure 2.9 *Representation of a water molecule and dipole formation.*

Carbon, with four outer shell electrons, is capable of forming four covalent bonds, and the shared orbitals point towards the corners of a regular tetrahedron. The diamond form of carbon shows this tetrahedral structure with each carbon atom being bonded covalently to four other carbon atoms (Figure 2.10).

The covalent bond is the form of bonding in organic molecules. ('Organic chemistry' is the chemistry of carbon and its compounds. Organic molecules, composed principally of carbon and hydrogen, form the basis of polymer materials and of all living matter.) The most simple hydrocarbon compound is methane, CH_4, in which a carbon atom is bonded covalently to four hydrogen atoms, thus

Figure 2.10 *Tetrahedral structure of diamond.*

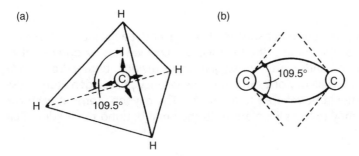

Figure 2.11 *(a) Tetrahedral form of methane, CH₄. (b) Double covalent bond between carbon atoms. Bonds under strain.*

The molecule is tetrahedral in shape with the carbon nucleus at the centre of a regular tetrahedron (Figure 2.11(a)). The angle between adjacent bonds is 109.5°. Not all bonds in organic molecules need be of the single covalent type. Multiple bonding is possible also. Consider the three compounds ethane, C_2H_6, ethene (or ethylene), C_2H_4, and ethine (or acetylene), C_2H_2. Ethane has single covalent bonds only:

$$
\begin{array}{cc}
\text{H} & \text{H} \\
| & | \\
\text{H}-\text{C}-\text{C}-\text{H} \\
| & | \\
\text{H} & \text{H}
\end{array}
$$

Ethene has a double covalent bond linking the two carbon atoms:

$$
\begin{array}{c}
\text{H} \diagdown \qquad \diagup \text{H} \\
\text{C}=\text{C} \\
\text{H} \diagup \qquad \diagdown \text{H}
\end{array}
$$

Ethine has a triple covalent bond between the two carbon atoms:

$$\text{H}-\text{C}\equiv\text{C}-\text{H}$$

The separation distance between carbon nuclei is smaller in multiple covalent bonds than in a single covalent bond. This separation distance is 0.154 nm in ethane, but is 0.134 nm in ethene and 0.120 nm in ethine. Multiple covalent bonds are not stronger than a single covalent link. In fact the reverse is true. Normally, the covalent bonds of carbon are directed outwards towards the corners of a tetrahedron. A multiple bond can be thought of as two or three spring linkages which are curved and, hence, under considerable strain (Figure 2.11(b)). A multiple bond in organic molecules can be broken more easily than a single bond. Stated in another way this means that compounds such as ethene and ethine are more reactive chemically than ethane. The splitting of double covalent bonds is part of the mechanism of addition polymerisation (see Chapter 4).

EXAMPLE 2.7

What type of bonding occurs in silica, SiO_2?

SOLUTION

Silicon, like carbon, has four outer shell electrons, giving it a valency of 4. Also, as in carbon, the bonds are directed towards the corners of a tetrahedron. Oxygen has six outer shell electrons. A silicon atom forms single covalent bonds with four oxygen atoms giving a tetrahedron of SiO_4, as in Figure 2.12. As an oxygen atom has a valency of 2, and only one valency per oxygen atom is used in forming the tetrahedron, the unit should be referred to as SiO_4^{4-}. The oxygen atoms at each corner of the tetrahedron are bonded covalently to the silicon atoms in the neighbouring tetrahedra. (Refer also to section 5.7.)

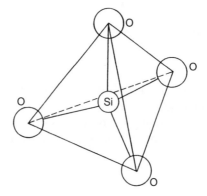

Figure 2.12 *SiO$_4$ tetrahedron.*

2.14 | The co-ordinate bond

The co-ordinate bond is a variation on the covalent bond. It involves the sharing of electrons but, in this case, the shared electrons are provided by one atom. The sulphate group, SO_4^{2-}, is a good example of co-ordinate bonding. In this group, the sulphur atom provides all the electrons required for covalent linkages with three oxygen atoms. The two additional electrons needed for the covalent link with the fourth oxygen atom are obtained from an ionic donor, for example, calcium, so making the SO_4 group an ion with a double negative charge (Figure 2.13).

The electron donor atom, calcium, becomes a calcium ion, Ca^{2+}, and forms an ionic bond with the sulphate ion to give the compound calcium sulphate. Thus there are two types of interatomic bonding

Figure 2.13 *Sulphate group, SO_4^{2-}.*

within the one molecule. There are many substances with interatomic bonding that is in part covalent and in part ionic.

2.15 | The metallic bond

In the bond types discussed so far, the electron distribution is rearranged to provide each nucleus with an outermost shell containing eight electrons. In the case of metals, where the number of valency electrons of each atom is small, usually 1, 2 or 3, it is not possible to fully satisfy the requirement of eight outer shell electrons per atom, and there is only a partial satisfying of this condition. In an assembly of metal atoms the principal forces acting will be an attraction due to gravity and a repulsion due to negatively charged electron shells in close proximity. For two adjacent atoms, the variation in energy with separation distance will be similar to that for ions (Figure 2.4, discussed earlier). (The separation distance, r_0, for minimum energy can be regarded as an atomic diameter. For the discussion of many properties of metals it is feasible to consider metal atoms as hard spheres of finite radius.) At the equilibrium separation distance, the outer, or valence, electron shells of the atoms can be regarded as being in contact or slightly overlapping. This means that the outer shell electrons, at certain points in their orbits, are attracted as much by one nucleus as by another, and the valence electrons follow complex paths around many nuclei. Thus, all valence electrons are shared by all the atoms in the assembly. This has similarities to a covalent bond, but the strength of the bond is weaker than that of a true covalent bond. One can liken the metallic state to an arrangement of positive ions permeated by an electron cloud or gas (Figure 2.14). The extreme mobility of the valence electrons accounts for the high electrical conductivity, and other properties of metals.

EXAMPLE 2.8

Calculate the number of electrons available to conduct an electrical current in 1 kg of aluminium. (For aluminium: $M = 26.98$, $Z = 13$. Avogadro's number $= 6.023 \times 10^{26}$/kmol.)

SOLUTION

Each aluminium atom has 13 electrons. The first two electron shells are filled with 2 and 8 electrons respectively, leaving 3 valence electrons in the third, or outer, electron shell. 1 kmol of aluminium has a mass of 26.98 kg, therefore 1 kg of aluminium contains

$$\frac{6.023 \times 10^{26}}{26.98} = 2.232 \times 10^{25} \text{ atoms}$$

so 1 kg aluminium has $(2.232 \times 10^{25}) \times 3 = 6.696 \times 10^{25}$ valence electrons.

Figure 2.14 *The metallic state – positive ions in an electron gas.*

2.16 | Secondary bonds

In addition to the types of primary bond discussed in the preceding sections, there also exist weaker secondary bonds. These are the hydrogen bond and van der Waal's bonds The bond energies of the secondary bonds, again are given by the Mie equation:

$$\text{P.E.} = \frac{A}{r^m} + \frac{B}{r^n}$$

For these bonds the values of the indices m and n are approximately 6 and 12 respectively and typical bond energies are of the order of 0.2 eV for a hydrogen bond and between 0.002 and 0.1 eV for van der Waal's bonds, as compared with 6.5 eV for the ionic bond in sodium chloride.

The bonding in the H_2O molecule was described in section 2.13 and it was shown that the molecule is polar with the two hydrogen atoms being positive relative to the two non-bonding orbitals of the oxygen atom. There is quite a strong force of attraction between the hydrogens and the negative ends of adjacent molecules (see Figure 2.15). The hydrogen bond can be written thus: H–O–H … O.

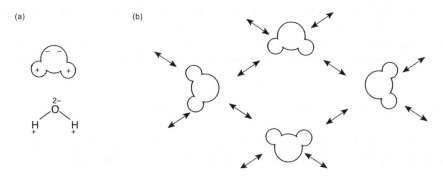

Figure 2.15 *(a) Representation of a polar H_2O molecule. (b) Attraction between neighbouring H_2O molecules – the hydrogen bond.*

The hydrogen bond does not occur only in water and ice but in a number of polymer materials also. In general terms, a hydrogen bond can be formed between an electronegative atom and a hydrogen atom which is also covalently bonded to another electronegative atom. Examples of hydrogen bonding in polymers are –N–H … O bonds between polyamide (nylon) molecules and –O–H … O bonds in cellulose and polyvinyl alcohols.

Many molecular compounds are polarised to some extent and electrostatic attractive forces exist between the molecular dipoles. These weak electrostatic attractive forces are termed van der Waal's bonds. Van der Waal's bonds can occur also between atoms. The monatomic inert gases, with full outer

Figure 2.16 *Momentary uneven electron distribution in atoms giving weak dipoles and weak interatomic attraction.*

electron shells, will condense into liquids and solids at extremely low temperatures. This indicates the existence of weak bonding forces. Within such atoms, owing to the continual movement of electrons, at any instant the centroid of negative charge may not necessarily coincide with the centroid of positive charge, that is, the nucleus. The atom becomes slightly polarised and may be weakly attracted to a similarly polarised atom (Figure 2.16). At very low temperatures, the kinetic energy of the atoms will be insufficient to overcome even these very weak attractions and the gas will condense. The energies of the van der Waal's bonds in inert gases is of the order of 0.002 eV but in some molecular substances with strong dipoles the magnitude of the bond energy may approach that of the hydrogen bond.

EXAMPLE 2.9

The energy, E, of a two-atom bond is given by $E = -\dfrac{A}{r^m} + \dfrac{B}{r^n}$ where A, B, m and n are all positive constants. Show that the requirement for stability of the bond is that $m < n$.

SOLUTION

In the Mie equation the potential energy due to attraction is $-A/r^m$ and that due to repulsion is B/r^n. At the stable atomic distance the net force is zero and the net potential energy will be negative (energy of attraction). This condition for bond stability will be achieved only if $A/r^m > B/r^n$ and the criterion for this is that $m < n$.

2.17 | Mixed bonds

The main types of primary and secondary bonds have been described in the preceding sections but, in very many cases, the bonding in substances cannot be described as purely ionic or purely covalent. There are a large number of compounds in which a mixture of bond types occurs. In many ceramics and glasses the bonding is part ionic and part covalent. Dipoles occur in a number of ceramic materials and permit the creation of van der Waal's bonds, examples being clay minerals and graphite. The interatomic bonding within polymers is predominantly covalent but van der Waal's intermolecular bonds occur in most of these and, as has been mentioned, hydrogen bonds occur also in some polymers. Within some metallic alloy systems intermetallic compounds can occur and this is an indication that some alloys contain bonds which are not entirely metallic.

SUMMARY

Atoms comprise a nucleus of protons and neutrons surrounded by electrons in orbits. An atom is characterised by an atomic number, the number of protons, and an atomic mass number, the combined number of protons and neutrons. The number of electrons balances the number of protons in a neutral atom. Each chemical element has a unique value of atomic number.

The unit of molar mass is a *mole*. A mole is the molecular mass number of a substance (the atomic mass number for a monatomic substance) expressed in grammes. A mole of any substance always contains the same number of molecules (same number of atoms for a monatomic substance), this number being Avogadro's number.

The electrons of an atom exist in a series of energy levels, or shells, surrounding the nucleus. The energy states of the electrons are described by a set of four quantum numbers and, in any atom, no two electrons can possess the same set of quantum numbers – the Pauli exclusion principle. The outer shell electrons are known as the valence electrons and the number of valence electrons determines the position of the element in the periodic table.

Some atomic nuclei are unstable and will decay spontaneously emitting α-particles or β-particles and γ-radiation. The rate of radioactive decay of an element decreases exponentially with time and the *half-life period* is an important characteristic of any radioactive substance.

Individual atoms bond together and, in general, there is interaction between the valence electrons in the creation of interatomic bonds. There are both forces of attraction and repulsion existing between adjacent atoms and when these opposing forces are in balance the potential energy of the atoms will have a minimum value. There are three types of primary bond: the ionic bond, based on electron transfer and the creation of positive and negative ions; the covalent bond, based on electron sharing between atoms; and the metallic bond, in which array of metal ions are permeated by a 'gas' of mobile valence electrons. Secondary bonds, based on electrostatic attraction, can occur between molecules with some electrical polarisation. The hydrogen bond and van der Waal's bonds are secondary bonds. These are weak and have low binding energies in comparison with the primary bonds. Many substances possess a mixture of bond types, for example, many ceramics have bonding which is part covalent and part ionic.

Questions

2.1 Silver, $Z = 47$, contains two isotopes in the following proportions: 56.5 per cent of atoms contain 60 neutrons, and 43.5 per cent of atoms contain 62 neutrons. Estimate the atomic mass number of silver.

2.2 Copper, $Z = 29$, has a mass number of 63.54 and contains two isotopes, Cu-63 and Cu-65. State the number of neutrons in each isotope and estimate the percentage of each isotope in copper.

2.3 Magnesium burns in oxygen according to the reaction $2Mg + O_2 = 2MgO$.

(a) What mass of oxygen is required for the full combustion of 1 kg of magnesium?
(b) How many molecules of MgO are produced in the reaction?
Assume: atomic mass numbers are Mg = 24.32, O = 16.00; $N_0 = 6.023 \times 10^{23}$/mol.

2.4 Pure silicon, $Z = 14$, $M = 28.09$, is a semiconductor. Assuming that the number of mobile electrons in 100 g of silicon is 5×10^{10}, what fraction of the total number of valence electrons present does this represent?

2.5 The following data are given for fluorine and magnesium:

	Symbol	Atomic no.	Atomic mass no.	Electron configuration
Fluorine	F	9	19.00	$1s^2, 2s^2, 2p^x$
Magnesium	Mg	12	24.31	$1s^2, 2s^2, 2p^y, 3s^z$

(a) What are the values of x, y and z?

(b) From the data, deduce the valencies of the two elements and give the formula for magnesium fluoride.

(c) What is the mass in grammes of one mole of magnesium fluoride?

(d) With simple reasoning, identify the possible properties of magnesium fluoride.

2.6 Indium, $Z = 49$, has all of its linear electron states filled, except for the 4f, which is empty. What would be the expected valency of indium?

2.7 The nucleus of an atom contains 13 protons and 14 neutrons. State
(a) the values of Z and M for the element,
(b) the electronic structure for atoms of this element, and
(c) the valency value.

2.8 Iron, $Z = 26$, may show a valency of either 2 or 3. Explain why this is so and state the electron configurations for the two valence states of iron.

2.9 Write down the values of Z and M for
(a) the element X formed from thorium-232 after six α-particle and four β-particle emissions, and
(b) the element Y formed from uranium-238 after six α-particle and two β-particle emissions.

2.10 The radioisotope iridium-192, which has a half-life period of 74.4 days, is used as a source material for industrial radiography. Determine the source intensity, as a percentage of the initial intensity, after (a) 25 days, (b) 50 days and (c) 100 days.

2.11 The remains of wooden artefacts were excavated from an archaeological site and were found to be emitting β-particles at a rate of 210 per second per kg wood. A living tree absorbs radioactive carbon-14 and emits 255 β-particles per second per kg wood. Estimate the age of the remains, assuming the half-life period of carbon-14 is 5600 years.

2.12 Complete the following disintegration table:

Nucleus	Symbol	Emission
Uranium	$^{235}_{92}U$	α
Thorium	?	?
Protoactinium	?	α
Actinium	?	?
Thorium	$^{227}_{90}Th$	α
Radium	?	?
Radon	$^{219}_{86}Rn$	α
Polonium	?	β
Astatine	$^{215}_{85}At$	?
Bismuth	?	?
Polonium	$^{211}_{84}Po$	?
Lead	$^{207}_{82}Pb$	stable

2.13 State, with reasons, the type of bonding that occurs in each of the following substances:
(a) common salt, NaCl,
(b) methane, CH_4,
(c) diamond, and
(d) brass containing 70 per cent copper and 30 per cent zinc.
Methane is a gas at ordinary temperatures and pressures while diamond is the hardest known solid with a melting temperature of 3800°C. Account for the properties of these substances on the basis of bonding.

2.14 The elements A, B, C, D and E have atoms with outer electron shell configurations containing 1, 2, 4, 7 and 8 electrons respectively. State and describe the type of bonding which is likely to occur in the following cases:
(a) between a large number of atoms of A,
(b) between a large number of atoms of C,
(c) between atoms of D,
(d) between atoms of E,
(e) between a mixture of atoms of A and B,
(f) between equal numbers of atoms of A and D, and
(g) between one atom of C and four atoms of D.

2.15 The curves showing the relationships between potential energy with interatomic separation distance for three substances are shown in Figure 2.17. One substance is a metal, another is an ionically bonded compound and the third is bonded by van der Waal's forces. Which curve relates to which type of material?

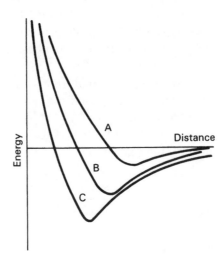

Figure 2.17

CHAPTER 3

Influence of bond type on structure and properties

How does the bonding affect structure and properties?

There are many factors that influence the final properties of a bulk material, and one of these is the nature of the bonding between atoms and molecules. The binding energies vary from one bond to another and the value of binding energy will influence the strength of a material. Also, it was stated that covalent bonds were highly directional whereas ionic and metallic bonds were not directional. This has an influence on the structures and, hence, on the properties of materials.

LEARNING OBJECTIVES

1. Know that the type of interatomic and intermolecular bonding is a major factor determining some basic materials properties including structure, density, elastic constants, electrical properties, stability and melting point.

3.1 | Structure

As stated in Chapter 2, both the ionic and the metallic bonds are non-directional. This permits ions and metal atoms to pack closely together in a regular crystalline array. All the atoms in a pure metal are identical and most metals form closely packed arrangements with *co-ordination numbers* of either 12 or 8.

The face-centred cubic and close-packed hexagonal crystal systems both have a co-ordination number of 12 while that for the body-centred cubic system is 8. (The co-ordination number is the number of other atoms or ions with which a particular atom or ion is in contact.) In an ionic compound there are at least two types of ion, of opposing electrical charge and which may be of differing size. This means that the co-ordination number for ionic crystals is usually 4, 6 or 8 and cannot be 12. On average, therefore, ionic crystalline materials have lower packing densities than metals.

Consider the two ionic compounds caesium chloride, CsCl, and sodium chloride, NaCl. The ions of caesium and chlorine are of comparable size and eight chlorine ions can pack around a caesium ion, whereas a sodium ion is small in relation to a chlorine ion and only six chlorine ions can pack around a sodium ion (Figure 3.1).

Why cannot a co-ordination number of 12 be achieved in ionic crystals? This is best illustrated by looking at a two-dimensional representation of the packing (Figure 3.2). In both caesium chloride and sodium chloride the negatively charged chlorine ions, while in contact with a positive ion, are not in direct contact with one another. However, to obtain a co-ordination number of 12 negative ions would need to be in contact with one another. Such a structure would not be stable as the ions of like sign would repel one another strongly.

Covalent bonds are directional and this will affect the structures of covalent crystals. It was shown in Chapter 2 (section 2.13, Figure 2.10) that the co-ordination number for diamond is four with each carbon atom bonded to four other atoms of carbon. A similar tetrahedral arrangement occurs in silicon

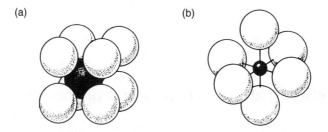

Figure 3.1 *(a) Caesium chloride structure. (b) Sodium chloride structure.*

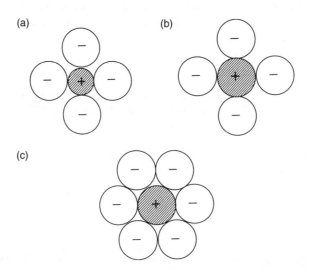

Figure 3.2 *Two-dimensional representation ion packing with co-ordination numbers of (a) 6, (b) 8 and (c) 12. Negative ions would need to be in contact in (c).*

and in silica, SiO_2 (Figure 2.12), where each silicon atom is bonded to four oxygen atoms, and each oxygen bonded to two silicon atoms.

Polymer materials are composed primarily of carbon and hydrogen and in a major class of polymers, linear polymers, the basic molecular structure is one composed of long chains of covalently bonded carbon atoms. The bond angle between adjacent covalent bonds is 109.5° (Figure 2.11) and so the 'linear chain' is far from straight. This means that the molecules may not be packed in a fully regular and symmetrical pattern and there may be void spaces of considerable size occurring within the structure.

Some materials possess structures which are either only partially crystalline or completely amorphous. This condition is known as the *glass* state (refer to Chapter 6). Because the atoms or ions are not packed in a regular and symmetrical pattern the total void space is greater in a glass than it would be in a crystalline material of the same composition.

3.2 | Density

The densities of materials are related to structure. The densities of crystalline metals range from $534 \, kg/m^3$ for lithium up to $22,500 \, kg/m^3$ for osmium. Crystalline ceramics show a wide variation of densities also, ranging from about 2000 to $17,000 \, kg/m^3$. However, the average density of metals is about $8000 \, kg/m^3$ while that of ceramics is about $5000 \, kg/m^3$. This is because the co-ordination numbers of metallic crystals are higher than those for ionically and covalently bonded crystals. Also, inorganic glasses have slightly lower densities than crystalline forms of the same material.

Carbon and hydrogen, the main components of most polymer materials, are of low density and, as stated above, there are considerable void spaces between molecules in many polymer structures. The densities of polymers are generally much lower than most metals and ceramics and range from about 900 to $3000 \, kg/m^3$.

EXAMPLE 3.1

The density of quartz, a fully crystalline form of silica, is $2660 \, kg/m^3$ but the measured densities of two samples of silica glass (vitreous silica) are $2210 \, kg/m^3$ and $2070 \, kg/m^3$ respectively. Why is this so?

SOLUTION

In the glass state the atoms are not packed in a fully symmetrical crystalline array and the total amount of void space is greater than in crystalline quartz giving the vitreous form a lower density. The vitreous sample with the higher density cannot be entirely glassy and must contain some crystalline areas.

3.3 | Stiffness and rigidity

The elastic moduli, E, G and K, are fundamental properties of materials. When a stress is applied to a material a dimensional change, or strain, is developed and this strain is termed elastic if the material returns to its original dimensions when the stress is removed. A direct tensile or compressive stress will produce an increase or a decrease in the length of the material. For a direct stress, the ratio of the stress applied to the elastic strain developed is termed the modulus of elasticity, E, or Young's modulus, and this constant is a measure of the stiffness of the material.

The primary bonds between atoms can be regarded as having the properties of tiny springs and when an external force is applied these springs will either extend, compress or twist, according to the type of force system. The effect of a tensile force on a line of atoms forming part of a crystal is shown schematically in Figure 3.3. In crystalline materials the amount of strain produced per unit of stress tends to be small and so the value of Young's modulus, E, is high. Crystalline ceramics and metals have high values of E, generally within the range 40–450 GPa.

The basic structure of many polymer molecules is a chain of carbon atoms with an angle of 109.5° between adjacent covalent bonds. Also, even in a so-called linear polymer, the total molecular chain is far from straight. The effect of a tensile force on a polymer molecule is not only to extend the covalent bonds but to straighten them and, also, to straighten the long chain molecule. These effects would cause a greater amount of extension per unit of stress than in a crystalline material (Figure 3.4) but, in addition the bonding between molecules consists of weak secondary bonds. Using the spring analogy, the secondary bonding springs are of lower stiffness than the primary bond springs. These effects combine to give polymer materials low values of Young's modulus, usually within the range 0.7–3.5 GPa.

Figure 3.3 *Effect of a tensile force extending the bonds between atoms.*

Figure 3.4 *Effect of a tensile force on a polymer. The bonds are straightened.*

EXAMPLE 3.2

A sample of lightly vulcanised natural rubber may be extended to up to seven times its original length by application of a tensile force. How can this very large extensibility be explained?

SOLUTION

The very long chain molecules in natural rubber (polyisoprene) tend to twist themselves into spiral forms and, when subject to a tensile force, these molecules behave like a series of miniature helical springs. Each little spring can be extended by a very large amount. The process of vulcanising (refer to section 4.6) produces some cross-links between molecules increasing the rigidity and stability of the material. Natural rubber is one of the group of materials referred to as *elastomers*.

The group of materials termed *elastomers*, of which natural rubber is one, possess extremely small elastic moduli, within the range 0.01–0.1 GPa. These materials are polymers which are at a temperature higher than their *glass transition temperature*. The glass transition temperature is discussed in Chapter 6. It is sufficient to say, at this stage, that this temperature is sufficiently high to break the van der Waal's intermolecular bonds. Elastomers lose their very low elastic moduli if cooled to a temperature below their glass transition temperature, thus allowing the van der Waal's bonds to reform. The glass transition temperature for natural rubber is about $-75°C$.

3.4 | Stability and melting point

The stability of a substance is related to the bond strength. One indicator of bond strength and stability is the melting point of a substance. When a substance is heated the absorption of heat energy causes an increase in the vibrational amplitude of the atoms. A point will be reached when the vibrational energy is sufficient to overcome the interatomic bonding energy of the atoms in the solid and the atoms become mobile, in other words the substance melts.

Both ionic bonds and covalent bonds are strong bonds that possess a similar range of bond energies. However, within both ionically and covalently bonded substances there are major variations in melting point. For example, consider the two substances sodium chloride, NaCl, and magnesium oxide, MgO. Because of the larger ionic charge the force of attraction between Mg^{2+} and O^{2-} ions is greater than that between Na^+ and Cl^- ions. The stronger bonding in MgO is reflected in the melting temperatures of the two substances, 2800°C for MgO as compared with about 800°C for NaCl.

In covalently bonded crystals of elements the strength of the bonding lessens as the atomic number of the element increases and the distance of the outer shell, or valence, electrons from the nucleus becomes greater. This is illustrated by comparing the melting points of the Group IV elements carbon ($Z = 6$), silicon ($Z = 14$), germanium ($Z = 32$) and tin ($Z = 50$). The respective melting points are carbon (diamond) 3800°C, silicon 1420°C, germanium 937°C and tin 232°C. The melting temperatures of the major ceramic materials, generally oxides, carbides and nitrides of the lower atomic number elements such as aluminium, boron, magnesium and silicon, are all high being in the general range 1000–3000°C.

Bonding in polymer molecules is covalent but, generally, polymers have low melting points, mainly within the range 100–400°C. Although the interatomic bonding is covalent only weak secondary bonds, mainly of the van der Waal's type, exist between molecules and it is this factor that influences the melting temperature. Some polymer materials have higher melting temperatures than others with broadly similar structures. An example is the polyamides (nylons) in which the intermolecular bonding is due to the presence of hydrogen bonds which are considerably stronger than van der Waal's bonding (refer to section 2.16). The melting temperatures of polymers are also influenced by molecular weight and increase as the molecular weight of the substance increases. The expression relating the melting temperature, T_m, in Kelvin, to the average molecular weight of polymer molecules is of the general form

$$T_m = \{A + (B/n)\}^{-1}$$

where n is the number of carbon atoms in the polymer chain and A and B are constants.

EXAMPLE 3.3

What is the difference between the melting temperatures of two samples of polyethylene if the average molecular chain lengths are 500 carbon atoms in sample A and 5000 carbon atoms in sample B. The melting temperature (K) for polyethylene is given by $T_m = \{0.0024 + (0.017/n)\}^{-1}$.

SOLUTION

For sample A, $n = 500$, so $T_m = \{0.0024 + (0.017/500)\}^{-1} = \{0.0024 + 0.000034\}^{-1}$.

$$= 0.002434^{-1} = 410.8\,K$$

For sample B, $n = 5000$, so $T_m = \{0.0024 + (0.017/5000)\}^{-1} = \{0.0024 + 0.0000034\}^{-1}$

$$= 0.0024034^{-1} = 416.1\,K$$

The difference in melting temperature between the two samples is $416.1 - 410.8 = 5.3\,K$.

In metals, although the strength of the metallic bond tends to be less than that of ionic and covalent bonds, the close packing of atoms in metal crystals leads towards relatively high melting points in many cases. There is a very wide range of melting temperatures for metals, from $-39°C$ for mercury to $3380°C$ for tungsten. However, some general trends are discernible. One trend is that melting temperatures tend to increase as the number of valence electrons increases. For example, looking at elements with atomic numbers of 11, 12 and 13, sodium, magnesium and aluminium with 1, 2 and 3 valence electrons respectively, the melting points are: $98°C$ for sodium, $649°C$ for magnesium and $660°C$ for aluminium. Another trend is for the melting point to decrease as the atomic number increases for metals within the same group of the periodic table. For example, the Group I metals lithium ($Z = 3$), sodium ($Z = 11$), potassium ($Z = 19$) and caesium ($Z = 55$) have melting points of $180°C$, $98°C$, $64°C$ and $30°C$, respectively. The metals of highest melting temperature are those of the three transition series and here the bonding may not be purely metallic but show some aspects of covalent bonding.

3.5 | Electrical properties

In both covalent and ionic bonding the valency electrons are localised in orbitals around one or a pair of atoms. In the metallic bond the valency electrons are not localised and a metal crystal can be likened to an array of positive ions permeated by a cloud of electrons, the *'electron gas'*. In an electric field the electron gas will move preferentially in one direction, thus constituting an electric current. All metals are electrical conductors. Ionic and covalent solids are not electrically conductive and are insulators, or dielectrics, because their valency electrons are bound to particular atoms. However, the ions in ionic compounds become mobile when either the substance is dissolved in a solvent or is melted. The mobile ions can move in a preferential direction when influenced by an electric field. This form of conduction, by the movement of ions, is termed *electrolytic conduction* and the solution or molten salt is termed an *electrolyte*. It may be possible for ions of small size to migrate, or diffuse, through a solid in a strong electric field causing the solid material to have a very small, through measurable, conductivity.

Some crystalline dielectric materials show a special effect, the *piezoelectric* effect. Piezoelectric materials polarise when strained mechanically, thus generating a small electrical voltage or field. This phenomenon is reversible and such a material will strain mechanically when subjected to an electrical field. The piezoelectric effect is used in *transducers* which convert electrical fields into mechanical vibrations, and vice versa. This phenomenon is dealt with in more detail in Chapter 13.

Although most materials can be regarded as either electrical conductors or insulators, there is a small but very important group of materials including, notably, silicon and germanium, which are known as semiconductors. The mechanism of semiconduction is covered in Chapter 13.

SUMMARY

Metal atoms tend to pack more closely together than is the case with ionic and covalently bonded structures and so the average density of metals is higher than that for ceramics. An inorganic glass has a lower density than the crystalline form of the same material. There are void spaces between molecules in polymer materials giving them lower densities than other materials.

Crystalline metals and ceramics tend to have high stiffness and rigidity. Polymer materials have much lower elastic moduli than metals and ceramics because they have weak intermolecular bonds. The group of materials termed elastomers have extremely low elastic moduli.

Ionically and covalently bonded crystals tend to have high melting temperatures because of the high bond strengths. Polymer materials tend to have low melting temperatures because of the weak secondary intermolecular bonds. Metals show a wide range of melting points ranging from below 0°C to more than 3000°C.

In metal structures the valence electrons are highly mobile and permeate the whole array of atoms permitting metals to be electrically conductive. The valence electrons in covalent and ionic solids are bound to particular atoms or pairs of atoms, so these materials are electrical insulators (dielectrics). When an ionic substance dissolves in a solvent or melts the charged ions have mobility. In these states electrical conduction occurs through migration of ions.

Questions

3.1 Why is it that the average density of metals is higher than the average density of ceramics?

3.2 On the basis of primary bond type, what differences would be expected between (a) the melting temperatures and (b) the elastic moduli of magnesium and magnesia (MgO)?

3.3 Beryllium ($Z = 4$) and magnesium ($Z = 12$) are both in Group IIA of the periodic table. The melting temperature of beryllium is 1280°C and that of magnesium is 649°C.
(a) Explain this, on the basis of bond energies.
(b) How would the values of E be expected to differ for the two elements?

3.4 How would properties such as Young's modulus and melting temperature be expected to differ between boron and aluminium, both metals in Group IIIB of the periodic table?

3.5 For the assembly of products involving the joining of a polymer to a metal, or the joining of two polymer components, designers make much use of 'clip-fit' assembly techniques. What major property of polymers is responsible for the usefulness of this method of assembly? How is the property related to the bonding forces within polymers?

3.6 As a class of materials, polymers tend to possess low values of Young's modulus and low densities. What can be done to produce materials based on polymers which will have much higher values of stiffness combined with relatively low densities?

4

The formation of polymers

Why learn about the formation of polymers?

Polymers form the basis of a very important range of engineering materials. Living organisms generate large and complex molecules from simple ingredients. These large molecules are composed principally of carbon and hydrogen. The long, fibrous molecules produced by the various organisms help to determine the properties of naturally occurring materials such as timbers, wool, cotton, leather and natural rubber. The whole range of plastics materials is based also on very large molecular structures but, generally, the large molecules are produced synthetically. The nature of the raw materials and the way in which the ingredients combine influences the molecular structure and, hence, the properties of the resulting polymer material.

LEARNING OBJECTIVES

1. Know the significance of the terms *thermoplastic, thermoset* and *elastomer*.
2. Understand the mechanisms of *addition* and *condensation* polymerisation.
3. Distinguish between *linear* and *non-linear* polymer structures.
4. Know how branching and cross-linking reactions can modify the structures and properties of linear polymers.
5. Know how the degree of polymerisation can influence properties.
6. Know the main methods used for polymerisation.

4.1 | Introduction

Polymer materials are composed of very large molecules and the name *polymer* is derived from *poly*, meaning many, and *mer*, meaning unit. The chemical compounds which are reacted together to create polymer molecules are termed *monomers*. These monomers, the building blocks of plastics materials, can be reacted together, or polymerised, in several ways and these are described in subsequent sections of this chapter. The polymers produced may possess *linear* or *non-linear* structures.

The plastics materials produced by polymerisation can be classified under three headings: as *thermoplastics*, *thermosets* and *elastomers*. A thermoplastic material is one which becomes more plastic when heated and becomes rigid again on cooling. A thermoset is a material which, on first heating, becomes plastic and can be moulded and then 'cures' at temperature, setting into a hard rigid structure. Once curing and setting has occurred the material cannot again be made plastic. Some materials are *cold-setting*. This type usually involves mixing a resin with a 'hardener' which causes curing to occur at ordinary temperature to produce a hard, rigid material. Elastomers are a group of polymer materials of very low elastic moduli giving them great flexibility and the ability to suffer major elastic deformations.

4.2 Addition polymerisation

Many covalently bonded compounds of carbon and hydrogen exist in nature. The simplest group of hydrocarbon compounds is the paraffin series, many members of which are found in petroleum. The first few compounds in the paraffin series are

$$
\text{methane, } CH_4 \quad \text{or} \quad
\begin{array}{c}
\text{H} \\
| \\
\text{H}-\text{C}-\text{H} \\
| \\
\text{H}
\end{array}
$$

$$
\text{ethane, } C_2H_6 \quad \text{or} \quad
\begin{array}{c}
\text{H} \quad \text{H} \\
| \quad\ | \\
\text{H}-\text{C}-\text{C}-\text{H} \\
| \quad\ | \\
\text{H} \quad \text{H}
\end{array}
$$

$$
\text{propane, } C_3H_8 \quad \text{or} \quad
\begin{array}{c}
\text{H} \quad \text{H} \quad \text{H} \\
| \quad\ | \quad\ | \\
\text{H}-\text{C}-\text{C}-\text{C}-\text{H} \\
| \quad\ | \quad\ | \\
\text{H} \quad \text{H} \quad \text{H}
\end{array}
$$

The paraffins are known as saturated hydrocarbons as every linkage is a single covalent bond, and the carbon chain is fully saturated with hydrogen. Other hydrocarbons may be unsaturated and contain double or triple covalent bonds. An example is ethylene, or ethene:

$$
C_2H_4 \quad \text{or} \quad
\begin{array}{c}
\text{H}\!\!\diagdown \qquad \diagup\text{H} \\
\text{C}=\text{C} \\
\text{H}\!\!\diagup \qquad \diagdown\text{H}
\end{array}
$$

An unsaturated compound may be made to react with itself, or with other compounds, to produce large complex molecules. This is termed *polymerisation*. In the case of ethylene, the double bonds can split allowing the separate molecules to link up, forming polyethylene (polythene).

$$
\begin{array}{c}
\text{H} \ \text{H} \\
| \ \ | \\
\text{C}=\text{C} \\
| \ \ | \\
\text{H} \ \text{H}
\end{array}
+
\begin{array}{c}
\text{H} \ \text{H} \\
| \ \ | \\
\text{C}=\text{C} \\
| \ \ | \\
\text{H} \ \text{H}
\end{array}
+
\begin{array}{c}
\text{H} \ \text{H} \\
| \ \ | \\
\text{C}=\text{C} \\
| \ \ | \\
\text{H} \ \text{H}
\end{array}
\rightarrow
\begin{array}{c}
\text{H} \ \text{H} \ \text{H} \ \text{H} \ \text{H} \ \text{H} \\
| \ \ | \ \ | \ \ | \ \ | \ \ | \\
-\text{C}-\text{C}-\text{C}-\text{C}-\text{C}-\text{C}- \\
| \ \ | \ \ | \ \ | \ \ | \ \ | \\
\text{H} \ \text{H} \ \text{H} \ \text{H} \ \text{H} \ \text{H}
\end{array}
$$

$$
\text{or } n\left(
\begin{array}{c}
\text{H} \ \text{H} \\
| \ \ | \\
\text{C}=\text{C} \\
| \ \ | \\
\text{H} \ \text{H}
\end{array}
\right)
\rightarrow
\left(
\begin{array}{c}
\text{H} \ \text{H} \\
| \ \ | \\
-\text{C}-\text{C}- \\
| \ \ | \\
\text{H} \ \text{H}
\end{array}
\right)_n
\quad
\begin{array}{l}
\text{where } n \text{ is a very large} \\
\text{number}
\end{array}
$$

This is the process termed *addition polymerisation*. According to the reaction equation shown above there is a free valence at each end of the molecular chain. This is not allowable and there must be a small quantity of some other reactant present to provide a logical ending to the chain, for example chlorine, giving

$$
\begin{array}{ccccc}
& H & H & H & H \\
& | & | & | & | \\
- - - - - & C & C & C & C - Cl \\
& | & | & | & | \\
& H & H & H & H
\end{array}
$$

The reactant used to provide a chain ending is termed the *initiator*. Certain peroxides, including hydrogen peroxide, H_2O_2 or HO – OH, are frequently used as initiators. The – OH or hydroxyl groups provide chain terminations. The amount of initiator added will affect the average molecular weight of the resultant polymer and this, in turn, will influence the properties of the polymer. In polymers of the addition type, the number of atoms in each molecular chain will be extremely large running into thousands.

EXAMPLE 4.1

Hydrogen peroxide, H_2O_2, is used as an initiator in the polymerisation of ethylene, C_2H_4. Assuming that all the hydrogen peroxide is used to form terminal groups, what would be the degree of polymerisation in the polyethylene if the amount of hydrogen peroxide added was 1 kg per 1000 kg of ethylene? Assume atomic mass numbers are: $C = 12$, $H = 1$ and $O = 16$.

SOLUTION

When hydrogen peroxide is used as an initiator one molecule of H_2O_2 splits and two –OH groups are available to terminate one linear polymer molecule.

Molar mass of $H_2O_2 = (2 \times 1) + (2 \times 16) = 34$

So, 1 kg $H_2O_2 = 1/34$ kmol

Molar mass of $C_2H_4 = (2 \times 12) + (4 \times 1) = 28$

So, 1000 kg $C_2H_4 = 1000/28$

Average number of C_2H_2 mers per linear chain $= \dfrac{1000 \times 34}{28}$

$$= 1214$$

So the degree of polymerisation in the polyethylene is 1214.

There are many unsaturated compounds that can be polymerised by an addition reaction and these include

vinyl chloride

$$
\begin{array}{cc}
H & H \\
| & | \\
C = C \\
| & | \\
H & Cl
\end{array}
$$

giving polyvinyl chloride (PVC)

tetrafluoroethylene

$$
\begin{array}{cc}
F & F \\
| & | \\
C = C \\
| & | \\
F & F
\end{array}
$$

giving PTFE

styrene

$$
\begin{array}{cc}
H & H \\
| & | \\
C & = C \\
| & | \\
H & C_6H_5
\end{array}
$$

giving polystyrene (PS)

propylene

$$
\begin{array}{cc}
H & H \\
| & | \\
C & = C \\
| & | \\
H & CH_3
\end{array}
$$

giving polypropylene (PP)

EXAMPLE 4.2

What mass of hydrogen peroxide, H_2O_2, would be needed as an initiator for the polymerisation of 100 kg of propylene, $CH_2 = CH.CH_3$, to produce an average degree of polymerisation of 750? Assume atomic mass numbers are $C = 12.01$, $H = 1.008$, $O = 16.00$.

SOLUTION

Molar mass of $H_2O_2 = (2 \times 1.008) + (2 \times 16) = 34.016$
Molar mass of $CH_2 = CH.CH_3 = (3 \times 12.01) + (6 \times 1.008) = 42.078$
Molar mass of polypropylene with an average degree of polymerisation of 750
$$= 42.078 \times 750 = 31{,}558.5$$
One molecule of H_2O_2 is needed for each long chain molecule. Therefore 34.016 kg of H_2O_2 would be needed to make 31,558.5 kg of polypropylene.

So, 100 kg propylene requires $\dfrac{100 \times 34.016}{31{,}558.5} = 0.108$ kg H_2O_2 for an average degree of polymerisation of 750.

It will be seen from the above that unsaturated monomers produce polymers that are fully saturated. In some addition polymerisation reactions, however, the resultant polymer is unsaturated. An example of this is the polymerisation of butadiene. Butadiene has two double bonds per molecule and the polymerisation reaction can be written

$$n(CH_2 = CH-CH = CH_2) \rightarrow (-CH_2-CH = CH-CH_2-)_n$$

The product, polybutadiene, is a rubber-like material and, being unsaturated, is capable of being further reacted. The type of reaction that can be brought about in unsaturated polymers will create cross-links between molecules. A good example of this is the vulcanisation of rubbers. Refer to section 4.6.

The addition polymerisation of a single monomer produces a *homopolymer* but it is also possible to polymerise a mixture containing two or more monomers. *Copolymerisation* is the term used for the addition polymerisation of two or more monomers. Two monomers, represented by A and B, when polymerised together, may produce a copolymer with either a random or ordered linear structure, as shown:

–A–A–B–A–B–B–B–A–A–B–A–B–B–A–A–A–B–A–B–B– (random)
–A–A–B–B–B–A–A–B–B–B–A–A–B–B–B–A–A–B–B–B– (ordered)

It is also possible to produce a *block copolymer* in which the structure is composed of large blocks of A alternating with large blocks of B, as

$-(-A-)_m-(-B-)_n-(-A-)_m-(-B-)_n-$, where m and n are large integers.

Examples of copolymers are SAN (styrene/acrylonitrile) and ABS (a terpolymer of acrylonitrile, butadiene and styrene).

4.3 | Condensation polymerisation

Another type of polymerisation reaction is condensation polymerisation. A condensation reaction is one in which two molecules combine to form a larger, more complex molecule, together with a small molecular substance as a by-product. Frequently, the by-product in condensation reactions is water, H_2O. The reaction between an alcohol and an organic acid to form an ester is an example of a condensation reaction. Ethylene glycol, a di-ol, and teraphthalic acid, a di-acid, can be reacted together to form a linear polyester, as shown in the equation below:

The product of the above reaction is used as a synthetic fibre, one trade name for which is *terylene*. Linear polyamides (PA), also known as *nylons*, are produced by the condensation of a di-acid and a di-amine or, in some cases by self condensation of an amino acid, a compound with an acid grouping at one end of the molecule and an amine group at the other end. A typical reaction is that between hexamethylene diamine and adipic acid. The product of the reaction is PA 6.6, so termed because both of the reacting monomers contain six carbon atoms:

EXAMPLE 4.3

PA 6 (nylon 6) is made by the self-condensation of amino caproic acid, $NH_2.(CH_2)_5.COOH$. Determine the degree of polymerisation in a batch of PA 6 with an average molecular weight of 175 kg/mole. Assume atomic mass numbers are $C = 12.01$, $O = 16.00$, $N = 14.008$, $H = 1.008$.

SOLUTION

A molecule of amino caproic acid contains 6 carbon atoms, 2 oxygen atoms, 1 nitrogen atom and 13 hydrogen atoms. Its molecular weight, therefore, is

$$(12.01 \times 6) + (16.00 \times 2) + 14.008 + (1.008 \times 13) = 131.172$$

When the compound condenses with itself to give a linear polymer molecule one molecule of water, H_2O, is created for each link between molecules so that the molecular weight of the repeating unit in the chain molecule is

$$131.172 - \{(1.008 \times 2) + 16.00\} = 113.156$$

If the average molecular weight of the PA6 is 175 kg then it must contain

$$\frac{175,000}{113.156} = 1547 \text{ repeating units}$$

The degree of polymerisation of the PA 6 is 1547.

If one of the reactants in condensation polymerisation has more than two functional groups, or reacts at more than two points, then a non-linear polymer structure can be created. A good example of this is the condensation between phenol, C_6H_5OH, and formaldehyde, HCHO, creating $-CH_2-$ linkages between adjacent phenol molecules, as shown below. The carbon atoms in the phenol structure shown are numbered and condensation reactions can occur involving the hydrogen atoms at the 2, 4 and 6 positions. Only one $-CH_2-$ linkage is shown in the equation but they can form from all three reactive positions in each phenol to create a three-dimensional network molecular structure. This type of structure is shown schematically in Figure 4.1.

Within a cyclic carbon compound the sites may be numbered thus:

Other condensation reactions of importance in the formation of non-linear network polymers are those between formaldehyde and amines (giving urea-formaldehyde and melamine-formaldehyde materials), and between acids and alcohols to give non-linear polyesters.

Epoxides are another valuable group of polymer materials with non-linear network structures but the epoxy reaction is not a condensation reaction as there is no reaction by-product evolved. Uncured epoxy resin consists of polymer molecules with a reactive epoxy group at both ends of each molecular chain:

It is necessary to mix a hardener compound with an epoxy resin to convert, or cure, it into a fully rigid network structure. Several types of hardener may be used, but the more commonly used types contain at least two amino, $-NH_2$, groups. The amine hardener will form cross-links between the epoxy molecules. This cross-linking reaction is shown below:

$$NH_2-\cdots-NH_2 + CH_2 \overset{O}{-} CH -\cdots \rightarrow NH_2-\cdots NH-CH_2-\overset{OH}{CH}-\cdots$$

Both hydrogen atoms in each amino group can react with an epoxy group in this way and, in consequence, considerable cross-linking can occur resulting in the formation of a very hard and rigid material.

4.4 Linear and non-linear polymers

In general, the polymers produced by addition polymerisation are linear polymers with long fibrous molecules. As has been stated, some polymers produced by condensation polymerisation are also linear in structure. Several types of non-linear structures are also possible in polymers. These are branched, cross-linked and full network (see Figure 4.1).

Linear, branched and lightly cross-linked polymers are thermoplastic, while heavily cross-linked polymers and those with a full network structure are thermosets. Most elastomers are linear polymers with light cross-linking and it is the presence of these cross-links which causes the elastomer to return to its original dimensions after the removal of a deforming stress.

4.5 Branching

Branching is the linking together of some polymer molecules to create side branches to the long chain molecules. Branching reactions require a certain input of energy (activation energy) before they can

Figure 4.1 *(a) Linear polymer, (b) branched polymer, (c) cross-linked polymer, (d) network polymer.*

occur and the necessary energy can be provided by high-energy photons. Irradiation of a polymer such as polyethylene with X- or γ-radiation can cause branching. The formation of branch chains stiffens the polymer and increases its melting point.

Branching reactions can be used to good effect in the manufacture of some products. For example, polyethylene, with low softening and melting points, can be processed and moulded with ease but the moulded product is flexible and can soften and lose its shape in boiling water or steam. Branching reactions can be initiated in the already moulded products to give stronger and more rigid articles and which possess higher softening points than untreated polyethylene.

EXAMPLE 4.4

Give an example of a type of product for which the use of branched polyethylene would be most suitable.

SOLUTION

Polyethylene is non-toxic, has good resistance to most chemicals, is of relatively low cost and can be moulded into shapes easily. Its main disadvantage is its low softening temperature. Its general properties would make it very suitable for many applications in the medical field including urine bottles, bedpans, etc., were it not for its low softening temperature which precludes the use of steam sterilisation. This can be overcome by generating branching reactions in the mouldings to greatly increase softening points and so give products that can withstand the temperatures of sterilisation.

Some copolymers can be produced by branching. These, known as *graft copolymers* are made by initiating branching reactions in a mix of two linear polymers. Chains of polymer A can be grafted on at points along the linear chains of polymer B.

4.6 | Cross-linking

A good example of a cross-linking reaction is the vulcanisation of natural rubber using sulphur. Natural rubber, cis-polyisoprene, is an unsaturated polymer with the formula $(-CH_2-C(CH_3)=CH-CH_2-)_n$. During vulcanising flexible cross-links are formed between adjacent polyisoprene chains with a consequent increase in the hardness and stiffness of the material. Each link may be either a single atom of sulphur or a short chain of sulphur atoms:

Generally, between 1 and 5 parts sulphur per 100 parts of rubber are used for vulcanisation, the number of cross-links formed and, hence, the hardness of the vulcanised rubber increasing as the amount of sulphur is increased. If the sulphur content is increased to about 40 parts per 100 parts rubber, complete vulcanisation, with all possible reaction sites used, will take place creating a three-dimensional network structure. The product of this full vulcanisation of natural rubber is the hard, rigid, elastic solid, *ebonite*.

EXAMPLE 4.5

A butyl rubber, BR, formula $(-CH_2-CH=CH-CH_2-)_n$, has an average degree of polymerisation of 600. What mass of sulphur per kg of BR is needed to lightly vulcanise by cross-linking at one-twentieth of the available sites, assuming that the cross-links are composed of S_2 short chains: Take atomic mass numbers as $H = 1.008$, $C = 12.01$, $S = 32.066$.

SOLUTION

Molar mass of BR = $\{(12.01 \times 4) + (1.008 \times 6)\} \times 600 = 32458.2$
There is one possible cross-link site per mer, or 600 sites per polymer molecule, so one-twentieth of the possible sites = 30.
Molar mass of sulphur needed for 60 links of $S_2 = (32.066 \times 2 \times 60) = 3848$
Mass of sulphur needed for 1 kg of BR = $3848/32458.2 = 0.12\,kg$

4.7 | Stereoregularity

A linear homopolymer which has side groups attached to the linear chain, for example, $-CH_3$ groups in polypropylene or $-C_6H_5$ groups in polystyrene, can show structural variations. These variations are termed *stereoregularity* or *tacticity*. When all the attached groups are on the same side of the carbon chain the arrangement is termed *isotactic*. A *syndiotactic* configuration is one in which the attached groups alternate in a regular manner along both sides of the linear chain. The third type of tacticity is the *atactic* structure in which there is a random arrangement of the attached groups. These three configurations are shown schematically in Figure 4.2 for the general case of a vinyl polymer with the repeating unit $-CH_2-CH(R)-$.

The tacticity of polymers has implications for the microstructures and properties of polymers as it affects the way in which adjacent molecules, or parts of molecules, may fit together. A regular isotactic homopolymer can have a strong tendency to crystallise whereas atactic polymers possess insufficient regularity to form crystals and, instead, give glassy polymers (see section 6.5). Some syndiotactic polymers

Figure 4.2 *(a) Isotactic structure, (b) syndiotactic structure, (c) atactic structure.*

possess sufficient regularity to give some crystallisation. Commercial polypropylene is isotactic while commercial polystyrene is atactic.

4.8 Degree of polymerisation

The molecular chain lengths within many linear polymers is very large, each molecule containing thousands of carbon atoms. Within any polymer the chain lengths of the constituent molecules are not equal and, frequently, a figure is quoted for the average molecular size. The average number of repeating units, or mers, in molecules of a linear polymer is referred to as the *degree of polymerisation*.

It was stated in section 4.2 that the process of addition polymerisation requires an initiator. The function of an initiator is to provide terminal groups for the polymer molecules. Therefore, the amount of initiator used will determine the degree of polymerisation and, hence, affect some of the properties of the resulting material.

For any particular polymer an increase in the degree of polymerisation will cause an increase in its melting temperature and also increase strength and hardness. The variation of melting temperature, T_m (in K), with the degree of polymerisation, n, is given by an empirical relationship of the form

$T_m = (A + B/n)^{-1}$ where A and B are constants.

4.9 Polymerisation methods

There are four basic methods used for the production of polymers from one or more monomers. These are:

(a) mass, or bulk, polymerisation,

(b) solution polymerisation,

(c) emulsion polymerisation,

(d) suspension polymerisation.

The choice of polymerisation method used for any particular material will depend on either the nature of the reaction chemistry or the nature of the resultant polymer, whether it is a low or high viscosity liquid or a flexible, friable or rigid solid.

Mass polymerisation is the simplest method. Monomer, with the appropriate amount of initiator added, is allowed to polymerise at a predetermined temperature. The process rate may be accelerated by means of a catalyst. The main disadvantage of this process is that, as polymerisation proceeds, the viscosity of the mass increases and stirring of the mass so as to maintain a uniform temperature becomes more difficult. Polymerisation is an exothermic process and the heat evolved during the reaction can cause an uneven distribution of temperature in the reactive mass and render it difficult to achieve full control of process temperature. A consequence of this could be a product of variable quality. Polystyrene and PVC may be made by mass polymerisation methods.

Solution polymerisation, as the name implies, is polymerisation with the monomer dissolved in a suitable solvent. Control of process temperature is much easier in this process than in mass polymerisation. However, there are disadvantages. Additional processes are required to separate solvent and any unreacted monomer from the polymer to give a material of marketable quality. These additional purification procedures may be complex, expensive and also hazardous. Polystyrene and SAN copolymer may be made by solution polymerisation, a suitable solvent being ethyl benzene.

A variation on solvent polymerisation is interfacial condensation and this can be used for the condensation polymerisation of an organic acid and an amine in the manufacture of a polyamide (nylon). The two reactants are dissolved in different solvents. The diamine is dissolved in water and the diacid dissolved in a chlorinated solvent giving a solution of high density. The aqueous amine solution is carefully poured over the other solution to form two immiscible layers. The condensation reaction occurs at the interface between the two liquid layers and the resulting polymer is drawn off continuously as nylon fibre.

Some monomers can be highly dispersed in water, forming an emulsion, and it may be possible to polymerise from such an emulsion. A water emulsion may possess a viscosity sufficiently low to permit good control of process temperature. In the same way as for solution polymerisation, it is necessary to have additional procedures for the separation of pure polymer from water and unreacted monomer. The advantage offered over solvent polymerisation is that the purification processes are less hazardous. Polyvinyl acetate and PVC may be produced by emulsion polymerisation.

Suspension polymerisation is a process in which the monomer is broken up into small particles or droplets and maintained in suspension in water by continuous agitation, but without an emulsion being formed. Polymer particles form and these tend to coalesce into larger globules which are removed from the suspension by screening or centrifuging. Styrene is almost insoluble in water and some polystyrene is produced by the polymerisation of a suspension of one part styrene in two parts water.

SUMMARY

Organic compounds, monomers, may take part in reactions to form very large molecular substances known as polymers. An unsaturated compound, in the presence of an initiator, may react with itself or with another unsaturated compound by addition polymerisation to give a linear polymers. Both linear and non-linear polymers can be generated by the process of condensation polymerisation. Non-linear polymer structures give rise to materials which are harder and more rigid than linear polymers. Some linear polymers can be converted into non-linear structures by means of branching and cross-linking reactions.

One characteristic of linear polymers is the average molecular chain length within the material. This is termed the degree of polymerisation and can be controlled by regulating the amount of initiator used in the polymerisation reaction.

Thermoplastic materials become more plastic on heating but return to the more rigid state on cooling. Thermoplastics tend to have linear, branched or lightly cross-linked structures. Thermosets become plastic on first heating, when they can be moulded, but then cure, setting into a hard rigid structure, at temperature. Some materials are cold-setting and cure at ordinary temperatures after mixing the ingredients – resin and hardener.

Questions

4.1 What is meant by the term *polymerisation*? Distinguish clearly between addition and condensation polymerisation.

The structural formulae of some chemical compounds are given below. For each of the compounds, state whether or not it can be polymerised and, if so, by what type of reaction.

4.2 What is meant by the following terms as applied to polymer structures?
(a) homopolymer,   (b) copolymer,   (c) linear polymer,
(d) branched polymer,   (e) cross-linked polymer,   (f) network polymer.
Describe briefly how branching and cross-linking can occur and explain how these reactions modify the properties of a polymer.

4.3 Determine the average molecular mass of polypropylene if the degree of polymerisation is 1000. Use atomic mass numbers $C = 12.01$, $H = 1.008$.

4.4 The mean molecular mass of a sample of PTFE $(-CF_2-CF_2-)_n$ is 3500. What is the degree of polymerisation? Use atomic mass numbers $C = 12.01$, $F = 19.00$.

4.5 Calculate the degree of polymerisation and the average molecular mass number of a sample of polyvinyl acetate, $(-CH_2-CH(COO.CH_3)-)_n$, polymerised using 100 g hydrogen peroxide, H_2O_2, as an initiator per 1000 kg vinyl acetate. Assume that all of the peroxide is used to provide terminal groups. Use atomic mass numbers $C = 12.01$, $H = 1.008$, $O = 16.00$.

4.6 Determine the ratio of styrene, $CH_2 = CH.C_6H_5$, to acrylonitrile, $CH_2 = CH.CN$, in the copolymer SAN having a degree of polymerisation of 600 and an average molecular mass of 50,000 kg/kmol. Use atomic mass numbers $C = 12.01$, $H = 1.008$, $N = 14.008$.

4.7 (a) Estimate the proportion of available cross-linking sites used up during the vulcanisation of polyisoprene (NR), $(-CH_2-CH = C(CH_3)-CH_2-)_n$, if the amount of sulphur used is 6 kg per 100 kg of polyisoprene.
(b) How much sulphur would be needed to convert the polyisoprene into ebonite?
Assume in each part of the question that the short cross-link chains are S_2. Use atomic mass numbers $C = 12.01$, $H = 1.008$, $S = 32.066$.

4.8 The melting temperature, T_m, of polyethylene, in K, is given by the expression $T_m = \{0.0024 + (0.034/n)\}^{-1}$, where n is the degree of polymerisation. Estimate the melting temperature for degrees of polymerisation of 50, 500 and 1000.

4.9 What general class (thermoplastic, thermoset etc.) of polymer material would be selected for each of the following applications? Give reasons for the choice.
(a) A drinking cup top for a vacuum flask, (b) an electrical wall socket, (c) a sandwich box, (d) insulation covering for electrical wire, (e) boots to cover automotive steering linkages, (f) a kitchen worktop surface, (g) the casing for a PC keyboard, and (h) the sole of a trainer shoe.

CHAPTER 5

Crystalline structures

Why study crystal structures?

Very many solid materials are crystalline in nature, that is their atoms or ions are arranged in regular, repetitive and symmetrical array. Many different crystalline structures exist and the properties of a material are, to a large measure, determined by the type of crystal structure. This is particularly true for metals.

LEARNING OBJECTIVES

1. Know the difference between crystalline and amorphous solids.

2. Know the names and definitions of the seven crystal systems.

3. Understand Miller notation and be able to derive the Miller indices for planes and directions within crystal unit cells.

4. Understand the concept of unit cells and be able to sketch the unit cells of the body-centred cubic, face-centred cubic and close-packed hexagonal systems.

5. Know how atoms are packed within planes and how the differing crystal systems are produced by the stacking of planes of atoms.

6. Know the siting of tetrahedral, octahedral and cubic interstices within crystal structures.

7. Know how atoms and ions are arranged in some of the simpler ceramic crystals, including the arrangements of SiO_4 tetrahedra in silica and some silicates.

8. Derive the relationships between lattice parameter and atomic diameter for body-centred cubic, face-centred cubic and close-packed hexagonal systems and be able to compute packing fractions and densities for crystalline arrangements.

9. Know the Bragg law for X-ray diffraction and be able to compute crystal lattice parameters and crystal type from X-ray diffraction data for cubic crystals.

5.1 Introduction

The terms crystalline, glassy or amorphous may be used to describe solid materials. What do these terms mean? The crystalline state is that in which the constituent atoms or molecules of the substance are arranged in a regular, repetitive and symmetrical pattern. The regularity of a structure may be termed long range, meaning that the same symmetrical pattern of atoms or ions exists over large distances within the material, or short range, in which local groupings of atoms or ions may be in a symmetrical pattern but the relationships between these local groupings may not be regular. This short range order is found in many inorganic glasses. Amorphous, meaning literally without form, is the term used to describe non-crystalline structures, even though these may possess a short range order.

Many solids are fully crystalline, some are partially crystalline, and others are amorphous. Metals and ceramics are crystalline solids, but in many polycrystalline ceramics there is frequently a glassy phase in the spaces between crystal grains and a new range of materials, metallic glasses is under development. Many polymer materials show a greater or lesser degree of crystallinity but others are amorphous. Inorganic glasses may have the same, or similar, compositions as some ceramic materials yet are amorphous rather than being crystalline.

If a material is allowed to cool slowly from the liquid state the shape of the cooling curve obtained may be either continuous (Figure 5.1(a)), or show discontinuities (Figure 5.1 (b), (c)). As the temperature falls, there is a steady decrease in the kinetic energy of the atoms or molecules that make up the liquid, and hence a steady increase in the viscosity of the fluid. Materials that show cooling curves of the type shown in Figure 5.1(a) change during cooling from true fluids, through stages of steadily increasing viscosity, into an apparently solid condition. The atoms or molecules present still have the same type of random arrangement that existed in the true fluid state. Such materials are amorphous solids or glasses.

Other solid materials are crystalline in nature, that is their constitutent atoms or molecules are arranged in a definite symmetrical pattern. Figure 5.1(b) shows the type of cooling curve obtained for the freezing of a pure crystalline substance. There is a very definite freezing or solidification point,

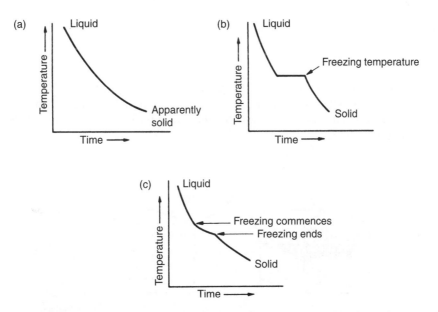

Figure 5.1 *(a) Cooling curve for a glassy material – no definite solidification temperature. (b) Cooling curve for a pure crystalline material. (c) Solidification and crystallisation occurring over a temperature range.*

shown by a marked arrest on the cooling curve. At this temperature the atoms cease their random movement and tend to 'stick' together in relatively fixed positions in a regular pattern. Atomic motion does not cease abruptly upon solidification, and the atoms or molecules in a crystalline solid vibrate about fixed positions.

The change from random fluid motion to vibration about a point within a crystal is a change to a much lower energy state. Energy, the latent heat, is emitted from the material at the freezing temperature. Figure 5.1(c) shows the type of cooling curve applicable to some crystalline mixtures. Freezing is not completed at constant temperature, but takes place over a definite temperature range. The latent heat energy emitted during freezing shows up as a less steep gradient on the cooling curve within the freezing range.

The remainder of this chapter contains details of the various classes of crystals and analysis of crystal structures. The glass state and crystallinity in polymers is dealt with in Chapter 6.

5.2 Crystal classes

A shape is said to be symmetrical if it possesses one or more of the elements of symmetry. The elements of symmetry are planes, axes, and points of symmetry. A shape of low symmetry may possess only one plane of symmetry, whereas a highly symmetrical shape, such as the cube, will contain several planes and axes of symmetry (Figure 5.2). A shape is said to be symmetrical about a plane if the plane divides the shape into either two identical halves, or into two halves that are mirror images of one another. If a shape can be rotated about an axis so that the shape occupies the same relative position in space more than once in a complete revolution then such an axis is termed an axis of symmetry. Such axes may be either 2, 3, 4 or 6-fold.

There are seven crystal systems. These are

triclinic
monoclinic
rhombohedral
hexagonal
orthorhombic
tetragonal
cubic

It is possible to define each of these systems by reference to three prinicipal axes.

Consider three axes OA, OB, OC, of lengths a, b and c respectively, inclined to one another at angles α, β and γ (Figure 5.3). Each crystal may now be defined in terms of the lengths of the principal axes, and the angles between them, as follows:

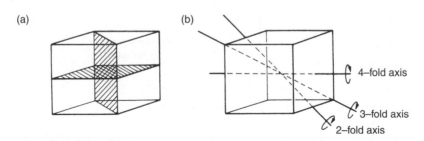

Figure 5.2 *Elements of cubic symmetry: (a) planes; (b) axes.*

triclinic	$a \overset{\neq}{=} b \overset{\neq}{=} c$	$\alpha \neq \beta \neq \gamma \neq 90°$
monoclinic	$a \overset{\neq}{=} b \overset{\neq}{=} c$	$\alpha = \beta = 90° \neq \gamma$
rhombohedral	$a = b = c$	$\alpha = \beta = \gamma \neq 90°$
hexagonal*	$a = b \neq c$	$\alpha = \beta = 90°, \ \gamma = 60°$
orthorhombic	$a \neq b \neq c$	$\alpha = \beta = \gamma = 90°$
tetragonal	$a = b \neq c$	$\alpha = \beta = \gamma = 90°$
cubic	$a = b = c$	$\alpha = \beta = \gamma = 90°$

* Frequently, a four reference axis system, the Miller–Bravais system, is used to define the hexagonal form. In this system there are three coplanar axes, a_1, a_2 and a_3, all of equal length and inclined at 120° to each other, with a fourth axis, c, mutually perpendicular to the other three axes, as shown in Figure 5.4. The four indices are h, k, i and l, the first three indices relating to the a axes and the fourth relating to the c axis.

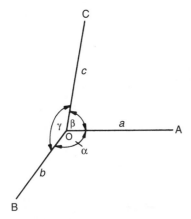

Figure 5.3   *Principal axes of reference.*

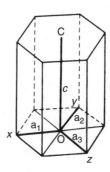

Figure 5.4   *Miller–Bravais reference axes.*

Figure 5.5   *Representation of part of a space lattice with a unit cell outlined.*

A crystal contains myriads of atoms arranged in a regular, repetitive pattern known as a *space lattice*. The various crystal systems, as defined above, indicate the type of symmetry to be found within the space lattice, but give no indications of the relative positions of atoms within the crystal. To show this the device known as a unit cell is used. The unit cell can be regarded as the smallest grouping of atoms still showing the symmetry elements of the type (Figure 5.5).

5.3 | Miller notation

As will be seen later, it is sometimes necessary to refer to specific planes and directions within a crystal lattice. The notation used is the Miller indexing system.

Referring to the principal axes OA, OB, and OC of the unit cell (Figure 5.6), the plane PQR can be described by the Miller indices h, k and l, where these are the reciprocals of the intercepts of the plane with the principal axes, in terms of the lengths of the axes.

$$h = \frac{OA}{OP}, k = \frac{OB}{OQ} \text{ and } l = \frac{OC}{OR}$$

The describing indices are enclosed in brackets, thus (h, k, l). Any fractions are cleared to give the smallest integers.

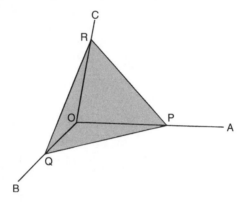

Figure 5.6 *Plane PQR in relations to crystal axes.*

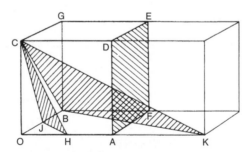

Figure 5.7 *Identification of planes (refer to text). Plane ADEF = (100), plane CJH = (231), plane BCK = (122).*

Figure 5.7 shows three planes in the cubic system. These planes would be described as follows: Plane ADEF intercepts the principal axes at A, ∞, and ∞.

Indices of plane ADEF are $\frac{OA}{OA}, \frac{OB}{\infty}, \frac{OC}{\infty}$ or $(1,0,0)$

On the same basis plane CDEG would be $(0,0,1)$ and plane BFEG would be $(0,1,0)$. All face planes of a cube are similar and are termed a family of planes. The use of curly brackets, thus, $\{1,0,0\}$ means reference to the whole family of face planes rather than to one specific plane.

Plane CJH intercepts the principal axes at H, J and C.

Indices of plane CJH are $\frac{OA}{OH}, \frac{OB}{OJ}, \frac{OC}{OC}$ or $(2,3,1)$

Plane BCK intercepts the principal axes at K, B, and C.

Indices of plane BCK are $\dfrac{OA}{OK}, \dfrac{OB}{OB}, \dfrac{OC}{OC}$ or ($\frac{1}{2}$, 1, 1)

Removal of fractions gives (1, 2, 2) as the indices for this plane

There are four axes of reference in the Miller–Bravais system and, therefore, four indices (*hkil*) must be quoted to identify planes and directions. A special feature of the Miller–Bravais system is that the sum of the rationalised intercepts on the three *a* axes must be zero. Refer to Figure 5.8 for examples.

The four reference axes are OX, OY, OZ and OC. Plane PQRSTU, the top plane of the cell, is parallel to the axes OX, OY and OZ and intercepts them at infinity, but it intercepts axis OC at C. The indices for this plane, therefore, are (0001). Now consider the plane PSXK. This plane intercepts the OX axis at X, the OY axis at K (which is at unit distance from O but in the negative direction) and the OC axis at C. This plane is parallel to the OZ axis, intercepting it at infinity. So, the indices of plane PSXK are ($1\bar{1}01$). For the prismatic face plane TSLX, this plane intercepts the OX axis at unit distance, is parallel to the OY axis and intercepts the OZ axis at unit distance in the negative direction. The indices for plane TSLX are ($10\bar{1}0$).

Figure 5.8 *Identification of planes using Miller–Bravais system (refer to text). Plane PQRSTU = (0001), plane PSXK = ($1\bar{1}01$) and plane TSLX = ($10\bar{1}0$).*

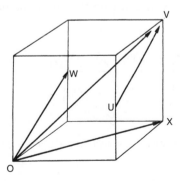

Figure 5.9 *Directions in crystals (refer to text). Direction UV = direction OW = [021], direction OV = [111], direction OX = [110].*

To describe a direction within a crystal, substract the coordinates of the start of the line from the coordinates of the end of the line (Figure 5.9). Again, fractions are cleared to give the smallest integers. The following examples illustrate this.

For direction UV: coordinates of V = 1, 1, 1, coordinates of U = 1,0,$\frac{1}{2}$. So direction UV = 1, 1, 1 − 1, 0, $\frac{1}{2}$ = 0, 1, $\frac{1}{2}$. Removal of fractions gives the indices as 0, 2, 1. Indices for crystallographic directions are enclosed in square brackets, thus [021]. It will be noticed that line OW is parallel to UV and has the same direction indices. The reciprocal direction, WO, would be 0, 0, 0 − 0, 1, $\frac{1}{2}$ =0,− 1, − $\frac{1}{2}$.

Removal of fractions would make this 0, −2, −1, which would be written as [0$\bar{2}$ $\bar{1}$]. Direction OV becomes [111] and direction OX becomes [110].

Directions in the hexagonal system are expressed as vector sums, as shown in the example in Figure 5.10. Consider the direction LX. This direction is the same as the sum of three vectors in the base plane, namely, a vector of length +2 in the OX direction, a vector of −1 in the OY direction and a vector of length −1 in the OZ direction. There is no vector in the c axis, so the direction is quoted as [2$\bar{1}$ $\bar{1}$0]. The reciprocal direction XL would be [$\bar{2}$110]. Similarly the directions MY and NZ are quoted as [1$\bar{2}$10] and [$\bar{1}$ $\bar{1}$20] respectively. The direction OC, normal to the base plane, is direction [0001].

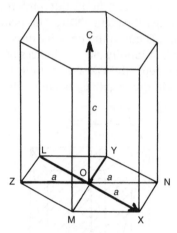

Figure 5.10 *Directions in the hexagonal system. Direction LX = [2$\bar{1}$ $\bar{1}$0], direction MY = [$\bar{1}$2$\bar{1}$0] direction NZ = [$\bar{1}$ $\bar{1}$20], direction OC = [0001].*

5.4 | Metallic crystals

The concept of the unit cell can be used to indicate the relative positions of atoms within a space lattice. Figure 5.11 shows three of the variations on the cubic theme, with the circles representing the positions of atom centres.

Figure 5.11 *Unit cells in the cubic system (circles represent atomic centres): (a) simple cubic; (b) body-centred cubic (b.c.c.); (c) face-centred cubic (f.c.c.).*

EXAMPLE 5.1

How many atoms are contained within the the three unit cells illustrated in Figure 5.11?

SOLUTION

Reference to Figure 5.5 will show that eight cells meet at the corner of a unit cell and that a face of a unit cell is common to two cells. Therefore, an atom in a corner position will be shared between the eight cells that meet at that point and an atom positioned in a cell face will be shared between two cells.

For the simple cubic system there are atoms at each cell corner, therefore the number of atoms per unit cell $= (8 \times 1/8) = 1$

For the body-centred cubic system there are atoms at each corner and one atom in the centre of the cell, so the number of atoms per unit cell $= \{(8 \times 1/8) + 1\} = 2$.

For the face-centred cubic cell there are atoms at each corner and one atom positioned in each of the six cube faces, so the number of atoms per unit cell $= \{(8 \times 1/8) + (6 \times 1/2)\} = 4$.

As stated earlier (section 2.15), atoms and ions in the bound state can sometimes be regarded as solid spheres in contact. In the case of the crystalline form of a pure element all the spheres will be identical. However, when considering a chemical compound or a metallic alloy, atoms of two or more elements will be present, and the different elements will possess differing atomic, or ionic, diameters. In the former case, with identical atoms, the structure tend to have high symmetry. Consequently very many metals crystallise in the cubic or hexagonal forms.

Consider uniform spheres. These may be packed together in more than one way. For example, the packing of spheres within a single plane could be square, or hexagonal (Figure 5.12), this latter being the closest possible packing for uniform spheres. A crystal lattice is three dimensional, so that the planes of the type shown in Figure 5.12 must be stacked above one another.

A series of planes with square packing, as in Figure 5.12(a), if stacked upon one another so that the centres of atoms in successive layers all lie on a line normal to the base, would give a three-dimensional space lattice in the simple cubic pattern. In this system, each atom is in contact with six other atoms, that is, a co-ordination number of six. The spheres in a simple cubic pattern occupy 52 per cent of the available space.

In the body-centred cubic system, the atoms are packed more closely than in the simple cubic form. The most densely packed planes within the body-centred cubic system are those of the (110) type (Figure 5.13). The three-dimensional crystal is built up by stacking a series of identical planes of the (110) type on one another in the manner shown in Figure 5.13(c). It will be seen that the atoms in alternate planes, or layers, are all in line, giving an ABABAB stacking pattern. Figure 5.16(a) shows the unit cell of the body-centred cubic system as a series of spheres in contact. It can be seen that the co-ordination number for this type of packing is 8. This is a higher density of packing than for simple cubic, and it can be calculated that the spheres occupy 68 per cent of the available space. A number of metals, including iron and chromium, crystallise in this form.

The closest possible packing of uniform spheres in a plane is hexagonal packing, and the stacking of planes of this type upon one another gives rise to the closest possible packing in three dimensions. It is, however, possible to build up a symmetrical three-dimensional structure from close-packed planes in two different ways. Figure 5.14 shows the centres of atoms (plane A) arranged in a hexagonal plane pattern. When a second plane of atoms is stacked adjacent to the first, or A, plane, the centres of atoms in the second layer may be in either the position marked B or that marked C. If three relative spatial

Figure 5.12 *Two possible packing arrangements for spheres in a plane: (a) square packing; (b) hexagonal packing.*

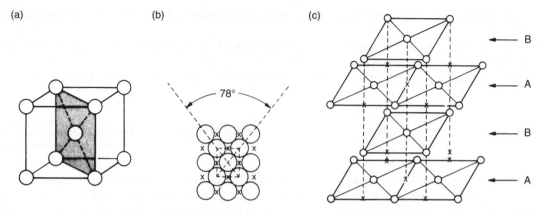

Figure 5.13 *Body-centred cubic system: (a) unit cell with (110) plane shaded; (b) plan view of a (110) plane, the most densely packed in the b.c.c. system. The positions of atomic centres in the second layer are marked x. (c) ABABAB type stacking sequence of 110 planes.*

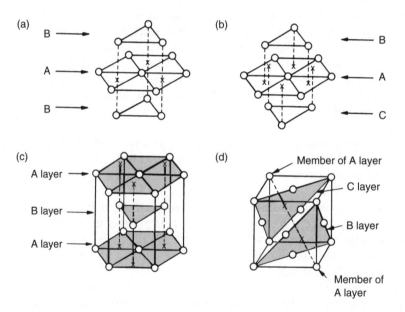

Figure 5.14 *Close-packed lattices: (a) ABAB type stacking sequence; (b) ABC type stacking sequence; (c) close-packed hexagonal crystal structure: ABAB stacking; (d) face-centred cubic crystal system: ABCABC stacking.*

positions exist for close-packed layers then symmetry in three dimensions can be obtained if the layers are stacked in either an ABABAB sequence (Figure 5.14(a), or else in an ABCABC sequence (Figure 5.14(b)).

In both cases, the density of packing is the same, with 74 per cent of the total space filled with spheres, and in both cases the co-ordination number is 12. The first type of stacking sequence produces the crystal form known as close-packed hexagonal, and the second sequence gives rise to the face-centred cubic type of crystal (Figure 5.14(c),(d)).

EXAMPLE 5.2

How many atoms are contained within the close-packed hexagonal unit cell?

SOLUTION

Six hexagonal prisms meet at a corner point and, so, a cell corner atom is shared between six cells. Referring to Figure 5.14(c), it will be seen that in the close-packed hexagonal cell there are twelve atoms situated at corner positions, an atom at the centre of both the top and bottom faces, plus three atoms within the cell, so the number of atoms per unit cell = $\{(12 \times 1/6) + (2 \times 1/2) + 3\} = 6$.

Many metals crystallise in these close-packed systems, including magnesium and zinc (close-packed hexagonal), and aluminium and copper (face-centred cubic).

5.5 Interstitial sites

In the structures considered so far we have considered that all the atoms have been spheres of uniform diameter. In such a build-up there are gaps, or interstices, between these uniform spheres into which may be fitted atoms or ions of smaller diameter. The co-ordination number of the various types of interstice or space in crystals, that is the number of atoms that an inserted atom will be in contact with, may be 4, 6 or 8 and the names given to the three types of interstitial site are tetrahedral, octahedral and cubic, respectively. In the first type the interstice is the central space between four atoms forming a tetrahedron, in the second the space is at the centre of six atoms which occupy the corners of an octahedral figure, and in the third type the space is the central space of a simple cubic structure. The locations of these spaces in some crystal systems is shown in Figure 5.15.

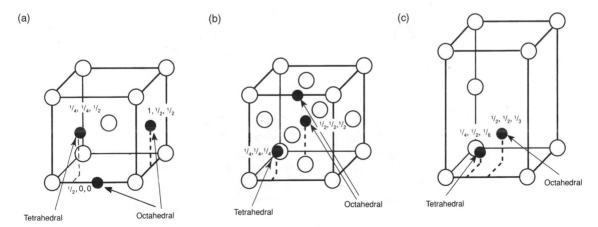

Figure 5.15 *Location of interstitial sites in some crystal types: (a) body-centred cubic; (b) face-centred cubic; (c) hexagonal close-packed.*

5.6 | Ceramic crystals

With the exception of diamond, all ceramic materials are compounds and, therefore, ceramic crystals contain atoms, and/or ions, of more than one type. Consider the case of compounds of the alkali metals (Group IA of the periodic table) and the halogen elements (Group VIIB). These are all ionic compounds between monovalent elements, for example LiF, NaCl, CsCl and KBr. These compounds possess symmetrical crystalline structures but, although all similar chemically, they do not all form in the same crystal pattern. In the case of caesium chloride, CsCl, the diameters of the caesium and chlorine ions are comparable in size and the caesium chloride structure is a body-centred cubic pattern (Figure 5.16(a),(b)), with each caesium ion in contact with eight chlorine ions, and vice versa. (This is the highest possible co-ordination number for an ionic crystal, as explained in section 3.1.)

In the case of sodium chloride, NaCl, the sodium ion is small in comparison with the chlorine ion and there is room for only six chlorine ions to surround and be in contact with a sodium ion, as in Figure 5.16(c). This arrangement also builds up into a symmetrical cubic pattern (Figure 5.16(d)) but of a different type from caesium chloride.

Calcium fluoride, known also as fluorite or fluorspar, with the formula CaF_2, is an example of an ionic compound between a divalent metallic ion and a halogen ion (Figure 5.17).

Many compounds with a composition of the type AB_3 crystallise in a face-centred cubic structure with A atoms occupying cell corner sites and the B atoms in the face centre sites. One such compound is tungsten trioxide, WO_3 (Figure 5.18(a)). The *perovskite* structure is similar but in this case the central site in the unit cell is occupied by a small third ion. The mineral perovskite, calcium titanate, has the formula $CaTiO_3$ and the crystal structure is shown in Figure 5.18(b). A number of ceramic materials possess the perovskite structure and an important one is barium titanate. The structure of the barium titanate crystal, at temperatures below 120°C, is distorted. The TiO_6 octahedra are not perfectly regular as the Ti ion is displaced from the body-centre position by a small amount, about 0.01 nm. This causes a spontaneous polarisation of the crystal. Barium titanate is a *ferroelectric* material in which the polarisation of the crystal can be affected by an external electric field, giving it piezoelectric

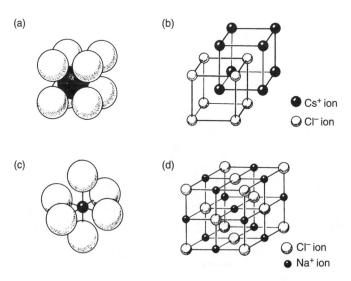

Figure 5.16 *Caesium chloride and sodium chloride structures. (a) Caesium chloride unit cell showing Cs$^+$ ion in contact with eight Cl$^-$ ions. (b) Interpenetration of cubes showing Cl$^-$ ion in contact with eight Cs$^+$ ions. (c) Sodium chloride showing Na$^+$ ion surrounded by six Cl$^-$ ions. (d) Cubic crystal of sodium chloride.*

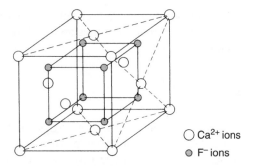

Figure 5.17   *The structure of calcium fluoride, CaF$_2$.*

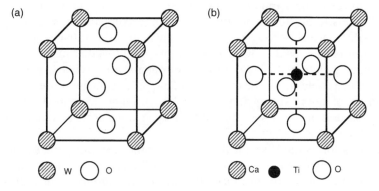

Figure 5.18   *(a) Structure of tungsten trioxide, WO$_3$. (b) Structure of perovskite, CaTiO$_3$.*

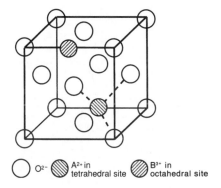

Figure 5.19   *Spinel structure. Only a small number of the available tetrahedral and octahedral sites are occupied.*

characteristics. It is used as a piezo crystal in transducers, for example ultrasonic inspection probes. Some materials which possess the perovskite strcture are also ferromagnetic.

There are some oxide ceramics with more complex crystal structures and an important group is the *spinels*. Spinels have the general formula AB$_2$O$_4$ where A is a divalent metal ion and B is a trivalent metal ion. The normal spinel structure has a structure in which oxygen ions are arranged in a face-centred cubic structure with A^{2+} ions in tetrahedral interstices and B^{3+} ions in octahedral interstices. Not all the available interstitial sites are occupied. A related structure is that of *inverse spinels*. In an inverse spinel half of the B^{3+} ions occupy octahedral sites while the other half, together with the A^{2+} ions, are in tetrahedral interstices. The arrangement for spinel and inverse spinel structures is shown in Figure 5.19.

Inverse spinels in which B is the Fe^{3+} ion are termed *ferrites*, ceramic materials which are ferromagnetic but also dielectrics and so electrical eddy currents are not established in them when they are subject to an alternating magnetic field. (Refer to section 13.9.)

5.7 | Silica and silicates

Silicon is, next to oxygen, the most abundant element in the earth's crust. It occurs in combination with oxygen as silicon dioxide, or silica, SiO_2, and in combination with oxygen and other metallic elements in a wide range of silicates. Silica and silicates are the basis of many ceramic materials and glasses.

The silicon atom is tetravalent and forms bonds with four oxygen atoms in a tetrahedral structure (Figure 5.20). The SiO_4 unit tetrahedron is the basis of silica and silicate structures. Bonds are formed between the corner oxygen atoms and other silicon atoms to give silica crystals (or with atoms of other metals in silicates). The size of the silicon atoms is small in comparison with oxygen and the silicon atom fits easily into the space between oxygen atoms giving a comparatively open structure. There are several allotropic forms of silica as the SiO_4 tetrahedra may add together in more than one way. In one form of silica, tridymite, the tetrahedra build up into a hexagonal array while in another form, cristobalite, the tetrahedra form into a cubic array. The build-up of SiO_4 tetrahedra gives rise to a fairly open network. One possible build-up of tetrahedra giving a three-dimensional structure is shown in Figure 5.21. The bonding within silica is principally of the covalent type and this means that silica is very hard and possesses a high melting point (1710°C).

The structure of silica forms the basis for that of many other ceramics, the silicates. It is possible for some substitution to occur with other metal atoms taking the place of some of the silicon atoms in the structure forming silicates. In many instances the substituted atoms bond ionically to the oxygen

Figure 5.20 *SiO₄ tetrahedron.*

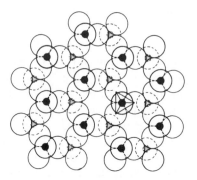

Figure 5.21 *Plan view of a silica structure.*

atoms. The crystal structures of silicates may or may not be distorted, depending on the relative sizes of substituted metal ions.

Some silicates possess chain structures with the tetrahedra adding in one plane only, and in some cases the chains could be linked forming a sheet or layer structure (Figure 5.22). Sheet or layer lattices are found in clays, talc and mica. In layer lattice structures, there is strong covalent and ionic bonding within the layer, but comparatively weak van der Waal's secondary bonds between layers. This allows for easy cleavage of the mineral into thin sheets, as is the case with mica. Kaolin, the chief mineral in china clay, is a hydrated aluminium silicate and is a good example of a mineral with a layer lattice (Figure 5.23). Because the structure of kaolin is not symmetrical, the crystal layers become highly polarised. The polarisation of layer crystals allows the clay to absorb water molecules, which are also polar, between the layers. This gives good intercrystalline lubrication and contributes significantly to the plasticity of the clay.

Figure 5.22 *Silicate chain and sheet structures: (a) single chain; (b) double chain; (c) sheet structure. Metal ions may be attached at points marked M.*

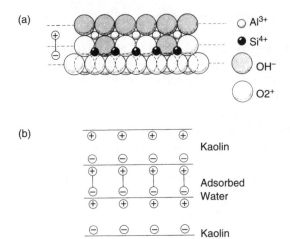

Figure 5.23 *(a) Structure of kaolin. The asymmetric layers are polarised. (b) Adsorbed water between layers acts as a lubricant aiding plasticity.*

Silica and many silicates can form glasses. The structures discussed so far exhibit crystal symmetry, that is, the same geometrical arrangements of groups of atoms or ions occur throughout the structure. In other words, the structure possesses long range order. Glasses are not crystalline, in other words they do not possess a full and regular symmetrical structure, but they cannot be said to be entirely structureless. Glasses based on silica and silicates have the same composition as the ceramic from which they are derived and, therefore, will still contain SiO_4 tetrahedra. However, these tetrahedra are not assembled in an ordered and symmetrical manner. This is termed short range order as, in silica glass for example, each silicon atom is bonded to four oxygen atoms. Glasses are dealt with in greater detail in Chapter 6.

5.8 | Analysis of crystals

The lattice constants, or parameters, are the lengths of the principal axes of the unit cell. For cubic crystals, therefore, there is only one constant to be quoted. The lattice constants of pure metal crystals can be directly related to the atomic diameter of the element, as below.

For face-centred cubic (Figure 5.24): the length of a face diagonal $= 2D$, where $D =$ an atomic diameter. If the lattice constant is a, then, by Pythagoras

$$(2D)^2 = a^2 + a^2$$

$$4D^2 = 2a^2$$

$$a = \frac{2D}{\sqrt{2}} = D\sqrt{2}$$

Similarly, for body-centred cubic (Figure 5.25): the length of the cubic diagonal $= 2D$. In this case, by Pythgoras

$$(2D)^2 = a^2 + 2a^2$$

$$4D^2 = 3a^2$$

$$a = \frac{2D}{\sqrt{3}}$$

In the close-packed hexagonal system there are two lattice constants, a and c (Figure 5.26). From Figure 5.26 it will be seen that parameter a is equal to one atomic diameter D. It can be calculated that the parameter c is equal to $2h$, where h is the height of the regular tetrahedron of edge length a:

$$c = 2h = \frac{D2\sqrt{2}}{\sqrt{3}}$$

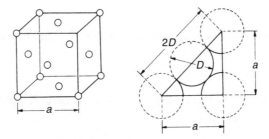

Figure 5.24   *Face-centred cubic unit cell. Relation between a and D.*

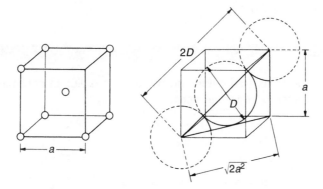

Figure 5.25   *Body-centred cubic unit cell. Relation between a and D.*

Figure 5.26   *Close-packed hexagonal structure with unit cell outlined, and showing relationship between a, c and h, the height of a tetrahedron.*

EXAMPLE 5.3

Calculate the packing fraction for uniform size spheres in each of the following crystal types: simple cubic, body-centred cubic, face-centred cubic and close-packed hexagonal.

SOLUTION

The packing fraction (PF) is the volume of spheres in a unit cell divided by the unit cell volume.

(a) For simple cubic, with atoms of diameter D in contact, the lattice parameter $a = D$, and the unit cell contains one atom only. Therefore

$$PF = \pi D^3/6 \div D^3 = 0.52$$

(b) For body-centred cubic the lattice parameter $a = 2D/\sqrt{3}$, and the unit cell contains 2 atoms. Therefore

$$PF = 2 \times (\pi D^3/6) \div (2D/\sqrt{3})^3 = 0.68$$

(c) For face-centred cubic the lattice parameter $a = D\sqrt{2}$, and the unit cell contains 4 atoms. Therefore

$$PF = 4 \times (\pi D^3/6) \div (D^3 \times 2\sqrt{2}) = 0.74$$

(d) For close-packed hexagonal the lattice parameters are a and c, where $a = D$ and $c = D\, 2\sqrt{2} \div \sqrt{3}$. The unit cell contains 6 atoms and its volume is the base of a regular hexagon of side D times c, namely $\{6(D/2 \times D\sqrt{3}/2) \times D\, 2\sqrt{2}/\sqrt{3}\} = 6\, D^3/\sqrt{2}$. Therefore

$$PF = 6 \times (\pi D^3/6) \div 6\, D^3/\sqrt{2} = 0.74$$

EXAMPLE 5.4

Chromium crystallises as body-centred cubic and the atomic radius of chromium atoms is 0.125 nm. Calculate the lattice parameter, a, of the unit cell.

SOLUTION

For the b.c.c. unit cell: $a = 4r/\sqrt{3}$
For chromium, $r = 0.125$ nm, therefore $a = 4 \times 0.125 \div \sqrt{3}$
$$= 0.289 \text{ nm}$$

EXAMPLE 5.5

Copper crystallises as face-centred cubic, its atomic mass number is 63.54 and its atomic radius is 0.1276 nm. Calculate the theoretical density of copper. Avogadro's number, $N_0 = 6.023 \times 10^{26}$/kmol.

SOLUTION

Lattice parameter for a f.c.c. cell, $a = 2r\sqrt{2}$
so lattice parameter for copper: $a = 2 \times 0.1276 \times 10^{-9} \times \sqrt{2}$ m
One unit cell of the f.c.c. system contains 4 atoms.
Volume of a copper unit cell $= a^3 = (2 \times 0.1276 \times 10^{-9} \times \sqrt{2})^3$ m^3 (equivalent to 4 atoms)

Mass of 4 copper atoms $= \dfrac{63.54 \times 4}{N_0} = \dfrac{63.54 \times 4}{6.023 \times 10^{26}}$ kg

$$\text{Density of copper} = \frac{\text{Mass}}{\text{Volume}} = \frac{63.54 \times 4}{6.023 \times 10^{26} \times (2 \times 0.1276 \times 10^{-9} \times \sqrt{2})^3}$$

$$= 8.98 \times 10^3 \, \text{kg/m}^3$$

EXAMPLE 5.6

(a) Calculate the diameter of the octahedral interstitial sites, as a fraction of an atomic diameter, for the body-centred cubic and face-centred cubic crystal systems.

(b) The crystal structure of iron is body-centred cubic at ordinary temperatures but changes to face-centred cubic at high temperatures. The diameter of a carbon atom is approximately 0.6 of that of an iron atom. What implications do the results from (a) have on the possibility of carbon atoms fitting in to interstitial sites with iron crystal lattices?

SOLUTION

(a) Referring to Figure 5.15 it will be seen that in both the body-centred cubic and face-centred cubic systems the centre of a unit cell edge is an octahedral site.

For body-centred cubic the cell edge length, $a = 2D/\sqrt{3}$

so the diameter of the octahedral site is $2D/\sqrt{3} - D = 0.155D$

And for face-centred cubic the cell edge length, $a = D\sqrt{2}$

so the diameter of the octahedral site is $D\sqrt{2} - D = 0.414D$

(b) The size of a carbon atom is about 50 per cent larger than the size of octahedral interstices in face-centred cubic but is about four times larger than the interstices in body-centred cubic. This means that if a carbon atom occupies an interstitial site b.c.c. iron (*ferrite*) it will cause a lot of strain in the lattice. Some carbon, but very little, can be accepted into interstices in ferrite – approximately one carbon atom for every 3300 iron atoms – before the total strain becomes unacceptably high. The f.c.c. form of iron (*austenite*), the stable form of iron at high temperatures, can accept more carbon into these interstitial sites before the strain in the lattice becomes too great.

This has major implications for the development of properties in steels which are, basically, alloys of iron and carbon. Refer to Chapter 16 for more details.

Within the cubic system the spacing between parallel planes can be readily calculated. The relationship between interplanar spacing, d, and the lattice constant, a, is

$$d = \frac{a}{\sqrt{(h^2 + k^2 + l^2)}}$$

where h, k, and l are the Miller indices of the plane.

The wavelengths of X-radiation are of the same order of magnitude as interplanar spacings in crystals. Consequently, when X-radiation is directed at a crystalline material there will be some diffraction effects. The radiation will be reflected by atomic planes, including both surface and sub-surface planes. This is illustrated in Figure 5.27. For radiation striking atomic planes at some angle, θ, there will be effective reflection only when the various ray paths differ by an amount that is equal to one wavelength, or an integral number of wavelengths. Under these conditions the reflected rays will be exactly

Figure 5.27 *Reflection of X-rays by crystal planes.*

in phase. From the diagram this condition will be satisfied when distance LMN is equal to $n\lambda$, where λ is the wavelength of the radiation, and n is a small integer.

OM = d, the interplanar spacing

therefore MN = $d \sin\theta$

and   LMN = $2d \sin\theta$

The condition for reflection is, therefore

$n\lambda = 2d \sin\theta$

This is the Bragg equation.

By using monochromatic X-radiation and measuring the angles at which Bragg reflections occur it is possible to determine the lattice constants for crystals. In the case of cubic crystals of metals the planes that give Bragg reflections are

body-centred cubic – (110), (200), (211), (220), (310), and (222) planes
face-centred cubic – (111), (200), (220), (311), and (222) planes

It will be seen from the above that different types of crystal give different diffraction patterns. For the less symmetrical crystal types, and for those containing more than one type of atom, the analysis is similar, but considerably more complex.

One method which is used to obtain X-ray diffraction patterns, and hence determine crystal interplanar spacings, is the Debye–Scherrer method. A powdered metal sample is placed in the centre of a circular diffraction camera and monochromatic X-radiation is directed at it. A strip of film, sensitive to

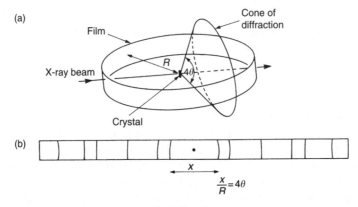

Figure 5.28 *(a) Principle of Debye–Scherrer camera. (b) Diffraction pattern.*

X-radiation, is mounted around the circumference of the camera. The incident X-ray beam is diffracted by crystal planes, according to Bragg's law, giving a series of emerging cones of radiation and a series of lines are recorded on the film. The included angle of each radiation cone is equal to $4\theta_n$, where n is an integer denoting the number of the Bragg diffraction. After exposure and development of the film the strip of film is held flat and the distances between diffraction lines measured. If the distance between a pair of lines is x then $x/R = 4\theta$ (in radians), where R is the radius of the camera (see Figure 5.28).

EXAMPLE 5.7

Copper atoms are 0.2552 nm in diameter and form a f.c.c. structure. X-radiation of wavelength 0.152 nm is used for the analysis of two samples of copper. For sample A the first order Bragg reflection from (111) planes occurred at an angle of 21°00′, while for sample B the first-order Bragg reflection from (111) planes was at 21°23′. Calculate the lattice parameter of the two samples and give an explanation for the difference between them.

SOLUTION

For f.c.c. crystal the lattice parameter, $a = D\sqrt{2}$
Lattice parameter for pure copper, $a = 0.2552 \times \sqrt{2}$
$$= 0.3609 \text{ nm}$$

For sample A

$$d_{(111)} = \frac{\lambda}{2 \sin \theta} = \frac{0.152}{2 \sin 21°00'}$$

$$= 0.212 \text{ nm}$$

so

$$a = d\sqrt{(h^2 + K^2 + l^2)} = 0.212 \times \sqrt{3}$$
$$= 0.3673 \text{ nm}$$

For sample B

$$d_{(111)} = \frac{\lambda}{2 \sin \theta} = \frac{0.152}{2 \sin 21°23'}$$

$$= 0.208 \text{ nm}$$

so

$$a = 0.208 \times \sqrt{3}$$
$$= 0.361 \text{ nm}$$

The lattice parameter for sample B equates with that calculated for pure copper but that for sample A has a higher value. Sample B is high purity copper, while sample A is either impure or is an alloy and the presence of atoms of another element with a larger atomic radius than that of copper has imposed strain in the copper crystal lattice.

EXAMPLE 5.8

X-radiation of wavelength 0.171 nm is directed at a cubic crystalline metal. The first two Bragg reflections occur at angles of 30°00 and 35°17 respectively.

Determine (a) whether the crystal type is b.c.c. or f.c.c., (b) the lattice parameter of the unit cell, and (c) the atomic diameter (assuming that all atoms are identical).

SOLUTION

If the metal is b.c.c. the first two reflections will be from the (110) and (200) planes respectively, whereas if the metal is f.c.c. the first two reflecting planes are (111) and (200).

Assuming that the metal is b.c.c.:

$$d_{(110)} = \frac{0.171}{2 \sin 30°00'} \quad \text{and} \quad d_{(200)} = \frac{0.171}{2 \sin 35°17'}$$

$$= 0.171 \, \text{nm} \qquad\qquad = 0.148 \, \text{nm}$$

The lattice parameter, $a = d\sqrt{(h^2 + K^2 + l^2)}$

Therefore $a = 0.171 \times \sqrt{2}$ or $a = 0.148 \times 2$
$\qquad\qquad = 0.242 \, \text{nm} \qquad\qquad = 0.296 \, \text{nm}$

The two computed values of a are not the same, therefore the metal is not b.c.c.

Assuming that the metal is f.c.c.:

$$d_{(111)} = \frac{0.171}{2 \sin 30°00'} \quad \text{and} \quad d_{(200)} = \frac{0.171}{2 \sin 35°17'}$$

$$= 0.171 \, \text{nm} \qquad\qquad = 0.148 \, \text{nm}$$

giving $a = 0.171 \times \sqrt{3}$ or $a = 0.148 \times 2$
$\qquad\qquad = 0.296 \, \text{nm} \qquad\qquad = 0.296 \, \text{nm}$

These two values of a are the same, showing that the metal is f.c.c. with a lattice parameter of 0.296 nm.

In the f.c.c. system the relationship between atomic diameter, D, and the lattice parameter is given by $D = a/\sqrt{2}$. Therefore

$$D = \frac{0.296}{\sqrt{2}} = 0.2093 \, \text{nm}$$

[In producing this solution the author deliberately made an initial assumption that the metal was b.c.c. and this resulted in a lengthy answer. Close inspection of the data shows that such an assumption was unlikely. The first two Bragg angles were 30°00' and 35°17'. The relative closeness of these two angles is an indication that they are for planes 3 (111) and 4 (200) rather than for planes 2 (110) and 4 (200). If the metal had been b.c.c. an angle of 30° for plane 2 (110) would mean an angle of about 45° for plane 4 (200).]

EXAMPLE 5.9

Bragg reflections may occur from the following planes in b.c.c. crystals: (110), (200), (211), (220), (310), (222), (321), (400), (411), (420), (332) and (422). Which of these planes will give reflections when X-radiation of wavelength 0.154 nm is directed at a sample of chromium? Assume the atomic diameter of chromium is 0.2494 nm.

SOLUTION

$\lambda = 2d \sin \theta$ or $\lambda/2d = \sin \theta$

$\sin \theta \not> 1$ therefore $d \not< \lambda/2$

$$d_{(hkl)} \not< \frac{0.154}{2}$$

But $d_{(hkl)} = \dfrac{a}{\sqrt{(h^2 + k^2 + l^2)}}$

therefore $\sqrt{(h^2 + k^2 + l^2)} \not> \dfrac{2a}{\lambda}$

In b.c.c. crystals $a = \dfrac{2D}{\sqrt{3}}$

so a for chromium is $a = \dfrac{2 \times 0.2494}{\sqrt{3}} = 0.288$ nm

Therefore

$$\sqrt{(h^2 + k^2 + l^2)} \not> \frac{2 \times 0.288}{0.154}$$
$$\not> 3.74$$

and so $h^2 + k^2 + l^2 \not> 13.98$

The highest possible values of (hkl) are therefore (222).

Bragg reflections will occur from the first six reflecting planes, that is, up to and including plane (222).

SUMMARY

In a crystalline material the atoms or ions are arranged in a regular, repetitive and symmetrical pattern. There are seven main classes of crystal but most metals form in densely packed arrangements with co-ordination numbers of 8 or 12 as body-centred cubic, face-centred cubic or close-packed hexagonal. Ceramic crystals containing more than one type of ion or atom tend to form structures which are less densely packed than metals.

Specific directions and planes within crystals can be identified using Miller indices.

Monochromatic X-radiation is diffracted by the regular planes of atoms in a crystal according to Bragg's law. Analysis of the X-ray diffraction patterns obtained from crystals permits identification of the crystal type and the determination of crystal lattice parameters.

Questions

5.1 Give the indices for the planes A, B, C, D and E in Figure 5.29.

5.2 Give the indices for the directions A, B, C, D and E in Figure 5.30.

5.3 (a) Sketch a cubic unit cell and mark the positions of the following planes: (220), (010) and (111).

   (b) Calculate the separation distance, in terms of the lattice parameter, a, between adjacent planes of each of the three types.

5.4 Considering atoms in a single plane to be represented by discs of uniform diameter, show, by calculation, that the packing density of (111) planes in f.c.c. is greater than that of (110) planes in b.c.c.

Figure 5.29

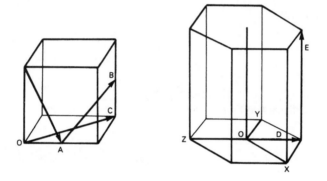

Figure 5.30

5.5 (a) Calculate the theoretical density of nickel from the following data: nickel has an atomic mass number of 58.69, an atomic radius of 0.124 nm and crystallises as f.c.c. (Avogadro's number is 6.023×10^{26}/kmol.)

    (b) Give reasons why this calculated density differs from the actual measured density of 8900 kg/m³.

5.6 Brass containing 70 per cent copper and 30 per cent zinc (both percentages by weight) has a face-centred cubic crystal structure. Calculate the average lattice parameter of the alloy unit cell if the atomic diameters are 0.2552 nm for copper and 0.2658 nm for zinc. Atomic mass numbers for copper and zinc are 63.54 and 65.38 respectively.

5.7 Calculate the fractional volume change when α-iron (b.c.c.) transforms into γ-iron (f.c.c.). Assume that the radius of the iron atoms remains constant.

5.8 Sodium chloride, NaCl, crystallises in a cubic pattern and the unit cell can be regarded as Cl^- ions in an f.c.c. arrangement with Na^+ ions occupying octahedral sites. Given that the ionic radii of sodium and chlorine ions are 0.097 nm and 0.181 nm respectively, calculate (a) the lattice parameter of the unit cell, and (b) the packing fraction.

5.9 Gold crystallises as f.c.c. and the lattice parameter of the unit cell is 0.407 nm. The atomic mass number of gold is 197.2 and Avogadro's number is 6.023×10^{26}/kmol. Calculate the theoretical density of gold.

5.10 (a) Derive the Bragg equation which relates the interplanar spacing, d, for a set of planes to the wavelength of incident radiation, λ, and the diffraction angle, θ.

    (b) Describe a diffraction camera which can be used for obtaining X-ray diffraction patterns from wire or powder metal specimens.

5.11 A crystal of high-purity sodium chloride is used for determining the wavelength of monochromatic radiation. Determine the wavelength of the radiation if a first-order Bragg reflection from the (111) planes of sodium chloride occurs at an angle of 5.2°. The lattice parameter, a, for sodium chloride is 0.563 nm.

5.12 X-radiation of wavelength 0.1542 nm is directed at a sample of an unknown metal. The first four Bragg reflections occur at angles of 19.25°, 22.38°, 32.58° and 39.16°. Determine (a) the crystal type, (b) the lattice parameter, and (c) the atomic diameter of the metal. (Assume that the metal crystallises in a cubic form.)

5.13 Bragg reflections may occur from the following planes in b.c.c. crystals: (110), (200), (211), (220), (310), (222), (321), (400), (411), (420), (332) and (422). Which of these planes will give reflections when X-radiation of wavelength 0.154 nm is directed at a sample of vanadium? Assume that the atomic diameter of vanadium is 0.2626 nm.

5.14 When X-radiation of wavelength 0.179 nm strikes a sample of pure copper a first-order Bragg reflection from the (111) set of planes occurs at an angle of 25°25′. Calculate the atomic radius of copper atoms, given that copper crystallises as face-centred cubic.

5.15 The X-ray diffraction pattern from a cubic crystalline metal, using radiation of wavelength 0.154 nm, shows lines at the following angles: 21.66°, 31.47°, 39.74°, 47.58°, 55.63°, 64.71° and 77.59°. Determine (a) the structure of the crystal, (b) the Miller indices of the reflecting planes, (c) the lattice parameter, and (d) the diameter of the metal atoms.

5.16 An X-ray diffractometer recorder chart for an element with a cubic structure shows diffraction peaks at the following 2θ angles (in degrees): 40, 58, 73, 86.8, 100.4, 114.7 and 129.6. If the wavelength of the X-rays used was 0.154 nm, determine (a) the crystal structure of the element, and (b) the lattice parameter.

5.17 X-ray diffraction analysis is conducted on two samples of copper (f.c.c.) using X-radiation of wavelength 0.152 nm and a Debye–Scherrer camera with a film diameter of 180 mm. For sample A, the measured distance between the first pair of lines was 132.0 mm while for sample B the distance between the first pair of lines was 134.4 mm. Determine the lattice parameter, a, for each sample and give an explanation for the difference between the two samples.

5.18 A sample of a metal M is subjected to X-rays of wavelength 0.1786 nm in a Debye–Scherrer camera with a diameter of 50.8 mm. The distances between the arcs for the first six reflected cones of radiation are 41.9, 60.9, 77.7, 94.2, 112.3 and 139 mm respectively. Determine (a) whether the metal is f.c.c. or b.c.c. in structure, (b) the lattice parameter, and (c) the density of M if the atomic mass number is 185.35. (Take Avogadro's number as 6.023×10^{26}/kmol.)

5.19 (a) A diffraction pattern has been obtained from a pure b.c.c. metal using a Debye–Scherrer type camera. Sketch the appearance of the developed film for low angle lines.

(b) Using the following data, calculate the atomic radius for the metal atoms:
diameter of camera 57.5 mm
wavelength of X-radiation 0.1541 nm
distance between first pair of lines on film 44.52 mm

(c) It is proposed to alloy the original b.c.c. metal with an f.c.c. alloying element which has an atomic radius of 0.135 nm. Assuming that the two metals form a substitutional solid solution with a solubility limit of 5 atomic per cent, discuss whether the X-ray diffraction method could be used to estimate alloy compositions. Suggest what the likely appearance of the X-ray diffraction pattern would be if the amount of added element exceeds 5 atomic per cent.

5.20 High-density polyethylene, $(-CH_2-CH_2-)_n$, shows a high degree of crystallinity with the hydrocarbon chains packing themselves into an orthorhombic form. The parameters of the unit cell are $a = 0.741$ nm, $b = 0.494$ nm and $c = 0.255$ nm. Given that the density of HDPE is 965 kg/m³, atomic mass numbers are C = 12, H = 1, and Avogadro's number is 6.023×10^{26}/kmol, determine the number of atoms in a unit cell.

CHAPTER 6

Glasses and partial crystallinity

What are glasses?

Glasses are non-crystalline or amorphous elastic solids formed by cooling from the fully molten state at a rate sufficiently rapid to prevent crystallisation occurring. The glass, or *vitreous* state can be achieved in all classes of materials. Inorganic glasses may have compositions which are the same as or similar to those of ceramics. Many polymeric materials exist in the glass state and metallic glasses have been created. A glass is not necessarily structureless. For example, the bonding between silicon and oxygen atoms is the same in vitreous silica as in the crystalline forms of silica, with covalent bonding forming a series of SiO_4 tetrahedra. The difference is that in the crystalline forms the tetrahedra are joined together to form large regular symmetrical arrays with long-range order, whereas in vitreous silica there is only short range order and no long-range symmetry.

LEARNING OBJECTIVES

1. Understand what is meant by the term *glass*.
2. Know the significance of the glass transition temperature, T_g.
3. Appreciate the terms *network formers* and *network modifiers* used in connection with inorganic glasses.
4. Know that many polymer materials may show partial crystallinity with glassy zones separating crystalline zones.

6.1 | Formation of glasses

When a substance is cooled from the fully molten state to, or just below, T_m, the melting point, the random fluid atoms or molecules tend to rearrange themselves into an ordered crystalline structure. This process of freezing into a crystalline solid does not occur instantaneously but by a process of nucleation and growth. Small crystals grow from nuclei through the liquid. Outward growth of a crystal ceases when it meets the advancing boundary of another crystal growing from a neighbouring nucleus and the result, when the freezing process is complete, is a polycrystalline solid. Generally, the change from a random fluid into an ordered solid is accompanied by an increase in density and a consequent decrease in volume. The rate at which this crystallisation process occurs in a liquid at a temperature just below T_m depends on the viscosity of the liquid and this parameter is dependent on the size and complexity of the molecules that compose the substance. Liquid metals tend to be of low viscosity and the atoms form into a crystalline structure readily. Many inorganic ceramic materials contain complex molecular arrangements but can still crystallise with relative ease. Polymeric materials tend to have very large complex molecules and have relatively high viscosities at T_m and, generally, are more slow to crystallise than other classes of materials.

When a liquid is cooled rapidly to below T_m, the viscosity of the liquid in the supercooled state may be so high that crystallisation is prevented. As the temperature falls further the viscosity steadily increases until, eventually, it is so high that the material is effectively solid but is amorphous. It is then termed a glass. A material is generally classed as being a solid when the viscosity increases beyond 10^{16} Pa s. There is no sharp change in density at T_m during the formation of a glass and, on cooling of the supercooled liquid below T_m, the volume continues to reduce with the same rate of thermal contraction as applied above T_m. At some lower temperature, there is a change in slope of the volume–temperature curve and this temperature is termed the *glass transition* or *fictive temperature*, symbolised as T_g (see Figure 6.1). T_g is considered to be the temperature at which, on cooling, the substance changes from a supercooled liquid to a glass.

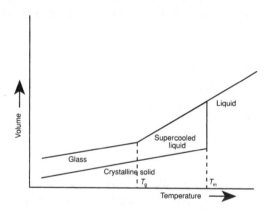

Figure 6.1 *Change in volume with temperature for both a crystalline solid and a glass.*

Polymers and some inorganic ceramic materials can be cooled sufficiently rapidly from the molten state to produce a supercooled liquid and a glass by rates of cooling which are readily achieved during normal commercial processing but to achieve the glass state in metals extremely high cooling rates, of the order of 10^6 °C/s, are required.

EXAMPLE 6.1

A rod of a thermoplastic material of length 100.00 mm at a temperature of 10°C was heated slowly and its length measured accurately at intervals. The following data was obtained:

Temp. (°C)	30	40	50	60	65	70	75	80
Length (mm)	100.155	100.245	100.330	100.410	100.450	100.515	100.610	100.705

Plot a graph showing the relationship between temperature and length and explain its significance.

SOLUTION

The graphical plot is shown in Figure 6.2.

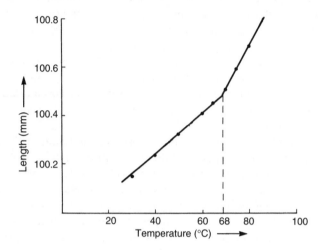

Figure 6.2

It will be seen that there is a change in slope at a temperature of 68°C. This temperature is the value of T_g for the material. At temperatures below T_g the material is in the glass state and elastic, while above T_g the material is viscoelastic.

6.2 | Inorganic glasses

Many inorganic glasses are based on silica, SiO_2, and the basic structural unit is the SiO_4 tetrahedron. The glass form of silica, *vitreous silica*, consists of a three-dimensional network of SiO_4 tetrahedra but with these tetrahedra arranged at random with no long-range order. Figure 6.3 is a schematic representation of the structural difference between crystalline silica and silica glass. For simplicity in this two-dimensional representation each silicon atom is shown bonded to three oxygen atoms only.

In addition to silica, SiO_2, some other oxides can form glass network structures, the principal ones being B_2O_3 and P_2O_5. All these oxides are referred to as *glass formers* or *network formers*. Some metallic oxides, while not being network formers in their own right, can form a joint glass network with silica. Oxides which are capable of forming joint glass networks include alumina, Al_2O_3, titania, TiO_2, and lead oxide, PbO. There is another group of metallic oxides which are important constituents in many commercial glasses even though they can neither form glass networks in their own right nor form joint full network structures with silica. This group, which includes soda, Na_2O, potash, K_2O, lime, CaO, magnesia, MgO and another lead oxide, PbO_2, are referred to as *glass* or *network modifiers*. The oxygen atoms of the oxides enter the silica network but, as there are now insufficient silicon atoms to bond fully with all the oxygen atoms present, the network begins to break up. The metal cations, such as Na^+, enter holes within the glass network. This is shown schematically in Figure 6.4. If too great an amount of a network modifier is added then it becomes difficult to form a glass at all. This tends to be the case when the ratio of oxygen to silicon atoms increases beyond a value of about 2.5/1.

Silica has a high melting point, $> 1700°C$, but the addition of network modifier oxides lowers the melting temperature. In this way commercial glasses with various melting temperatures from about

Figure 6.3   *Schematic representation of (a) crystalline silica, (b) silica glass.*

Figure 6.4   *Schematic diagram of soda glass structure. Network begins to break up and cations enter holes.*

700°C upwards can be made. The main commercial glasses, apart from silica glass, are common glass (soda–lime–silica), leaded glass, borosilicate glass (heat resistant, one trade name for which is 'Pyrex'), and aluminosilicate glass.

EXAMPLE 6.2

The mix for the manufacture of a batch of common glass contains, by weight, 73 per cent silica, 12 per cent lime and 15 per cent soda. Calculate the oxygen/silicon ratio for this mix. Take the atomic mass numbers as $Ca = 40$, $Na = 23$, $Si = 28$, $O = 16$.

SOLUTION

The molecular weights for the ingredients of the glass are

for lime, CaO, $M = 40 + 16 = 56$
for soda, Na_2O, $M = (2 \times 23) + 16 = 62$
for silica, SiO_2, $M = 28 + (2 \times 16) = 60$

Consider 1000 kg of mix:

730 kg SiO_2 contain $730/60 = 12.12$ kmol
120 kg CaO contain $120/56 = 2.14$ kmol
150 kg Na_2O contain $150/62 = 2.42$ kmol

Therefore, in this mix, for every 12.12 Si atoms there are $(12.12 \times 2) + 2.14 + 2.42 = 28.8$ oxygen atoms. So the oxygen/silicon ratio is $28.8/12.12 = 2.38$.

6.3 | Metallic glasses

A metallic glass may be made by cooling a liquid metal extremely rapidly. Cooling rates of about $10^6\,°C/s$ or greater are required. This can be achieved by melt spinning from a chilled disc to form very thin tapes or filaments. Typical tape thicknesses are of the order of $40\,\mu m$. Some metallic glasses will devitrify on heating to moderate temperatures (350°C+) and become brittle, but for some a controlled crystallisation to form a ductile, high-strength microcrystalline structure can be achieved. High-strength metallic glass tapes and filaments offer potential as reinforcing elements in ceramics and polymers.

6.4 | Polymer glasses

Many polymers cool from the molten state into glassy elastic solids. As the temperature reduces molecular movement lessens and the viscosity increases. Polymer molecules are extremely large and, often, complex in shape and it is not possible in a supercooled liquid for the complete molecule to be in fluid motion at any one time. The molecular motion which occurs is largely due to rotation of part of a molecule around a covalent bond (Figure 6.5). As the temperature falls, the lower molecular kinetic energy means that bond rotation cannot occur to any significant extent and adjacent molecules cannot easily move relative to one another. The *glass transition temperature*, T_g, is the temperature at which bond rotation largely ceases.

Figure 6.5 *A single covalent bond permits bond rotation.*

The value of T_g varies considerably from one polymer to another and is much lower for simple flexible chain molecules than for those with bulky side groups attached to the chain. For example, T_g is of the order of $-90°C$ for polyethylene while for polystyrene, with bulky $-C_6H_5$ groups attached, it is 95°C. Below T_g the polymer is an elastic solid but at temperatures between T_g and T_m the material is viscoelastic (refer to Chapter 9). This is sometimes termed the *rubber* state.

6.5 | Crystallinity in polymers

X-ray analysis of polymer materials shows that many polymer materials give diffraction patterns which indicate a regular array, or crystalline structure. The diffraction patterns obtained also show diffuse regions indicating that there is not full crystallinity but a mixture of crystalline and amorphous zones. It is easy to visualise how in polyethylene, for example, one can readily have regular arrangements of the $-CH_2-$ chains, as shown in Figure 6.6.

An early theory was that the polymer structure contained small high-density areas, termed *crystallites*, where sections of molecular chains were aligned together, and other, less dense amorphous areas (Figure 6.7(a)). In this model, one individual polymer chain could form part of both crystalline and amorphous zones. At a later stage, single crystals of polymer were separated out from solution and analysis of these showed that they were small platelets with a thickness of about 10 nm and a plate size of about $10\,\mu m$. Within these platelets the linear polymer chain was repeatedly folded (Figure 6.7(b)).

Many crystalline polymers show a structure containing areas termed *spherulites* when examined under a polarising microscope. These spherulites may vary in size from several millimetres across to less than a micrometre. A spherulite is not a single crystal but is believed to consist of many crystalline

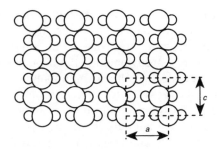

Figure 6.6 *Crystalline packing within polyethylene, with unit cell outlined.*

(a)

(b)

Figure 6.7 *(a) Crystallites. (b) Crystal platelets formed by folding.*

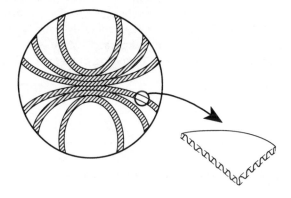

Figure 6.8 *Structure of a spherulite.*

platelets which grow from a nucleus, and that these platelets are twisted (see Figure 6.8). Within the spherulite there are both platelets and amorphous regions.

The structure of a crystalline polymer can also be affected by deformation. During the cold drawing of a polymer there can be a major alignment, or orientation, of the structure in the direction of drawing. This orientation produces an anisotropic material which is much stronger in the direction of the draw than in a transverse direction. A similar orientation takes place during film blowing.

The degree of crystallinity in polymers varies. It is about 50 per cent in commercial polyethylene and may be considerably less than this in some other materials. Nevertheless, polymers which show partial crystallinity are termed *crystalline polymers*. The main crystalline polymers are linear homopolymers in which there are strong van de Waal's intermolecular forces, and some block copolymers. Frequently polymers with isotactic and syndiotactic structures (refer to section 4.7) possess sufficient regularity for them to form crystals. Polyethylene, polypropylene, PTFE, polyamides, polycarbonate, PEO, POM and PEEK are crystalline. A branched polymer or a linear homopolymer with bulky side groups, particularly if the structure is atactic (see section 4.7), is unlikely to form ordered crystalline zones. This is also true for random copolymers. The main amorphous polymers are polystyrene, ABS, PMMA, polyurethanes and the thermosets. Some materials, including PVC, are borderline cases.

Crystalline polymers show viscoelasticity and possess a glass transition temperature, T_g. This is associated with the amorphous regions of the structure, which may make up 50 per cent or more of the volume.

SUMMARY
••

The glass state is a solid state in which the material is not fully crystalline in structure. As the material cools from the fully molten state it does not freeze at a specific temperature but becomes a supercooled liquid of steadily increasing viscosity. At some lower temperature, when the viscosity is of the order of 10^{16} Pa s, there is a change in gradient of the volume–temperature curve and this temperature is that at which the material changes during cooling from a supercooled liquid to a glass. This temperature is termed the *glass transition temperature*, symbolised as T_g. Inorganic glasses, often based on silica, show a short-range order but no long-range order and symmetry. Many commercial inorganic glasses are formed from mixtures of oxides, a glass former, such as silica, to give a network structure, and network modifiers, such as soda, which lower the melting temperature but which will break up the network structure if present in too great a quantity.

Many thermoplastic polymer materials cool into glassy elastic solids. Some thermoplastic materials are termed *crystalline*. These are not completely crystalline but have numerous small areas which are crystalline. Generally, the crystalline areas in a so-called crystalline polymer form up to about 50 per cent of the total volume of the material. The area between crystals is amorphous. Crystalline polymers also show a glass transition temperature and this is associated with the amorphous zones separating the crystalline areas.

Generally, metals are crystalline but the process of crystallisation which occurs when a metal solidifies may be prevented and a metallic glass formed if the rate of cooling of the metal is extremely rapid and of the order of 10^6 °C/s.

Questions

6.1 What is the significance of the temperature T_g for a polymer material? Explain how the value of T_g could be determined.

6.2 A sample of a thermoplastic material is heated through a range of temperature and changes in its length are measured using a dilatometer. The data obtained are given below:

Temperature (°C)	10	20	30	40	50	60	70	80	90
Dilatometer reading (mm $\times 10^{-2}$)	5.5	11.0	16.5	22.0	27.5	34.1	41.5	49.0	56.6

Plot a graph showing the relationship between temperature and change of length and explain its significance. What changes occur within the material and what effects do any changes have on the properties of the material?

6.3 What are the effects of adding the following oxides to a glass forming oxide such as silica? (i) Al_2O_3, (ii) B_2O_3, (iii) Na_2O.

In the case of Na_2O, is there a maximum limit to the amount that can be added to silica and, if so, what are the reasons for this?

6.4 Boric oxide, B_2O_3, and silica, SiO_2, are used to make a borosilicate glass. A requirement is that the oxygen/silicon ratio should not exceed 2.4. Calculate the maximum percentage of boric oxide, by weight, that can be in the mix. Take the atomic mass numbers as B = 10.8, Si = 28.1, O = 16.0.

6.5 Calculate the oxygen/silicon ratio for a soda/silica mix containing 18 per cent soda by weight. State, giving reasons, whether or not this mix would be suitable for the manufacture of a glass. Take the atomic mass numbers as Na = 23.0, Si = 28.1, O = 16.0.

CHAPTER 7

Elastic behaviour

What is elasticity?

When a solid material is subject to a stress the material will change its shape in response to that stress. This response may be *elastic* or *plastic*. If, when the stress is removed, the material immediately returns to its original dimensions then the deformation is said to be elastic. Many materials, including ceramics, glasses and many thermosetting polymers, exhibit elasticity at all levels of stress up to their fracture stress. Metals and alloys are elastic up to a certain level, after which they deform plastically. When the stress is removed the elastic portion of the deformation is recovered but any plastic deformation remains. Many thermoplastic materials possess little true elasticity but show the property of viscoelasticity in which much of the deformation is not completely recovered after the removal of a deforming stress. In an elastic material the amount of deformation, or *strain*, is directly proportional to the stress producing it and the ratio of stress to strain will be a constant for the material. There are several elastic constants for a material, depending on the type of stress system applied.

LEARNING OBJECTIVES

1. Understand the concepts of stress and strain.

2. Know the differences between direct and shear stresses and strains.

3. Be able to calculate the stress within a material of given dimensions for a specified externally applied force.

4. Know the various elastic constants, how they are derived, and the relationships between them.

5. Understand the way in which the applied force is distributed between the components in a composite material.

6. Be able to calculate the stress distribution in composite materials with continuous, parallel, fibre reinforcement.

7. Understand the development of thermally induced stresses within materials.

8. Be able to calculate the thermal stresses developed in duplex materials of simple geometry.

7.1 | Stress and strain

When a load is applied to a material a balancing force is set up within the material, and this internally acting force is termed a *stress*. The stress acting upon a material is defined as the force exerted per unit area.

Figure 7.1 *Stresses and strains: (a) tensile force F, cross-sectional area A, tensile stress $\sigma = F/A$, tensile strain $\varepsilon = \delta l/l$; (b) compressive force F, cross-sectional area A, compressive stress $\sigma = F/A$, compressive strain $\varepsilon = -\delta l/l$; (c) shear force F, cross-sectional area A, shear stress $\tau = F/A$, shear strain $\gamma = \tan \phi = y/x$.*

For a force of 10 newton acting on a surface of 10 mm square the stress is

$$\frac{\text{force}}{\text{area}} = \frac{10}{10 \times 10 \times 10^{-6}} = 10^5 \text{ newton/metre}^2 \text{ (N/m}^2) = 10^5 \text{ pascal (Pa)}$$

The types of stress normally considered are tensile, compressive and shear stresses. When a material is in a state of stress its dimensions will be changed. A tensile stress will cause an extension of the length of the material, while a compressive stress will shorten the length. Tensile and compressive stresses are termed *direct stresses*. A shear stress imparts a twist to the material. The dimensional change caused by stress is termed *strain*. In direct tension or compression the strain is the ratio of the change in length to the original length. As a ratio strain has no units and is simply a numerical value.

A bar of length l and cross-sectional area A, when subjected to a tensile force F, will increase in length by an amount δl (Figure 7.1(a)). The direct stress, σ, in the material is given by $\sigma = F/A$. The direct strain, ε, developed is given by $\varepsilon = \delta l/l$. Similarly, a compressive force F will cause a reduction in length of δl, so that a direct compressive stress of $\sigma = F/A$ will cause a strain $\varepsilon = -\delta l/l$ (Figure 7.1(b)). The action of a shear force F will cause a block of material to twist through a small angle ϕ. Referring to Figure 7.1(c), the shear force F acting on a plane of area A creates a shear stress, τ, given by $\tau = F/A$. The shear strain, γ, which is generated is given by $\gamma = y/x = \tan \phi$.

In elastic behaviour the strain developed in a material, when the material is subjected to a stress, is fully recovered immediately the stress is removed. Some materials, notably ceramics, glasses and many thermosets, show elastic properties up to their fracture stress. Metals and alloys exhibit both elasticity and plasticity, with non-recoverable or plastic strain occurring when the applied stress exceeds a critical level, the *elastic limit* or *yield stress*. Many thermoplastic materials possess little, if any, true elasticity but show viscoelasticity (see Chapter 9) and much of the strain developed is not fully recoverable.

Robert Hooke, in 1678, enunciated his law stating that the strain developed in a material is directly proportional to the stress producing it. This law holds, at least within certain limits, for many materials.

Figure 7.2 is a force–extension diagram for a metal stressed in tension. The first portion of the curve, OA, shows that the length of the specimen increases in direct proportion to the applied load, hence strain will be proportional to stress. Hooke's law does not hold beyond point A. Behaviour within the region OA will be elastic. Beyond point A the extension of the material ceases to be wholly elastic and some permanent strain is developed. The non-elastic permanent strain is termed *plastic strain*. If the

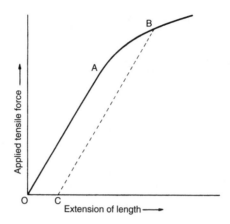

Figure 7.2 *Force–extension curve for a metal.*

material were loaded from O to B and the load then removed the material would not revert to its original length, but follow the unloading path BC showing a permanent extension, or plastic deformation, of OC when completely unstressed. Point A is known as the *limit of proportionality* or *elastic limit*.

7.2 Elastic constants

For an elastic material, namely one that obeys Hooke's law, the ratio of stress to strain will be a constant of the material. For direct stresses acting in tension or compression:

$$\frac{\text{direct stress } (\sigma)}{\text{direct strain } (\varepsilon)} = E \text{ (a constant for the material)}$$

Strain, being the ratio of change in dimension to original dimension, has no dimension, therefore the constant E has the dimensions of stress, namely pascals (Pa) (or newtons/metre^2 (N/m^2)). E is termed the *modulus of elasticity*, or *Young's modulus*, and is a measure of the stiffness of a material.

There is a similar relationship between shear stress and shear strain within the elastic limit

$$\frac{\text{shear stress } (\tau)}{\text{shear strain } (\gamma)} = G$$

G is a second constant for a material and is termed the *modulus of rigidity* or *shear modulus*.

The modulus of elasticity can be determined in a tensile or compressive test (refer to Chapter 27) and the modulus of rigidity may be determined from the results of a torsion test but vibration test methods (refer to Chapter 28) are used for the most accurate evaluation of these elastic constants.

There is a third elastic modulus for a material, this being the *bulk modulus of elasticity*, K. This is the elastic response of the material to equilateral tension or compression

$$\frac{\text{stress}}{\text{volume strain}} = K$$

If a material is subjected to a direct longitudinal stress, a longitudinal strain will be developed extending the length of the specimen. This will be accompanied by a certain lateral contraction, compensating for the increase in length (Figure 7.3). The ratio of lateral to axial strain is termed *Poisson's ratio, v*. The value of Poisson's ratio for many materials lies between 0.28 and 0.35.

Figure 7.3 *Poisson's ratio. Longitudinal strain = $\delta l/l$, lateral strain = $\delta x/x$, Poisson's ratio v = lateral strain/longitudinal strain.*

The various elastic constants are related to one another as follows:

$$G = \frac{E}{2(1 + v)} \quad \text{or} \quad E = 2G(1 + v)$$

$$K = \frac{E}{3(1 - 2v)} \quad \text{or} \quad E = 3K(1 - 2v)$$

EXAMPLE 7.1

A direct axial force is applied to a cylindrical metal bar of diameter 15 mm and the extension over a 100 mm length is measured as the force is increased. The data obtained are given below.

Force (kN)	60	120	150	160	170
Extension (mm)	0.40	0.80	1.00	1.20	1.50

From this data determine: (a) E for the metal, (b) the elastic limit, and (c) the strain at the elastic limit.

SOLUTION

Produce a graphical plot of force against extension (see Figure.7.4). The plot is linear up to a force of 150 kN, where it departs from linearity.
(a) At a force of 120 kN the elastic extension is 0.80 mm.
   Stress, σ = force/area = $120 \times 10^3/\pi \times 7.5^2$ = 679 N mm^{-2} = 679 MPa
   Strain, ε = extension/original length = 0.80/100 = 8×10^{-3} $E = \sigma/\varepsilon = 679 \times 10^6/8 \times 10^{-3}$
   = 84.9×10^9 Pa (84.9 GPa)
(b) The elastic limit is the stress at the point where the plot ceases to be linear, when the force is 150 kN.
   Stress, σ = force/area = $150 \times 10^3/\pi \times 7.5^2$ = 849 N mm^{-2} = 849 MPa
(c) At a force of 150 kN, the elastic limit, the extension is 1.00 mm.
   Strain, ε = extension/original length = 1.00/100 = 1×10^{-2}

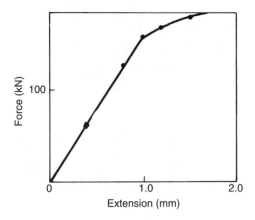

Figure 7.4

EXAMPLE 7.2

Given that E for aluminium is 70.5 GPa, and Poisson's ratio, v, is 0.34, estimate the value of G for aluminium.

SOLUTION

The relationship between E, G and v is $E = 2G(1 + v)$
So

$$G = \frac{70.5 \times 10^9}{2(1 + 0.34)}$$

$$= 26.3 \times 10^9 \, \text{Pa} = 26.3 \, \text{GPa}$$

The values of the elastic constants for a material will be independent of the direction of stressing for some materials. Such materials are termed *isotropic* (possessing the same properties in all directions). However, in an anisotropic material the values of the elastic constants differ according to the directions in which they are measured. Crystals show anisotropy and a single crystal will possess differing values of E, G and Poisson's ratio, depending on the direction of stressing relative to the crystal axes. However, most commercial ceramics and metals are polycrystalline, their structures consisting of many small crystals with a random orientation. In consequence, they approximate in behaviour to isotropic materials.

Some bulk materials, for example, timbers and fibre reinforced composites, particularly composites with uniaxial fibre reinforcement, exhibit marked anisotropy and possess different elastic constants in different directions. In timber, for example, the values of E and G are different in the longitudinal, radial and tangential directions relative to the grain. Thus, for a timber, there are three values of Young's modulus, three of shear modulus and six values of Poisson's ratio.

7.3 Fibre reinforced composites

Fibre reinforced composites are a very important group of engineering materials. Many of those in current use are composed of a thermoset or thermoplastic matrix material reinforced with thin glass, carbon or aramid fibres. The greatest reinforcing effect is obtained when the fibres are continuous and parallel to one another and the maximum strength is obtained when such a composite is stressed in tension in a direction parallel to the fibre axis.

Consider a uniaxial continuous fibre composite subjected to an axial tensile force F (Figure 7.5). The force will be divided between fibres and matrix such that

$$F = F_f + F_m \tag{1}$$

where F_f and F_m are the forces carried by fibres and matrix respectively.

The same longitudinal strain, ε, will be developed in both fibres and matrix so that

$$\varepsilon_c = \varepsilon_f = \varepsilon_m \tag{2}$$

Stress, σ, = force F/area A, giving $F = \sigma A$ so, from equation (1):

$$\sigma_c A_c = \sigma_f A_f + \sigma_m A_m$$

giving

$$\sigma_c = \sigma_f \frac{A_f}{A_c} + \sigma_m \frac{A_m}{A_c} \tag{3}$$

where A_f, A_m and A_c are the cross-sectional areas of fibre, matrix and composite respectively. For fibres of uniform cross-section, the area fractions A_f/A_c and A_m/A_c will equal the fibre and matrix volume fractions V_f and V_m so that, from equation (3) we have

$$\sigma_c = \sigma_f V_f + \sigma_m V_m \tag{4}$$

From Hooke's law $\sigma = \varepsilon E$ giving, from equation (4)

$$\varepsilon_c E_c = \varepsilon_f E_f V_f + \varepsilon_m E_m V_m$$

As, in equation (2) there is equal strain in composite, fibre and matrix, the value of E for the composite, E_c, is given by

$$E_c = E_f V_f + E_m V_m \tag{5}$$

where E_f and E_m are the values of Young's modulus for fibre and matrix respectively.

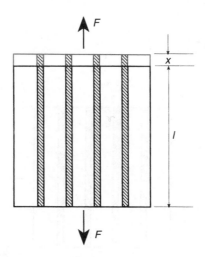

Figure 7.5 *Strain in composite, ε_c, strain in fibres, ε_f, and strain in matrix, ε_m, are all equal to x/l.*

Equations (4) and (5) show that the properties of the composite are proportional to the volume fractions of the components of the composite. This is the *Rule of Mixtures* and this rule can also be used to calculate other properties such as the density or the thermal conductivity of a composite in the fibre direction.

The values of E and tensile strength are greatest in the fibre axial direction and will be very much less in the transverse direction. When subjected to a force, F, normal to the fibre axis (see Figure 7.6) the effective cross-sectional areas of fibre and matrix can be assumed to be equal and so the stress developed in both fibres and matrix can be assumed as equal, giving

$$\sigma_c = \sigma_f = \sigma_m \tag{6}$$

The total transverse deformation of the composite is y_c and will be the sum of the deformations of fibres and matrix. The total strain, $\varepsilon_c = y_c/t_c$, so that

$$y_c = \varepsilon_c t_c$$

Therefore

$$\varepsilon_c t_c = \varepsilon_f t_f + \varepsilon_m t_m \tag{7}$$

where t_f and t_m are the respective thicknesses of the fibre and matrix elements.

From equation (7)

$$\varepsilon_c = \varepsilon_f \frac{t_f}{t_c} + \varepsilon_m \frac{t_m}{t_c}$$

But the thickness ratios t_f/t_c and t_m/t_c equate to the respective volume fractions, so

$$\varepsilon_c = \varepsilon_f V_f + \varepsilon_m V_m \tag{8}$$

From Hooke's law $\varepsilon = \sigma/E$

so $$\frac{\sigma_c}{E_c} = \frac{\sigma_f V_f}{E_f} + \frac{\sigma_m V_m}{E_m}$$

But, from equation (6) $\sigma_c = \sigma_f = \sigma_m$

so $$\frac{1}{E_c} = \frac{V_f}{E_f} + \frac{V_m}{E_m}$$

or $$E_c = \frac{E_f E_m}{E_m V_f + E_f V_m} \tag{9}$$

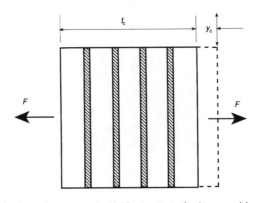

Figure 7.6 *A transverse force, F, causes the composite thickness, t_c, to be increased by an amount y_c.*

EXAMPLE 7.3

Calculate the values of E in both the longitudinal and transverse directions for an epoxy composite reinforced with continuous, parallel glass fibres, the volume fraction of glass in the composite being 0.35. Assume E for epoxy is 4 GPa and E for glass fibre is 70 GPa.

SOLUTION

$$E_{cL} = E_f V_f + E_m V_m = (70 \times 0.35) + (4 \times 0.65) = 27.1 \text{ GPa}$$

$$E_{cT} = \frac{E_f E_m}{E_m V_f + E_f V_m} = \frac{70 \times 4}{(4 \times 0.35) + (70 \times 0.65)} = 5.97 \text{ GPa}$$

In practice, the use of equation (9) does not give an accurate value for E in the transverse direction. One reason for this is that there may not be an equal distribution of force between the fibres and matrix. Several equations of an empirical nature have been proposed to reduce inaccuracies. Two such equations are the Halpin–Tsai and the Brintrup equations:

Halpin–Tsai equation $\qquad E_{cT} = E_m \dfrac{1 + 2\beta V_f}{1 - \beta V_f}$ where $\beta = \dfrac{(E_f/E_m) - 1}{(E_f/E_m) + 2}$

Brintrup equation $\qquad E_{cT} = \dfrac{E_f E_m'}{E_f V_m + V_f E_m'}$ where $E_m' = \dfrac{E_m}{(1 - v_m^2)}$

EXAMPLE 7.4

Determine values of E_{cT} for the glass-filled epoxy composite quoted in Example 7.3 on the basis of both the Halpin–Tsai and Brintrup equations. Assume Poisson's ratio for the epoxy matrix $= 0.35$.

SOLUTION

Using the Halpin–Tsai equation

$$\beta = \frac{(E_f/E_m) - 1}{(E_f/E_m) + 2} = \frac{70/4 - 1}{70/4 + 2} = \frac{16.5}{19.5} = 0.846$$

$$E_{cT} = E_m \frac{1 + 2\beta V_f}{1 - \beta V_f} = 4\left[\frac{1 + (2 \times 0.846 \times 0.35)}{1 - (0.846 \times 0.35)}\right] = 4\left[\frac{1.592}{0.704}\right] = 9.05 \text{ GPa}$$

Using the Brintrup equation

$$E_m' = \frac{E_m}{(1 - v_m^2)} = \frac{4}{(1 - 0.35^2)} = 4.56 \text{ GPa}$$

$$E_{cT} = \frac{E_f E_m'}{E_f V_m + V_f E_m'} = \frac{70 \times 4.56}{(70 \times 0.65) + (4.56 \times 0.35)} = 6.78 \text{ GPa}$$

As stated earlier, a composite with uniaxial fibre reinforcement will be highly anisotropic and have a low modulus value and be relatively weak in a direction transverse to the fibre direction. In order to obtain some degree of isotropy in, say, flat sheet material, reinforcing fibres can be arranged in the form of a two-dimensional mat. This gives a greater uniformity of properties in the plane of the sheet but at the expense of the maximum strength obtainable. The maximum strength in any direction in such sheet is only about one-third of the maximum strength parallel to the fibre direction in a composite with uniaxial fibre reinforcement. If a fibre-composite has reinforcing fibres arranged in a random three-dimensional manner then the maximum strength in any direction is only about one-sixth of that for a uniaxial composite.

Another method for reducing anisotropy effects in sheet material is to compile a series of thin uniaxially reinforced laminae into a laminate. Generally, these laminates are symmetrical in build-up and contain an odd number of laminae with each lamina orientated at some angle $\pm\phi$ to a particular direction (see Figure 7.7). Plywood is a timber product with reduced anisotropy produced in similar manner by bonding an odd number of laminae, or plies, together. The analysis of laminates is beyond the scope of this volume.

Figure 7.7 *Build-up of a laminate to reduce anisotropy effects.*

7.4 | Thermal stresses

Under certain circumstances stresses can be induced in materials and components through a change in temperature. The cooling of a casting, forging or moulding from the processing temperature may cause residual thermal stress to be generated and, in some cases, the level of internal stress developed may be so high as to exceed the fracture stress of the material.

When a material is heated it will, if completely unrestrained, expand. The change in length, δl, can be calculated using the expression $\delta l = l_0 \alpha \delta\theta$ where

l_0 = original length,
α = the coefficient of linear expansion, and
$\delta\theta$ = the temperature interval.

If the material is so constrained that its dimensions cannot change and it is then heated or cooled through a temperature interval of $\delta\theta$ a stress will be set up within the material. The length of the sample should have been changed by an amount $l_0 \alpha \delta\theta$. The length is forced to remain as l_0 and this is equivalent to a direct strain of $l_0 \alpha \delta\theta / l_0 = \alpha \delta\theta$. If Young's modulus, E, for the material is known then the stress induced in the material may be calculated, the stress, σ, being given by $\sigma = E\alpha\delta\theta$.

EXAMPLE 7.5

A 1000 m length of welded steel railway track is heated to 85°C before being firmly secured to the track bed. Calculate the stress in the track at a temperature of 15°C. *E* for steel is 208 GPa, α for steel is 12×10^{-6}/°C.

SOLUTION

The tensile strain developed in the track on cooling from 85°C to 15°C is given by

$\alpha\delta\theta = 12 \times 10^{-6} \times (85 - 15)$.

The tensile stress induced in the rail is

$12 \times 10^{-6} \times 70 \times 208 \times 10^{9} = 174.7 \times 10^{6}\,\text{Pa} = 174.7\,\text{MPa}$

Thermal stresses can be generated in composite structures through the unequal expansion or contraction of each element. The bimetallic strip (Figure 7.8) is a good example showing the effects of differing expansion coefficients. Consider a bimetallic strip made of copper and steel. Copper has a larger coefficient of expansion than steel and so, if the initially flat strip is heated through a temperature interval of $\delta\theta$, the copper, if unrestrained, would expand to a greater extent than the steel. However, the two elements of the strip are fully bonded to one another and the greater expansion of the copper forces the strip into a curved profile. Full expansion of the copper is prevented by the restraining presence of the steel with the result that the copper will be in a state of compression while the steel contains a tensile stress.

Figure 7.8 *Bimetallic strip. The upper element, of higher expansion coefficient, is not allowed to expand to the full extent, and the strip is forced to adopt a curved shape.*

If the various elements in a composite construction are arranged in a symmetrical manner bending or twisting of the composite will not occur when thermal stresses are induced. An example of such a symmetrical construction is a high voltage electrical transmission cable consisting of an aluminium conducting element surrounding a reinforcing steel core.

EXAMPLE 7.6

A high voltage electrical conducting cable with an external diameter of 25 mm consists of a steel core of 5 mm diameter surrounded by aluminium. Assuming that there is no relative movement between the steel and aluminium, calculate the thermal stresses developed in both elements if the conductor is heated through a temperature interval of 40 °C.
Assume: α for steel $= 12 \times 10^{-6}$/°C; α for aluminium $= 26 \times 10^{-6}$/°C; E for steel $= 208$ GPa; E for aluminium $= 70$ GPa.

SOLUTION

Refer to Figure 7.9. When heated through some temperature interval $\delta\theta$, if completely unrestrained, the aluminium would expand by an amount $L\alpha_{A1}\delta\theta$ and the steel expand by an amount $L\alpha_{st}\delta\theta$. But both of the elements are constrained to move together so that they will both expand by some amount δL. This means that the aluminium is effectively subject to a compressive strain of $(\alpha_{A1}\delta\theta - \delta L/L)$ at the higher temperature while the steel is subject to a tensile strain of $(\delta L/L - \alpha_{st}\delta\theta)$. The assembly is in equilibrium at the higher temperature, therefore the compressive force in the aluminium must be equal to the tensile force in the steel.

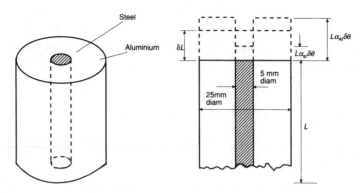

Figure 7.9 *Thermal strains induced in an alumimum/steel composite.*

Force = stress × cross-sectional area

Therefore $F = \sigma_{Al}A_{Al} = \sigma_{st}A_{st}$

Stress, $\sigma = \varepsilon E$ so $\varepsilon_{Al}E_{Al}A_{Al} = \varepsilon_{st}E_{st}A_{st}$

Substituting for strain gives

$$(\alpha_{Al}\delta\theta - \delta L/L)E_{Al}A_{Al} = (\delta L/L - \alpha_{st}\delta\theta)E_{st}A_{st}$$

$$\alpha_{Al}\delta\theta E_{Al}A_{Al} - (\delta L/L)E_{Al}A_{Al} = (\delta L/L)E_{st}A_{st} - \alpha_{st}\delta\theta E_{st}A_{st}$$

$$\frac{\delta L}{L} = \frac{\alpha_{Al}\delta\theta E_{Al}A_{Al} + \alpha_{st}\delta\theta E_{st}A_{st}}{E_{st}A_{st} + E_{Al}A_{Al}}$$

$$\frac{\delta L}{L} = \frac{\{26 \times 10^{-6} \times 40 \times 70 \times 10^9 \times ((25 \times 10^{-3})^2 - (5 \times 10^{-3})^2) \times \pi/4\} + \{12 \times 10^{-6} \times 40 \times 208 \times 10^9 \times (5 \times 10^{-3})^2 \times \pi/4\}}{\{208 \times 10^9 \times (5 \times 10^{-3})^2 \times \pi/4\} + \{70 \times 10^9 \times ((25 \times 10^{-3})^2 - (5 \times 10^{-3})^2) \times \pi/4\}}$$

$$= 9.783 \times 10^{-4}$$

So $\varepsilon_{A1} = (\alpha_{Al}\delta\theta - \delta L/L) = (26 \times 10^{-6} \times 40) - 9.783 \times 10^{-4} = 61.7 \times 10^{-6}$

and $\varepsilon_{st} = (\delta L/L - \alpha_{st}\delta\theta) = 9.783 \times 10^{-4} - (12 \times 10^{-6} \times 40) = 498.3 \times 10^{-6}$

$\sigma = \varepsilon E$

so compressive stress in aluminium, $\sigma_{A1} = 61.7 \times 10^{-6} \times 70 \times 10^9 = 4.32\,MPa$

and tensile stress in steel, $\sigma_{st} = 498.3 \times 10^{-6} \times 208 \times 10^9 = 103.6\,MPa$

7.5 | Toughened glass

In some cases, residual thermal stresses are deliberately induced to produce some specific characteristics in a material. A good example of this is the manufacture of thermally toughened glass.

A plate of glass to be thermally toughened is heated to a temperature which is sufficiently high to allow for some thermal movement of molecules and adjustment of stresses and then the surfaces are cooled quickly by means of an air blast or quenching in oil. The surface layers contract and become rigid while the centre section remains at a high temperature. As the centre section subsequently cools down, the associated contraction causes the generation of compressive stresses in the surface layers and

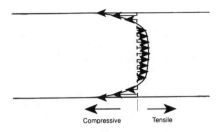

Compressive   Tensile

Figure 7.10   *Stress distribution across a section of thermally toughened glass.*

tensile stresses in the centre. The stress distribution across a section of the glass plate is shown in Figure 7.10. Glasses are stronger in compression than in tension and the residual stress distribution in toughened glass means that a considerable tensile force must be applied before the residual compressive stress in the surface layers is overcome and the net stress moves into the tensile mode. The major disadvantage of thermally toughened glass is that if the surface is damaged and a surface chip or scratch goes through the compressive layer and into the tension zone then failure may be sudden and catastrophic. This type of toughened glass was used to a considerable extent at one time for the manufacture of car windscreens. However, impact with a stone or other object thrown up by the wheels of a preceding vehicle could be sufficient to penetrate the compressive skin leading to immediate total failure of the screen. Today, the majority of vehicle windscreens are made of laminated glass rather than of thermally toughened glass. Catastrophic failure does not occur in glass which is strengthened by lamination.

SUMMARY

Many materials, notably ceramics, glasses and thermosets are elastic solids, that is the strain developed is proportional to the stress acting, and the strain is fully recovered when the stress is removed. Generally, these materials are fully elastic at all stresses up to the fracture stress. Metals show elastic behaviour up to a point but, for stress levels beyond the elastic limit, the strain produced is no longer proportional to the stress and is non-recoverable or plastic strain. The ratios of stress to strain, for elastic materials, give elastic moduli which are important constants for the materials. They are measures of the stiffness and rigidity of the materials.

Fibre reinforced materials in which continuous parallel axial strong fibres are embedded in a softer matrix material are highly anisotropic and have maximum strength when stressed in the fibre direction.

Thermal stresses can be established in a homogeneous material when it is subjected to a change in temperature when its dimensions are constrained and in composite materials when the components of the composite have differing thermal expansion coefficients.

Questions

7.1 When a steel bar of 12 mm × 12 mm cross-section is subjected to a tensile force of 11 kN the measured elastic extension over a length of 150 mm is 5.5×10^{-2} mm. Determine: (a) the stress in the bar, (b) the elastic strain developed, and (c) the value of Young's modulus for steel.

7.2 Determine the shear stress in the coupling pin when the coupling (shown in Figure 7.11) is subject to a tensile force of 5 kN. The pin is 10 mm in diameter.

Figure 7.11

7.3 (a) Estimate the shear modulus of bronze, given that its modulus of elasticity is 80 GPa and Poisson's ratio is 0.34. (b) Estimate the value of Poisson's ratio for nickel, given that the values for E and G for nickel are 208 GPa and 79 GPa respectively.

7.4 A fibre composite rod of 10 mm diameter, with a 0.3 volume fraction of continuous axially aligned glass fibres, is subject to an axial force of 1.8 kN.

Calculate: (a) the force carried by the fibres, (b) the stress in the fibres, and (c) the stress in the matrix. Assume E for matrix is 3 GPa and E for fibres is 85 GPa.

7.5 A fibre-reinforced composite is made from a polyester resin and long uniaxially aligned glass fibres with a fibre volume fraction of 45 per cent. Estimate the maximum strength of the composite and the values of E for the composite in both the longitudinal and transverse directions using the following data: tensile strength of fibre = 1800 MPa; tensile strength of polyester = 65 MPa; E for fibre = 70 GPa; E for matrix = 2.8 GPa.

7.6 A uniaxially aligned carbon fibre/epoxy composite with a fibre volume fraction of 0.4 is in the completely unstressed condition at a temperature of 15°C. Calculate the stresses in both fibre and matrix when the temperature is raised to 50°C.

Coefficients of thermal expansion are: epoxy 60×10^{-6}/°C; carbon 5.4×10^{-6}/°C. E for epoxy = 3 GPa; E for carbon fibre = 200 GPa.

7.7 A component is to be manufactured from a uniaxially-aligned continuous fibre composite material. Two composites are available, one containing 40 per cent, by volume, of glass fibre and the other containing 40 per cent, by volume, of carbon fibre. A minimum axial tensile strength of 700 MPa and a minimum value of E of 50 GPa parallel to the fibre direction is required. Using the property values in the table below, which of the two composite materials is the most suitable?

Material	Tensile strength (MPa)	E (GPa)
Matrix for glass composite	60	3
Matrix for carbon composite	15	3
Glass fibre	3500	62
Carbon fibre	2000	415

CHAPTER 8

Dislocations and plasticity in metals

Why learn about dislocations?

Metals are crystalline but, unlike crystalline ceramics, show both elasticity and plasticity. Crystalline structures contain imperfections termed dislocations and point defects. These exert major influences on the properties of the materials. The plastic deformation of metal crystals is the result of the movement of dislocations under the action of shear stresses, while the point defects in crystal lattices make the movement of dislocations more difficult and, thus, have a strengthening effect. A knowledge of these crystal defects and their behaviour is most necessary for the full understanding of the properties of metals and alloys, the ways in which the properties can be modified, and the effects of heat treatments on the properties of the materials.

LEARNING OBJECTIVES

1. Know that plastic slip in metal crystals occurs by shear across specific planes and in specific directions.
2. Know the specific slip planes and directions for the b.c.c., f.c.c. and h.c.p. crystal systems.
3. Be able to calculate the resolved shear stress in a plane inclined at a known angle to the axis of direct stress.
4. Be able to describe both edge and screw type dislocations and know the significance of the Burger's vector of a dislocation.
5. Be able to describe the various types of point defect that can occur in crystals.
6. Understand the nature of crystal grain boundaries and know the Petch equation relating crystal grain size to the yield strength of a metal.
7. Understand the principles of work hardening and recrystallisation.
8. Understand the principles of solution strengthening and dispersion hardening.
9. Understand the principles of diffusion via vacancies and dislocation climb.

8.1 | Plastic flow in metals

Many materials possess an elastic limit and when stressed they strain in an elastic manner up to a certain point. Beyond this point the strain developed is no longer directly proportional to the applied stress, and also, the strain developed is no longer fully recoverable. If the stress is removed elastic strain is recovered but the material will be left in a state of permanent, or plastic, strain. The mechanism of plastic deformation is not the same for all classes of materials and it is necessary to consider the various materials groups separately.

Metals, in general, are characterised by possessing high elastic modulus values, and also the ability to be strained in a plastic manner. Some metals will begin to deform plastically at very low values of stress and will yield to a very considerable extent before fracture occurs. Other metals and alloys show little plastic yielding before fracture. These latter materials are termed brittle. Plastic deformation in metals may take place by the process of slip, or by twinning. Slip is the more common deformation process encountered and will be dealt with first.

An early theory evolved to explain plastic deformation in metals was the 'block slip' theory. In this it was postulated that when the yield stress of the metal was exceeded plastic deformation took place by the movement of large blocks of atoms sliding relative to one another across certain planes – slip planes – within the crystal (Figure 8.1).

The 'block slip' theory accounted for many of the observed phenomena but possessed a number of drawbacks. The theoretical shear yield strength of a metal, τ_c, calculated on the basis of this theory is given by $\tau_c = G/2\pi$, where G is the modulus of rigidity. This calculated value of τ_c is many times greater than the experimentally observed yield strengths of pure metals. Present day theories of plastic flow in metals are based on the existence of small imperfections, or defects, within crystals. These are structural defects, termed dislocations, and plastic deformation is due to the movement of dislocations across the slip planes of a crystal under the action of an applied stress. The calculated stress required to bring about the movement of dislocations is of the same order of magnitude as observed yield stresses in metals. In recent years, it has been possible to produce small single crystals that are virtually defect free. These small perfect crystals, termed whiskers, possess properties close to the very high theoretical strengths predicted for perfectly crystalline metals.

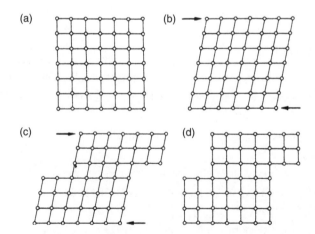

Figure 8.1 *Plastic deformation according to the 'block slip' theory. (a) Original unstressed condition. (b) Elastic strain only. Specimen will revert to (a) when the stress is removed. (c) Elastic and plastic deformation. When stress is removed the specimen will appear as in (d). (d) Unstressed. No elastic strain but permanent plastic strain remains.*

EXAMPLE 8.1

An annealed single crystal of a metal with an h.c.p. crystal structure shows an elastic extension of 0.1 per cent when subjected to a stress of 200 MPa. Estimate the theoretical shear yield strength of the metal, given that $G = E/2(1 + v)$ and that v for the metal is 0.31.

SOLUTION

$E = \sigma/\varepsilon$, $\sigma = 200\,\text{MPa}$, $\varepsilon = 0.1$ per cent $= 0.001$, so $E = 200\,\text{GPa}$

$G = E/2(1 + v) = 200/\{2(1 + 0.31)\} = 76.3\,\text{GPa}$

The theoretical shear yield strength, τ_c, is given by $\tau_c = G/2\pi$

so $\tau_c = G/2\pi = 76.3/2\pi = 12.14\,\text{GPa}$.

8.2 | Slip planes

The crystal planes on which slip generally takes place are those that possess the highest degree of atomic packing. The direction of slip within a crystal plane is the direction of greatest atomic line density. The majority of metals crystallise as close-packed hexagonal, face-centred cubic, or body-centred cubic, and slip within these three systems will be discussed.

In the close-packed hexagonal system, the planes with the greatest density of atomic packing are the basal planes (0001) and within planes of this type there are three possible slip directions (Figure 8.2). Within any one hexagonal crystal space lattice there is only one series of parallel planes of this type.

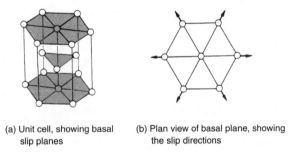

(a) Unit cell, showing basal slip planes
(b) Plan view of basal plane, showing the slip directions

Figure 8.2 *Slip planes and directions in the close-packed hexagonal system.*

The results of research work in the 1930s into the effects of tensile stresses on single-crystal specimens of zinc, a close-packed hexagonal metal, showed that the behaviour of the sample was dependent on the angle at which the slip planes were inclined to the axis of the applied stress. If the slip planes were aligned either normal to, or parallel to, the stress axis, failure occurred in a brittle manner with negligible plastic deformation (Figure 8.3). If however, the slip planes were inclined at some angle, θ, other than 0° or 90°, plastic yielding took place before failure. The amount of plastic yielding and the value of the direct tensile stress necessary to commence yielding were not constant but varied with angle θ (Figure 8.4).

Analysis of the results indicated that, for single crystals of zinc, plastic deformation was due to slip initiated by the action of a shearing stress on the slip plane and also that the magnitude of the critical shearing stress (namely, that necessary to initiate slip), was constant. This has been enunciated as

Figure 8.3 *Effect of stress on single crystals of a hexagonal metal. (a) Slip planes normal to the applied stress; brittle fracture; no slip. (b) Slip planes in line with the applied stress; brittle fracture; no slip. (c) Slip planes inclined to the applied stress; slip and plastic deformation caused by shear force acting on slip planes.*

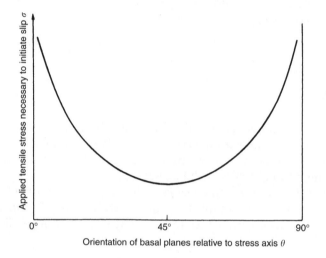

Figure 8.4 *Variation of tensile yield stress with slip plane orientation for hexagonal single crystal.*

Schmid's law: *slip in a metal crystal begins when the stress resolved on the slip plane in the slip direction reaches a certain value termed the critical resolved shear stress.*

The magnitude of a resolved shear stress, τ, across a plane inclined at some angle θ to the axis of direct stress, σ (see Figure 8.5), can be determined as follows.

Consider the prism XYZ. Figure 8.5(b) shows the stresses acting on this prism. The forces F, F_θ and F_s (Figure 8.5(c)) must be in equilibrium. Force is stress times area, so that if the thickness of the prism XYZ is w then

$$F = \sigma \times XY \times w, \; F_\theta = \sigma_\theta \times XZ \times w, \text{ and } F_s = \tau \times XZ \times w.$$

Resolving along $\quad$ XZ: $\tau \times XZ \times w = \sigma \times XY \times w \times \cos \theta$

$$\tau = \sigma \times (XY/XZ) \times \cos \theta$$

$$= \sigma \times \sin \theta \cos \theta$$

$$= \tfrac{1}{2}\sigma \sin 2\theta$$

For a constant value of critical shearing stress, τ_c, the applied tensile stress necessary to initiate slip will be at a minimum when $\theta = 45°$, that is, when $\sin 2\theta = 1$, and infinite when $\theta = 0°$ or $90°$, that is when $\sin 2\theta = 0$.

There is a similar expression, namely $\tau = \sigma \times \cos \theta \cos \varphi$, where θ is the angle between the axis of direct stress and the slip direction within a slip plane and φ is the angle between the axis of direct stress and the normal to the slip plane.

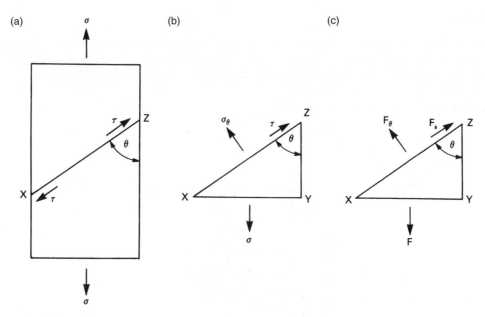

Figure 8.5 *Resolution of shear stress.*

EXAMPLE 8.2

The annealed single crystal sample cited in Example 8.1 is of cylindrical cross-section of 5 mm diameter. The alignment of the crystal is such that both the (0001) slip plane and the [$2\bar{1}\bar{1}0$] slip direction are at an angle of 45° to the axis of tensile force. It is found that plastic slip begins when the magnitude of the applied tensile force is 8 kN. Calculate the critical resolved shear stress on the slip plane in the slip direction and give a brief explanation of the difference between this result and that calculated in Example 8.1.

SOLUTION

The tensile stress, σ, acting is F/A, where $F = 8$ kN and $A = \pi r^2$, where $r = 2.5$ mm.

So $\sigma = 8000/(\pi \times 2.5^2) = 407.4$ N mm^{-2} = 407.4 MPa

The critical resolved shear stress, $\tau_{c'} = \sigma \cos \theta \cos \varphi = 407.4 \cos 45° \cos 45°$
$$= 203.7 \text{ MPa}$$

The value of critical resolved shear stress calculated in Example 8.1 is about 60 times greater than the experimentally observed value in this example. The reason for the discrepancy is that in the theoretical shear strength calculation in Example 8.1 the assumption is made that the crystal lattice is perfect. Crystals contain defects known as dislocations and the presence of these permits slip through the movement of dislocations at a much lower level of stress than that required to cause block slip.

EXAMPLE 8.3

The critical shear stress, $\tau_{c'}$ for the <110> {111} slip system of a pure face-centred cubic metal is found to be 1.6 MPa. What direct stress, σ, must be applied in the [001] direction to produce slip in the [101] direction on the ($\bar{1}$11) plane?

SOLUTION

Refer to Figure 8.6. It will be seen that the angle θ between the direction of direct stress [001] (line AB) and the direction of slip [101] (line OB), is 45°. The angle ϕ between the direction [001] and the normal to the slip plane (line AC) is sin^{-1} (BC/AC). But BC = $a\sqrt{2}$ and AC = $a\sqrt{3}$, where a is the lattice parameter, so that $\varphi = \sin^{-1} (2/3)^{\frac{1}{2}} = 54.74°$.

From the expression $\tau_c = \sigma \times \cos \theta \cos \varphi$

$$\sigma = 1.6 \div (\cos 45° \cos 54.74°)$$
$$= 3.92\,\text{MPa}$$

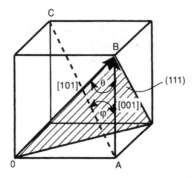

Figure 8.6

For the face-centred cubic system the most densely packed planes of atoms are those of the {111} family and within each (111) plane there are three possible directions of slip (Figure 8.7). The combination of a slip plane and a slip direction is termed a *slip system*. There are four sets of planes of this type, each set occurring at different inclinations, within the face-centred cubic structure. Consequently, no matter from what direction relative to the crystal a direct stress is applied there will be resolved shearing stresses acting on several slip planes and at least one slip system will be inclined in such a way that plastic deformation can occur. This means that face-centred cubic crystals are comparatively soft and ductile.

Figure 8.7 *Slip plane and directions in the f.c.c. system.*

In the body-centred system, the most densely packed planes are those of the {110} type. Within this type of plane there are two possible slip directions (Figure 8.8).

Slip in body-centred cubic metals is more complex than in the other two systems already discussed. In addition to slip taking place on (110) planes, slip may also occur on planes of the (112) and (123) types. Although there are numerous possible slip directions within body-centred cubic crystals, metals crystallising in this form are generally harder and less ductile than face-centred cubic metals. This is thought to be due to the body-centred cubic lattice being less densely packed than the face-centred

(a) Unit cell showing a (110)
slip plane

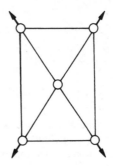

(b) Plan view of (110) plane showing the
slip directions

Figure 8.8 *Slip plane and directions in the b.c.c. system.*

cubic lattice. An increase in temperature will bring more slip planes into action, and also reduce the value of critical shearing stress. This latter statement also applies to other forms of metallic crystal.

The main slip planes and directions for the three principal types of crystal found in metals are summarised in Table 8.1.

Table 8.1 *Slip planes and directions in metal crystals.*

Crystal type	Slip planes	No. of slip planes	Slip directions	No. of slip directions per plane	Total no. of slip systems
h.c.p.	(0001)	1	$[2\bar{1}\bar{1}0][\bar{1}2\bar{1}0][\bar{1}\bar{1}20]$	3	3
f.c.c.	{111}	4	<110>	3	12
b.c.c.	{110}	6	<111>	2	12

8.3 | Dislocations

Crystal space lattices are not perfectly crystalline but contain certain structural defects. These may be classified as either line or point defects. The line defects are known as *dislocations*, and these may be classified as either *edge-type* or *screw-type*. In practice, dislocation lines are rarely of the pure-edge or pure-screw type but are mixed dislocations, that is, lines of dislocation containing both an edge and a screw component. An important property of a dislocation is its *Burger's vector, b*, which indicates the extent of lattice displacement caused by the dislocation. The Burger's vector also indicates the direction in which slip will occur. Figure 8.9 is a representation of an edge dislocation. It can be considered that the edge dislocation is due to the presence of an additional half-row of atoms within the lattice. If an atom-to-atom circuit is described within a portion of regular lattice, as shown at the bottom left of Figure 8.9(a), it will be a complete closed circuit. If, however, a similar circuit is described around the dislocated portion of lattice the start and finish will not be coincident. The distance *sf* in Figure 8.9(a) will be the Burger's vector, *b*. In an edge dislocation, the Burger's vector is normal to the dislocation line.

When a shearing stress is applied to a section of crystal lattice, and this stress is beyond the elastic limit, minor atomic movements will occur causing the dislocation line to move through the lattice (see Figure 8.9(b)). The magnitude of the stress necessary to initiate the movement of a dislocation across a slip plane is considerably less than that which would be required to bring about block slip. It will be seen that plastic flow occurs in the direction of the Burger's vector.

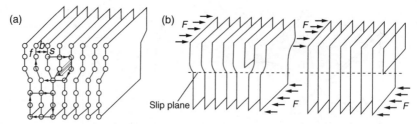

Figure 8.9 *Edge dislocation: (a) representation of a portion of crystal lattice containing an edge dislocation, dislocation shown as ⊥ and Burger's vector as b; (b) application of shear stress F causes the dislocation to move along the slip plane until it leaves this section of lattice causing an increment of plastic deformation.*

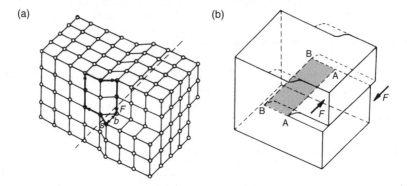

Figure 8.10 *Screw dislocation: (a) representation of a lattice containing a screw dislocation; (b) application of a shear stress F will cause the dislocation to move from AA to BB (slipped area is shaded).*

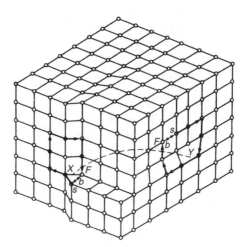

Figure 8.11 *Line of dislocation that is screw at X and edge at Y.*

A diagrammatic representation of a screw dislocation is shown in Figure 8.10. In this case, a Burger's circuit describes a helical, or screw, path and the Burger's vector, *b*, is in the same direction as the line of the dislocation. It should be noted, however, that plastic flow under the action of a shearing stress still occurs in the direction of the Burger's vector.

As stated earlier, many dislocations are mixed dislocations with both an edge and a screw component. The magnitude and direction of the Burger's vector are constant for all points on any one dislocation line. Referring to Figure 8.11 it will be seen that the dislocation line XY which is pure screw at

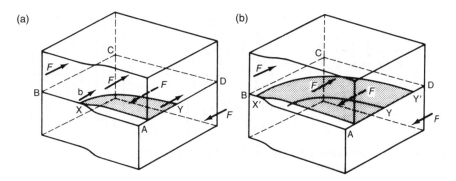

Figure 8.12 *Slip caused by movement of curved dislocation line across a slip plane ABCD. The dislocation moves from XY to X'Y' (slipped region is shown shaded).*

X and pure edge at Y has the same vector, *b*, at each end. When a shear stress, *F*, is applied to such a curved dislocation line lying within a slip plane ABCD it will cause slip to occur, as shown in Figure 8.12. All slip will be in the direction of the Burger's vector, *b*, but it will be noted that the curved dislocation line XY has now moved to position X'Y', the pure screw portion at X having moved normal to the applied force and the pure edge portion at Y having moved parallel to the applied force (see Figure 8.12(b)).

EXAMPLE 8.4

Copper is f.c.c. in structure and the lattice parameter of the unit cell is 0.361 nm. A dislocation line lies in a (111) plane and its Burger's vector, *b*, is in the [110] direction. What is the magnitude of the Burger's vector?

SOLUTION

The Burger's vector has the magnitude of one atomic spacing. The [110] direction is a face diagonal of a unit cell and it can be seen (refer to Figure 8.7) that the length of the face diagonal is equal to two atom spacings. The length of the diagonal, in terms of the lattice parameter, *a*, is $a\sqrt{2}$ and so an atomic spacing is equal to $a/\sqrt{2}$. Therefore, the magnitude of the Burger's vector in copper $= 0.361/\sqrt{2} = 0.255$ nm.

It should be appreciated that there will be strain fields associated with dislocations. In the example of an edge-type dislocation, considered as an extra half-row of atoms inserted into the structure, the crystal lattice in the region of the dislocation will be strained. There will be a compressive strain above the dislocation and a tensile strain below it, as shown in Figure 8.13(a).

It can be said that dislocations of like sign repel one another. Consider two similar dislocations on the same slip plane (Figure 8.13(b)). These will tend to repel one another in an effort to reduce the strain in that portion of the lattice, and it would require an increase in the level of applied stress to move them closer together. Conversely, dislocations of opposite sign attract one another. If the two dislocations shown in Figure 8.12(c) were brought together they would cancel one another out leaving a perfect unstrained section of lattice.

Crystals invariably contain numerous point defects, in addition to the structural defects mentioned above. The principal types of point defect which occur in metal crystals are the vacancy, the substitutional defect, and the interstitial type defect (Figure 8.14). The latter two types are due to the presence of 'stranger' atoms of other elements which are present either as impurities or are deliberately added

Figure 8.13 *(a) Strain field surrounding a dislocation. (b) Dislocations of like sign tend to repel one another. (c) Dislocations of opposite sign tend to attract one another.*

as alloying elements. The interstitial type occurs when the stranger atoms are very small in comparison with the atoms of the parent metal and fit into interstitial sites (see section 5.5), while the substitutional type occurs when parent and stranger atoms are comparable in size. It will be noticed from Figure 8.14 that strain is developed within a crystal lattice in the neighbourhood of point defects.

All crystal lattices contain a number of vacancies and the number of vacancies present increases as the temperature is increased. The rate at which the number of vacancies increases with temperature is exponential, following an Arrhenius type expression, $N = N_0 \exp(-Q/RT)$, where N is the number of vacancies per m³ at some temperature T (in Kelvin), N_0 is the number of lattice points (atoms) per m³ in the crystal structure, Q is the activation energy for vacancy formation (J/kmol) and R is the universal constant (8314 J/kmol K).

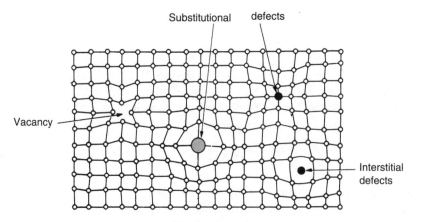

Figure 8.14 *Lattice containing several types of point defect. Note that substitutional atoms may be larger or smaller than the parent atoms causing positive or negative lattice strain.*

EXAMPLE 8.5

(a) Estimate the number of vacancies per m³ in copper at a temperature of 15 °C, given that the activation energy for vacancy formation in copper is 84 MJ/kmol and the lattice parameter, *a*, of the f.c.c. unit cell of copper is 0.361 nm.

(b) What would be the effect on the number of vacancies present of raising the temperature of the material by 100 °C?

SOLUTION

(a) The volume of one unit cell $= (0.361 \times 10^{-9})^3$ m$^3 = 4.705 \times 10^{-29}$ m^3
So the number of unit cells / m$^3 = 2.126 \times 10^{28}$
As there are 4 atoms per unit cell in the f.c.c. system, the number of lattice points per m$^3 =$ $4 \times (2.126 \times 10^{28}) = 8.502 \times 10^{28}$

The number of vacancies per m^3, $N = N_0 \exp(-Q/RT)$

At 15 °C (288 K) $N = 8.502 \times 10^{28} \exp(-84 \times 10^6/(8314 \times 288))$
$= 4.94 \times 10^{13}$/m^3

This is a very large number but, looking at it in perspective, because of the extremely large number of unit cells per m^3 it works out at approximately one vacancy in 4×10^{14} unit cells.

(b) At 115°C (388 K) $N = 8.502 \times 10^{28} \exp(-84 \times 10^6/(8314 \times 388))$
$= 4.17 \times 10^{17}$

A temperature rise of 100 °C increases the number of vacancies by a factor of about 10^4.

EXAMPLE 8.6

Calculate the theoretical density of iron, given that its atomic diameter is 0.2476 nm, it crystallises as b.c.c., and its atomic mass number is 55.85. Take Avogadro's number as 6.023×10^{26}/kmol.
The measured density of iron at 15°C is 7870 kg/m^3. Assuming that the difference between the calculated and experimental values of density are due to the presence of vacancies in the crystal lattice, estimate the number of vacancies per 1000 unit cells.

SOLUTION

For the b.c.c. system the lattice parameter, $a = 2D/\sqrt{3}$, where D is the atomic diameter.
so a for iron $= (2 \times 0.2476 \times 10^{-9})/\sqrt{3} = 0.2859 \times 10^{-9}$ m
and the volume of a unit cell, $a^3 = 2.337 \times 10^{-29}$ m^3
1 kmol of Fe has a mass of 55.85 kg and contains 6.023×10^{26} atoms
so the mass of a unit cell (2 atoms) $= (2 \times 55.85)/(6.023 \times 10^{26}) = 1.8546 \times 10^{-25}$ kg
Density is mass/volume $= (1.8546 \times 10^{-25})/(2.337 \times 10^{-29}) = 7936$ kg/m^3
The measured density of 7870 kg/m^3 is 0.83 per cent less than the theoretical density.
This means that in 1000 unit cells (2000 atoms) the number of vacancies will be 0.83 per cent of 2000, or approximately 17 vacant sites.

EXAMPLE 8.7

An aluminium alloy contains 2 per cent (by weight) of magnesium held in substitutional solid solution within the f.c.c. aluminium lattice. Calculate the average volume strain developed in the aluminium lattice.
Data: for aluminium – atomic diameter $= 0.2856$ nm, atomic mass number $= 26.97$
for magnesium – atomic diameter $= 0.3180$ nm, atomic mass number $= 24.32$.

SOLUTION

Consider 100 kg of alloy:

98 kg of aluminium contain 98/26.97 = 3.6337 kmol
2 kg of magnesium contain 2/24.32 = 0.0822 kmol

Therefore molar percentage of aluminium in alloy = $(100 \times 3.6337)/(3.6337 + 0.0822)$
= 97.79 per cent

and the molar percentage of magnesium in alloy = 100.0 − 97.79
= 2.21 per cent

The average atomic diameter of the alloy = $(0.2856 \times 0.9779) + (0.3180 \times 0.0221)$
= 0.2863 nm

The lattice parameter for an f.c.c. crystal, $a = D\sqrt{2}$
a for pure aluminium = $(0.2856 \times 10^{-9}) \times \sqrt{2} = 0.4039 \times 10^{-9}$ nm $a^3 = 6.589 \times 10^{-29}$ m^3
and a for alloy = $(0.2863 \times 10^{-9}) \times \sqrt{2} = 0.4049 \times 10^{-9}$ nm $a^3 = 6.638 \times 10^{-29}$ m^3

The volume strain developed is $(6.638 − 6.589)/6.589 = 0.0074$.

8.4 | Deformation by twinning

Another mechanism of plastic deformation, namely deformation by twin formation, may sometimes occur in close-packed hexagonal, body-centred cubic or tetragonal metal crystals. This type of deformation may occur if the slip planes within the crystals are not aligned to the axis of stress in a favourable manner. Twin formation in these crystals often occurs as the result of impact or shock loading conditions. In this type of deformation, the atoms in each plane undergo some displacement and a revised lattice arrangement is formed which will produce planes more favourably placed for slip to occur. The new lattice arrangement is a mirror image of the original in the twin, or shear, plane (Figure 8.15). When a sample of the metal tin, a body-centred tetragonal structure, is deformed plastically it is sometimes possible to hear high-pitched sounds emanating from the metal. These distinctive sounds, known as *tin cry*, are emitted during mechanical twin formation within the material. Deformation by mechanical twinning does not occur in crystals of the face-centred cubic type but twinned crystals, known as annealing twins, commonly occur in face-centred cubic metals that have been annealed following cold deformation.

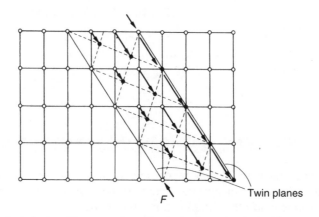

Figure 8.15 *Deformation by twinning. The new twinned lattice is shown by dashed lines.*

8.5 | Polycrystalline metals

In the foregoing sections, the mechanisms of plastic deformation within metallic single crystals have been discussed. Metallic engineering materials, however, are not normally in the form of single crystals, but are composed of many small crystals, or grains, and in many cases these small crystals have a random orientation. For a polycrystalline sample of a close-packed hexagonal metal, such as zinc, it will be apparent that while the slip planes in some crystals might be favourably inclined for slip under the action of an applied stress other crystals may not be aligned in a suitable direction. Plastic deformation of the favourably positioned crystals will be hindered, or even completely prevented, by unfavourably placed adjacent crystals (Figure 8.16(a)).

The crystal boundaries (grain boundaries) will also hinder plastic deformation. These boundaries are not simple planes but are transition zones between adjacent crystals of differing orientation (Figure 8.16(b)). These transition zones, which may be of several atoms in thickness, do not possess regular crystal planes and consequently act as barriers to the movement of dislocations. From this it follows that the grain boundaries in a pure metal tend to be stronger than the crystal grains and that, consequently, a sample composed of a large number of very small crystals will be stronger than another sample of the same metal containing a smaller number of larger crystal grains. These statements do not hold true at all times. For example, in some alloys and impure metals there may be segregation of alloy constituents, or impurities, to grain boundaries. In some cases this renders the crystal boundaries weak; for example, copper is embrittled by the segregation of traces of bismuth impurity to the grain boundaries. At very high temperatures close to the melting temperature, the grain boundaries become weaker than the crystals and plastic deformation may occur by a process of viscous flow of the grain boundary material. The turbine blades for aircraft gas turbines, which operate in a very high temperature environment, are manufactured by a special process which produces a single crystal structure with a specific orientation.

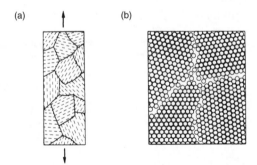

Figure 8.16 *Polycrystalline metals. (a) Random orientation of slip planes (shown dashed). Some crystals are more favourably placed than others for plastic deformation. (b) Representation of grains and boundaries.*

The grain size of a material is determined by counting the number of grains observed in a known area during the microexamination of a prepared sample. The results may be expressed either as the number of grains/mm² or as an index number. The grain size index, G_E, is based on the number of grains/mm² and is given by $m = 2^{(G_E + 3)}$ where m is the number of grains/mm² (refer to BS 4490), while the ASTM (American Society for Testing and Materials) index is based on the number of grains/in², as counted at a magnification of $\times 100$. The ASTM index number, n, is given by $N = 2^{(n-1)}$, where N is the number of grains/in² at $\times 100$ magnification. Numerically, these two index numbers are the same and the higher the index number, the greater will be the number of grains per unit area.

EXAMPLE 8.8

(a) A grain count on a sample of aluminium gives a value of 450 grains/mm^2. Calculate the grain size index number, G_E, for this material.

(b) Microexamination of a mild steel sample at a magnification of $\times 100$ gave a grain count of 160 grains/in^2. Calculate the ASTM index for this sample.

SOLUTION

(a) 450 grains/mm$^2 = m = 2^{(G_E + 3)}$

$\log 450 = (G_E + 3) \log 2 \qquad (G_E + 3) = 2.653/0.301 = 8.81$

So $G_E = 5.81$.

(b) 160 grains/in^2 at $\times 100$ magnification $= N = 2^{(n-1)}$

$\log 160 = (n - 1) \log 2$ so $(n - 1) = 2.204/0.301 = 7.32$

giving the ASTM index, n, as 8.32.

EXAMPLE 8.9

What is the average diameter, in mm, of the crystal grains in a metal with an ASTM index number of 7?

SOLUTION

For $n = 7$, the number of grains/in^2. at $\times 100 = 2^6 = 64$

The number of grains/mm$^2 = 64(100^2/25^2) = 1024$

The number of grains/mm $= \sqrt{1024} = 32$

thus giving an average grain diameter of $1/32 = 0.031$ mm.

The effect of grain size on tensile yield strength has been quantified by Petch. The Petch relationship relates the yield strength, σ_y, of a polycrystalline metal to the crystal grain size and is: $\sigma_y = \sigma_0 + K\,d^{-1/2}$, where d is the average grain diameter in mm and σ_0 and K are constants for the material.

EXAMPLE 8.10

Annealed samples of a steel with varying grain sizes were tested and the results obtained are shown in the table below:

	Average grain size (grains/mm)	Tensile yield strength, σ_y (MPa)
Sample A	13	250
Sample B	34	300
Sample C	45	320

Using the Petch relationship, estimate the grain size which would be necessary to give the steel a tensile yield strength of 360 MPa.

SOLUTION

The average grain sizes, d, for the three samples are:
Sample A – $1/13 = 0.077$ mm; B – $1/34 = 0.030$ mm; C – $1/45 = 0.022$ mm, from which $d^{-1/2}$ values are: Sample A – 3.6; B – 5.8; C – 6.75.
   A plot of σ_y against $d^{-1/2}$ yields a straight line (Figure 8.17). From the graph $\sigma_y = 360$ MPa corresponds to a value of $d^{-1/2} = 8.5$ from which $d = 0.014$ giving an average grain size of 72 grains/mm.

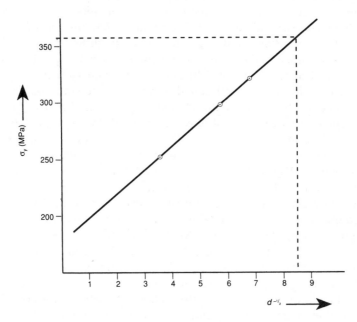

Figure 8.17   *Graph of σ_y against d^{-1}.*

A single crystal will exhibit anisotropy, that is, it will possess differing properties in different directions. This is very evident in the case of hexagonal metals that possess only one series of parallel slip planes. However, a polycrystalline material in which the orientation of the individual crystals is a purely random arrangement will, on a macro scale, be virtually isotropic.
   The deformation behaviour of the principal crystal types in polycrystalline samples of pure metals may be summarised as follows:

(a) Close-packed hexagonal metals are generally brittle, or of low ductility only. If ductility occurs, it usually exists only within a narrow range of temperatures.

(b) Face-centred cubic metals are invariably highly ductile over a wide range of temperatures.

(c) The body-centred cubic system is less favourable to ductility than the face-centred cubic system. The body-centred cubic system contains both ductile and brittle metals, but even the ductile metals in this system often show good ductility only over a fairly small range of temperature.

8.6 | Plastic deformation and strain hardening

When a metal is stressed beyond its elastic limit, dislocations within the metal begin to move and plastic deformation occurs. The movement of one dislocation through a lattice will give one increment of plastic deformation, namely a displacement of the order of one atomic spacing. The movement of extremely large numbers of dislocations would be necessary to give a visible amount of plastic slip and the number of dislocations that would need to be present to account for the large amount of plastic deformation that takes place in commercial metal-working processes such as rolling or forging would be so great as to mean that there would be virtually no regular sections of lattice at all. In fact, the number of dislocations present in a fully annealed metal is comparatively small and additional dislocations are generated during plastic deformation. The dislocation density in a metal can be considered as N, the number of dislocations that intersect a unit area of 1 mm^2. For fully annealed metals the dislocation density, N, is usually between 10^4 and 10^6/mm^2. The number of atoms per unit area in a typical metal crystal is of the order of 10^{13}/mm^2 and so it can be seen that the dislocations in an annealed metal are widely separated. There is a much higher dislocation density in plastically deformed metals and typical values for heavily cold worked material would be between 10^9 and 10^{10}/mm^2.

One suggested mechanism for the generation of dislocations is the Frank–Read source. This supposes a length of dislocation line firmly anchored at each end. Application of a shear force will cause the dislocation line to bow. The successive stages of movement of a dislocation line AB, fixed at both A and B, is shown in Figure 8.18. As movement and growth continues a kidney-shaped loop (Figure 8.18(iv)) develops. The two sections of the loop advancing towards one another are of opposing sign and will cancel out when they meet (Figure 8.18(v)) forming a complete loop and a new dislocation line between A and B, thus allowing the process to be repeated.

Frank–Read sources and dislocation generation have been observed experimentally by means of transmission electron microscopy of thin metal foils. The effect of expanding dislocation loops from a Frank–Read source giving plastic deformation is shown in Figure 8.19.

As plastic deformation occurs, the level of applied stress must be continually increased if the process is to be continued. In other words, as more plastic deformation is given to the metal, the greater will be the force required to continue deforming the material, and the metal can be said to become strengthened. Dislocations cannot move freely throughout the whole of a crystalline material. Their progress will be impeded by barriers such as grain boundaries. A continued movement of the dislocations, that is, further plastic deformation, can occur only if the applied stress is raised so that the dislocations can move across the potential barriers. Also, in cubic crystals, where slip may be occurring simultaneously on several intersecting slip-planes, the dislocations may interfere with one another's movement. After severe cold working, the array of dislocations within the metal will be in a highly tangled state, but there will be some areas that are almost dislocation free while other areas will have a very high dislocation density. Figure 8.20 is a diagrammatic representation of the dislocation distribution in a sample of heavily cold worked metal. This type of pattern can be seen using transmission electron microscopy.

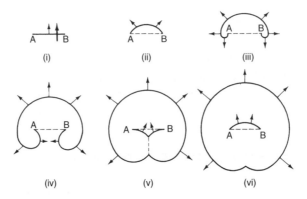

Figure 8.18 *Dislocation generation from a Frank–Read source.*

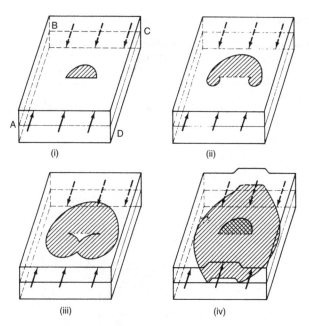

Figure 8.19 *Dislocation loops developing from a Frank–Read source on slip plane ABCD and causing plastic deformation.*

Figure 8.20 *Sketch showing formation of dislocation tangles in heavily cold worked metal (from an electron micrograph). The dislocation-free areas become small crystals.*

There is a high degree of structural disorder in those areas with a high dislocation density and these eventually become new crystal grain boundaries. One of the results of cold working is that the crystal grains of the original annealed metal become distorted and fragmented.

This phenomenon of strain, or work, hardening is utilised in practice for the strengthening of metals. By performing cold deformation operations, such as rolling or drawing, the strength of ductile metals may be increased considerably (see Table 8.2). Strain hardening is accompanied by a decrease in ductility and an increase in the electrical resistivity of the material.

If attempts are made to continue cold working beyond the point where maximum hardness is achieved, cracks will develop within the material and failure will occur. The start of failure may be due to a number of dislocations of like sign on the same slip plane being forced together at a major barrier, such as a grain boundary, by a large applied stress, thus forming a void or internal crack (see Figure 8.21).

Table 8.2 *Effect of cold rolling on the properties of commercial purity aluminium.*

Reduction in thickness (per cent)	Condition	Hardness (VPN)	Tensile strength (MPa)	Elongation (per cent)
0	Annealed	20	92	40
15	Quarter hard	28	107	15
30	Half hard	33	125	8
40	Three-quarters hard	38	140	5
60	Fully work hardened	43	155	3

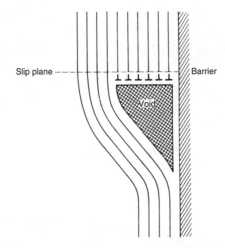

Figure 8.21 *The pile-up of several dislocations of like sign can form a void which can be a starting point for total failure.*

8.7 | Recrystallisation

A metallic material that has been subjected to straining beyond the elastic limit becomes harder and stronger, but more brittle, and its crystal structure becomes deformed. When the external deforming stress is removed, the material will still be in a state of some internal stress. Each individual crystal of the original material will have been strained both elastically and plastically. Removal of the external force should allow for the recovery of all elastic strain, but for any individual crystal complete elastic recovery will be hindered by the surrounding rigid crystals and there will be some locked-in elastic strain.

When a material in this condition is heated, changes will begin to take place. These changes may be classified under three headings:

(a) stress relief,

(b) recrystallisation,

(c) grain growth.

As the temperature of the material is raised so the vibrational energies of the individual atoms are increased and atomic movements can occur. Comparatively minor atomic movements result in the removal of the residual stresses associated with the locked-in elastic strains. This change, which occurs at comparatively low temperatures, has a negligible effect on the strength and hardness of the material, and the microstructure of the metal is unchanged in its appearance. When the temperature is

Figure 8.22 *Recrystallisation: (a) original cold worked structure; (b) nuclei forming; (c) and (d) new grains growing from nuclei; (e) recrystallisation complete.*

raised further the process of recrystallisation begins. New unstressed crystals begin to form and grow from nuclei until the whole of the material has a structure of unstressed polygonal crystals (Figure 8.22). This change in structure is accompanied by a reduction in hardness, strength, and brittleness to the original values prior to plastic deformation. The driving force for the recrystallisation process is the release of the strain energy stored in the zones of high dislocation density. The temperature at which recrystallistion occurs is, for a pure metal, within the range from one-third to one-half of the melting temperature (K). The recrystallisation temperature is not, however, constant for any material as its value is affected by the amount of plastic deformation prior to heating, and it is lower for very heavily cold worked metals than for samples of the same material which have received small amounts of plastic deformation. The recrystallisation temperature is also affected by composition. The presence of impurities or alloying elements will increase the recrystallisation temperature of the material. If the temperature is raised further, grain growth may occur following the completion of recrystallisation, with some crystal grains growing in size at the expense of others by a process of grain boundary migration. Again, the driving force for grain growth is the release of boundary surface energy as the amount of total grain boundary surface is reduced.

The industrial process of *annealing* is a heat treatment allowing recrystallisation to take place with consequent softening of work hardened materials. Some metals, such as lead, have recrystallisation temperatures below room temperature. This means that they cannot be work hardened at ordinary temperatures. Note that many of the various annealing treatments for steels, while still being softening treatments, are based on a different principle (refer to Chapter 16).

Plastic deformation processes carried out at temperatures above the recrystallisation temperature are referred to as hot working. There is no strain hardening effect due to hot working as there is almost immediate recrystallisation. Hot working processes such as hot rolling, forging and extrusion are used for the initial major plastic deformation of large ingots in industrial metal processing.

8.8 | Solution hardening

As mentioned in section 8.3, the presence of interstitial and substitutional point defects introduces strain in the lattice and disturbs the regularity of crystal planes. This has the effect of increasing the stress necessary to cause movement of dislocations and, therefore, strengthens the metal. The presence of stranger atoms within the parent metal space lattice also affects the movement of valency electrons and increases the electrical resistivity of the metal.

Atoms can move, or diffuse, through a crystal lattice because there are vacancies within the lattice. An atom can move from a position adjacent to a vacancy into that vacant site. Conversely, it could be considered that the vacancy had moved in the opposite direction to the atom jump. There is a tendency for solute atoms within a crystal lattice to be associated with vacancies. In this way the extent of the strain associated with a solute atom will be reduced to a minimum. Movement of solute atom/vacancy pairs can result in diffusion, a movement of solute atoms through the crystal lattice. (There is further coverage of diffusion in Chapter 12.) Solute atoms will tend to diffuse or migrate to sites near dislocation lines as a means of reducing the lattice strain to a minimum. When solute atoms are larger than the parent metal atoms they will diffuse to sites below the dislocation line, where the strains are tensile (Figure 8.23(a)). Solute metal atoms smaller than parent metal atoms will migrate to sites above the dislocation line where compressive strains occur (Figure 8.23(b)). These concentrations of solute atoms close to dislocations are termed *Cottrell atmospheres*. The Cottrell atmospheres have the effect of pinning dislocations in position. A relatively large force will be required to cause the dislocations to move, thus increasing the strength of the alloy.

Interstitial solute atoms also tend to migrate to sites near dislocations (see also section 8.10). The alloying of metals to produce interstitial or substitutional solid solutions is a useful technique for increasing the strengths of metals (see also section 11.5). Table 8.3 shows the hardness and strength increases obtained in copper–nickel solid solutions.

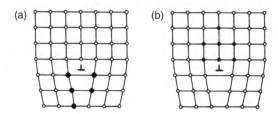

Figure 8.23 *Cottrell atmospheres. These tend to pin the dislocations. (a) Large solute atoms migrate to sites below the dislocation. (b) Small solute atoms migrate to sites above the dislocation.*

Table 8.3 *Solid solution hardening. Properties of annealed copper–nickel solid solutions.*

Composition (per cent)		Hardness (VPN)	Tensile strength (MPa)	Elongation (per cent)
Cu	Ni			
100	—	60	210	65
95	5	65	230	50
90	10	70	250	45
80	20	75	315	45
70	30	80	370	45
60	40	90	430	45
30	70	120	520	45
—	100	95	450	50

8.9 | Dispersion hardening

A third principle for the strengthening of metallic materials is termed *dispersion hardening*. A very finely dispersed second phase distributed throughout the crystal lattice will provide a series of barriers to the movement of dislocations, and considerably higher stresses will be necessary to cause yielding and plastic deformation than would be the case for a single-phase material. This dispersed second phase may be metallic or non-metallic. An example of a non-metallic dispersion is the use of thorium oxide to increase the high-temperature strength of tungsten wire in electric lamp filaments. Sintered aluminium powder (SAP) is another example of an oxide dispersion strengthened material. The compacting and sintering of finely powdered aluminium produces aluminium containing a very fine dispersion of aluminium oxide.

Precipitation hardening (age hardening) is another dispersion hardening process, but in this case the finely divided second phase is produced by giving an alloy a specific type of heat treatment. The types of alloys that may respond to this type of treatment are those in which partial solid solubility occurs (refer to section 11.7). The first stage of the heat treatment produces a supersaturated solid solution (Figure 8.24(a)). Such a supersaturated solution is metastable and there will be a tendency for diffusion of solute atoms to occur leading to the formation of a precipitate. Such diffusion will be accelerated if the temperature is increased. The first results of the diffusion of solute atoms is to increase their concentration in some localised areas of the crystal lattice. At this stage the solute atom aggregates are termed a *coherent* precipitate (Figure 8.24(b)). As the process continues the solute metal atoms separate from the parent metal lattice to form small crystals with their own lattice structure. At this stage the precipitate is a true precipitate, non-coherent with the parent metal (Figure 8.24(c)).

During the diffusion stage leading up to the formation of a coherent precipitate, areas of high strain are created within the parent lattice. These high-strain areas will hinder the movement of dislocations, thus strengthening the metal. Once true precipitation begins the strain in the parent lattice is released and the alloy loses the increased strength brought about by diffusion.

Certain commercial alloy systems, of which one of the best known is aluminium–copper, may be strengthened in this way. The dispersion strengthening of aluminium–copper alloys, also known as age hardening or precipitation hardening, is discussed in sections 12.6 and 15.3.

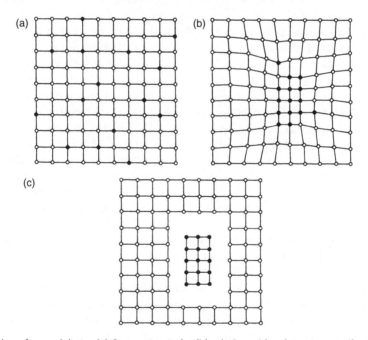

Figure 8.24 *Formation of a precipitate. (a) Supersaturated solid solution with solute atoms uniformly distributed through the lattice. (b) Diffusion of solute atoms has occurred producing severe strain in area around concentration of solute element. The solute atoms are still coherent with the parent metal lattice, and termed a coherent precipitate. (c) Solute has now separated as a true non-coherent precipitate. Both parent metal and precipitate lattice are unstrained.*

8.10 Yield point in mild steel

In most metallic materials, there is a gradual transition from elastic to plastic strain with increasing stress, but in the case of mild steel there is a sharp discontinuity in the tensile stress–strain curve (Figure 8.25).

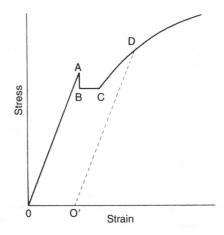

Figure 8.25 *Tensile stress–strain curve for mild steel showing sharp yield point.*

The material behaves in an elastic manner up to a certain point, A, and then suddenly yields. Point A is termed the *upper yield point*. The stress level necessary to continue plastic straining falls to level B, the lower yield stress, and a considerable amount of plastic strain takes place at this lower level. After this sudden yielding has occurred, the stress has to be increased again to bring about further plastic strain and, beyond point C on the curve normal strain hardening occurs. If the metal is stressed to some point, D, on the curve and the stress then removed, elastic strain will be recovered along the path DO′. Immediate reloading of the test-piece will give elastic strain to point D with continued plastic strain beyond point D, but with no sudden yielding. However, if after stressing to point D and unloading, the sample is left at ordinary temperatures for about a week before retesting (or heated at 100°C for about one hour) it is found that the sudden yield phenomenon returns. This is termed strain ageing.

Mild steel contains a very small percentage of carbon present in interstitial solid solution within the iron space lattice. The carbon atoms, small though they are, are somewhat larger than the interstices between iron atoms, and so the lattice is in a state of strain. The interstices along a dislocation line are slightly larger than the normal lattice interstitial sites and carbon atoms tend to migrate by diffusion through the lattice to positions along dislocations. This has the effect of partially locking the dislocation in position and requiring a relatively high stress, the upper yield stress, to initiate movement of the dislocation from the neighbourhood of carbon atoms.

Once the dislocations have been freed, their movement can be sustained at a lower stress level, the lower yield stress. During the ageing period, after stressing beyond the yield point and unloading, the interstitial atoms have a chance to diffuse through the lattice and return to positions on lines of dislocation. The diffusion process is speeded up if the temperature is raised. Pure iron completely free from interstitial impurities does not show a pronounced yield point.

8.11 Diffusion and dislocation climb

As stated earlier (section 8.3) crystal lattices contain vacancies and it is possible, through the medium of vacancies, for a certain amount of atomic movement, or diffusion, to take place. An atom occupying a position next to a vacancy in a crystal lattice may jump from its original position into the vacant site. Alternatively, it could be said that the vacancy has moved one atomic position in the opposite direction. In the case of a pure metal with all atoms identical, this atomic movement cannot be observed, but observations and measurements have been made of diffusion in pure metals containing

Figure 8.26 *Dislocation climb. (a) Lattice containing vacancy and dislocation. If numbered atoms diffuse into the vacancy the vacancy will move to position 4 causing dislocation to climb. (b) Dislocation pile-up at barrier. The climb of a dislocation will allow further dislocation movement.*

a small number of atoms of a radioactive isotope. The movements of these 'marked' atoms can be registered. Diffusion becomes of consequence in an alloy, or in a metal containing impurities, as the atoms of an alloying element or impurity may migrate through the parent lattice. Any individual atom, in moving into a vacant site, has to cross a potential barrier, and must possess an energy equal to, or greater than, q, in order to break away from the attraction of its neighbours, q is the activation energy for the movement of one atom. Diffusion is a thermally activated process and is governed by the Arrhenius rate law. The Arrhenius law and the laws of diffusion are dealt with in Chapter 12.

The migration of vacancies through a crystal lattice can lead to interaction with dislocations. When a vacancy reaches a point at a dislocation line, there will be a collapse of the vacancy at that point and the dislocation will move to the next lattice plane (Figure 8.26). The collapse of one vacancy will only cause the dislocation line to be moved to the next plane at one point. This is termed a *jog*. If many vacancies migrate to the same area then a large portion, or the whole, of a dislocation line will move to the next plane. This is known as the climb of dislocations. This has significance in connection with the creep of metals. Metallic creep is the continued slow straining of the material under constant stress. In most metals, creep occurs at significant rates only at elevated temperatures, although lead will creep at ordinary temperatures. Figure 8.26(b) shows a series of dislocations, that have been moved by the action of a stress, piled up at a barrier. No further movement (plastic strain) can take place unless the level of stress is increased. However, if thermally activated diffusion leading to dislocation climb occurs, then some further dislocation movement may take place at constant stress. This has been postulated as a possible mechanism of metallic creep (see also Chapter 25).

8.12 | Superplasticity

Plastic deformation through the medium of dislocation movement has been discussed in earlier sections but some alloys, under certain circumstances may be plastically deformed to an extremely large extent (elongation values of the order of a thousand per cent) under low deformation forces. This phenomenon is termed *superplasticity*. The superplastic alloys possess an extremely fine grain structure with grain diameters of less than 5 μm and, generally, possess a duplex microstructure. The strain rate must be closely controlled for superplastic deformation to occur. Neither very low nor high rates of strain will permit superplastic deformation. Superplastic behaviour occurs at high temperatures and a temperature of 0.5–0.7 T_m (K) is generally required. The mechanism of deformation is largely by grain boundary sliding. A good example of a superplastic alloy is the zinc–aluminium alloy of eutectoid composition (22 per cent Al) (see section 15.11) which will show very large deformations when formed at 275°C. Other superplastic alloys of commercial importance include the titanium alloy containing 6 per cent Al and 4 per cent V, which can be deformed at about 930°C, and some of the Ni–Cr–Fe superalloys.

SUMMARY

Metals, when stressed, deform elastically up to a certain level of stress, the *yield stress*, and deform plastically beyond this point. Generally, the plastic deformation of metals is the result of a shear stress causing dislocations to move across specific crystal planes, the slip planes, within metal crystals. The slip planes are the planes of greatest atomic density and slip takes place across these planes in a direction of greatest atomic line density. Cubic crystalline metals have numerous slip planes and slip directions. Face-centred cubic metals tend to be relatively soft and ductile. Body-centred cubic metals tend to be harder and less ductile than f.c.c. metals, probably because they have less densely packed lattices. Hexagonal close-packed metals possess only one set of parallel slip planes and in a polycrystalline material the slip planes in many crystals may not be aligned in a manner favourable for slip. In consequence, commercial h.c.p. metals tend to be non-ductile.

A characteristic of a dislocation is its Burger's vector, as this gives the direction in which slip will take place and the magnitude of one increment of slip. Dislocations may be described as edge-type or screw-type but many dislocations are mixed, with both an edge and a screw component. During the plastic deformation process new dislocations are generated and the dislocation density in a heavily plastically deformed metal is greater by a factor of 10^4 or 10^5 than that of the undeformed metal. Metal crystals also contain point defects, namely vacancies, interstitial and substitutional defects. These point defects cause local strain in the lattice and distortion of the slip planes, thus hindering the movement of dislocations and, thereby, strengthening the metal.

Most commercial metallic materials are polycrystalline and the boundaries between crystal grains act as barriers to the movement of dislocations. From this it follows that a metal with a very fine grain size will be stronger than a coarse-grained sample of the same metal. The Petch equation relates the yield strength of a polycrystalline metal to its grain size.

There are three main ways in which the strength of a metal may be increased, these being *strain hardening, solution hardening* and *dispersion hardening*. Strain hardening is the result of severe plastic deformation as, for example, in the cold rolling of metal sheet. During the process the original crystal structure is distorted and broken up into small, highly strained grains with a high dislocation density and the hardness and strength of the metal is increased, with a corresponding reduction in ductility. The effects of strain hardening may be removed by a heat treatment known as *annealing* at a temperature of between one-third and one-half of the melting temperature (in Kelvin) in which new unstrained crystals form by a process of recrystallisation. Solution hardening is the result of making an alloying addition to a metal such that the atoms of the added element enter into either interstitial or substitutional sites within the original crystal lattice. This generates lattice strain and disturbs the regularity of the slip planes, thus hindering the movement of dislocations and strengthening the metal. Dispersion hardening occurs when a very finely dispersed second phase is distributed widely throughout the crystal lattice, thus providing many barriers to the movement of dislocations and strengthening the metal. Dispersion hardening may be achieved either by the mechanical dispersion of a finely divided insoluble phase through the metal, as with the distribution of thorium oxide in tungsten, or by generating coherent precipitate sites throughout a metal structure by a *precipitation* or *age hardening* heat treatment.

Vacancies in a crystal lattice play an important role in diffusion processes. An atom in a site adjacent to a vacancy can move into the vacant site. Atoms of a second element occupying substitutional positiions can, through interaction with vacancies, diffuse through the crystal lattice. When an atom moves into a vacant site it means that the vacancy moves through one increment in the opposite direction. When, during the movement or migration of vacancies, a vacancy reaches a dislocation line it will cause the collapse of the vacancy and the movement of part of the dislocation to the next lattice plane. The migration of many vacancies to the same area will cause a dislocation line to *climb* to the next plane. Dislocation climb is of significance in the creep of metals.

Questions

8.1 A direct stress of 100 MPa acts on a sample of a metal. Calculate the shear stress acting within planes at the following angles to the axis of direct stress: (a) 20°, (b) 40°, (c) 60°, (d) 80°.

8.2 The critical shear stress for the <111> {110} slip system of a pure b.c.c. metal is determined to be 23 MPa. What direct stress is required to be applied in the [001] direction to initiate slip in the [1$\bar{1}$1] direction on the (110) plane?

8.3 A single crystal sample of a h.c.p. metal of circular cross-section, with a diameter of 4 mm and of 50 mm length, is subjected to an axial tensile force.
  (a) When the tensile force is 1.25 kN the measured elastic extension is 0.047 mm. Given that Poisson's ratio for the metal is 0.35, estimate the theoretical shear strength of the metal.
  (b) Plastic deformation of the sample begins when the tensile force is increased to 2.8 kN. Given that the [2$\bar{1}$$\bar{1}$0] slip direction and the (0001) slip plane are both aligned at 35° to the axis of the tensile force, calculate the critical shear stress for the metal.
  (c) Why do the values obtained in (a) and (b) differ by so much?

8.4 Describe with the aid of sketches: (a) an edge dislocation, (b) a screw dislocation, and (c) a dislocation loop. In each case, show the relationship between the dislocation line and its Burger's vector.

8.5 Given that the operative slip systems in f.c.c. crystals are <110> {111}, calculate the magnitude of the Burger's vector of a dislocation in a sample of silver, given that the lattice parameter for the silver lattice is 0.420 nm.

8.6 The density of dislocations in soft iron is approximately 10^5/mm^3. Estimate the total length of dislocations in a 1 kg sample of iron, given that the density of iron is 7870 kg/m^3.

8.7 Describe with the aid of sketches three types of point defect which may occur in metal crystals and indicate the ways in which they affect the lattice and the properties of the material.

8.8 Calculate the number of vacancies per 10^6 atoms for a sample of niobium, given that the atomic radius of niobium is 0.1426 nm and the lattice parameter of the b.c.c. niobium crystal is determined as 0.3294 nm.

8.9 Nickel atoms and copper atoms have radii of 0.1243 nm and 0.1276 nm respectively and both metals crystallise as f.c.c. Calculate the proportion of nickel atoms in a copper–nickel solid solution alloy if the lattice parameter of the alloy lattice is determined as 0.3594 nm.

8.10 Estimate the amount of lattice strain in the copper lattice when it contains in solid solution: (a) 5 molar per cent nickel, (b) 5 molar per cent tin, and (c) 5 molar per cent zinc. Assume that the atomic radii for the metals are copper 0.1276 nm, nickel 0.1243 nm, tin 0.1508 nm, zinc 0.1329 nm.

8.11 In f.c.c. crystals the largest interstitial spaces are at sites where the unit cell coordinates are of the $a/2$, $a/2$, $a/2$ type, while in b.c.c. crystals the largest spaces are in sites with coordinates of the $a/2$, $a/4$, 0 type. Given that the radius of an iron atom is 0.124 nm, calculate the radii of these interstitial sites in both the b.c.c. and f.c.c. forms of iron.
  The radius of a carbon atom is 0.077 nm. Explain how this influences the interstitial solid solubility of carbon in both crystalline forms of iron.

8.12 (a) Describe a mechanism for the generation of dislocations under the action of an applied shear stress.
  (b) Give brief accounts of the following phenomena:
    (i) the yield point in mild steel,
    (ii) dislocation climb and creep,
    (iii) the hindering effect on dislocation movement of coherent and non-coherent precipitates.

8.13 The Petch equation $\sigma_y = \sigma_0 + Kd^{-0.5}$ relates the yield strength of a metal to its grain size. A steel sample with an ASTM index of 6 has a tensile yield strength of 275 MPa, one with an ASTM index of 8 has a yield strength of 318 MPa and one with an ASTM index of 10 has a yield strength of 375 MPa. Evaluate the constants K and σ_0 and explain the significance of σ_0. What would be the expected yield strength of a steel sample with an average grain diameter of 0.03 mm?

8.14 The tensile yield strength of annealed pure aluminium with a grain size index, G_E, of 6.5 is 36 MPa and that for a single-crystal sample of the metal is 25 MPa. Estimate the tensile strengths of samples of aluminium with measured average grain diameters of (a) 0.01 mm and (b) 0.10 mm.

8.15 Samples of cold worked and annealed commercial purity aluminium were subjected to tensile testing and the average grain size in each sample determined. The results obtained were:

Sample A   Tensile yield strength 97.5 MPa   Grain size 800 grains/mm²
Sample B   Tensile yield strength 85.0 MPa   Grain size 190 grains/mm²
Sample C   Tensile yield strength 95.0 MPa   Grain size 600 grains/mm²

(a) Estimate the maximum grain size allowable if the minimum tensile yield strength required in the aluminium product is 87.5 MPa.

(b) Determine the value of σ_0 in the Petch relationship and explain its significance.

CHAPTER 9

Viscoelastic behaviour

What is viscoelasticity?

For an elastic solid the strain developed is directly proportional to the level of stress applied, but for an ideal viscous fluid it is the rate of strain, or flow, that is directly proportional to the level of stress. In the latter case the constant relating strain rate to stress is termed the *viscosity*. Some materials, when subject to stress, behave in a manner in which there is some elasticity but also some time-dependent deformation or creep. This type of behaviour is termed *viscoelastic* and is shown by thermoplastic materials at temperatures between their glass transition and melting temperatures. The viscosity of materials decreases exponentially as the temperature is increased and so the viscoelastic properties of a material are highly temperature-dependent.

LEARNING OBJECTIVES

1. Know the meaning of the term *viscoelasticity*.
2. Understand how viscoelastic behaviour can be simulated by combining the simple elements of a Hookean spring, for elasticity, and a dashpot, for viscosity, in various ways.
3. Know the characteristics of both the Maxwell and the Voigt–Kelvin models for viscoelasticity.
4. Understand the significance of the terms *creep, recovery* and *relaxation*.
5. Be able to perform simple calculations based on the Maxwell and Voigt–Kelvin models.

9.1 | Introduction

The behaviour of polymeric materials, when subjected to stress, varies very considerably from one material to another and also is greatly affected by temperature. Thermoplastic materials, at temperatures below their *glass transition* temperatures, and thermosets are brittle elastic solids but thermoplastics, above T_g, show *viscoelastic* behaviour. In a viscoelastic material the deformation under stress is time-dependent and the phenomenon of *creep* can occur. Also, in such a material, there can be *relaxation*, that is, a reduction in the level of stress with time at constant strain. A consequence of viscoelasticity is that parameters such as modulus, strength and ductility are functions of time and rate of strain.

9.2 Viscoelasticity

For an ideal elastic material the strain developed, ε, is directly proportional to the stress applied, σ, and this may be expressed as $\varepsilon = \sigma \times$ constant. For this Hookean elasticity, this relation is usually rewritten as $\sigma = \varepsilon \times E$, the constant E being the *modulus of elasticity*.

In the case of a Newtonian fluid, an ideal viscous fluid, the rate of strain, $d\gamma/dt$, is directly proportional to the magnitude of the shear stress, τ, and this may be expressed as $\tau = d\gamma/dt \times$ constant. In this case, the constant is the *viscosity*, η.

Thermoplastic materials at temperatures above T_g possess properties which lie somewhere between these two ideal cases and show some aspects of elasticity and also some viscosity, hence the term *viscoelastic solids*. In a viscoelastic solid, stress, σ, is a function of both strain, ε, and time, t, and can be expressed by an equation: $\sigma = f(\varepsilon, t)$ or $\sigma = \varepsilon f(t)$. It is also temperature-dependent as viscosity decreases with increase in temperature according to the Arrhenius law (see Chapter 12).

Viscoelastic behaviour can be simulated using a series of models. These models are composed of two elements, a Hookean spring, representing ideal elasticity for which strain $\varepsilon = \sigma/E$, and a dashpot, representing Newtonian viscosity for which strain rate $d\varepsilon/dt = \sigma/\eta$, combined in various ways. Several models have been proposed and some of these are considered below.

9.3 The Maxwell model

The Maxwell model for viscoelasticity consists of a Hookean spring and a dashpot combined in series (Figure 9.1).

Consider the behaviour of this model when subjected to some constant stress, σ_c. There will be an instantaneous strain, ε_1, developed in the spring, such that

$$\varepsilon_1 = \sigma_c/E \tag{1}$$

Figure 9.1 *Maxwell model.*

and a continuous straining at a constant rate of the dashpot occurs according to the equation

$$d\varepsilon/dt = \sigma_c/\eta \qquad (2)$$

The total strain developed after some time t, ε_t, is given by

$$\varepsilon_t = \sigma_c \left(\frac{1}{E} + \frac{t}{\eta} \right) \qquad (3)$$

This behaviour is shown graphically in Figure 9.2(a) (the section between zero time and t_1). If the strain is then maintained constant until some time t_2 the phenomenon of stress relaxation will occur (Figure 9.2(b)), the stress reducing in this time interval from σ_c to σ_1.
From equation (1), the strain rate is

$$\frac{d\varepsilon}{dt} = \frac{1}{E} \times \frac{d\sigma}{dt}$$

Combining this with equation (2), we have

$$\frac{d\varepsilon}{dt} = \left(\frac{1}{E} \times \frac{d\sigma}{dt} \right) + \left(\frac{1}{\eta} \times \sigma \right) \qquad (4)$$

At constant strain $d\varepsilon/dt = 0$. The solution of the differential equation (4) when $d\varepsilon/dt = 0$ and the condition that when $t = 0$ then $\sigma = \sigma_c$ becomes

$$\sigma_t = \sigma_c \exp \left(\frac{-E}{\eta} \right) t \quad \text{where } \sigma_t \text{ is the stress at some time } t \qquad (5)$$

In other words stress relaxation is exponential, as shown in Figure 9.2(b) with a time constant of η/E.
If the stress relaxes from σ_c to σ_1 by time t_2 and the stress is then removed completely, there will be a corresponding *recovery* of elastic strain of amount ε_2, as shown in Figure 9.2(a), such that $\varepsilon_2 = \sigma_1/E$.
The Maxwell model gives a reasonable approximation to the stress relaxation phenomenon in thermoplastics but does not correspond particularly well to either the creep behaviour or the elastic recovery of these materials.

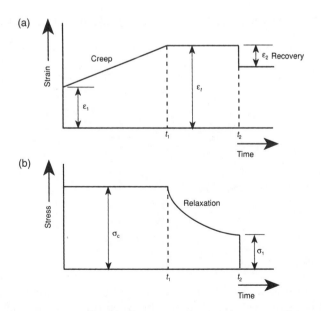

Figure 9.2 *(a) Strain response of a Maxwell model. Stress is maintained constant until time t_2. (b) Stress response of a Maxwell model showing relaxation between t_1 and t_2.*

EXAMPLE 9.1

A coupling in a fluid line is fastened using bolts of PA6. The joint may be subject to leakage if the tensile stress in the bolts is less than 30 MPa. Initially, the bolts are tightened until the tensile stress in them is 45 MPa. What is the maximum time that should be allowed to elapse before the bolts are retightened if leakages are not to occur? Assume that E for PA6 is 2.5 GPa, that the viscosity of PA6 is 7.5×10^{17} Pa s, and that the relaxation of PA6 is as predicted by the Maxwell model of viscoelasticity.

SOLUTION

According to the Maxwell model, stress relaxation is according to equation (5):

$$\sigma_t = \sigma_c \exp\left(\frac{-E}{\eta}\right)t \text{ or } \ln\left(\frac{\sigma_c}{\sigma_t}\right) = \frac{Et}{\eta}$$

Substituting:

$$\ln\left(\frac{45 \times 10^6}{30 \times 10^6}\right) = 0.4055 = \frac{2.5 \times 10^9 t}{7.5 \times 10^{17}} = 3.333 \times 10^{-9}t$$

from which

$$t = 1.22 \times 10^8 \text{ s} \approx 3.85 \text{ years}$$

According to this, if the Maxwell model accurately reflects the stress relaxation of PA6, no leakages should occur within a time of 3.85 years. In practice, retensioning of the holding bolts would occur during regular maintenance at more frequent intervals than this maximum time.

9.4 | Voigt–Kelvin model

The Voigt–Kelvin model for viscoelasticity comprises a Hookean spring and a dashpot arranged in parallel (Figure 9.3).

Figure 9.3 *Voigt–Kelvin model.*

Consider the response of this model when subject to a stress, σ. As the two elements of the model are arranged in parallel the stress will be distributed between the elements such that

$$\sigma = \sigma_s + \sigma_d$$

where σ_s is the stress in the spring and σ_d is the stress in the dashpot, so that, from equations (1) and (2)

$$\sigma = E\varepsilon + \eta \mathrm{d}\varepsilon/\mathrm{d}t \qquad (6)$$

which may be rewritten as

$$\frac{\mathrm{d}t}{\eta} = \left(\frac{\mathrm{d}\varepsilon}{\sigma - E\varepsilon}\right)$$

Integration gives

$$\frac{t}{\eta} = -\frac{1}{E}\ln\left(\sigma - E\varepsilon\right) \quad \text{or} \quad \exp\left(\frac{-Et}{\eta}\right) = \sigma - E\varepsilon$$

from which

$$\varepsilon = \frac{\sigma}{E}\left\{1 - \exp\left(\frac{-Et}{\eta}\right)\right\} \qquad (7)$$

This equation indicates that when a stress, σ, is applied to the model strain will increase exponentially from zero to a value of σ/E, that is, the instantaneous strain which would have been developed in the spring alone in the absence of a dashpot. This behaviour is illustrated in Figure 9.4(a).

The Voigt–Kelvin model does not show any stress relaxation as when the maximum strain has been developed, at time t_1, the dashpot is fully extended and the elastic spring element bears all the stress. Maintaining constant strain for a time interval between t_1 and t_2 causes the stress to remain constant also. (See Figure 9.4.)

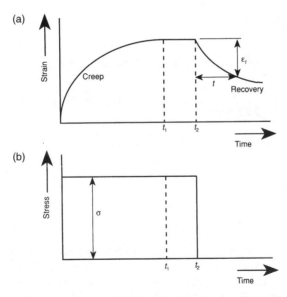

Figure 9.4 *(a) Strain response of a Voigt–Kelvin model. Strain is maintained constant until t_2, when stress is removed. (b) Stress response of a Voigt–Kelvin model. No relaxation occurs between t_1 and t_2.*

If the stress is removed there will be strain recovery. With zero stress, equation (6) becomes

$$E\varepsilon + \eta \mathrm{d}\varepsilon/\mathrm{d}t = 0$$

and the solution of this differential equation gives

$$\varepsilon_t = \varepsilon \exp\left(\frac{-E}{\eta}\right)t$$

where ε_t is the reduced strain at some time t after the removal of stress. This exponential recovery is shown in Figure 9.4(a).

The Voigt–Kelvin model gives a response which corresponds to both the creep and recovery behaviour of thermoplastics but is deficient in terms of relaxation.

EXAMPLE 9.2

A plastic material for which E is 1 GPa and viscosity, η, is 250 GPa s is subjected to a tensile stress of 12 MPa. Assuming that the material behaves according to the Voigt–Kelvin model, calculate the value of tensile strain after periods of 5, 10 and 20 minutes. If the stress is removed after 20 minutes how long will it take for 50 per cent of the strain to be recovered?

SOLUTION

For the Voigt–Kelvin model $\varepsilon = \dfrac{\sigma}{E}\left\{1 - \exp\left(\dfrac{-Et}{\eta}\right)\right\}$

So, for 5 minutes (300 s)

$$\varepsilon = \frac{12 \times 10^6}{1 \times 10^9}\left\{1 - \exp\left(\frac{-1 \times 10^9 \times 300}{250 \times 10^9}\right)\right\}$$

$$= 0.012 \times (1 - 0.301) = 8.38 \times 10^{-3}$$

Similarly for 10 minutes (600 s) $\varepsilon = 10.9 \times 10^{-3}$
and for 20 minutes (1200 s) $\qquad \varepsilon = 11.9 \times 10^{-3}$

For strain recovery $\varepsilon_t = \varepsilon \exp\left(\dfrac{-E}{\eta}\right)t$ where ε_t is the strain at time t.

When ε_t is 0.5ε $\qquad 0.5 = \exp\left(\dfrac{-1 \times 10^9}{250 \times 10^9}\right)t$ or $0.693 = 4 \times 10^{-3} \times t$

giving $t = 173$ s.

9.5 | Other models

In the preceding sections it was shown that the Maxwell model was realistic in terms of relaxation but does not accurately relate to creep and recovery while the Voigt–Kelvin model approximates to the creep and recovery of many thermoplastics but shows no relaxation phenomenon. Various complex models comprising combinations of these two simple models have been proposed. One such is the Maxwell–Kelvin model in which a Maxwell model and a Voigt–Kelvin model are combined in series as shown in Figure 9.5. The response of this model shows creep, recovery and stress relaxation and is a closer approximation to the actual behaviour of a thermoplastic than either model singly. The response of a Maxwell–Kelvin model is shown in Figure 9.6.

The Maxwell–Kelvin model, while showing creep, recovery and relaxation, is only an approximation and does not truly reflect the complex responses of real thermoplastic materials. More accurate simulations can be achieved by making other models with more Maxwell and Kelvin elements added in various combinations. However, in these more complex models the mathematical analysis becomes very much more involved and complicated.

Figure 9.5 *Maxwell–Kelvin model.*

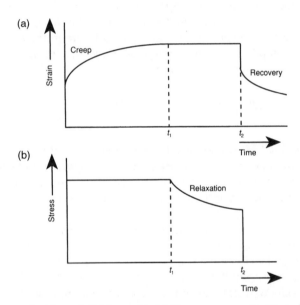

Figure 9.6 *(a) Strain response of a Maxwell–Kelvin model. (b) Stress response of a Maxwell–Kelvin model.*

EXAMPLE 9.3

The strain–time response for a thermoplastic material when subjected to a force for a given time followed by complete removal of the external force, is shown in Figure 9.7.

Using the notation given in the figure:

(a) Draw the full Maxwell–Kelvin model, indicating the state of the springs and dashpots, to describe the situation at A (with no load applied) and at B (with maximum strain).

(b) Similarly, indicate the state of the springs and dashpots in the Maxwell–Kelvin model at points C, D and E in the figure.

(c) Describe briefly what occurs in the polymer structure to give rise to:
  (i) the instantaneous changes, and
  (ii) the retarded and the irrecoverable changes.

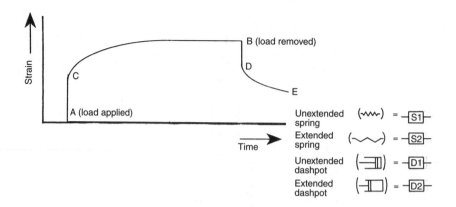

Figure 9.7

SOLUTION

(a) The Maxwell–Kelvin model comprises a Maxwell element (a Hookean spring and a dashpot in series) in series with a Voigt–Kelvin element (a Hookean spring and a dashpot in parallel).

At A there is no strain and both springs and both dashpots are in the unextended state.

At B, maximum strain, creep has occurred with time and both springs and both dashpots are in the extended state. These configurations are given below.

(b) At C the application of load causes an instantaneous extension of the elastic spring of the Maxwell element. This initial loading causes no extension of the dashpots and the spring of the Voigt–Kelvin element, which is restrained from extending by the dashpot in parallel. After some time under constant load both dashpots and the spring in the Voigt–Kelvin element creep and extend fully to the state in B.

At point D the removal of the load allows the spring in the Maxwell element to return fully to its unextended state. Both dashpots are in the extended state and instantaneous recovery of the strain in the spring in the Voigt–Kelvin element is prevented by the parallel dashpot. With time the force exerted by the spring in the Voigt–Kelvin element on its parallel dashpot causes the dashpot to return slowly to its unextended state, giving recovery of strain to point E in the figure, but the dashpot of the Maxwell element remains in its extended state. The configurations for points C, D and E are shown below.

(c) (i)   When a force is applied to a thermoplastic material an instantaneous elastic strain is generated. This elastic strain is a combination of bond extension and bond straightening within the individual polymer molecules (refer to section 3.3). When the force is removed this elastic strain is recovered.

   (ii) Some effects in polymer materials are time-dependent. Continued slow straining, creep, occurs under load. There is a continued slow straightening of the polymer chains and also a relative sliding movement between molecules as the weak van der Waal's bonds between molecules are broken. This relative sliding movement is akin to a viscous flow. When an applied load is removed, there is some instantaneous elastic recovery, and there is also some further recovery of strain with time. Many of the molecules cannot return to their totally unstrained condition immediately the load is removed because they will be constrained by the configurations of surrounding molecules. The material, therefore, will contain residual internal stresses and these will cause, over time, a slow recovery of some strain. However, much of the viscous flow which occurs during the creep stage is irrecoverable.

SUMMARY
••

Polymer materials, at temperatures above their glass transition temperatures, behave when stressed in a manner which is part elastic and in part similar to a highly viscous fluid. The strain developed in a viscoelastic solid is a function both of stress and of time, as the material slowly deforms by viscous flow, or creep. Similarly, when the stress is removed, the elastic portion of strain is recovered instantaneously but a further recovery of strain occurs with time. The phenomenon of stress relaxation also occurs in viscoelastic solids, this being a reduction of stress with time when the material is held in a state of constant strain. Also, the properties of viscoelastic solids are highly temperature-dependent as viscosity decreases with an increase in temperature according to an exponential law. Various models, using a combination of Hookean elastic springs and Newtonian flow dashpots, have been developed to simulate viscoelasticity but these, while giving approximations, do not reflect completely the complex responses to stress of real thermoplastic materials.

Questions

9.1 Describe both the Maxwell and the Voigt–Kelvin models for viscoelasticity and explain both the good and bad features of each of these models in relation to the actual viscoelastic characteristics of real thermoplastic materials. How would it be possible to design a viscoelastic model that would simulate more closely the behaviour of real materials?

9.2 A thermoplastic material with an elastic modulus of 1.5 GPa and a viscosity of 100 GPa s is subject to a constant stress of 15 MPa. Calculate the predicted values of strain after (a) 2 minutes and (b) 5 minutes on the basis of (i) the Maxwell model and (ii) the Voigt–Kelvin model.

9.3 A cylindrical plastic rod, of 12 mm diameter and length 300 mm, is subjected to an axial tensile force of 1400 N for a time of 20 minutes. Assuming that the material behaves as a Voigt–Kelvin model, estimate the length of the rod after (a) 10 minutes, (b) 20 minutes, and (c) 20 minutes after the force has been removed. Assume that the spring and dashpot constants are 900 MPa and 300 GPa s respectively.

9.4 A rubber band with an instantaneous modulus of elasticity of 4 MPa is stretched around a sheaf of papers and the instantaneous stress developed in the rubber is 2 MPa. The stress within the rubber is determined to be 1.9 MPa after 40 days. Estimate the stress in the rubber after a period of 1 year, assuming that relaxation follows the pattern of the Maxwell model.

9.5 A thermoplastic material which can be represented by a Maxwell model has a tensile modulus of 1 GPa. When in a state of constant strain it is found that the stress reduces to 0.6 of its original value after 300 days. Estimate the viscosity of the material.

CHAPTER 10

Toughness and fracture of materials

Why should we learn about toughness and fracture?

We use the terms *tough* and *brittle*, sometimes loosely, to describe the manner in which a material may fracture. When a material suffers considerable plastic deformation before fracture it is deemed to be tough, while a brittle material shows little or no plastic deformation before fracture. However, *toughness* is an intrinsic property of a material.

Materials invariably contain flaws and defects and these will have a major influence on the level of applied stress necessary to cause fracture of a material. It is essential that an engineer should understand the various factors that affect both the manner of failure and the levels of stress necessary for failure, in order that he or she can design components and structures that should not fail within their projected life cycle. In some circumstances components made from materials which would be expected to plastically deform if overstressed may fracture suddenly in a brittle manner, the fracture growing rapidly from some existing flaw, even though the average level of stress in the material is well below its yield stress. The resistance of a material to fast crack propagation is termed *fracture toughness*.

This chapter contains an introduction to the subject of *Fracture Mechanics*, which is the study of the relationships between the strength and toughness of a material, the stress systems and the geometry of flaws and defects. The aim of fracture mechanics is to determine the critical size of a crack or other defect necessary for the occurrence of fast fracture.

LEARNING OBJECTIVES

1. Distinguish between ductile and brittle fracture modes.
2. Understand the stress raising effect of small defects.
3. Distinguish between *toughness* and *fracture toughness* and their relationship to flaw size and fracture stress for plane stress conditions.

4. Understand the influence of component geometry on critical stress intensity.

5. Use the appropriate equations to calculate critical flaw sizes and fracture stresses for some cases.

6. Appreciate the effect of temperature and know the effect of strain rate on the ductile–brittle transition temperature of steels.

10.1 Introduction

The terms *ductile, tough, brittle, fatigue* and *creep* are used frequently to describe the mode of failure of a material. Ceramics and glasses are brittle in that there is little or no plastic deformation prior to failure and the energy needed to bring about fracture is relatively low. Many thermoplastic materials deform plastically and require a considerable energy input to bring about tearing or fracture. For metals and alloys there is a wide spectrum of failure mode from tough to brittle. The type of failure which actually occurs will be dependent largely on the nature and condition of the material but it is also affected by a number of other factors including the type of stress applied, the rate of application of stress, temperature and environmental conditions, the size and shape of any internal flaws or surface defects that may be present and the geometry and surface condition of the component or structure.

The influence of some of the above factors will be considered in this chapter, together with an introduction to the subject of fracture mechanics and way in which this can be applied to calculate critical defect sizes and fracture stresses in some cases.

Failure by fatigue and creep is not discussed in this chapter but is dealt with in Chapter 25.

10.2 Crack propagation and failure

Ductile failure is preceded by plastic deformation. In the case of most metals, as the level of applied stress is increased, firstly the metal deforms in an elastic manner. Plastic deformation begins when the yield stress value is exceeded and, eventually, the component fractures. Let us examine what happens when a metal test-piece is subjected to a direct tensile stress, as in a tensile test. Firstly, the stress causes elastic strain in the test-piece (Figure 10.1(a)) but when the level of stress exceeds the yield strength plastic deformation begins. This is manifested by a necking effect in one part of the test-piece (Figure 10.1(b)). The level of stress increases in the necked region, sufficiently so to create small voids within the material (Figure 10.1(c)) and then, through a tearing action, these voids increase in size and fuse into one

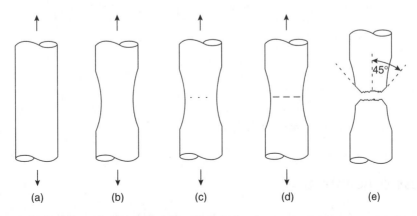

(a)   (b)   (c)   (d)   (e)

Figure 10.1 *Stages in the failure of a ductile metal tensile test-piece.*

large internal crack (Figure 10.1(d)). The level of stress in the outer annulus of the section increases rapidly, eventually causing failure by shear, giving rise to the 'cup and cone' fracture (Figure 10.1(e)).

During the propagation of a crack in a ductile metal, plastic deformation takes place near the tip of a crack and this plastic flow tends to blunt the crack tip, thus reducing its stress concentrating effect. However, continued plastic flow in the strain hardened region ahead of the crack causes the creation of further internal cavities. (See also section 8.6 (Figure 8.21)). As plastic flow continues, the cavities ahead of the crack tend to link up through a tearing mechanism (see Figure 10.2). The plastic deformation and tearing absorb considerable energy. The tearing and failure process is similar in many thermoplastic materials.

In brittle materials, on the other hand, little, if any, plastic deformation can occur ahead of the crack tip to blunt the crack, so the actual local stress at the sharp crack tip is intensified to a very high level. When the local stress at the crack tip exceeds the fracture stress of the material the interatomic bonds are ruptured and the crack develops catastrophically as a fast brittle fracture. Ceramics, glasses, brittle metals, thermosets and thermoplastics at temperatures below their T_g values behave in this way.

Many fibre reinforced composite materials possess higher toughness than either the matrix or reinforcing material on their own. The growth of a crack through the matrix in a direction normal to the line of reinforcing fibres can be stopped at a fibre. The high stress concentrated at the crack tip can cause a rupture of the bonding between matrix and fibre, thus blunting the crack tip and reducing the local stress so that fracture cannot proceed into the fibre (Figure 10.3). This crack stopping effect only applies to crack growth in a direction normal to the fibre direction. Crack growth in a direction parallel to the fibre direction can proceed without hindrance. A good example of this is the fracture characteristics of timber, a naturally occurring composite fibrous material. Timbers split easily along the grain but give a tough fracture across the grain.

(a)   (b)

Figure 10.2 *Crack propagation in a ductile metal: (a) blunting of crack tip and creation of voids ahead of tip; (b) voids linking up through a tearing action.*

Figure 10.3 *Crack in a fibre composite matrix stopping at a fibre.*

10.3 | Stress concentration

The theoretical strength of a material, σ_t, is given by $\sigma_t = (E\gamma/b)^{\frac{1}{2}}$ where E is the modulus of elasticity, γ is the surface energy and b is the interatomic spacing. The calculated value of σ_t for a material is

Figure 10.4 *(a) Enclosed elliptical defect. (b) Edge defect.*

approximately one-tenth of the value of E. The actual strengths of most materials are very much lower than this theoretical value, differing by some factor between 10^3 and 10^5. Real materials contain structural imperfections and small cracks. The presence of a crack will cause a high local concentration of stress. The classical equation which may be used for calculating the stress concentrating effect of an elliptical defect lying normal to the stress axis is given by the expression: $\sigma_{loc} = \sigma(1 + 2(a/r)^{\frac{1}{2}})$, where σ_{loc} is the local stress at the defect tip, σ is the nominal stress, $2a$ is the size of the defect and r is the radius at the defect tip (refer to Figure 10.4). (Note that $2a$ is the size of a wholly enclosed flaw while for a defect which breaks the surface the size is quoted as a.)

For a defect of high aspect ratio with a crack tip radius which is small in relation to the defect size, then

$$\sigma_{loc} = \sigma(1 + 2(a/r)^{\frac{1}{2}}) \approx 2\sigma(a/r)^{\frac{1}{2}}$$

The quantity $(1 + 2(a/r)^{\frac{1}{2}})$ or $2(a/r)^{\frac{1}{2}}$ is termed the *stress concentration factor, K_t.*

EXAMPLE 10.1

Estimate the stress concentrating effects of
(a) a circular hole of 1 mm diameter and
(b) a small enclosed crack of length 2 mm and with a crack tip radius of 1 μm.

SOLUTION

(a) In this case $a = r = 1$ mm so $(a/r)^{1/2} = 1$
giving $\sigma_{loc} = \sigma(1 + 2) = 3\sigma$
Note that for a circular hole the actual diameter is immaterial as a/r is always 1.
(b) In this case a is 1 mm (crack length is $2a$)
so $\sigma_{loc} = \sigma(2(1 \times 10^{-3}/1 \times 10^{-6})^{1/2}) = 63\sigma$

10.4 | Fracture mechanics

Fracture mechanics is the study of the relationships between crack geometry, material strength and toughness, and stress systems, as they affect the fracture characteristics of a material. The aim of fracture mechanics is to determine the critical size of a crack or other defect necessary for the occurrence of fast fracture, that is, catastrophic crack propagation and failure, under service loading conditions.

The analysis of the critical stress to cause crack propagation in brittle materials was made by Griffith who proposed the relationships

$$\sigma_c = \left(\frac{2\gamma E}{\pi a}\right)^{\frac{1}{2}} \text{ (for plane stress)}$$

$$\sigma_c = \left\{ \frac{2\gamma E}{\pi(1 - v^2)a} \right\}^{\frac{1}{2}} \text{ (for plane strain)}$$

where σ_c is the critical stress for fracture, γ is the fracture surface energy per unit area, E is the modulus of elasticity, a is one half the crack length and v is Poisson's ratio. The plane stress version is used for crack propagation in thin materials.

EXAMPLE 10.2

(a) Determine the fracture stress of a glass if it contains a series of microcracks with a maximum dimension of 10 μm. Assume that E for glass is 70 GPa and the surface energy is 0.6 J/m².

(b) What would be the likely fracture stress for the glass if a fine scratch with a depth of 0.3 mm was made on the surface?

SOLUTION

(a) Using the Griffith equation for plane stress, $\sigma_c = (2\gamma E/\pi a)^{\frac{1}{2}}$ and substituting the values for γ, E and a ($2a = 10$ μm, $a = 5 \times 10^{-6}$ m) we have

$$\sigma_c = \left(\frac{2 \times 0.6 \times 70 \times 10^9}{\pi \times 5 \times 10^{-6}} \right)^{\frac{1}{2}} = 73.1 \times 10^6 \text{ Pa} = 73.1 \text{ MPa}$$

(b) For a surface scratch the depth is taken as a, so $a = 3 \times 10^{-4}$ m, so

$$\sigma_c = \left(\frac{2 \times 0.6 \times 70 \times 10^9}{\pi \times 3 \times 10^{-4}} \right)^{\frac{1}{2}} = 9.44 \times 10^6 \text{ Pa} = 9.44 \text{ MPa}$$

The presence of a small surface scratch causes a major reduction in the fracture stress of the glass.

The Griffith relationship only applies to brittle materials and was modified by Orowan and Irwin to take account of the plastic flow which occurs at a crack tip at the beginning of crack development. With the assumption that the size of the plastic zone at the crack tip is very small, the Orowan–Irwin modification becomes

$$\sigma_c = \left(\frac{2(\gamma + \gamma_p)E}{\pi(1 - v^2)a} \right)^{\frac{1}{2}} \text{ for plane strain}$$

where γ_p is the work done in plastic deformation per unit area of crack extension. The equation is generally written as

$$\sigma_c = \left(\frac{G_c E}{\pi(1 - v^2)a} \right)^{\frac{1}{2}} \text{ for plane strain or } \sigma_c = \left(\frac{G_c E}{\pi a} \right)^{\frac{1}{2}} \text{ for plane stress}$$

The term G_c ($G_c = 2(\gamma + \gamma_p)$) is a property of the material and is termed the *toughness*, or *critical strain energy release rate*, and has the units J/m².

There are three possible ways to apply stress to a crack so as to cause a crack extension. These are in plane strain (opening mode I), in plane shear (opening mode II), and in anti-plane shear (opening mode III). These modes are shown in Figure 10.5. The respective toughnesses for each opening mode are written as G_{Ic}, G_{IIc} and G_{IIIc}.

The toughness value G_{Ic} is the value generally quoted for materials as mode I opening is the easiest mode for crack extension and, hence, represents the lowest toughness value. Values of toughness, G_{Ic}, for a range of materials are given in Table 10.1.

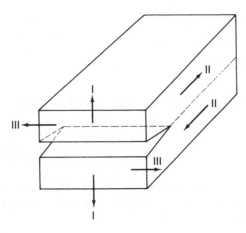

Figure 10.5 *Crack opening modes.*

Table 10.1 *Toughness and fracture toughness values for some materials.*

Material	Toughness G_{lc} (kJ/m²)	Fracture toughness K_{lc} (MPa m$^{\frac{1}{2}}$)	Material	Toughness G_{lc} (kJ/m²)	Fracture toughness K_{lc} (MPa m$^{\frac{1}{2}}$)
Steels	30–135	80–170	acrylic (PMMA)	0.4–1.6	0.9–1.6
Cast irons	0.2–3	6–20	nylons (PA)	1–4	3–5
Aluminium alloys	0.4–70	5–70	rigid PVC		
			(UPVC)	1.3–1.4	1–4
Copper alloys	10–100	30–120	polyester resins	0.1–0.2	0.5–1
Nickel alloys	50–110	100–150	epoxy resins	0.1–0.3	0.3–0.5
Titanium alloys	20–100	50–100	GFRP	20–100	20–60
Polyethylene (LDPE)	6–10	1–2	CFRP	5–30	30–45
Polyethylene (HDPE)	3–7	1–5	soda glass	0.01	0.7
Polypropylene (PP)	7–10	3–4	alumina	0.02	3–5
Polycarbonate (PC)	0.4–5	1–2.5	silicon nitride	0.1	4–5
Polystyrene (PS)	0.4–0.8	0.7–1.1	concrete	0.03	0.2

The plane stress equation above can be rewritten as $\sigma_c(\pi a)^{\frac{1}{2}} = (G_c E)^{\frac{1}{2}}$. Both G_c and E are material properties, that is, constants for the material. It follows that the particular combination of stress and defect size at which fracture will occur must also be a material constant.

As stated above, the aim of fracture mechanics is to determine the critical values of stress and crack size necessary for fast fracture. The following examples should aid understanding.

EXAMPLE 10.3

If the fracture stress of a sheet of high-tensile steel containing a central crack of length 30 mm is 500 MPa, calculate the fracture stress in the case of a 50 mm crack in a similar sheet.

SOLUTION

$\sigma_c = (G_{lc}E/\pi a)^{\frac{1}{2}}$ which may be written as $\sigma_c = \text{constant} \times a^{-\frac{1}{2}}$
When $2a = 30$ mm (0.03 m) $\sigma_c = 500$ MPa.

so constant $= \sigma_c \times a^{\frac{1}{2}} = 500 \times 0.015^{\frac{1}{2}}$ MPa m$^{\frac{1}{2}}$

For a 50 mm crack $a = 25$ mm (0.025 m)

therefore $\sigma_c = 500 \times (0.015/0.025)^{\frac{1}{2}}$

$\quad\quad\quad\quad = 387$ MPa

For a crack of length 50 mm the critical fracture stress will be 387 MPa.

EXAMPLE 10.4

Determine the maximum crack size that can be tolerated in a sheet of high-tensile material with a tensile yield stress of 1100 MPa and a toughness of 25 kJ/m^2, if the design stress safety factor is 1.5. Assume that E for the material is 208 GPa.

SOLUTION

The maximum design stress = yield stress ÷ safety factor = 1100/1.5 = 733.3 MPa.

Since $\sigma = \left(\dfrac{G_{Ic}E}{\pi a} \right)^{\frac{1}{2}}$

so $733.3 \times 10^6 = \left(\dfrac{25 \times 10^3 \times 208 \times 10^9}{\pi a} \right)^{\frac{1}{2}}$

and hence

$$a = \frac{25 \times 208 \times 10^{12}}{(733.3 \times 10^6)^2 \, \pi} = 3.08 \times 10^{-3} \, \text{m} = 3.08 \, \text{mm}$$

The maximum crack size that can be tolerated is a crack of length $2a$, namely 6.16 mm.

Irwin noted that stresses at a crack tip were proportional to $(\pi a)^{-\frac{1}{2}}$ and proposed a parameter, K, the *stress intensity factor*, where $K = \sigma(\pi a)^{\frac{1}{2}}$. K has the units MPa m$^{\frac{1}{2}}$. It should be noted that this expression is based on an elliptical flaw under plane stress in a plate of infinite size. The general form of the equation can be written as $K = \sigma M(\pi a)^{\frac{1}{2}}$, where M is a *geometry factor* to take account of varying flaw shapes and finite boundaries. Stress intensity factors for some geometries are given in Table 10.2. This factor, K, while it is associated with stress distribution close to the tip of a crack, should not be confused with the *stress concentration factor*, K_t, described in section 10.3.

Sudden fast fracture will occur when K reaches some critical value, K_c. The parameter $K_c = (G_cE)^{\frac{1}{2}}$ and, hence, is a constant for the material. K_c, the *critical stress intensity factor*, is generally termed the *fracture toughness* of the material and has the units MPa m$^{\frac{1}{2}}$. The fracture toughness value generally quoted for a material is that for K_{Ic}, the value for mode I crack opening. Values of K_{Ic} for a range of materials are given in Table 10.1.

10.5 | Applications of K_{Ic}

No material is wholly free from defects but it is essential that any crack-like defects that are present are relatively harmless. Using K_{Ic} values of fracture toughness it is possible to calculate the defect size

Table 10.2 *Stress intensity factors for some geometries.*

Type of crack	Stress intensity factor, K
Centre-crack of length $2a$ in infinite plate	$K = \sigma(\pi a)^{\frac{1}{2}}$
Centre-crack of length $2a$ in plate of width W^*	$K = \sigma(\pi a)^{\frac{1}{2}}\{W/\pi a \times \tan(\pi a/W)\}^{\frac{1}{2}}$ or $K = \sigma(\pi a)^{\frac{1}{2}}\{\sec(\pi a/W)\}^{\frac{1}{2}}$
Single edge crack of length a in semi-infinite plate	$K = 1.12\sigma(\pi a)^{\frac{1}{2}}$
Two symmetrical edge cracks, each of length a in plate of width W	$K = \sigma(\pi a)^{\frac{1}{2}}\{[W/\pi a \times \tan(\pi a/W)] + [0.1\,W/\pi a \times \sin(\pi a/W)]\}^{\frac{1}{2}}$
Central penny-shaped crack of diameter $2a$ in infinite body	$K = \sigma(\pi a)^{\frac{1}{2}}(2/\pi)$

* Of the two functions listed the sec function is probably the more accurate.

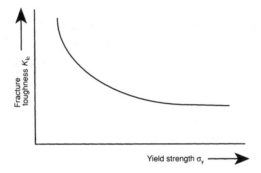

Figure 10.6 *General relationship between yield strength and K_{Ic}.*

necessary to initiate fast fracture failure at a given value of applied stress, or the value of applied stress required to cause failure in a component containing a defect of a known size. A high value of fracture toughness, K_{Ic}, is an indication that the resistance to fracture will be high. Generally, as the hardness and tensile yield stress of a material increases so its fracture toughness decreases (see Figure 10.6). This is because materials with high yield strengths tend to have low values of ductility and their ability to plastically deform ahead of a crack with a consequent blunting of the crack tip is negligible.

It can be seen from the general expression $\sigma_c = K_{Ic}/(\pi a)^{\frac{1}{2}}$ that σ_c, the critical stress for fast fracture, reduces as the size of a flaw, denoted by a, increases. This relationship is shown graphically in Figure 10.7. Many materials, particularly metals and alloys, can plastically deform when stressed beyond a certain level, the yield stress. The value of yield stress, σ_y, is marked in the figure showing that there is a critical flaw size, below which plastic yielding of a material will occur. If the material contains a flaw larger than this critical size it will fail by fast fracture before the yield stress is reached. Quantitative non-destructive testing methods can be used to determine the location and sizes of flaws in many materials. Thus it is possible to decide whether the dimensions of a discovered defect are below the critical value and thus 'safe', or above the critical value and initiate catastrophic failure in service. Flaws in components which are subject to fluctuating stresses in service can grow in size due to *fatigue*. Similarly, corrosion effects could produce crack growth with time. Thus, a flaw which was considered 'safe' initially may become unstable and reach critical dimensions after a period of time in service. Many engineering fabrications, such as pressure vessels, aircraft structures and bridges, are examined at regular intervals, using quantitative non-destructive testing techniques, and components taken out of service or the damage rectified before a crack has reached the critical size.

Figure 10.7   *Plot of σ_c and σ_y for a material against defect size.*

EXAMPLE 10.5

A component manufactured from a steel with a tensile yield strength of 1200 MPa and a fracture toughness of 150 MPa $m^{\frac{1}{2}}$ is found to contain a small crack of length 4.5 mm. Determine whether or not the component is safe in terms of its resistance to fast fracture assuming that the maximum applied stress is limited by the yield stress.

SOLUTION

From the expression $\sigma_c = K_{Ic}/(\pi a)^{\frac{1}{2}}$, $a = (K_{Ic}/\sigma_c)^2 \div \pi$
Making $\sigma_c = \sigma_y$ we have

$$a = \left(\frac{150 \times 10^6}{1200 \times 10^6}\right)^2 \div \pi = 4.97 \times 10^{-3}\,m = 4.97\,mm$$

from which the critical crack size, $2a = 9.94$ mm.

 The critical crack size for fast fracture is just over twice the size of the detected crack and it would be safe to use the component.

EXAMPLE 10.6

Low alloy steel plate with a yield strength of 1200 MPa and a fracture toughness of 70 MPa $m^{\frac{1}{2}}$ is to be used in a structure with a maximum design stress of $\sigma_y/1.5$. Full examination of the plate, as received, reveals the presence of an edge crack with a depth of 2 mm.

 Can the plate be used in this condition and, if not, what steps could be taken to render the plate usable? Assume the plate size to be such as to be classified as semi-infinite.

SOLUTION

For a single edge crack in a semi-infinite plate use $K = 1.12\sigma(\pi a)^{\frac{1}{2}}$
Design stress $= \sigma_y/1.5 = 1200/1.5 = 800$ MPa
The critical fracture stress, $\sigma_c = K_{Ic}/1.12(\pi a)^{1/2} = (70 \times 10^6)/1.12(\pi \times 2 \times 10^{-3})^{\frac{1}{2}}$
$$= 788\,MPa$$

This is fractionally less than the design stress of 800 MPa and it would be unwise to use the plate in this condition. As the plate is semi-infinite in size it could be made usable by grinding out the 2 mm deep crack. This would not significantly change the overall dimensions of the plate.

EXAMPLE 10.7

An aluminium alloy with a yield strength of 475 MPa and a fracture toughness of 21 MPa $m^{\frac{1}{2}}$ is suggested for the manufacture of a structural component with a maximum design stress of $\sigma_y/2$. The inspection facilities are capable of detecting defects of 3 mm length or greater. Determine whether the inspection capability is sufficiently sensitive to detect internal flaws before they reach the critical dimensions for fast fracture.

SOLUTION

Using the expression $a = (K_{Ic}/\sigma_c)^2 \div \pi$ and making $\sigma_c = \sigma_y/2$ we have

$$a = \left(\frac{2 \times 21 \times 10^6}{475 \times 10^6}\right)^2 \div \pi = 2.49 \times 10^{-3}\,m = 2.49\,mm$$

from which the critical crack size, $2a = 4.98$ mm, which is well within the capabilities of the inspection system.

The failure of a pressure vessel in service by the mechanism of fast fracture because of the unmonitored slow growth of a flaw could be extremely dangerous. One measure which can be adopted to prevent this happening is to design the pressure vessel on the *leak before break* principle, that is, to design the vessel such that the wall thickness is less than the critical crack size for fast fracture. In this way, a crack could develop through the full wall thickness, with a consequent leakage and loss of internal pressure, but without being large enough to reach the critical size for fast fracture. The wall thickness, t, needs to be less than the critical crack size $2a$ but in practice, for added safety, generally cylinders are designed so that $t \approx a$.

For a cylinder in which wall thickness, t, is very much smaller than the radius, r, the hoop stress, σ, is given by $\sigma = Pr/t$, where P is the internal pressure in the cylinder. It will be seen that although designing on the leak before break principle will give added safety there may be some disadvantages in other respects. If the critical flaw size is relatively small then the cylinder wall thickness will have to be equally small and this can only be achieved by either reducing the internal pressure P or by increasing the radius r, or altering both parameters.

EXAMPLE 10.8

A pressure cylinder is to be constructed from a steel with a yield strength of 1100 MPa and a fracture toughness of 110 MPa $m^{\frac{1}{2}}$. Safety factors of 2 on the yield strength and ratio of wall thickness to critical flaw size are to be used. If the maximum working pressure is to be 150 bar (1 bar = 10^5 Pa), calculate the maximum wall thickness and minimum cylinder diameter required if failure by leak before break is to occur.

SOLUTION

Maximum design stress $= \sigma_y/2 = 550\,\text{MPa}$
Using $a = (K_{Ic}/\sigma)^2/\pi$
we have

$$a = \left(\frac{110}{550}\right)^2 \div \pi = 12.7 \times 10^{-3}\,\text{m} = 12.7\,\text{mm}$$

Using $\sigma = Pr/t$, $P = 150\,\text{bar} = 15\,\text{MPa}$ and making $t = a = 12.7\,\text{mm}$
we have

$$r = \frac{550 \times 10^6 \times 12.7 \times 10^{-3}}{15 \times 10^6} = 0.466\,\text{m}$$

To satisfy the leak before break criteria with a maximum working pressure the cylinder should have a maximum wall thickness of 12.7 mm and a minimum diameter of $(2 \times 0.466) = 0.932\,\text{m}$.

10.6 | Effect of temperature

A change in temperature can have a significant effect on the fracture toughness of a material, a reduction in temperature increasing the liability to fail by brittle fracture. In the case of thermoplastic materials, there is a major change in the fracture toughness as the temperature falls through the glass transition temperature, T_g. There is also a major change in the behaviour of many metals.

An increase in temperature will render dislocations in metals more mobile and, hence, will cause a reduction in the yield strength. Conversely, a decrease in temperature will be reflected in an increased yield strength. Metals with face-centred cubic structures remain ductile at all temperatures but metals with body-centred cubic and hexagonal crystal structures undergo a change from ductile to brittle behaviour as the temperature is lowered. For example, zinc (h.c.p. structure) is brittle at ordinary temperatures but will deform plastically at temperatures in excess of 100°C. Low carbon steel is ductile at low strain rates at all temperatures above a value of about −170°C, but when a low carbon steel is subjected to impact loading conditions the transition from a tough, ductile fracture to a brittle cleavage fracture occurs at about 0°C. The energy absorbed in fracturing a low carbon steel test-piece in a Charpy impact test may reduce from about 2 MJ/m² at +15°C to less than 200 kJ/m² at −10°C.

The cleavage fracture stress, σ_f, of a metal does not vary as much with temperature as does the yield stress, σ_y. In Figure 10.8(a), curve (i) shows the variation of σ_f with temperature, while curves (ii) and (iii) show the variation of σ_y with temperature for a body-centred cubic metal. Curve (ii) is the effect

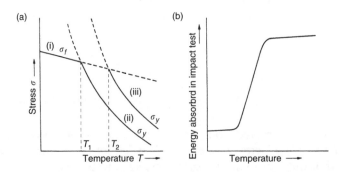

Figure 10.8 *(a) Variation of σ_f and σ_y with temperature for a b.c.c. metal: (i) σ_f–T, (ii) σ_y–T for slow strain rate, (iii) σ_y–T for fast strain rate. (b) Notch-impact energy against temperature showing ductile–brittle transition.*

for slow rates of strain and curve (iii) applies under conditions of high rates of strain (impact loading conditions). At low rates of strain $\sigma_f < \sigma_y$ at all temperatures below T_1, while for impact conditions $\sigma_f < \sigma_y$ at temperatures below T_2. In other words, at temperatures between T_1 and T_2, the metal will show ductile behaviour at low rates of strain but will be brittle under impact conditions.

Figure 10.8(b) shows the relationship between impact strength and temperature, as measured in a Charpy notch-impact test, for a body-centred cubic metal. The transition from tough to ductile behaviour occurs over a narrow range of temperature but, generally, a specific transition temperature is quoted. In the case of the transition in steels, the transition temperature is that at which the energy for fracture of a standard size Charpy test-piece is 40 J ($200\,kJ/m^2$). The ductile–brittle transition temperature, which is around $0\,°C$ for a low carbon steel, is affected by alloying and impurity elements. It is reduced by manganese and nickel but increased by carbon, nitrogen and phosphorus.

It should be noted that the parameter of toughness, as measured in a notch-impact test of the Charpy or Izod type, is not the same as the fracture toughness of a material. The units of notch-impact toughness are J/m^2 while those of fracture toughness are $MPa\,m^{\frac{1}{2}}$.

10.7 | Determination of fracture toughness

Standard tests for fracture toughness are designed to determine reproducible values for K_{Ic} (mode I crack opening). Two types of standard test-piece may be used. These are the single edge-notched bend test-piece (SEN) and the compact tensile test-piece (CTS) (see Figures 10.9 and 10.10). Both types of test-piece are prepared with a notch conforming to standard dimensions and with a crack emanating from the base of the notch. The crack is generally developed from the notch by a fatigue process to a pre-determined size. During a fracture toughness test the values of applied force required to cause given amounts of crack extension are measured. The full details of the test procedures and the manner

Width = W
Thickness = $B = 0.5W$
Half loading span L = $2W$
Notch width N = $0.065W$ max. (if W is over 25 mm) or $= 1.5\,mm$
max. (if W is less than or equal to 25 mm)
Effective notch length M = $0.25W$ to $0.45W$
Effective crack length a = $0.45W$ to $0.55W$

All dimensions are in millimetres.

Figure 10.9 *Details of standard bend test-piece (SEN) from BS 5447: 1987.*

Net width W
Total width C $= 1.25\,W$ min.
Thickness B $= 0.5\,W$
Half height H $= 0.6\,W$
Hole diameter D $= 0.25\,W$
Half distance between
hole outer edges F $= 1.6\,D$
Notch width N $= 0.065\,W$ max.
Effective notch length M $= 0.25\,W$ to $0.40\,W$
Effective crack length a $= 0.45\,W$ to $0.55\,W$

All dimensions are in millimetres.

Figure 10.10   *Details of standard tension test-piece (CTS) from BS 5447: 1987.*

in which K_{Ic} can be determined from the force–displacement results is given in BS 5447 'Methods of Test for Plane-Strain Fracture Toughness (K_{Ic}) Testing of Metallic Materials' and in ASTM E 399-83.

10.8 | Yielding fracture mechanics

The linear elastic fracture mechanics principles outlined earlier in this chapter are used in connection with high-strength materials where valid K_{Ic} values can be obtained using relatively small test-pieces. However, with low-strength, high fracture toughness materials such as low-carbon plain carbon steels, it would be necessary to use an extremely large test-piece in order to satisfy the standard conditions for the determination of K_{Ic} and this would be impracticable.

If the influence of plasticity on the stress field at a crack tip is considered then, for a plastic zone of length $2r_y$ at the tip of a crack of length $2a$, the relationship $K_{Ic} = \sigma\{\pi(a + r_y)\}^{\frac{1}{2}}$ is valid, where r_y is given by $\frac{1}{2}a(\sigma_c/\sigma_y)^2$. The use of this modified relationship is not appropriate if the ratio of applied stress to yield stress is large and the size of the plastic zone becomes large, but it does permit the use of a smaller sized test-piece.

In the case of materials that show a high degree of plasticity, a small test-piece could show general yielding before fracture, even though a large structure could fail by fast fracture before general yield takes place. In these circumstances the plane strain fracture toughness is estimated from *crack opening displacement* (COD) measurements. A test-piece of the SEN type (Figure 10.9) is used and displacement is measured by means of a clip gauge attached above the crack. The crack opening displacement, δ, is related to K and the yield stress, σ_y according to the following equations for conditions in which total yield is fairly small:

$$\delta = \frac{\sigma_a^2 \pi a}{\sigma_y E}$$

where σ_a is the applied stress.
This can also be written as

$$\delta = \frac{K^2}{\sigma_y E} \text{ or } \delta = \frac{G}{\sigma_y}$$

as K, G and E are related by $G = K^2/E$.

In COD testing, a critical displacement value, δ_c, can be measured and this can be used to calculate the fracture toughness of a material, but it should be emphasised that the relationships $\delta = K^2/\sigma E = G/\sigma$ hold only for small σ_a/σ_y ratios.

10.9 | Conclusion

Failure of a material by fast or brittle fracture can have disastrous consequences. Fracture mechanics is useful for assessing the risk potential of cracks in structures and for determining a 'safe maximum' or permissible crack size in a component. This is invaluable when designing structures to be 'fracture safe'. The techniques can be used with confidence for very high strength materials where fracture occurs under conditions that are almost totally elastic, but greater care must be taken where plastic deformation occurs. If the amount of plastic deformation is limited the crack opening displacement methods are applicable but the use of K_{Ic} or δ_c factors does not have sufficient reliability if major plastic deformation occurs.

SUMMARY
. . . .
Failure of materials when overstressed may be in a tough manner, characterised by plastic deformation prior to fracture, or in a brittle manner, with little or no plastic deformation.

Materials contain defects and flaws and these will have a significant effect on the fracture stress of the material, possibly permitting fast brittle fracture in a material that would be expected to fail by plastic yielding.

By means of fracture mechanics, which is the study of the relationships between material properties of strength and toughness and the geometry of flaws and defects, it becomes possible to determine the critical dimensions of cracks and other flaws necessary for the occurrence of fast or brittle fracture. Using these techniques it becomes possible to design to minimise the risk of failure in service. The parameters of *toughness*, G_{Ic}, and *fracture toughness*, K_{Ic}, are constants for a material.

Metals with face-centred cubic structures are tough and ductile at all temperatures but metals with body-centred cubic or hexagonal structures show a transition from ductile to brittle behaviour as the temperature is lowered.

Questions

10.1 (a) What is meant by the term *toughness* as applied to materials? Distinguish clearly between notch-impact toughness and fracture toughness.
  (b) Explain how both notch-impact toughness and fracture toughness may be determined.

10.2 A material is subjected to a direct tensile stress of 160 MPa. An internal crack of length 0.3 mm and with a crack tip radius of 3 μm lies normal to the axis of direct stress. Estimate the level of local stress at the tip of the crack.

10.3 (a) Estimate the theoretical strength of a brittle material from the following data: interatomic spacing = 0.3 nm, modulus of elasticity = 320 GPa, surface energy = 0.6 J/m^2.
   (b) Estimate the fracture strength of the material if it is known to contain some internal defects, the largest having a length of 3 mm.

10.4 An aluminium alloy has a fracture toughness, K_{Ic}, of 27 MPa m$^{1/2}$. A large plate of this alloy was subject to stress and failed through fast fracture at a stress of 250 MPa. It was found that failure had commenced at a surface crack. What was the size of the crack that initiated failure?

10.5 A thick plate of a high-strength aluminium alloy of width 0.24 m and which contains a centre crack of length 40 mm fails by fast fracture when the level of applied stress reaches 160 MPa. Determine the fracture toughness of the material and compare the results obtained using both the tan and sec functions.

10.6 Calculate the fracture stress for a glazed ceramic material with a modulus of elasticity of 300 GPa and containing a glazing crack 150 μm deep. To what value would the fracture stress be reduced if the depth of crack increased to 0.4 mm? Assume that the mean surface energy is 20 J/m^2.

10.7 What will be the limiting design stress, in both cases, for a steel possessing a tensile yield stress of 550 MPa and a fracture toughness of 40 MPa m$^{\frac{1}{2}}$, using a safety factor of 1.5, if no plastic deformation is permitted and the maximum tolerated crack is (a) 3.0 mm and (b) 5.0 mm in length?

10.8 Calculate the critical crack length for brittle fracture in a steel with a tensile yield stress of 1200 MPa, and for which the modulus of elasticity is 208 GPa, and the toughness, G_c is 110 kJ/m^2, if the maximum stress is limited by the yield stress.

10.9 The fracture stress of a piece of Perspex is 57 MPa. The fracture toughness of Perspex is 1.2 MPa m$^{\frac{1}{2}}$. Estimate the probable size of internal defects within the material.

10.10 Assess the relative merits of the materials described in the table below to resist failure by fast fracture in structures in which the design stress is 0.5 × yield stress. The materials test and inspection facilities are such that the minimum size of internal defect for which detection can be guaranteed is 8 mm.

Material	Yield stress (MPa)	Fracture toughness (MPa m$^{\frac{1}{2}}$)
High-strength steel	1730	110
Aluminium alloy	590	35
Titanium alloy	960	86

10.11 What should be the appropriate wall thickness and diameter for an aluminium alloy vessel if it is to operate at a maximum pressure of 100 bar and be designed to 'leak before break' criteria using safety factors of 2.5 on yield strength and 1.5 on ratio of wall thickness to critical flaw size? σ_y for aluminium alloy is 500 MPa and K_{Ic} is 23 MPa m$^{-\frac{1}{2}}$.

10.12 Charpy notched impact tests carried out on a steel over a range of temperatures gave the results in the table below:

Test temperature (°C)	Energy absorbed (J)	Test temperature (°C)	Energy absorbed (J)
95	164	5	50
80	164	0	47
70	157	−10	34
60	149	−20	25
50	149	−30	18
40	123	−40	12
20	84	−50	12

Plot these data and estimate the ductile–brittle transition temperature for the steel.

10.13 Samples of a plain carbon steel are subjected to Charpy notch-impact testing at temperatures between −100 °C and +100 °C and the results are given below. Estimate the ductile–brittle transition temperature for the steel.

Temperature (°C)	Energy absorbed (J)	Temperature (°C)	Energy absorbed (J)
100	135	−20	30
80	135	−40	17
60	130	−60	9
40	115	−80	5
20	90	−100	5
0	55		

Phase diagrams and alloy formation

Why should phase diagrams be studied?

What is a phase diagram and why should an engineer have some knowledge and understanding of such diagrams? Basically, a phase diagram is a map of an alloy system and it indicates phases which should exist under conditions of equilibrium for any particular combination of composition and temperature. Such diagrams are important for the proper understanding of metallic and ceramic alloy systems and they provide information relating to the heat-treatment processes that are used to modify the structures and properties of alloys.

LEARNING OBJECTIVES

1. Understand the concepts of binary temperature–composition phase diagrams.

2. Know the principles of two-component systems where the components are soluble in the liquid state but insoluble in the solid state and form eutectic mixtures.

3. Know the basic rules for the interpretation of phase diagrams.

4. Understand the factors which control the formation of solid solutions and how the presence of solid solubility influences the form of phase diagrams.

5. Know how the formation of compounds within a binary system can influence the shape of a phase diagram.

6. Know how an allotropic modification of an element can affect the form of a phase diagram.

7. Understand how the form of a phase diagram can influence the microstructure and, hence, the properties of alloys.

11.1 | Introduction

I have noticed during my teaching that some students experience difficulties in understanding phase diagrams. I will deal with this section of the subject in easy stages, illustrating each stage with appropriate examples and hope that, in this way, the concepts will be understood fully by the reader.

Firstly, what do we mean by the term *phase*? A phase may be defined as a portion of matter which is homogeneous. A pure substance, for example, water, is a single phase. However, a pure substance can exist in three states – solid, liquid and vapour, each of these states being a single phase. The relationships between these states, or phases, of a pure substance for various conditions of pressure, p, and temperature, t, can be shown on a simple diagram. The p–t diagram, or phase diagram, for water is shown in Figure 11.1. The lines in the diagram divide the space into three areas or *fields*, and each field shows the ranges of pressure and temperature over which a phase, solid, liquid or vapour, is stable. In this diagram the line OA indicates the variation in the vapour pressure of water with change of temperature. A liquid is said to boil when its vapour pressure is equal to the external pressure. Water boils at 100°C at standard atmospheric pressure ($101.5\,kN/m^2$), point P in Figure 11.1. The line OC indicates the influence of pressure on the solid to liquid change or melting point. This line is almost vertical (the slope has been exaggerated in the diagram) and an increase in pressure of 1 atmosphere ($10^5\,N/m^2$) will reduce the freezing temperature by less than 0.01°C.

Now consider the effect of adding a second substance, or *component*, to the system, for example the addition of common salt, NaCl, to water. Salt will dissolve in water to give a homogeneous solution. This then is a single phase. However, water can hold just over 30 per cent by weight of salt in solution. If more salt than this is added then we have two phases present, a fully saturated solution of salt in water plus the excess undissolved salt.

The phase diagram for a one-component system, such as that for water in Figure 11.1, has axes for the two variables pressure and temperature. In two-component, or *binary*, systems there is a third variable, namely composition. Phase diagrams for metallic and ceramic alloy systems are used to show liquid/solid phase changes and some changes wholly within the solid state. The effect of changes in pressure upon the melting point of a substance is extremely small and so the pressure variable may be ignored. This permits us to use temperature and composition as the two axes in a t–c or binary phase diagram. If it is required to show phase relationships for a three-component, or ternary, system then it becomes necessary to construct a three-dimensional diagram. The composition of any three-component

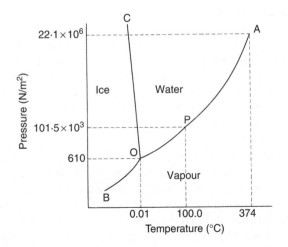

Figure 11.1 *A one-component phase diagram – H_2O.*

mixture can be shown by a point on an equilateral triangle. Temperature is represented on an axis orthogonal to such a composition plane (refer to section 11.12).

11.2 | Alloy systems

The term *alloy* is used to describe a mixture of two or more metals or of two or more ceramic substances. Certain metallic alloys may be a mixture of a metal and a non-metallic element, a good example being alloys of iron and carbon, or steels. Many alloys are created by mixing substances in the liquid phase and it is convenient to consider the formation of alloy structures on the basis of the solidification of liquid alloys. Alloy structures and, hence, alloy properties, will depend on the nature of the solid phase or phases formed.

When two liquid substances are mixed they may dissolve in one another in all proportions, forming a homogeneous liquid solution, or they may be insoluble in one another, either partially or totally, and separate into two liquid layers on density grounds like oil and water. Most alloys of commercial importance are based on systems in which the liquids are completely soluble in one another and, so, partially soluble or insoluble liquid mixes will not be considered in this text.

Although two substances may be completely soluble in one another in the molten state it does not necessarily mean that the liquid solution will solidify to give a homogeneous solid phase, or solid solution. There are several possibilities and the two substances may:

(a) be totally insoluble in one another when solid,

(b) be totally soluble in one another when solid,

(c) be partially soluble in one another when solid, or

(d) combine with one another to form a compound.

The binary phase diagrams and the types of solid structure which can occur for each of the above possibilities are described in the following sections. Much of this relates to metallic systems but a few ceramics examples are quoted. The phase diagram for a binary system composed of the two components, A and B, is drawn with composition represented on a base and temperature on the ordinate. The base line will range from 100 per cent A, 0 per cent B on the left to 0 per cent A, 100 per cent B on the right. Compositions may be expressed either as weight percentages or as atomic, or molar, percentages but the former is the more usual method of expression.

11.3 | Total solid insolubility

Consider the case of two hypothetical metals, A and B, which are completely soluble in one another in the liquid state, but are totally insoluble in one another in the solid state. If a composition base line and temperature scale is drawn (Figure 11.2), certain information can be plotted. The melting point of pure metal A can be marked off as point A on the left-hand temperature axis. Similarly point B on the right-hand axis represents the melting point of pure metal B. At high temperatures, any mixture of the two liquid metals will be a single-phase liquid solution.

In the same way as the presence of dissolved salt depresses the freezing point of water, so the freezing point of a liquid metal will normally be depressed if the liquid metal contains some other metal in solution. Line AL in Figure 11.2 indicates the depression of freezing point of pure metal A containing dissolved metal B. Similarly, line BM is the depression of the freezing-point curve for pure metal B containing dissolved metal A. It is important to note that at any point on line AL it is pure metal A that is freezing; that is, during cooling the solid that is forming is crystals of pure metal A. For example, X per cent of B dissolved in liquid A will depress the freezing point of A by an amount δT. The liquid solution will begin to solidify at a temperature of $A - \delta T$ (point S on curve AL in Figure 11.2) but it is crystals of

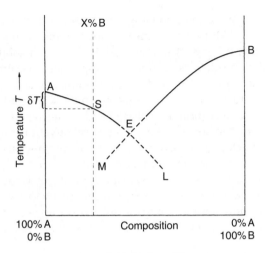

Figure 11.2 *Freezing point curves for two metals insoluble in the solid state.*

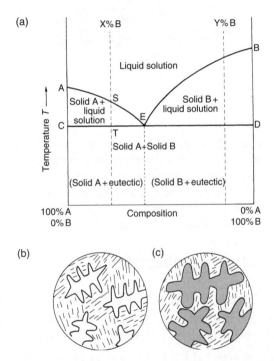

Figure 11.3 *(a) Binary phase diagram for solid insolubility (simple eutectic). (b) Solid structure of alloy containing X per cent of B. Dendrites of A in eutectic. (c) Solid structure of alloy containing Y per cent of B. Dendrites of B in eutectic.*

pure A that will begin to solidify. The two curves AL and BM intersect at point E. The section of curves EL and EM are hypothetical as no liquid can exist at a temperature lower than that of point E.

Figure 11.3 shows the completed phase, or equilibrium, diagram for the alloy system of metals A and B. Consider again an alloy containing X per cent of B. At a high temperature, the alloy exists as a single-phase liquid solution. On cooling, the liquid will commence freezing at a point denoted by S on curve AE. Crystals of solid pure A will begin to form. If pure A is rejected from the solution, the composition of the remaining liquid must become enriched in metal B, that is, the composition of the liquid varies toward the right. This means that as the freezing of metal A continues, the temperature and composition of the liquid remaining follow the curve AE toward point E. Point E, which is the only point common to both freezing-point curves, represents the lowest temperature that a liquid solution can exist at, and at this point all remaining liquid solution solidifies, forming a fine-grained crystal mixture of both solid metals, A and B. Point E is termed the *eutectic* point and the fine-grained crystal mixture formed is termed the eutectic mixture. The final structure of the solid alloy containing X per cent of B will, therefore, be composed of large crystals of pure metal A (primary crystals) and a eutectic mixture of A and B. The primary crystals will probably be *dendritic* in form. (Many metals solidify from liquid in a dendritic manner. Solidification commences at a nucleus and outward growth from the nucleus occurs preferentially in three directions. Subsequently, secondary, and then tertiary arms grow, producing a skeleton-type crystal, as in Figure 11.4. Outward growth ceases when the advancing dendrite arms meet an adjacent crystal. When outward growth has ceased the dendrite arms thicken and eventually the whole mass is solid and no trace of the dendritic formation remains, except where shrinkage causes interdendritic porosity, or in alloy systems where the final liquid to solidify is of a different composition from the primary dendrites.)

Figure 11.4 *Representation of a dendrite.*

If a liquid alloy containing Y per cent of metal B is allowed to solidify, solidification will follow a similar pattern, but in this case primary crystals of pure metal B will solidify first. It is important to note that the composition of the eutectic mixture remains constant.

In the phase diagram, line AEB is termed the *liquidus* and line CED is termed the *solidus*. At all points above the liquidus the alloy is always liquid, and below the solidus the alloy is always wholly solid. Between liquidus and solidus, in the solidification range, the alloy is in a pasty stage.

11.4 | Interpretation of phase diagrams

It is appropriate to bring in at this stage some simple rules for the interpretation of binary phase diagrams.

(a) A phase diagram consists of lines that divide it into a number of areas, or fields. These fields may be single phase, as is the area above the liquidus in Figure 11.3, or two phase, as is area ACE in the same figure.

(b) Single-phase areas are always separated by a two-phase zone, and three phases can only coexist at a point, for example, the eutectic point.

(c) When a vertical line representing the composition of some alloy in the system crosses a line in the phase diagram, it indicates that some change is taking place in the alloy. For example, in Figure 11.3 the X per cent B line cuts two other lines, AE at S and CD at T, indicating that during cooling the alloy is beginning to freeze at point S and that solidification is complete at point T.

(d) For any point in a two-phase region the composition of the two phases in equilibrium with one another can be determined. If a horizontal tie-line is drawn through the point the intersections of this line with the phase boundary lines denote phase compositions.

(e) The relative proportions of the phase present can be determined using the lever rule. The quantities of phases present are in proportion to the lengths of the lever lines.

EXAMPLE 11.1

State the phases present and their relative proportions for both point U and point V in Figure 11.5.

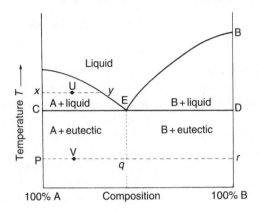

Figure 11.5 *Application of lever rule.*

SOLUTION

Consider point U in the figure. This is in a two-phase field, the phases being solid metal A and liquid solution. Draw a horizontal tie-line, xUy. This intercepts the field boundaries at x and y showing that pure solid metal A is in equilibrium with a liquid solution containing y per cent of metal B.

The relative quantities of the two phases are proportional to the lengths of the lever lines so that

$$\frac{\text{quantity of solid A}}{\text{quantity of liquid (composition } y)} = \frac{Uy}{Ux}$$

Similarly, at point V the phases present are solid metals A and B in the ratio

$$\frac{\text{quantity A}}{\text{quantity B}} = \frac{Vr}{Vp}$$

Alternatively it could be considered that the phases present are solid A plus eutectic mixture in the ratio

$$\frac{\text{quantity of A}}{\text{quantity of eutectic}} = \frac{Vq}{Vp}$$

or percentage of eutectic mixture in the solid alloy of composition V is

$$\frac{Vp}{pq} \times 100$$

11.5 ▌ Solid solubility

It is possible for a mixture of solid substances to form as a single phase; what is termed a *solid solution*. This concept may sound strange but it simply means that the atoms or molecules of the two substances have taken up positions in a common crystal lattice forming a single phase. Solid solutions may be of

Figure 11.6 *Schematic representation of solid solutions: (a) substitutional (random); (b) substitutional (ordered); (c) interstitial.*

either the substitutional or interstitial type (refer also to sections 5.5 and 8.3) and a schematic representation of substitutional and interstitial solid solutions is shown in Figure 11.6.

Generally, the arrangement of solute in a substitutional solid solution is random, but in some instances solid solutions of an ordered type may be formed. An ordered solid solution (also known as *superlattice*) can exist only at either one fixed composition or over a narrow range of composition.

As stated previously (section 8.3), atoms in interstitial or substitutional solid solution cause strain to be developed in the parent lattice. As there is an upper limit to the amount of strain that can be tolerated in a crystal lattice, it follows that there will be some restrictions to solid solution formation. The nature of solid solution in metal systems was studied extensively by Hume-Rothery and his work is summarised in the following rules:

(a) Relative size. If the sizes of the atoms of two metals do not differ by more than 14 per cent, conditions are favourable for the formation of substitutional solid solutions. If the relative sizes of atoms differ by more than 14 per cent, solid solution formation (if it occurs at all) will be extremely limited. Interstitial solid solutions may be formed if the atoms of the solute element are very small in comparison with those of the solvent metal.

(b) Chemical affinity. When two metals have a high affinity for one another the tendency is for solid solubility to be severely restricted and intermetallic compounds to be formed instead. This occurs when one element is electronegative and the other is electropositive.

(c) Relative valency. If a metal of one valency is added to a metal of another valency the number of valency electrons per atom, the electron ratio, will be altered. Crystal structures are very sensitive to a decrease in the electron ratio. Consequently, a metal of high valency can dissolve very little of a metal of low valency, although a metal of low valency might be able to dissolve an appreciable amount of a high-valency metal.

(d) Crystal type. If two metals are of the same crystal lattice type and all other factors are favourable, it is possible for complete solid solubility to occur over the whole composition range. (It is also necessary that the relative sizes of atoms differ by not more than 7 per cent for complete solid solubility.)

Some examples of binary systems in which a continuous range of solid solutions is formed are Cu–Ni, Cu–Au, Al_2O_3–Cr_2O_3 and MgO–NiO.

11.6 | Phase diagram for total solid solubility

The two possible types of phase diagram shapes for a binary alloy system where a continuous range of solid solution is formed, as exemplified by the Cu–Ni and Cu–Au systems, are shown in Figure 11.7. It should be noted that point C in the Cu–Au diagram (Figure 11.7(b)) is not an eutectic point.

An alloy containing X per cent of nickel (Figure 11.7(a)) would solidify in the following manner. Freezing of the liquid solution would commence at temperature t_1. At this temperature, liquid of composition 1 would be in equilibrium with a solid solution of a composition corresponding to point p on the solidus, so the first solid solution crystals to form are of composition p. Consequently, the composition of the remaining liquid becomes enriched in copper and the freezing temperature falls slightly. As the temperature falls so the composition of the solid solution tends to change by a diffusion process following the solidus line toward B. At some temperature t_2 liquid of composition m is in equilibrium with solid solution of composition q.

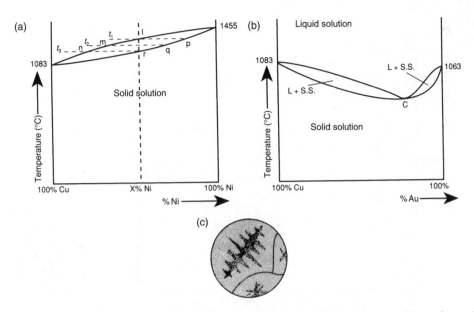

Figure 11.7 *Phase diagrams for complete solid solubility: (a) Cu–Ni system; (b) Cu–Au system; (c) cored crystal structure.*

Solidification of this alloy will be complete at temperature t_3 when the last drops of liquid, of composition n, solidify, correcting the composition of solid solution crystals to r. If the solidification rate is very slow, allowing for the attainment of equilibrium at all stages during the cooling process, the final solid solution crystals will be uniform in composition. In practice however, solidification rates are too rapid for full equilibrium to be attained and the crystals will be cored. In a cored crystal the compositional is not the same at all points. The crystal lattice is continuous but there will be a gradual change in composition across each crystal. The centre of a crystal will be rich in nickel while the outer edges will be rich in copper. In the Cu–Ni alloy system, the coring in crystals is clearly visible under microscopical examination of a prepared section as the alloy colour is dependent on composition. The centres of crystals are rich in nickel and silver in appearance while the outer edges of crystals are rich in copper and darker in colour. This colour shading clearly shows the dendritic manner of growth. Coring can occur in all solid solution alloys, even though it may not be visible.

Coring in alloys may be subsequently removed by heating this material to a temperature just below the solidus. During this treatment, *annealing* or *homogenising*, diffusion takes place, evening out composition gradients within the crystals.

11.7 Partial solid solubility

It is far more common to find that solid metals are partially soluble in one another rather than totally insoluble or totally soluble. The phase diagram for the Pb–Sn system, which shows partial solid solubility, is given in Figure 11.8. This diagram is, in effect, a combination of the two previous types and shows solid solubility sections and also a eutectic. The liquidus is line AEB and the solidus is ACEDB. Lines FC and GD are *solvus* lines and denote the maximum solubility limits of tin in lead and of lead in tin respectively. As there are two separate solid solutions formed, the Greek letters α and β are used to identify them.

Consider the solidification of three alloy compositions in this system. For alloy composition (1) solidification begins at temperature t_1 with the formation of a tin-rich β solid solution of composition O. As cooling continues, the composition of the liquid varies along the liquidus towards point E and the composition of the solid β phase varies according to the solidus toward point D. When the eutectic temperature is reached, there will be primary cored crystals of β and liquid solution of the eutectic

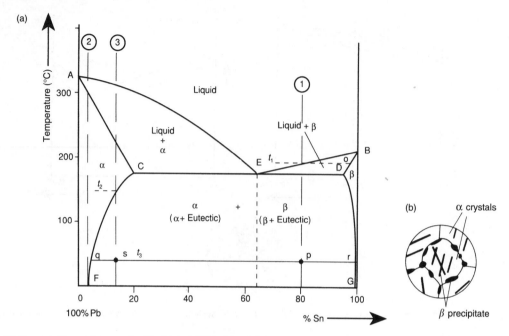

Figure 11.8 (a) Phase diagram for Pb–Sn (partial solid solubility with eutectic). (b) Structure of alloy (3).

composition. This liquid then freezes to form a eutectic mixture of two saturated solid solutions, α (lead-rich) of composition C and β of composition D. During further cooling the compositions of the α and β phases will adjust, following the solvus lines, until eventually at point p saturated α crystals of composition q will be in equilibrium with saturated β solid solution of composition r.

For alloy (2) solidification of the liquid solution takes place in the same manner as for a complete solid solution alloy (section 11.6) and when solidification is complete the structure will be one of cored α (lead-rich) solid solution.

In the case of alloy (3), a new concept emerges, namely the possibility of structural changes occurring within the solid state. The liquid alloy will freeze on cooling to give a cored α solid solution. During further cooling below the solidus, the α solid solution will remain unchanged until temperature t_2 is reached. At this temperature, the composition line meets the solvus and the solid solution is fully saturated with dissolved tin. As the temperature of the alloy falls below t_2 the solubility limit is exceeded and excess tin is rejected from α solid solution as a precipitate. In this case, it is not pure tin that forms as a second solid phase, but rather, saturated β (tin-rich) solid solution. Eventually, at temperature t_3 the structure is composed of crystals of α of composition q with precipitated β particles of composition r. Applying the lever rule in this case the proportion of phases present would be in the ratio

$$\frac{\text{quantity of } \beta \text{ solid solution}}{\text{quantity of } \alpha \text{ solid solution}} = \frac{sq}{sr}$$

The second phase, β, may be precipitated either at the α grain boundaries, within the α crystals, or at both types of site (Figure 11.8(b)).

Changes within the solid state take place slowly in comparison with changes between liquid and solid states. In consequence they may be suppressed by rapid cooling. Rapid cooling of alloy (3) from some temperature below the solidus may prevent the precipitation of β from taking place and giving at temperature t_3 an α solid solution of composition s supersaturated with dissolved tin. This is of significance in connection with precipitation hardening and age hardening and will be discussed further in Chapters 12 and 15.

11.8 Peritectic diagram

Another form of phase diagram that can occur for systems showing partial solid solubility is the *peritectic* type shown in Figure 11.9. The liquidus and solidus lines are AEB and ACDB respectively, and FC and GD are solvus lines. The horizontal line CDE is termed the peritectic line and point D is the peritectic point.

Consider the cooling of liquid alloy of composition (1). Solidification will commence at temperature t_1 with α solid solution of composition q forming. As freezing continues, the composition of the liquid follows the liquidus toward point E and the composition of the solid solution follows the solidus toward point C. When the peritectic temperature is reached liquid of composition E exists in equilibrium with α solid solution of composition C. At this temperature the two phases react together to form β solid solution according to the reaction

$$\alpha \text{ (composition C)} + \text{liquid (composition E)} \underset{\xleftarrow{\text{heating}}}{\xrightarrow{\text{cooling}}} \beta \text{ (composition D)}$$

If the reactants α and liquid were present in equivalent proportions, that is in the ratio

$$\frac{\text{amount}\,\alpha}{\text{amount liquid}} = \frac{DE}{CD}$$

they would both be totally consumed in the reaction producing β solid solution.

In the case of alloy (1) the reactants were present in the ratio

$$\frac{\text{amount}\,\alpha}{\text{amount liquid}} = \frac{XE}{CX} \text{ (by lever rule)}$$

where

$$\frac{XE}{CX} > \frac{DE}{CD}$$

so that the peritectic reaction will cease when all the liquid is consumed and there is some unreacted α remaining. The structure of this alloy below the peritectic temperature is, therefore, α and β. During

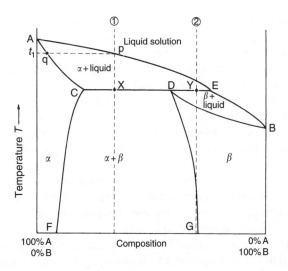

Figure 11.9 *Phase diagram for partial solid solubility with peritectic.*

further cooling, the compositions of both phases will vary according to the solvus lines. In the case of alloy (2) the ratio of reactants immediately before the peritectic reaction occurs is

$$\frac{\text{amount } \alpha}{\text{amount liquid}} = \frac{\text{YE}}{\text{CY}} < \frac{\text{DE}}{\text{CD}}$$

Consequently the reaction will cease when all the α has been consumed and there is some excess liquid remaining. During further cooling this liquid will solidify as β.

11.9 | Compound formation

As stated earlier, in some binary systems the components may combine to form compounds. Often these are not compounds in the true chemical sense, but are ordered crystal space lattices with atoms of the various elements taking up specific positions within the lattice. Compound formation occurs in both ceramic and metallic alloy systems and in many cases more than one compound may be formed within a single binary alloy system. Compounds possess crystal structures that differ from those of either constituent component and they often possess unique melting points. In the case of intermetallic compounds, their melting points are often higher than the melting points of the constituent metals and they are generally of high hardness and quite brittle. An example of an intermetallic compound is the compound between magnesium and tin Mg_2Sn.

	Melting point (°C)	Crystal structure
Mg	649	close-packed hexagonal
Sn	232	body-centred tetragonal
Mg_2Sn	783	complex cubic

Compounds are formed in many ceramic alloy systems, examples being mullite, $Al_6Si_2O_{13}$, in the Al_2O_3–SiO_2 system and the four different compounds formed in the Al_2O_3–CaO system (see Figure 11.10(a) and (b)). It will be seen from the figure that mullite and tricalcium aluminate ($3CaO.Al_2O_3$) are formed as a result of peritectic reactions but the other compounds in the Al_2O_3–CaO system possess unique melting points.

From many points of view an intermetallic compound can be regarded as a pure metal. Frequently the binary phase diagram for an alloy system in which an intermetallic compound is formed is effectively two simple diagrams linked together. Figure 11.11(a) shows the phase diagram for two metals, A and B, which form one compound, A_xB_y, and assumes no solid solubility. It is simply two binary eutectic diagrams joined together, one between metal A and the compound and the other between metal B and the compound. Figure 11.11(b) shows a variation on the above in which there is some solid solubility. As there are now three separate solid solution zones three Greek letters, α, β and γ, have to be used to describe these. Conventionally, these are used progressively, working from left to right. When, as in this case, a compound may exist over a small range of composition it is termed an *intermediate* phase.

In some binary alloy systems, several intermetallic compounds or intermediate phases may be formed. Although the complete phase diagram may look highly complex at first sight, it can usually be split into the small and comparatively simple elements discussed in earlier sections of this chapter, and it can be interpreted according to the same simple rules.

We have now covered the basic elements of phase diagrams, including total solid insolubility, complete and partial solid solubility and compound formation. Let us now consolidate this with the aid of some examples.

Figure 11.10 *(a) Al₂O₃–SiO₂ phase diagram. (b) Al₂O₃–CaO phase diagram.*

Figure 11.11 *Binary diagrams showing intermetallic compound (a) with no solid solubility, (b) with partial solid solubility.*

EXAMPLE 11.2

The phase diagram for the binary alloy system of two metals A and B is shown in Figure 11.12. Label all the unmarked phase fields.

SOLUTION

Referring to the rules for the interpretation of phase diagrams (section 11.4), single-phase areas are always separated by a two-phase field and three phases can only coexist at a unique point. There are four single-phase fields: liquid, α, β and γ, and each of these single-phase fields must be separated by a two-phase zone. This is shown in the solution diagram (Figure 11.13).

Figure 11.12

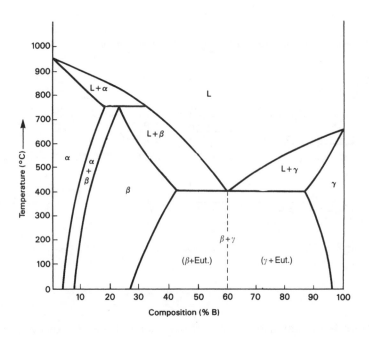

Figure 11.13

EXAMPLE 11.3

A number of different alloys of copper (melting point 1083°C) and silver (melting point 960°C) were allowed to cool and the thermal arrest points noted. These are given in the table.

Per cent copper	5	20	28	50	65	90
Per cent silver	95	80	72	50	35	10
1st arrest (°C)	945	860	780	900	950	1045
2nd arrest (°C)	860	780	—	780	780	780
3rd arrest (°C)	570	—	—	—	—	—

Plot the phase diagram for the copper silver system using the above data.
Describe the changes which occur when the following alloy compositions are allowed to cool slowly from the liquid state to room temperature:
(a) 5 per cent copper/95 per cent silver, (b) 50 per cent copper/50 per cent silver.

SOLUTION

The data in the table are plotted in Figure 11.14. The positions of the points indicate that it is a eutectic system, the eutectic composition being 28 per cent copper. The presence of a third arrest point for the 5 per cent copper alloy indicates a change in the solid state, namely the separation or precipitation of a second phase. There is sufficient information to complete the liquidus line but there is incomplete information about the solubility of copper in silver for the solidus and solvus lines to be drawn with certainty. Also, if there is some solubility of copper in silver it is likely that silver will be soluble to some extent in solid copper giving a β solid solution. The dashed lines in Figure 11.14 show the likely form of the diagram in these areas.

Figure 11.14

(a) The liquid alloy containing 5 per cent copper would begin to freeze when the liquidus temperature is reached at 945°C and a solid solution rich in silver, α-phase, separates out. Solidification will be completed when the temperature reaches 860°C, the solidus temperature for this alloy. At this stage the structure will consist entirely of crystals of α solid solution. On cooling to 570°C, the solvus temperature, the saturation limit for copper in silver is reached and precipitation of a copper-rich solid solution, β-phase, will occur when the temperature falls below this value. At room temperature the structure will consist of primary α crystals with precipitated β particles.

(b) Freezing of the 50 per cent copper alloy will begin at 900°C, the liquidus temperature, and the first crystals to form will be of a copper-rich solid solution, β. The composition of the remaining liquid will be enriched in silver and its freezing point will reduce, following the liquidus line to the eutectic point. At 780°C the alloy will consist of primary β crystals and liquid of the eutectic composition (28 per cent copper), and at this temperature all remaining liquid freezes to give a fine-grained eutectic mixture of α and β crystals. At room temperature the final alloy structure will be primary β crystals and $(\alpha + \beta)$ eutectic.

EXAMPLE 11.4

Some data for four hypothetical metals, Anglium (An), Celtium (Ct), Hibernium (Hb) and Londinium (Lo), are given in the accompanying table. From these data sketch the probable forms of the binary phase diagrams for the following alloy systems: (a) An–Lo, (b) An–Hb, (c) Ct–Lo, (d) Ct–Hb.

Element	M. pt. (°C)	Atomic radius (nm)	Crystal structure	Valency	Electro-negativity
An	710	0.124	f.c.c.	2	1.4
Ct	850	0.143	b.c.c.	3	1.9
Hb	610	0.127	h.c.p.	1	1.2
Lo	520	0.128	f.c.c.	2	1.5

SOLUTION

(a) An–Lo system. Referring to the Hume–Rothery rules for solid-solution formation, it will be seen that these two metals fulfil all the conditions for complete solid solubility. They are of the same crystal type, they have the same valency, they possess similar levels of electronegativity, and their atom sizes differ by less than 4 per cent. The expected form of the phase diagram is as shown in Figure 11.15(a).

(b) An–Hb system. Although the atom size difference is less than 3 per cent, complete solid solubility cannot occur as the elements are of different crystal type. The solubility of Hb in An is likely to be restricted because Hb has a lower valency than An although the solubility of An in Hb should be somewhat greater. The most likely form of phase diagram will be a eutectic type with limited solid solubility as shown in Figure 11.15(b).

(c) Ct–Lo system. With metals of differing crystal form and a difference of about 11 per cent in atom size solid solubility is likely to be very restricted. A possible phase diagram would be of the simple eutectic type (Figure 11.15(c)) but, as there is a fairly wide difference in melting temperatures, an alternative possibility would be the formation of a peritectic (Figure 11.15(d)).

(d) Ct–Hb system. The differences in crystal structure, size and valency all mean that solid solubility between the two metals will be severely restricted. However, the relatively large difference in the electronegativity of the elements suggests that one or more intermetallic compounds could be formed. For the solution (Figure 11.15(e)) it is assumed that one compound with the formula Ct_xHb_y is formed.

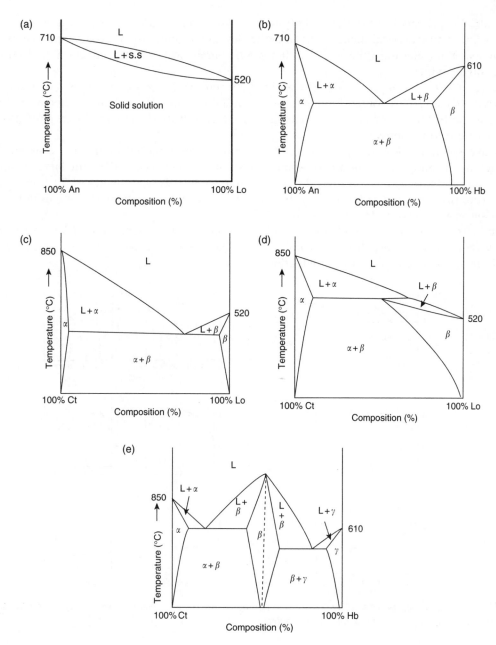

Figure 11.15

11.10 | Effect of alloy type on properties in metallic systems

For systems in which the component metals are completely insoluble in one another in the solid state, the structure of the solid alloy is simply a mixture of two pure metals. Consequently, the variation of properties with alloy composition should obey the *Rule of Mixtures* and be linear. In actual practice, there is a departure from linearity owing to a grain-size effect. A eutectic is a finely divided mixture of two metals. The primary crystals that solidify first on either side of the eutectic point are much larger in size. A fine-grained crystal metal tends to be harder and stronger than a coarse-grained sample of the same material (refer to section 8.5). Similarly, a fine grain size causes a reduction in electrical and thermal conductivities.

In Figure 11.16(a) the approximate relationship between two properties, hardness (H) and electrical conductivity (G), and alloy composition is shown for a simple eutectic alloy. The dashed lines show the expected property variation, neglecting the grain-size effect. Tensile yield and ultimate strengths follow a similar pattern to hardness.

In a solid solution alloy, the presence of the solute atoms imposes strain in the parent lattice, strengthening the alloy (see section 8.8). Maximum strengthening occurs when the lattice is subjected to maximum strain, that is, when there are equal numbers of both types of atoms. 50 atomic per cent is not necessarily the same as 50 per cent by weight. Property variations with composition for solid solution alloys are shown in Figure 11.16(b).

Figure 11.16(c), (d) shows the relationships between properties and composition for the partial solid solubility cases. As the phase diagrams contain some features from each of the previous type of diagram, so the property diagrams are combinations of the former two types.

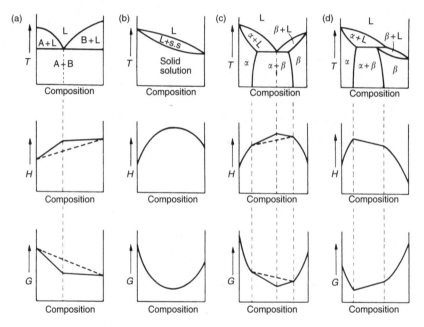

Figure 11.16 *Relationships between alloy composition and hardness (H) and electrical conductivity (G) for (a) simple eutectic, (b) solid solubility, (c) partial solubility with eutectic and (d) peritectic.*

11.11 | Allotropy

Certain elements and compounds can exist in more than one crystalline form.

The various separate forms are termed *allotropes* and the element or compound is said to exhibit allotropy. Diamond and graphite are two allotropic forms of the element carbon. A number of metals and ceramic compounds also exhibit allotropy and the presence of allotropes has an effect on the forms of phase diagrams. In the Al_2O_3–SiO_2 diagram, Figure 11.10(a), a horizontal line below the solidus indicates the solid state transformation from the cristobalite form of silica to the tridymite form on cooling. The most important allotropic metal is iron, which can exist in two crystal forms, body-centred cubic

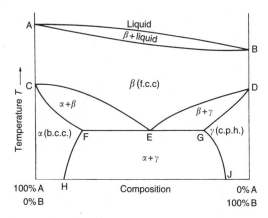

Figure 11.17 *Possible form of a binary phase diagram for two hypothetical metals, both being allotropic.*

and face-centred cubic. Pure iron is body-centred cubic in structure (α iron) at all temperatures up to 908°C, and also between 1388°C and the melting point of iron at 1535°C. (The high temperature b.c.c. form is known as δ iron.) Between 908°C and 1388°C the structure is face-centred cubic and this form is termed γ iron. Some phase diagrams involving iron appear in Chapter 16.

If a metal is allotropic, this will have an effect on the shape of equilibrium diagrams for alloy systems involving the metal. Consider a hypothetical alloy system between two allotropic metals, A and B. Suppose that metal A is body-centred cubic at low temperatures and face-centred cubic at high temperatures while metal B is hexagonal at low temperatures and face-centred cubic at high temperatures. If all factors are favourable, it is possible for a complete range of solid solutions to exist between the two metals at high temperatures. At lower temperatures only partial solid solubility can occur because the metals differ in crystal form. The complete phase diagram for the alloy system A–B could be as in Figure 11.17. It will be noticed in this diagram that, immediately below the solidus, there is a phase field containing one solid solution, β, and that below this there is a diagram apparently identical to the eutectic with partial solid solubility shown in Figure 11.8. This is a *eutectoid* diagram and point E is the eutectoid point. In a eutectic system it is a case of a single-phase liquid changing during cooling into two separate solid phases. A eutectoid is similar but it is a case of a single-phase solid solution changing during cooling into two differing solid phases. Line CED is termed the *liquidoid* and line CFEGD is termed the *solidoid*. The interpretation of a eutectoid diagram is fundamentally the same as the interpretation of a eutectic diagram. It must be remembered though, that reactions wholly within the solid state take place more slowly than liquid to solid changes. Variations in the rate of cooling through a eutectoid phase change can exert a profound effect on the structure and properties of the alloy. This will be discussed in greater detail in section 12.7 and in connection with the heat treatment of steels in Chapter 16.

Eutectoids may also appear in alloy systems between non-allotropic metals if several intermediate phases occur. There are examples of this in the copper–aluminium and copper–tin systems (see Chapter 15).

11.12 Ternary diagrams

When three components are present in a system the composition of any mixture cannot be represented by a point on a line. The composition abscissa of the binary diagram becomes an equilateral triangle (Figure 11.18(a)) in a ternary diagram. Temperature is represented on an axis orthogonal to the base triangle and the phase diagram becomes a three-dimensional solid figure. The ternary diagram for a hypothetical system of three components, A, B and C, all totally insoluble in one another in the solid state, is represented in Figure 11.18(b). In this system, each pair of components would give a simple

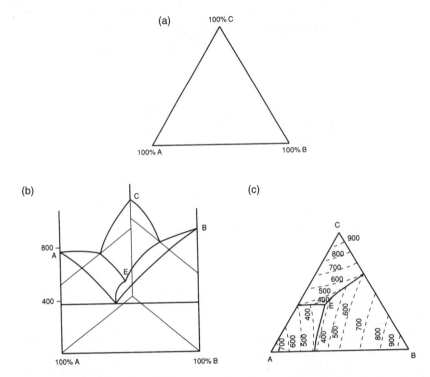

Figure 11.18 *(a) Ternary composition plane. (b) Representation of ternary diagram for eutectic system. (c) Liquidus projection of (b).*

eutectic system, and the three components give a ternary eutectic. It will be seen that the liquidus curve of a binary system becomes a curved surface in a ternary system, and the eutectic point of each binary diagram becomes a eutectic line, with the three eutectic lines intersecting at point E, the ternary eutectic point. One method of representing this in two dimensions is to show the figure as a liquidus projection, with thermal contours marked. This form of representation is shown in Figure 11.18(c).

SUMMARY
• •
A phase diagram for a binary system indicates the nature and compositions of the phases which should be present under equilibrium conditions for any particular temperature and composition. A phase diagram is, therefore, a useful tool for the understanding of the structures and, hence, the properties of alloys. The actual shape of any particular phase diagram will depend on whether the components are soluble or insoluble in one another, whether or not intermediate compounds are formed, or whether one or both components show allotropic modifications. When intermediate compounds are formed the phase diagram may look complex, but it is made up of several basic elements, each of which can be interpreted quite simply.

Questions

11.1 The phase diagram for a binary alloy system in shown in Figure 11.19.
  (a) Label all the phase fields.
  (b) Estimate the liquidus and solidus temperatures for the alloy containing 20 per cent B.

Figure 11.19

(c) For the alloy containing 40 per cent B, state what phases are present, and in what relative proportions, at (i) 600°C, (ii) 300°C and (iii) 100°C.

(d) What percentage of the microstructure is eutectic mixture in the alloy containing 70 per cent of B at room temperature?

11.2 A range of alloys is made from the two pure metals A, of melting point 750°C, and B, of melting point 1110°C, and the alloys are allowed to cool from the molten state to room temperature. During the cooling process, thermal arrest points are noted for each alloy. These are given in the table below:

Per cent B in alloy	8	30	45	55	78	90
1st arrest temperature (°C)	810	910	980	1010	1060	1090
2nd arrest temperature (°C)	760	830	910	910	940	1030
3rd arrest temperature (°C)	—	720	—	—	800	—

(a) Using these data draw and fully label the phase diagram for the alloy system of metals A and B.

(b) Describe the room temperature microstructure of an alloy containing 78 per cent of B in the following conditions: (a) sand cast, (b) cast, reheated to 900°C and quenched in water.

11.3 Two pure metals A and B, and a series of alloys of these two metals, were cooled from the liquid state and the information, in the table below, was obtained.

Per cent A in alloy	100	95	85	60	30	10	5	0
1st arrest point (°C)	600	575	535	425	300	400	425	450
2nd arrest point (°C)	—	500	300	300	—	300	350	—
3rd arrest point (°C)	—	40	—	—	—	—	200	—

(a) From the data, plot and fully label the thermal equilibrium diagram for the alloy system of metals A and B.

(b) For the alloy containing 75 per cent of A, what phases exist at the following temperatures: (a) 525°C, (b) 425°C, (c) 250°C?

(c) Explain how the properties of hardness and tensile strength of slowly cooled alloys would vary with composition from pure A to pure B.

11.4 Some data for the metallic elements absurdium, Ab, anglium, An, hypothetium, Hy, and Londonium, Ln, are given in the table below. On the basis of this information, sketch the likely forms of the phase diagrams for the following binary systems: (i) Ab–Hy, (ii) Ab–Ln, (iii) An–Ln, and (iv) Hy–Ln.

Metal	T_m (°C)	Atomic diameter (nm)	Structure	Valency	Electronegativity
Ab	855	0.255	h.c.p.	2	1.4
An	720	0.271	f.c.c.	3	1.8
Hy	1300	0.251	b.c.c.	2	1.5
Ln	950	0.242	b.c.c.	2	1.3

11.5 The cooling curves obtained when two pure metals and several of their alloys were cooled from the liquid phase are shown in Figure 11.20. Construct and label a phase diagram for the alloy system.

Figure 11.20

11.6 Two metals, A and B, have melting points of 1300°C and 1000°C respectively. In the binary alloy system between A and B there is a eutectic at 800°C and at a composition 20 per cent A/80 per cent B. Two solid solutions, α and β, are formed and the maximum solid solubility limits at various temperatures are given in the table

Temperature (°C)	0	200	400	600	800	900	950
Max. solubility of B in A (%)	3	10	20	32	50	40	35
Max. solubility of A in B (%)	2	2	3	5	10	5	3

(a) From the above information, draw the phase diagram for this alloy system.
(b) Describe the sequence of phase transformations that occur during the slow cooling of an alloy containing 40 per cent B from 1300°C to room temperature and calculate the percentage of β that would be present in the structure at 400°C.
(c) State what heat treatments would be required to produce the following room temperature structures in an alloy containing 30 per cent of B: (i) a homogeneous α phase structure, and (ii) a very fine dispersion of β in a matrix of α. Which of these two conditions would be the stronger?

11.7 A portion of the Cu–Al phase diagram is shown in Figure 11.21.
(a) Label the unmarked phase fields and identify the eutectic and eutectoid points.
(b) Calculate the relative amounts of phases present in: (i) a 10 per cent Al alloy at 700°C, and (ii) a 13 per cent Al alloy at room temperature.
(c) Sketch the expected room temperature microstructures of a 5 per cent Al alloy after (i) slow cooling from 1200°C, and (ii) rapid cooling from 1200°C, explaining the differences between them.
(d) Sketch the expected room temperature microstructures of a 12 per cent Al alloy after (i) slow cooling from 800°C, and (ii) rapid cooling from 800°C, explaining the differences between them.
(e) Suggest which alloy you would select for the manufacture of (i) cold drawn tubing, and (ii) a high strength casting, giving a brief justification of your choice.

Figure 11.21

CHAPTER 12

Phase transformations and diffusion

What are phase transformations and why should we study them?

In Chapter 11 we studied the phase change between liquid and solid states but saw as well that phase changes can occur within the solid state. Many of these phase changes have significant effects on the properties of materials. Examples are the martensitic transformation in steels which gives a major increase in hardness, the changes in martensite during tempering heat treatment to bring about increased toughness, and the microstructural changes that occur in some alloys causing precipitation hardening. It is necessary to study these phase changes in order to have a good appreciation of the development of the properties of engineering materials and the effects of the various heat treatments.

Many of the changes, the martensitic transformation being an exception, are the result of diffusion of atoms through crystal lattices. Diffusion itself is a function of temperature and is referred to as a *thermally activated process*. Many phase transformations are, therefore, termed diffusion controlled processes and are thermally activated. It is necessary then to study the mechanism of thermal activation and diffusion.

LEARNING OBJECTIVES

1. Know what is meant by the term *thermally activated process*.
2. Understand the significance of activation energy and temperature in connection with process rates.
3. Be able to perform calculations using the Arrhenius rate equation.
4. Know Fick's laws of diffusion.
5. Be able to perform calculations in connection with, for example, the case carburising of steel, using Fick's equations.

6. Understand the role of diffusion in precipitation hardening processes.

7. Understand the conditions necessary for diffusionless transformation, the martensitic transformation.

8. Know how information can be presented using time–temperature–transformation (T–T–T) diagrams.

12.1 | Introduction

The rate at which many processes occur is affected significantly by a change in temperature, reducing if the temperature is lowered and accelerated when the temperature is raised. Also, there is a certain minimum, or threshold, temperature below which the process will not be active. These principles apply to both physical processes, such as diffusion and diffusion related phenomena, and chemical processes. The quantity of energy which is necessary to initiate a process is termed the *activation energy* for the process.

A simple mechanical analogy can be used to explain the significance of activation energy. Consider a tetragonal prism of mass m (see Figure 12.1). This prism may have two stable rest positions, marked as A and C in the figure, although the potential energy of the prism is greater in position A than in position C owing to the centroid of the prism being at a higher level above datum. If the prism is translated from position A to position C by pivoting about point O, it is seen that the centroid will follow the path GG'G". When it is in an intermediate position B, the potential energy of the prism will be greater than at position A by an amount $mg\delta h$. This quantity of energy, $mg\delta h$, would be termed the activation energy for the process of translation of the prism from position A to position C.

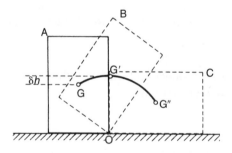

Figure 12.1 *Activation of a process – mechanical analogy.*

12.2 | Thermally activated processes

The atoms or molecules within any substance possess kinetic energy and are continually in motion. In a crystalline solid, the motion of atoms takes the form of vibration about fixed points. The total energy content of a system is determined by temperature and rises as the temperature increases. However, at any particular temperature, the atoms do not all possess the same value of kinetic energy. At any instant in time, some atoms may be momentarily at rest while others are in motion. The distribution of energy between all the atoms or molecules in a system is given by the Maxwell–Boltzmann distribution law. This is illustrated in Figure 12.2. $N(E)$ is the number of atoms per unit volume possessing an energy in the range from E to $E + \delta E$, where E is some value of energy and δE is a small increment of energy. Figure 12.2 shows the $N(E)$ distribution at three different temperatures T_1, T_2 and T_3, where $T_1 < T_2 < T_3$. As the temperature is raised, the total energy content of the system increases and the number of atoms possessing high values of kinetic energy increases. But from the

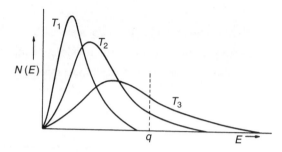

Figure 12.2 *Maxwell–Boltzmann distribution of molecular energies in a system at three temperatures ($T_1 < T_2 < T_3$).*

definition of $N(E)$:

$$\int_0^\infty N(E)\,dE = N$$

where N is the total number of atoms in the system. Therefore the areas under the three curves shown in Figure 12.2 are the same. If q is the activation energy of some process then the number of atoms per unit volume possessing an energy of q, or greater, will be given by $\int_q^\infty N(E)\,dE$. This function would have a value of zero at temperature T_1, but definite values at temperatures T_2 and T_3. It follows that the reaction would not take place at temperature T_1, but would occur at the temperature T_2 and T_3. Further, the rate at which the process would occur would be greater at T_3 than T_2 as a greater number of atoms possess energies above level q at this higher temperature.

The rate at which a process occurs is governed by the Arrhenius rate law, which can be written as

$$\text{Reaction rate} = v = A \exp\left(-\frac{q}{kT}\right)$$

where q is the activation energy for a single molecule, k is Boltzmann's constant, T is the temperature (K) and A is a constant. This may be rewritten as

$$v = A \exp\left(-\frac{Q}{R_0 T}\right)$$

where Q is the activation energy per kilomole, and R_0 is the universal gas constant (8.314 kJ/kmol K). The Arrhenius expression can be written in logarithmic form as

$$\ln v = \ln A - \frac{Q}{R_0 T}$$

This is a linear equation of the form $y = c + mx$, the two variables being the logarithm of rate and $1/T$. A graph plot of $\ln v$ against $1/T$ will give a straight line. The slope of the line is equal to $-Q/R_0$ and $\ln A$ is given by the intercept with the ordinate. If preferred, the logarithmic equation can be written as

$$\log_{10} v = \log_{10} A - \frac{Q}{2.3026 R_0 T}$$

The value of Q is of the order of 100 to 400 MJ/kmol for many physical processes and in the range from 20 to 100 MJ/kmol for many chemical reactions. In industrial chemical processes, catalysts are frequently used to accelerate the rate of a reaction. Their presence effectively reduces the activation energy for the reaction. Conversely, a reaction may be retarded with the aid of a negative catalyst which effectively raises the activation energy. Catalysts and negative catalysts are not permanently consumed in the reaction.

EXAMPLE 12.1

The rate of a process varies with temperature and the following table gives experimentally determined values:

Temperature (°C)	175	250	355	500
Rate (mm/s)	1.13×10^{-5}	5.93×10^{-5}	3.10×10^{-4}	1.46×10^{-3}

Show graphically that these data are consistent with the Arrhenius law and calculate the activation energy for the process. Also determine the temperature at which the process rate will be 1.5×10^{-2} mm/s ($R_0 = 8.314 \times 10^3$ J/kmol).

SOLUTION

The data should first be put into a form suitable for plotting as a graph. This is best done by producing a new table:

Temperature		1/T	ν	In ν
°C	K			
175	448	2.23×10^{-3}	1.13×10^{-5}	−11.39
250	523	1.91×10^{-3}	5.93×10^{-5}	−9.73
355	628	1.59×10^{-3}	3.10×10^{-4}	−8.08
500	773	1.29×10^{-3}	1.46×10^{-3}	−6.53
?	?	?	1.50×10^{-2}	−4.20

In ν can now be plotted against $1/T$ as shown in Figure 12.3.
The slope of the line is measured as $-5170 = -Q/R_0$.
Therefore the activation energy $Q = 5170 \times 8314 = 43 \times 10^6$ J/kmol.
From the graph a value of In $\nu = -4.20$ corresponds to $1/T = 0.83 \times 10^{-3}$.
Therefore the temperature at which ν is 1.50×10^{-2} is 1205 K or 932°C.

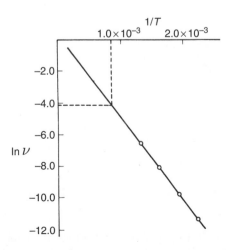

Figure 12.3 *Plot of In ν against $1/T$.*

EXAMPLE 12.2

The rates of linear growth of aluminium crystals during the recrystallisation of cold worked aluminium are given below:

Temperature (°C)	200	250	300	400
Rate (mm/s)	5.62×10^{-10}	1.38×10^{-7}	1.35×10^{-5}	1.82×10^{-2}

Determine if this is an example of a reaction that obeys the Arrhenius law and if so, calculate the activation energy.

SOLUTION

This problem can either be solved graphically as for example 1 or by a numerical method, as follows.
  If the reaction rates of temperatures T_1 and T_2 are v_1 and v_2 respectively, then

$$v_1 = A \exp\left(-\frac{Q}{R_0 T_1}\right) \text{ and } v_2 = A \exp\left(-\frac{Q}{R_0 T_2}\right)$$

Dividing we have

$$\frac{v_2}{v_1} = \exp\left(-\frac{Q}{R_0 T_2} + \frac{Q}{R_0 T_1}\right) = \exp\left[+\frac{Q}{R_0}\left(\frac{T_2 - T_1}{T_1 T_2}\right)\right]$$

From which Q can be evaluated.
  If this calculation is completed for several pairs of values and each calculation gives the same result for Q then the process satisfies the Arrhenius law.
Let $v_1 = 5.62 \times 10^{-10}$ mm/s, $T_1 = 200°C = 473$ K
  $v_2 = 1.38 \times 10^{-7}$ mm/s, $T_2 = 250°C = 523$ K
  $v_3 = 1.35 \times 10^{-5}$ mm/s, $T_3 = 300°C = 573$ K
and $v_4 = 1.82 \times 10^{-2}$ mm/s, $T_4 = 400°C = 673$ K

$$\frac{v_2}{v_1} = \frac{1.38 \times 10^{-7}}{5.62 \times 10^{-10}} = \exp\left\{\frac{Q}{8314}\left(\frac{523 - 473}{473 \times 523}\right)\right\}$$

$245.6 = \exp(5.5) = \exp(24.3 \times 10^{-9}\, Q)$
$Q = 226.2$ MJ/kmol

$$\frac{v_3}{v_1} = \frac{1.35 \times 10^{-5}}{5.62 \times 10^{-10}} = \exp\left\{\frac{Q}{8314}\left(\frac{573 - 473}{473 \times 573}\right)\right\}$$

$24.02 \times 10^3 = \exp(10.09) = \exp(44.38 \times 10^{-9}\, Q)$
$Q = 227.4$ MJ/kmol

$$\frac{v_4}{v_1} = \frac{1.82 \times 10^{-2}}{5.62 \times 10^{-10}} = \exp\left\{\frac{Q}{8314}\left(\frac{673 - 473}{473 \times 673}\right)\right\}$$

$32.38 \times 10^6 = \exp(17.29) = \exp(75.57 \times 10^{-9}\, Q)$
$Q = 228.8$ MJ/kmol

  The three calculated values of Q are in close agreement indicating that the process obeys the Arrhenius law. The average value of activation energy is 227.5 MJ/kmol.

EXAMPLE 12.3

A reaction is complete in 5 seconds at 600°C but requires 15 minutes for completion at 290°C. Calculate the time necessary for the reaction to be completed at 50°C.

SOLUTION

If the reaction requires 5 s for completion at 600°C then the reaction rate is proportional to $1/5\ \text{s}^{-1}$. At 600°C (873 K) $v = k/5$ and at 290°C (563 K) $v = k/(15 \times 60)$:

$$\frac{v_{873}}{v_{563}} = \frac{15 \times 60}{5} = \exp\left\{-\frac{Q}{8314}\left(\frac{563 - 873}{873 \times 563}\right)\right\}$$

$180 = \exp(5.19) = \exp(75.86 \times 10^{-9}\ Q)$
$Q = 68.4\ \text{MJ/kmol}$

$$v_{563} = \frac{1}{5} = A\exp\left(-\frac{68.4 \times 10^6}{8314 \times 563}\right)$$

$0.2 = A\exp(-14.62) = 4.49 \times 10^{-7}\ A$
Therefore $A = 445 \times 10^3$.
Let the time for the reaction at 50 °C (323 K) be t s:

$$\frac{1}{t} = 445 \times 10^3 \exp\left(-\frac{68.4 \times 10^6}{8314 \times 323}\right) = 3.86 \times 10^{-6}\ \text{s}^{-1}$$

$t = 259 \times 10^3\ \text{s} = 71.94\ \text{hours}$

12.3 | Diffusion

Many processes of significance in the materials world involve solid-state diffusion. How can material diffuse through a solid? It was stated in Chapter 8 that crystal lattices contain a number of point defects, one type of point defect being a vacancy. Consider a section of a crystal lattice containing some vacancies (Figure 12.4).

Atoms and vacancies can move through the lattice. For example, if the atom, marked A in Figure 12.4 and sited adjacent to a vacancy, possesses sufficient energy it can jump into the vacant site causing the vacancy to move one incremental distance in the opposing direction. As temperature rises not only is there an increase in the mean kinetic energy of atoms but also there is an increase in the percentage of vacancies in the lattice and so the frequency of atom/vacancy jumps increases exponentially according

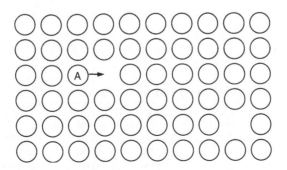

Figure 12.4 *Selection of crystal lattice containing vacancies.*

to Arrhenius' law. As explained in Chapter 8, the migration of vacancies to a dislocation can cause dislocation climb, one of the mechanisms postulated for creep in metals. It follows, then, that creep is a diffusion controlled process and steady-state creep rates vary with temperature according to an Arrhenius type expression.

In the atom/vacancy jump example illustrated in Figure 12.4, all atoms are identical. However, if the crystal structure contains a second component in solid solution, it will be evident that substitutional atoms can move through the lattice via the mechanism of atom/vacancy jumps. This migration of a second component through the lattice is diffusion. Atoms in interstitial solution can move by 'jumping' from one interstice to another. The diffusion of one substance into or through another is of importance in many aspects of materials. Examples are: diffusion of carbon in steels giving both carburising and decarburising effects, depending on condition, ageing and precipitation hardening effects in alloys, diffusion bonding, controlled diffusion of impurity elements into silicon to form p–n junctions.

12.4 Fick's laws of diffusion

Adolph Fick, in 1855, proposed a mathematical treatment for diffusion in one dimension at a constant temperature. For steady-state diffusion, in which the concentration gradient is constant, he stated that such steady-state diffusion could be described by the equation

$$J = -D\frac{dC}{dx}$$

where J is the flux (kg m^{-2} s^{-1}) (the quantity of material crossing a unit area normal to the flux direction in unit time), D is the coefficient of diffusion (m^2 s^{-1}) and dC/dx is the concentration gradient (kg m^{-4}). This is known as Fick's first law of diffusion. As stated earlier, the process of diffusion is a thermally activated process and, so, the diffusion coefficient, D, is not a constant but varies with temperature according to the Arrhenius equation.

A more common instance in practice is where the concentration of the diffusing substance changes with time. This means that both the flux and the concentration gradient also change with time. This dynamic condition is described by Fick's second law which can be written

$$\frac{dC_x}{dt} = D\frac{d^2C_x}{dx^2}$$

where C_x is the concentration of the diffusing species at some distance x from a surface and t is the time (s).

In the case of solid-state diffusion, where the concentration of the diffusing species at the surface of the material, C_s, remains constant as, for example, in the case carburising of steels, the solution to this differential equation may be written

$$\frac{C_s - C_x}{C_s - C_0} = \mathrm{erf}\left\{\frac{x}{2(Dt)^{\frac{1}{2}}}\right\}$$

where C_0 is the concentration at the start of the process ($t = 0$) and erf is the Gaussian error function or normalised probability (Table 12.1).

For values up to about 0.6, the Gaussian error function can be approximated as erf $y \approx y$, so that

$$\frac{C_s - C_x}{C_s - C_0} \approx \frac{x}{2(Dt)^{\frac{1}{2}}}$$

Table 12.1 *Gaussian error function.*

y	erf y	y	erf y
0.65	0.6420	0.85	0.7707
0.70	0.6778	0.90	0.7970
0.75	0.7112	0.95	0.8209
0.80	0.7421	1.00	0.8427

12.5 | Carburising and decarburising

Carbon can diffuse through iron and this phenomenon is utilised in the carburising processes which are used to increase the carbon content in the surface layers of low carbon steels to give surface, or case hardened components (see also Chapter 16). The reverse process of decarburisation can also occur if a steel is heated in an oxidising atmosphere. In these circumstances, carbon can diffuse out from the steel. Some surface decarburisation may occur during heat treatments but this effect is unwanted as it gives a softer surface layer than is generally required. In the preparation of carbon steel tools, the thin decarburised surface layer is removed during tool grinding.

During carburising operations, it can be assumed that a high carbon content is immediately established at the surface and that this surface carbon concentration remains constant during the process.

EXAMPLE 12.4

A low carbon (0.20% C) steel component is to be surface hardened by carburising at 950°C. Calculate the length of treatment needed to give a carbon content of 0.35 per cent at a distance of 1.0 mm below the surface. Assume that the carbon content at the surface is constant at 0.95 per cent throughout the process and the diffusion coefficient, D, at 950°C is $1.74 \times 10^{-11} \, m^2 \, s^{-1}$.

SOLUTION

This is a case of dynamic diffusion and Fick's second law applies. As the surface concentration is constant the solution

$$\frac{C_s - C_x}{C_s - C_o} = \mathrm{erf}\left\{\frac{x}{2(Dt)^{\frac{1}{2}}}\right\}$$

applies. Substituting in this expression we have

$$\frac{0.95 - 0.35}{0.95 - 0.2} = 0.8 = \mathrm{erf}\left\{\frac{1.0 \times 10^{-3}}{2(1.74 \times 10^{-11}t)^{\frac{1}{2}}}\right\}$$

When erf $y = 0.8$, $y = 0.9$ so that

$$0.9 = \frac{1.0 \times 10^{-3}}{2(1.74 \times 10^{-11}t)^{\frac{1}{2}}}$$

from which $t = 17738 \, s = 4$ hours 56 minutes.

EXAMPLE 12.5

A component, made from a high carbon steel containing 0.8 per cent carbon, is exposed to an oxidising environment at 1000°C for a period of one hour. At what depth below the surface will the carbon content of the material be reduced to 0.4 per cent? D, the diffusion coefficient for carbon in steel is $3.11 \times 10^{-11} \, m^2 \, s^{-1}$ at 1000°C.

SOLUTION

The assumption is made that the concentration of carbon at the extreme surface is zero throughout the exposure period. The expression

$$\frac{C_s - C_x}{C_s - C_0} \simeq \frac{x}{2(Dt)^{\frac{1}{2}}}$$

can be used as, for a value of 0.5, erf $y \simeq y$

$$\frac{0 - 0.4}{0 - 0.8} = 0.5 = \frac{x}{2(3.11 \times 10^{-11} \times 3600)^{\frac{1}{2}}}$$

from which $x = 0.33 \times 10^{-3}$ m $= 0.33$ mm.

12.6 ▌ Precipitation hardening

Precipitation hardening is the form of dispersion hardening (see also section 8.9) which can be brought about in certain alloys by heat treatment. The types of alloys in which precipitation hardening effects may be induced are those in which there is partial solid solubility (see section 11.7) and where there is a fairly large difference between the high-temperature and low-temperature limits of solubility. In alloy systems of this type the alloys within the narrow composition band shown in Figure 12.5 may be capable of being precipitation hardened.

Consider an alloy with a composition within this range. When heated to a high temperature just below the solidus, the alloy is in the α-phase field and the structure will consist of homogeneous α solid solution crystals. If the alloy is allowed to cool slowly from this high temperature the equilibrium structure of $\alpha + \beta$ will form by precipitation of particles of the β phase when the temperature falls below the solvus temperature. If, however, the alloy is cooled very rapidly from this temperature by, for example, quenching in water, there may be insufficient time for the equilibrium structure of β precipitate within the α crystals to form and, instead, the high-temperature structure will be retained at low temperature as a supersaturated α solid solution. This is a non-equilibrium state and, as such, is a metastable phase. In some alloy systems, such a supersaturated solid solution may remain indefinitely at room temperature without change while in other systems some diffusion and progression towards

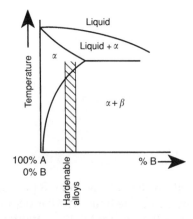

Figure 12.5 *Portion of partial solubility phase diagram showing the range of alloy compositions which may respond to precipitation hardening.*

the stable equilibrium $\alpha + \beta$ structure may occur at room temperature. As diffusion is a thermally activated process, any increase in temperature will cause an increase in the rate of diffusion and change.

Diffusion of solute within the crystals is a necessary prerequisite for precipitation as the supersaturated solution is, originally, homogeneous with a uniform distribution of solute atoms. There must be a diffusion of solute to build up the solute concentration in some areas of the lattice prior to precipitation of the second phase. This diffusion creates zones of high strain where there is a high solute concentration within the parent lattice. As the solution lattice is still fully coherent these zones are termed areas of coherent precipitate (see section 8.9). These regions of high strain offer considerable resistance to the passage of dislocations and, hence, have a major strengthening effect.

If diffusion of solute proceeds fully and the coherent precipitate zones separate out into true precipitate particles which are non-coherent with the parent lattice then the strain in the solute lattice begins to be released and the strength of the material begins to reduce. When this phenomenon occurs it is termed *overageing*.

There are many alloys which respond to precipitation hardening including aluminium–copper, copper–beryllium, magnesium alloys containing thorium, zirconium or zinc, some titanium alloys and maraging steels. The first precipitation hardening alloys to be developed and exploited were alloys of aluminium containing about 4 per cent copper, one proprietary name for which is *duralumin*.

The heat treatments necessary for a precipitation hardening alloy involve firstly, a *solution heat treatment*. This involves heating the alloy to a temperature in the single-phase region to create a homogeneous solid solution followed by rapid quenching to produce a supersaturated solution. The second stage of the heat treatment is heating at a relatively low temperature, the *precipitation treatment*, to permit diffusion and formation of coherent precipitate. Some alloys do not need this second heat treatment as diffusion and hardening will take place slowly at room temperatures. Such alloys are termed *age hardening* or *naturally ageing* alloys. However, the rate of hardening of these naturally ageing alloys can be accelerated by raising the temperature. Figure 12.6 shows the change of hardness with time at various temperatures for a naturally ageing alloy of aluminium and copper which has been solution heat treated. It will be seen that overageing will occur if this alloy is heated to a temperature of 200°C. If cooled to some temperature below room temperature the ageing of this alloy can be prevented. This phenomenon is put to good use in rivetted aluminium alloy structures. Aluminium–copper alloy rivets are placed in a refrigerator immediately after solution heat treatment to retard ageing and enable them to be kept for several days without hardening. They are then in the soft condition, permitting easy forming of the head during the rivetting operation, but subsequently age harden to full strength at room temperature.

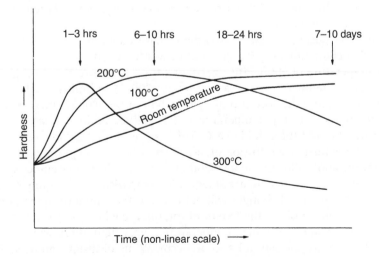

Figure 12.6 *Hardness–time relationships for an aluminium–copper age hardening alloy.*

EXAMPLE 12.6

A heat-treatable aluminium alloy will fully age harden in 10 days at 20°C but will reach maximum hardness after 10 hours when precipitation treated at 180°C. How long will this alloy take to harden if maintained at −25°C after solution heat treatment?

SOLUTION

The Arrhenius equation can be used for this example.
Hardening time at 20°C (293 K) is 240 hours and at 180°C (453 K) is 10 hours.

$$\frac{240}{10} = \exp\left\{\frac{B(453 - 293)}{453 \times 293}\right\} = 3.178$$

From which $B = 2636.3$.

Let x be the hardening time, in hours, at −25°C (248 K)
So

$$\frac{x}{10} = \exp\left\{\frac{2636.3(453 - 248)}{453 \times 248}\right\} = 122.8$$

From which $x = 1228$ hours ≈ 51 days.

12.7 | Martensitic transformations

In some alloys when the rate of cooling from a high temperature is extremely rapid and there is insufficient time for diffusion to occur it is still possible for a transformation to take place. In such a case the transformation will be of the type known as diffusionless or *martensitic*. In diffusionless transformation the movement of atoms will be small, though regular, and of less than one atomic spacing in magnitude. Martensitic transformations can take place in many types of crystal, both metallic and non-metallic, but the most important is that which occurs in many steels and is the mechanism for producing hardened steels.

When a steel, an alloy of iron and carbon, is cooled very rapidly and the transformation into the equilibrium state to give the eutectoid mixture of α iron and Fe_3C is prevented, a diffusionless transformation into martensite can occur.

Consider the eutectoid portion of the iron–carbon phase diagram (Figure 12.7). According to the phase diagram an alloy containing, say 0.6 per cent C, would be a homogeneous solid solution of carbon in γ iron at a temperature of 900°C. During cooling the γ structure should begin to transform into α iron when the liquidoid temperature is reached (about 800°C). α iron is capable of holding very little carbon in solid solution and the concentration of carbon in the γ phase increases as cooling continues. On cooling through the eutectoid temperature, 723°C, all remaining γ phase should transform into the eutectoid mixture of α and the carbide Fe_3C. This eutectoid mixture in steels is termed *pearlite*. It will be evident that there must be a diffusion of dissolved carbon through the γ-iron lattice to build up areas of carbon concentration prior to the formation of Fe_3C particles and eutectoid formation.

Transformations within the solid state are sluggish in comparison with liquid–solid transformations and in practical cases the structural changes will not occur at the equilibrium temperatures. There will be a certain amount of undercooling, the extent of the undercooling increasing as the rate of cooling is increased. If the rate of cooling from high temperature is very high, there will be insufficient time for any diffusion of carbon within the γ iron lattice to occur. While diffusion can be suppressed by a very rapid cooling from high temperature, the allotropic transformation from γ to α-iron is not prevented.

Consequently, when this change from the γ-form to the α-form occurs a large amount of carbon will be trapped in supersaturated interstitial solid solution creating extremely large strains. (The equilibrium solid solubility limit for carbon in α-iron is only 0.006 per cent at 20°C.) The body-centred cubic α-iron lattice is distorted into a body centred tetragonal structure and is referred to as the α′ structure, or martensite. This highly strained structure is extremely hard and also extremely brittle.

This transformation from γ into α′ only involves small movements of the atoms in the crystal lattice. Figure 12.8 shows, according to Bain, how a body-centred cubic lattice can be formed from a face-centred cubic structure. Carbon atoms present at interstitial sites in the f.c.c. lattice will remain, now in b.c.c. interstices, causing the body-centred cubic structure to be distorted into a body-centred tetragonal form.

The relationship between cooling rate and undercooling, and the conditions required for martensite formation can be shown on a time–temperature–transformation (T–T–T) diagram (Figure 12.9). (See also section 16.6.)

It will be noted that if the rate of cooling exceeds a particular value, the *critical cooling velocity*, the γ-phase will transform into martensite rather than into pearlite, the transformation beginning at the M_s temperature, about 340°C for a 0.6 per cent C steel.

As mentioned above, α′ or martensite is a non-equilibrium phase which is metastable. It will not undergo any changes at room temperature as there is insufficient thermal energy to permit diffusion, but if the temperature is raised diffusion of carbon can occur leading to the precipitation of particles of Fe_3C from the martensite. This process can begin to occur if the temperature is raised to about 200°C

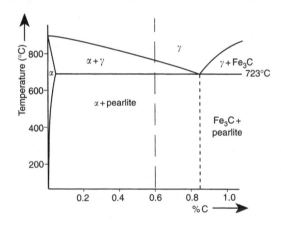

Figure 12.7 *Portion of iron–carbon phase diagram.*

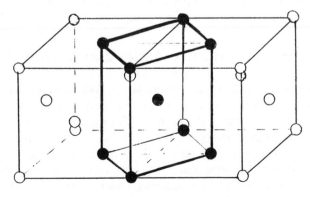

Figure 12.8 *Sketch showing a potential b.c.c. cell within an f.c.c. structure. The interstitial carbon atoms will distort the b.c.c. structure.*

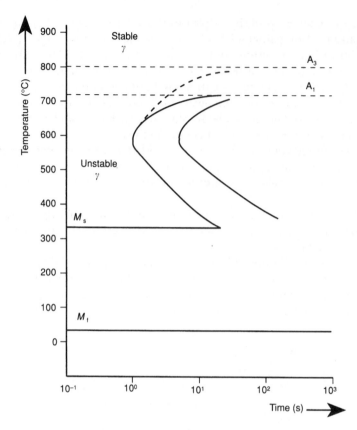

Figure 12.9 *T–T–T diagram for a 0.6% carbon steel.*

or above. Any precipitation of carbon in the form of Fe_3C from the α' lattice will release some of the lattice strain with a consequent reduction in the hardness and brittleness. This phenomenon is utilised in the process of *tempering* of hardened steels. The higher the tempering temperature, the greater will be the amount of strain release and the greater will be the reduction in hardness and increase in ductility. It is important to note that during tempering the structure does not change to the equilibrium structure of α and pearlite, as formed during slow cooling, but rather to one containing very fine precipitated particles of Fe_3C.

Martensitic transformation has been described with reference to steels but it can occur in other systems containing eutectoids, a good example being the copper–aluminium system.

12.8 Phase transformations in ceramics and glasses

Solid-state phase changes can occur in ceramic materials as well as in metals. Phase changes are generally accompanied by a volume change and in a non-ductile ceramic material such a volume change can lead to the formation of cracks within the material. For example, zirconia, the oxide of zirconium, ZrO_2, transforms on cooling from a tetragonal structure into a monoclinic structure with a volume change of about 3 per cent. For this reason pure zirconia is rarely used as a ceramic material and stabilised zirconia products are manufactured containing additions of lime, CaO, magnesia, MgO or yttria, Y_2O_3. These form solid solutions with a cubic structure which is stable at room temperature.

The glass state is metastable and a glass may slowly transform into a crystalline structure. This process is termed *devitrification*, and is accompanied by a decrease in volume. This could cause cracks to form within the glass or crazing in glazed coatings. Commercial glasses are formulated so that devitrification will not occur at ordinary temperatures, but in some materials, the *glass ceramics*, the material is first formed as a glass and then converted into a fine-grained crystalline material by controlled crystallisation heat treatments. Glass ceramics maintain their mechanical strength to much higher temperatures than glasses. LAS glass ceramics, based on $LiO–Al_2O_3–SiO_2$, have high strength coupled with very low thermal expansion coefficients.

SUMMARY

The rate of many processes is highly dependent upon temperature. In thermally activated processes the rate increases exponentially with temperature according to the Arrhenius law. Diffusion is one such process. A number of processes which can occur in materials, such as age hardening and the precipitation of a second phase within a solid, involve diffusion of atoms through a crystal lattice. These are termed *diffusion controlled processes* and, like diffusion, are thermally activated.

Diffusion of solute atoms through a crystal lattice is a phenomenon that leads to a number of important applications. These include the infusion of carbon and nitrogen into the surface layers of a steel to confer surface hardening, the infusion of impurity atoms into the surface of high-purity silicon in the manufacture of electronic components, and the migration of atoms across an interface between two solid components to bring about diffusion bonding.

Not all solid-state transformations involve diffusion. Diffusionless, or martensitic, transformations can take place in some materials through cooling at very rapid rates. This is the mechanism involved in the hardening of steels by means of a quenching process from high temperature. The martensitic structure created is extremely hard, but also brittle. Subsequent controlled heating, or tempering, will allow some diffusion of carbon and some precipitation of carbide particles to occur which will reduce the hardness to some extent but greatly reduce the brittleness of the hardened steel.

Questions

12.1 For a thermally activated reaction with an activation energy of 100 MJ/kmol, calculate by what factor the rate of reaction will be increased if the temperature is raised from 15°C to 25°C.

12.2 The time required to recrystallise aluminium varies with temperature according to the Arrhenius law. Determine the activation energy of the recrystallisation process if recrystallisation is complete in 6 minutes at 330°C but required 100 hours at 260°C. (Assume $R_0 = 8314$ J/kmol K.)

12.3 A thermally activated reaction is completed in 15 seconds at 600°C but requires 20 minutes for completion at 260°C. Calculate the time necessary for the reaction to be completed at 55°C and determine the value of the activation energy for the process.

12.4 The times required for a 50 per cent recrystallisation of cold worked pure copper are 1 minute at 162°C and 1000 minutes at 72°C. Estimate the time needed for 50 per cent recrystallisation at 20°C. Discuss what changes you would expect to occur in the mechanical properties of a cold worked copper cable over a period of 10 years.

12.5 The time required to soften a deformed metal by 50 per cent at 400 K is 1.5 minutes and at 430 K is 0.5 minute. Calculate the activation energy of this softening process.

12.6 The strength of a steel component designed for use at high temperatures is reduced to below its safe working limit after 50 per cent recrystallisation has occurred. This limit is reached after a time of 4250 s at a temperature of 580°C and after 3.6 s at 750°C. Assuming that the Arrhenius law applies, calculate the maximum temperature at which a working life of 10^4 hours is attainable.

12.7 The variation in thermionic emission from a tungsten filament with temperature is given in the table below:

Temperature (K)	Current density (mA/mm²)
1470	7.63×10^{-7}
1543	4.84×10^{-6}
1761	4.62×10^{-4}
1897	4.31×10^{-3}

  (a) Show graphically that the above data are consistent with the Arrhenius law and calculate the activation energy (work function) for tungsten.

  (b) Determine the thermionic current density for a filament temperature of 2000 K. (Use Boltzmann's constant, $k = 1.38 \times 10^{-23}$ J/K; 1 A = 1 Cs^{-1}.)

12.8 Joints made using an epoxy-based adhesive can be cured to maximum strength in 30 minutes at 150°C but requires 3 hours to reach maximum strength at 80°C. Assuming that the curing reactions obey the Arrhenius law, estimate the time required for complete curing at 25°C and determine the activation energy of the process.

12.9 The diffusion coefficient for carbon in steel is measured at two temperatures and the values found to be: 4.826×10^{-12} m² s^{-1} at 850°C and 1.805×10^{-11} m² s^{-1} at 950°C.

Calculate: (a) the activation energy for diffusion of carbon in steel for the temperature range 850–950°C, and (b) the value of diffusivity at 1000°C, assuming the activation energy is unchanged at this higher temperature.

12.10 The maximum hardness of aged aluminium alloy 2014 occurs after 30 hours at 150°C and after 3 minutes at 260°C. Assuming that the rate of reaction as a function of temperature is given by the Arrhenius relationship, calculate the time to achieve maximum hardness at both 180°C and 220°C.

12.11 The diffusion rates for carbon in titanium were determined at the following temperatures:

Temperature (°C)	736	782	835
Diffusion coefficient, D (m² s^{-1})	2×10^{-13}	5×10^{-13}	1.3×10^{-12}

  (a) Determine whether the equation $D = A \exp(-Q/RT)$ applies and, if so, evaluate the constant A and the activation energy for diffusion, Q.

  (b) Calculate the rate of diffusion at 500°C and comment on the value for 1000°C.

12.12 If the creep strain rate of a metal is 3×10^{-9} s^{-1} at 560°C, determine the creep strain rate at 650°C, assuming that the applied stress remains constant. Assume the creep process has an activation energy of 65 MJ/kmol and the universal constant, R, is 8.314 kJ/kmol K.

12.13 A mild-steel boiler plate (0.16 per cent C) was exposed to a very strongly oxidising atmosphere at 1000°C owing to the failure of a heat exchanger. Estimate the depth below the surface of a 10 mm thick plate at which the concentration of carbon decreases to one-half of its original value after exposure for (a) 1 hour and (b) 1 day. Assume the diffusivity of carbon in iron at 1000°C is 3.11×10^{-11} m² s^{-1} and discuss any assumptions made in the calculations.

12.14 A steel containing 1.2 per cent carbon is heated at 1000°C in a strongly oxidising environment. Estimate the depth below the surface at which the carbon concentration is reduced to 0.5 per cent after 1 hour's exposure. Take D for carbon in steel at 1000°C as $3.1 \times 10^{-11} \, m^2 s^{-1}$. Assume that the carbon concentration at the surface during oxidation is zero.

12.15 Calculate the percentage of carbon 0.3 mm below the surface of a 0.15 per cent carbon mild steel component which has been case carburised for 10 hours at 850°C. Assume that the surface carbon concentration has been maintained at 1.0 per cent during the process and that the diffusivity of carbon in mild steel at 850°C is $4.5 \times 10^{-12} \, m^2 s^{-1}$.

12.16 In a case carburisation process the surface of a 0.2 per cent carbon steel is maintained at 1.2 per cent carbon for 100 minutes at 800°C. Calculate the depth below the surface at which the carbon content is 0.6 per cent after this process. D for carbon in steel at 800°C is $2.4 \times 10^{-12} \, m^2 s^{-1}$.

12.17 A gear made from steel of nominal carbon content 0.2 per cent is to be gas carburised at 927°C. Calculate the time in minutes needed to increase the carbon content at a depth of 0.55 mm below the surface to 0.4 per cent. Assume that the carbon content at the surface is maintained at 0.9 per cent and that D at 927°C is $1.28 \times 10^{-11} \, m^2 s^{-1}$.

12.18 A case carburising operation for mild steel (0.15 per cent carbon) is conducted at 1000°C and gives a surface carbon content of 0.85 per cent and a carbon concentration of 0.75 per cent at a depth of 0.25 mm below the surface. It is suggested that economies may be affected if the temperature of the process is reduced to 950°C as furnace fuel costs per hour of furnace operation are 20 per cent less at the lower temperature. Is this suggestion practical if the same quality carburised product is to be obtained? Assume that the diffusivity of carbon in mild steel varies with temperature according to the Arrhenius law and that the activation energy is 150 MJ/kmol.

12.19 A steel component is to be nitrided and it is required, for optimum characteristics, that the concentration of nitrogen at 0.5 mm below the surface be 0.025 per cent. Calculate the time necessary for nitriding, assuming that the concentration of nitrogen at the surface of the steel is maintained at 0.125 per cent throughout and that the diffusion coefficient for nitrogen in steel is $3.48 \times 10^{-12} \, m^2 s^{-1}$.

12.20 There is a time delay, the incubation period, between water quenching and the commencement of hardening for a precipitation hardening aluminium alloy. Precipitation hardening tests show that, for a particular aluminium alloy, the incubation period is 30 s at a temperature of 150°C and 3 minutes at 100°C. To what temperature should the alloy be cooled, immediately after solution heat treatment, if it is desired to maintain the alloy for a period of three days without commencement of hardening?

CHAPTER 13

Electrical and magnetic properties

Why should we study the electrical and magnetic properties of materials?

It is likely that the majority of the readers of this book will be students of mechanical engineering, some of whom may question the need for studying the electrical and magnetic characteristics of materials. No sector of engineering exists in a vacuum and the divisions between the various engineering specialisms are largely artificial. The mechanical engineer should know the mechanisms of electrical conduction, the factors which control the conductivity of materials, and how materials respond to electrical and magnetic fields. The purpose of this chapter is to give the reader a brief insight into this aspect of materials science.

LEARNING OBJECTIVES

1. Describe the band structures for electron energies in aggregates of atoms and the concepts of valence bands and conduction bands.
2. Understand how excitation can create free electrons, and corresponding electron 'holes'.
3. Distinguish between *intrinsic* and *extrinsic* semiconductors.
4. Describe the electron/hole interaction at a *p–n* junction and the responses of this type of junction to forward-bias and reverse-bias applied voltages.
5. Know how *polarisation* can be induced in materials and distinguish between *ferroelectric* and *piezoelectric* materials.
6. Distinguish between paramagnetism, diamagnetism and ferromagnetism.

7. Describe the process of magnetisation and explain the terms *remanence*, *coercivity* and *hysteresis*.

8. Distinguish between magnetically *soft* and magnetically *hard* materials and have an understanding of the main factors controlling the magnetic properties of a material.

13.1 Conduction

The conduction of electricity through a material is the transference of an electrical charge from one position to another. The charge may be transferred by movement of electrons, or by the migration of ions. Conduction in metals is due to electron migration, but it is the movement of ions that is responsible for the conductivity of electrolytes and for the very low conductivities observed in some insulating materials. In the class of materials known as *semiconductors* charge is carried by the motion of electrons and the movement in the opposite direction of positive 'holes'.

Electrical conduction in metals is due to the migration of electrons through the crystal lattice and metals are classed as good conductors because of the high mobilities of electrons. The resistivity of a metal increases with increasing temperature. As the temperature is raised the vibrational amplitude of the atoms is increased and there is a greater probability of interference and collision between moving electrons and atoms. However, an increase in pressure effectively increases the number of electrons per unit volume of material and so the resistivity is reduced. This property is useful and electrical strain gauges are based on this principle.

In the case of ionic conductivity an increase in temperature increases the mobility of ions and causes a rise in the conductivity. This is the reverse of the effect of heat on an electron conductor.

There is a complete spectrum of resistivities from the extremely low resistivity of silver, the best metallic conductor, to the very high resistivities of plastics and ceramic materials (Table 13.1). Resistivity is expressed in ohm metres (Ωm). The resistance of a piece of conducting material increases with the length of the conductor and decreases with an increase in the cross-sectional area (c.s.a.):

$$\text{resistance } (\Omega) = \text{resistivity } (\Omega\,\text{m}) \times \frac{\text{length (m)}}{\text{c.s.a. (m}^2)}$$

Table 13.1 *Resistivities of some pure metals, alloys and other materials.*

Material	Resistivity Ω m @ 20°C	Material	Resistivity Ω m @ 20°C
Silver	1.6×10^{-8}	Silicon	8.5×10^{-4}
Copper	1.67×10^{-8}	Germanium	1.0×10^{-3}
Aluminium	2.66×10^{-8}	Urea-formaldehyde	
Magnesium	3.9×10^{-8}	(white Bakelite)	1×10^{6}
Sodium	4.3×10^{-8}	Fireclay	1.4×10^{8}
Zinc	5.92×10^{-8}	Alumina	1×10^{11}
70/30 brass	6.2×10^{-8}	Phenol-formaldehyde	
Nickel	6.84×10^{-8}	resins (Bakelite)	1×10^{12}
Iron	8.85×10^{-8}	Diamond	5×10^{12}
Tin	1.15×10^{-7}	Polyethylene	1×10^{13}
Mild steel	1.7×10^{-7}	Nylon	1×10^{14}
Lead	2.1×10^{-7}	Mica	9×10^{14}
50/50 cupronickel	5.5×10^{-7}	Quartz	9×10^{14}
18/8 stainless steel	7.0×10^{-7}	Pyrex glass	1×10^{16}
Mercury	9.6×10^{-7}	PTFE	2×10^{16}
Graphite	1.4×10^{-5}	Vitreous silica	1×10^{20}

EXAMPLE 13.1

What is the resistance/km length of cadmium–copper overhead power wires for an electric tramway if the wire cross-section is $5\,mm \times 3.5\,mm$ and the resistivity of the cadmium–copper at 20°C is $1.9 \times 10^{-8}\,\Omega$ m.

SOLUTION

The resistance of the wire = resistivity × (length/c.s.a.)
The cross-sectional area is $5 \times 10^{-3} \times 3.5 \times 10^{-3} = 17.5 \times 10^{-6}\,m^2$
Resistance/km $= 1.9 \times 10^{-8} \times (1000/(17.5 \times 10^{-6})) = 1.086\,\Omega/km$

13.2 | Band structure

It is necessary to consider the band structure of atoms in order to understand electronic conduction in conductors and semiconductors. The electronic structure of atoms was discussed in Chapter 2 and it was stated (section 2.8) that the energy levels of electrons become closer to one another as the distance from the nucleus increases. In the elementary description of metallic bonding (section 2.15) it was stated that when atoms are in close proximity the outer shell electrons tend to overlap. The Pauli exclusion principle that no two electrons can possess exactly the same energy still applies and, in an aggregate of atoms, the discrete energy level for an electron state, for example 3s, broadens into a band to accommodate all the 3s electrons in the aggregate, the energies of individual 3s electrons differing by only infinitesimal amounts. Let us see how this works out in practice by considering some examples. Figure 13.1 shows the broadening of energy levels for the element magnesium. The dashed line in the figure indicates the interatomic distance in crystalline magnesium. It will be seen that there is an overlap of the 3s and 3p in an aggregate of magnesium atoms.

In the case of the element sodium each atom contains one electron in the 3s state, although the 3s state could hold two electrons per atom, one with a spin quantum number $m_s = +\frac{1}{2}$ and the other with $m_s = -\frac{1}{2}$. In an aggregate of sodium atoms, if there are n atoms present there will be $2n$ available energy levels within the 3s energy band. With only one valence electron per atom, this means that the 3s band

Figure 13.1 *Broadening of the energy levels of the electron states of magnesium.*

in sodium is only half filled. The valence electrons tend to occupy the lowest available energy levels. Energy must be given to an electron in order to move it from one position to another, that is, to conduct electricity through the material. In the case of sodium the energies of the 3s, or valence, electrons can be readily increased within the part-filled 3s band, and the metal is a good electrical conductor. In magnesium each atom possesses two 3s electrons and the 3s energy band is completely filled. But, in this element the 3s and 3p bands overlap and valence electrons may have their energies increased within the limits of the 3p band, and so magnesium conducts electricity readily. Adjacent permitted electron energy bands do not overlap in all materials. In the diamond form of carbon, for example, each carbon atom is covalently bonded do four other carbon atoms (see section 2.13) thus effectively giving each carbon atom a complement of eight valence electrons. The valence band is completely filled and there is a large energy gap between the valence band and the next permitted energy range. The magnitude of this gap is a measure of the amount of energy an electron requires to break away from bonds and become a free electron. In diamond this energy is 8.33×10^{-19} J (5.2 eV). Consequently, diamond is an insulating material with a high resistivity. The other Group IV elements, silicon, germanium and tin, also crystallise in the diamond crystal form, but the energy gaps between the valence and conduction bands are smaller than in diamond. In silicon the energy gap is 1.76×10^{-19} J (1.1 eV), in germanium it is 1.5×10^{-19} J (0.72 eV) and in grey tin it is only 0.13×10^{-19} J (0.08 eV) (Figure 13.2). The resistivities of these elements (see Table 13.1) reflect the magnitudes of the energy gaps. A small electrical field is sufficient to cause the valence electrons of tin to cross the energy gap and so tin is a reasonably good conductor of electricity. It is possible, though more difficult, for valence electrons in silicon and germanium to cross the energy gap; hence, these elements are semiconductors.

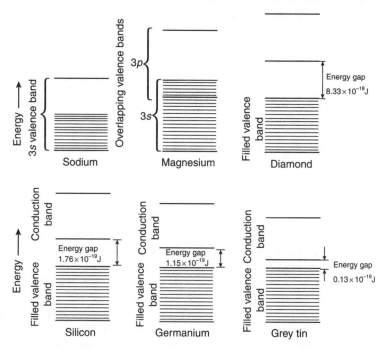

Figure 13.2 *Energy bands in some elements.*

13.3 | Conduction in metals

The good conductivity of a metal is due to the high mobility of the outer shell, or valence, electrons throughout the crystal lattice. Any factor that disturbs the regularity of a space lattice will, therefore, adversely affect the mobility of valence electrons and increase resistivity. The presence of lattice defects such as dislocations and interstitial or substitutional solute atoms will increase the probability of collisions between moving electrons and the cores of atoms and cause an increase in resistivity. Consequently, a solid solution of two elements will possess a higher resistivity than the pure metals. Refer to section 11.10 and also compare the resistivities of copper, nickel, and a cupronickel alloy quoted in Table 13.1.

If a metal or alloy is cold worked the grain structure is fragmented and distorted and the number of dislocations in the material is increased. This also leads to an increase in resistivity.

An increase in temperature will cause an increase in the amplitude of vibration of atoms and, hence, increase the probability of electrons colliding with an atom core. This is the reason for the increase in resistivity observed when the temperature of a metallic conductor is raised. The resistivity of a metallic conductor varies with temperature and this variation is almost linear over a wide range of temperatures. It can be expressed as follows:

$$\rho_2 = \rho_1 (1 + \alpha(T_2 - T_1))$$

where ρ_1 and ρ_2 are the resistivities at temperatures T_1 and T_2 respectively and α is the thermal coefficient of resistivity for the metal concerned.

EXAMPLE 13.2

The resistivity of nickel is measured at 20°C and 150°C, the two values being $6.84 \times 10^{-8}\,\Omega\,m$ and $1.24 \times 10^{-7}\,\Omega m$ respectively. Calculate the value of the temperature coefficient of resistivity over this temperature range.

SOLUTION

Substituting in the expression $\rho_2 = \rho_1 (1 + \alpha(T_2 - T_1))$ we have

$$1.24 \times 10^{-7} = 6.84 \times 10^{-8} (1 + \alpha(150 - 20))$$

So $\alpha = ((1.24 \times 10^{-7}/6.84 \times 10^{-8}) - 1)/130 = 6.25 \times 10^{-3}/°C$

13.4 | Semiconductors

It has already been stated that silicon and germanium are semiconductors. Their conductivity is due solely to the distribution of electron energies within the pure material. Materials of this type are termed *intrinsic* semiconductors. Silicon and germanium possess the diamond type of crystal structure with each atom covalently bonded to four other atoms. The freeing of one valence electron to cross the energy gap into the conduction band will create a situation with one atom within the crystal lattice possessing only three bonds. The gap in the bonding is termed an *electron hole*. The bonding electrons are constantly in motion and may switch from one atom to another. Movement of an electron in one direction will cause an electron hole to move through the lattice in the opposite direction (Figure 13.3). As the motion of a hole is opposite in direction to the drift of bonding electrons the hole can be regarded as a positive charge carrier. In intrinsic semiconductors conduction is due to both the movement of freed valence electrons in the conduction band and the motion of positive holes. The energy necessary to free an electron and allow it to cross the energy gap can be provided by thermal activation. From this it can be seen that increasing the temperature of an intrinsic semiconductor will increase the conductivity. This is the opposite of the effect of increasing the temperature of a fully conducting metal.

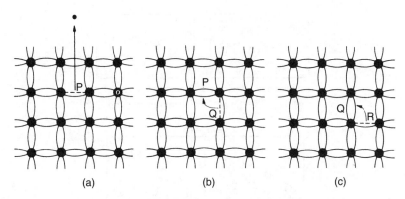

Figure 13.3 *Electrical conduction by movement of 'holes'. (a) An electron is freed from a covalent bond creating an electron hole at P. (b) Electron transfer to P from adjacent site. Hole effectively moves to Q. (c) Electron transfer from R to Q. Hole moves to R.*

Another type of semiconductor is the impurity, or *extrinsic*, semiconductor. Silicon and germanium have four outer shell electrons per atom and a balanced covalently bonded crystal structure is formed. If some atoms of an element with only three outer shell electrons are substituted for some of the parent atoms there will be a shortage of electrons for full covalent bonding and some electron holes will be present in the crystal lattice. Conduction due to the migration of positive holes can then occur. This type of semiconductor is known as *p-type*. A small amount of aluminium dissolved in silicon or germanium is an example of this type. Alternatively, if an element such as phosphorus with five outer shell electrons is introduced into the covalent silicon lattice there will be additional electrons present. These additional electrons can be freed to enter the conduction energy band relatively easily. This type of semiconductor, in which transference of electrical charge is by the motion of freed electrons, is termed *n-type*.

By the introduction of small controlled amounts of various elements into silicon or germanium, semiconductors can be 'designed' to meet specific requirements.

Other semiconductor materials may be ceramic compounds made by combining a Group III element and a Group V element, or a Group II element with one from Group VI, or oxides with a variety of structures. The important compound semiconductors are gallium arsenide, GaAs, indium phosphide, InP, indium antimony, InSb, zinc oxide, ZnO, zinc sulphide, ZnS, and cadmium sulphide, CdS. Three-component semiconductor materials have also been developed and some important ones are InGaAs, GaAsP and CdHgTe. Elements used as doping additions in Group III–Group V compound semiconductors are zinc or cadmium to make *p*-type, and selenium and sulphur for *n*-type. Semiconductor materials are used in integrated circuits and electronic devices.

13.5 | The *p–n* junction

A *p*-type semiconductor, although it contains positive holes due to the presence of atoms of a Group III element, is electrically neutral. Similarly, although an *n*-type crystal contains free electrons from the Group V donor atoms it is still electrically neutral. If a *p*-type semiconductor is joined to an *n*-type semiconductor* the result is a *p–n* junction and this is effectively a one-way valve. At the interface between the *p*-type and *n*-type materials there will be some interaction between positive holes and free electrons. The combination of a hole and an electron destroys both as mobile charge carriers. The small interface zone in which the charge carriers are lost is termed the depleted layer. This zone is very narrow, being only about 10^{-6} m in thickness. Within the depleted layer the *p*-type impurity atoms have each

---

* *p–n* junctions are normally made by introducing *p*-type and *n*-type impurities into opposite ends of a crystal of silicon or germanium, rather than by joining two separate crystals.

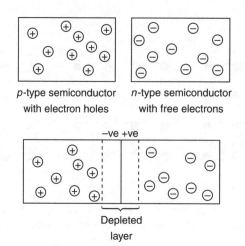

p-type semiconductor
with electron holes

n-type semiconductor
with free electrons

−ve +ve

Depleted
layer

Figure 13.4 *p–n junction showing depleted layer.*

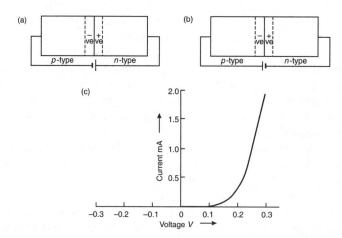

Figure 13.5 *p–n junction characteristics. (a) Applied voltage tending to increase height of potential barrier. (b) Applied voltage tending to decrease height of potential barrier. (c) Voltage–current relationship for silicon p–n junction.*

captured an electron and have become negatively charged ions, and the *n*-type impurity atoms have lost a free electron to become positively charged ions. These are termed acceptor and donor atoms respectively (Figure 13.4). The depleted layer is thus highly charged, carrying a negative charge on one side and a positive charge on the other. This charged zone is a potential barrier and effectively prevents charge carriers from passing freely across it.

Now let us consider what happens when a voltage is applied across the ends of a crystal containing a *p–n* junction. If the *n*-type end is made positive with respect to the *p*-type end (Figure 13.5(a)) the positive charge on the *n*-type side of the depleted layer will be increased. The height of the potential barrier at the depleted layer will thus be increased and it becomes extremely difficult for charge carriers to pass across the junction. Conversely, if an applied voltage is applied in the opposite sense (Figure 13.5(b)) the height of the potential barrier will be reduced and current will be able to flow fairly freely across the junction. Figure 13.5(c) shows the voltage–current relationship for a silicon *p–n* junction. The reverse current with *V* negative is extremely small and is of the order of one microampere (1 μA).

Application of a voltage in the sense shown in Figure 13.5(b) is referred to as *forward bias* and when applied in the opposite direction (Figure 13.5(a)) is termed *reverse bias*. It will also be seen that such a device, with the facility for an electric current to pass in one direction only, can be used as a rectifier or *diode*.

13.6 | Insulating materials

Insulating materials are those in which the bond structure is ionic, or covalent, or a mixture of both types. In these bonds the electrons are tightly bound and are not 'free' as in metals. Electron and ion mobility is very low and measured resistivities are 10^6 Ω m or greater (see Table 13.1).

The energy band diagram for an insulating material is similar to that for a semiconductor with a completely filled valence band and an energy gap between the valence band and the next permitted energy band. However, in the case of insulating materials the energy gap is so large that electrons are unable to cross it under normal conditions. An occasional electron may possess a sufficiently high energy to cross the gap, and it is this that accounts for the very high, but measurable resistivities of insulators. At high temperatures there will be a greater probability of an occasional electron possessing the necessary energy for conduction and so the resistivities of insulators show a decrease with increasing temperature. In the case of some materials containing ionic bonds there is also the possibility that small ions in a relatively open lattice could migrate, giving some small degree of conductivity. Again, at high temperatures ions become more mobile and can diffuse more readily, so reducing the measured resistivity.

Insulating materials are also known as *dielectrics*, and have the capacity to store an electrostatic charge. When a dielectric is placed in an electric field some polarisation occurs. If the material already contains polar molecules the effect is to cause the molecules to align themselves in the direction of the applied field. In non-polar materials the effect of the field is to create dipoles by causing small atomic movements leading to slight separation of positive and negative ions. When an applied electric field induces polarisation in a material there is a back effect which modifies the field. This effect is termed *permittivity*.

The relative permittivity, ε_r, or dielectric constant is a parameter of dielectric materials and is the ratio of the capacity of a capacitor (condenser) with the material filling the space between the plates, to the capacity with a vacuum between the condenser plates. The parameter is also known as relative permittivity or specific inductive capacity. The value of the dielectric constant, ε_r, is within the range of 2 to 10 for many polymers and ceramics, but the value tends to reduce as frequency increases. High permittivity ceramics based on barium titanates have been developed for the production of miniaturised capacitors. These possess values of ε_r, of the order of 1000.

Some energy is lost as heat when a dielectric material is polarised in an alternating electrical field. Energy losses are due to two factors, dipole friction and current leakage. The fraction of the energy lost during each reversal is termed the dielectric loss, W, and is given by $W = \omega C V^2 \delta$, where V is the root mean square of voltage, C is capacitance, δ is the power factor or loss angle and $\omega = 2\pi f$, where f is the frequency. A high power factor is an indication of greater power in an insulator being dissipated as heat. Low loss steatite ceramics were developed for high-frequency insulation, for example, in radio communication equipment.

Insulating materials, or dielectrics, can break down at very high voltages. The applied electrical field may be sufficiently high to raise the energies of some electrons above the energy gap and thus free them. The possibility of this occurring is increased if the material contains some impurities or defects. The high energy electrons may then disrupt covalent bonds within the insulator liberating more electrons to take part in a chain reaction. This type of breakdown is referred to as *cascading* and is irreversible. The *dielectric strength* of a material is the electrical stress in volts per unit volume required to cause breakdown. During the testing of the strength of dielectrics, the electrical stress is maintained for one minute. Far more common than cascading is surface breakdown. The presence of moisture or an accumulation of dirt on the surface of an insulator may provide a surface-conduction path. To minimise the incidence of surface breakdown, ceramic insulator surfaces are glazed and also made with a shaped or corrugated surface to greatly increase the surface path length.

13.7 | Ferroelectric materials

In many dielectric materials, the polarisation which occurs within an electrical field is directly proportional to the strength of the field and all polarisation is lost when the field decreases to zero. Some materials retain a net polarisation when the electrical field is removed and these are termed *ferroelectric*. In such a material the dipoles line up with the field when an electrical field is applied until all the dipoles are in alignment and a saturation polarisation is achieved. When the field is removed the degrees of dipole alignment largely remains and there is a remanent polarisation (compare with magnetic remanence – section 13.9). In order to remove the polarisation completely, the electrical field must be reversed and increased in strength to a specific value (compare with magnetic coercivity – section 13.9). In an alternating electric field, a ferroelectric hysteresis loop is described. The ferroelectric materials are ceramics, such as barium titanate, with a distorted perovskite structure (see section 5.6) and are dielectric materials with a high dielectric constant and are suitable for use in capacitors. Also, the ability to retain polarisation gives them the capability for holding and storing information, thus making them suitable for use in computer circuitry. Ferroelectric materials may also be *piezoelectric* (see section 13.8).

Ferroelectric behaviour is temperature-dependent and is lost above a critical value of temperature, the Curie temperature. The value of the Curie temperature varies from one material to another and, for example, is 120°C for barium titanate and 570°C for lead niobate.

Some dielectric materials show a different phenomenon, known as anti-ferroelectric behaviour. In anti-ferroelectric materials, such as lead zirconate and sodium niobate, individual dipoles occur but they arrange themselves so as to be antiparallel to adjacent dipoles. As a result, the net spontaneous polarisation is zero.

13.8 | Piezoelectric effect

Some crystalline dielectric materials polarise when strained mechanically and an electrical voltage or field is created. This is termed the *piezoelectric effect*. It is a reversible phenomenon and such a material, when subject to an electric field, will strain. Piezoelectric materials are crystals which do not possess a centre of symmetry. A good example is barium titanate which has a distorted perovskite structure (see section 5.6). Natural quartz is a piezoelectric material and, for many years, was used in the manufacture of transducers. Today, quartz has been largely superseded by inorganic crystalline materials such as barium titanate, lead niobate, lead zirconium titanate and lithium sulphate. Transducer crystals made from these materials are polycrystalline with all the crystals orientated in the same direction. This is achieved by allowing the material to crystallise within a strong electrical field.

There are two piezoelectric coefficients, d and g. The electrical field, ζ (V/m), caused by a stress σ (Pa) is given by: $\zeta = g\sigma$ and the strain, ε, produced by a field is given by: $\varepsilon = d\zeta$. The two coefficients are related: $g = 1/dE$, where E is the modulus of elasticity.

Piezoelectric materials are used in transducers which convert electrical fields into mechanical vibrations, generating acoustic waves, and vice versa.

13.9 | Magnetic behaviour

The motion of an electrical charge creates a magnetic field. Consequently, the motion of electrons within an atom produces a magnetic effect. In Chapter 2 it was stated that electrons are not just moving in orbitals around the nucleus but can also be regarded as spinning about their own axes. The fourth, or spin quantum number, m_s, for an electron is either $+\frac{1}{2}$ or $-\frac{1}{2}$ signifying one of two opposing spin directions. In most instances an element with an even number of electrons has as many electrons spinning in one sense as in the other, while an element with an odd number of electrons will just have one more electron spinning in one direction than in the other. Because of this, most elements possess very weak magnetic properties and are either *paramagnetic*, very weakly attracted to a magnet, or *diamagnetic*, very slightly repelled by a magnet. Four elements, however, possess structures in which several more electrons spin in one direction than in the opposing sense and these elements are strongly attracted by a magnet. This type of behaviour is termed *ferromagnetic* and the elements that behave in this manner are iron, cobalt, nickel and gadolinium. In these ferromagnetic elements each atom behaves as a small magnet as a result of the out-of-balance electron spins.

Within a ferromagnetic material there is a sub-structure known as a *domain* structure. A magnetic domain is a small section of a crystal in which the atomic magnets are aligned parallel to one another. The dimensions of an individual domain are very small and the volume of a domain is of the order of 1×10^{-4} mm^3. In the unmagnetised condition the domains possess random orientation, but when the material is placed in a magnetic field all the domains tend to realign themselves in the direction of the field. The material is now magnetised. When the external magnetic field is removed the domain alignment may remain and the material be permanently magnetised. When a material behaves in this way it is said to be magnetically *hard*. A *soft* magnetic material, on the other hand, will lose its magnetism when the external field is removed, with the domain structure reverting to a random orientation.

When a ferromagnetic material is magnetised the degree of magnetisation induced does not increase linearly with increasing magnetic field strength but follows the curved path OA in Figure 13.6 until magnetic saturation is reached at point A. If the intensity of the magnetising field is reduced the induced magnetisation falls off slowly, but still retains a positive value when the magnetising field has been completely removed. The extent of magnetisation remaining, B_r, is termed the *remanence*. To reduce the magnetisation, B, to zero the direction of the applied magnetic field must be reversed and its magnitude increased to a value H_c. This value, H_c, is termed the *coercivity*. If the magnetising force is increased in a reverse sense beyond the value of H_c the material will be magnetised in the opposite direction until saturation is reached at A'. Reducing the magnetising field from A' and then increasing it in the positive direction will complete the magnetisation, or B–H, curve. For a material to be a good permanent magnet the remanence, B_r, should be close to the saturation value, and the coercivity, H_c, should be large. The area enclosed within the B–H loop is proportional to the energy used to realign the magnetic domains during each magnetisation cycle. This energy is then dissipated as heat. If a ferromagnetic

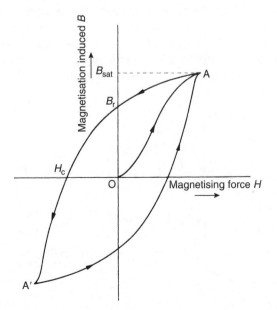

Figure 13.6 *Magnetisation curve for ferromagnetic material.*

material is used as the core of a coil through which an alternating electric current is passed it will be magnetised, first in one direction and then in the other, with every cycle of current. A magnetically soft material with a low remanence and low coercivity is necessary for this type of application. Figure 13.7 shows the difference in shape of B–H curves for two ferromagnetic materials.

A ferromagnetic material that is physically soft is likely to be magnetically soft, and vice versa. In a material such as soft annealed iron the magnetic domains will have random orientation but will readily become aligned on magnetisation. When the external magnetic field is removed, the majority of the domains will return to a state of randomness and the material will have a low remanence value. Any factor that will disturb the regularity of the iron lattice and strengthen the material physically, for example cold working or the introduction of alloying elements, will impede the rotation of domains and movement of domain boundaries and so will increase the coercivity of the material. Highly alloyed materials containing iron, cobalt, nickel, and other elements are used for the manufacture of permanent magnets. These alloys, which can be precipitation hardened to give a very highly dispersed precipitate, are most suitable because domain movement is very greatly restricted.

If a magnetised material is annealed to soften it the domains can revert to a random orientation. The characteristics of some magnetic materials are given in Tables 13.2 and 13.3.

In addition to metals and alloys some complex oxides can also be strongly magnetic. Magnetite, the lodestone of the ancient navigators, is magnetic. Magnetite has the formula Fe_3O_4 but of the three iron atoms per molecule one is divalent and the other two are trivalent. Magnetite is crystalline and possesses the spinel-type structure (refer to section 5.6). These are ceramic mixed oxides of iron and another metal, M, such as copper, cobalt, manganese, nickel or zinc with the general formula $MO.Fe_2O_3$. These materials are *ferrimagnetic* and are generally referred to as *ferrites*. Unlike metals, these materials are insulators and, thus, become suitable materials for use as magnetic cores at ultra-high frequencies as eddy currents cannot be established within them.

The characteristics required in a material for a permanent magnet are high values of both remanence and coercivity, and high power. The power of a magnet is related to the size of the hysteresis loop and is the maximum product of B and H. Such a material is termed magnetically hard. Originally, fully hardened high carbon steels were used for permanent magnet manufacture. These were followed by the development of alloy steels containing cobalt, chromium and tungsten. In the 1930s aluminium–nickel steel was developed in Japan as a magnet material and since then many alloy developments have taken place producing a series of Fe–Al–Ni–Co alloys with additions of copper, niobium and titanium. The coercivities and permeabilities of these alloys can be improved further by *magnetic annealing* which is the slow cooling of a ferromagnetic material from the annealing temperature within a strong magnetic field. This produces a permanent alignment of magnetic domains in a specific crystal direction. Other magnetic alloys include Co–Pt, Cu–Co–Ni and Cu–Ni–Fe materials, while other permanent magnet

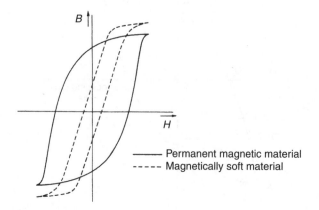

Figure 13.7 B–H *curves for two materials.*

Table 13.2 *Properties of some permanent magnet materials.*

Material	B_r (T)	H_c (A/m)	$B-H_{max}$
1% C steel	1.0	4,000	2,000
Alni (25% Ni; 13% Al; 4% Cu; 58% Fe)	0.56	46,000	9,500
Alnico I (21% Ni; 12% Al; 5% Co; 62% Fe)	0.71	35,000	14,000
Alnico V (14% Ni; 8% Al; 24% Co; 3% Cu; 51% Fe)	1.31	53,000	60,000
Alnico XII (18% Ni; 6% Al; 35% Co; 8% Ti; 33% Fe)	0.58	76,000	16,000
Cunico (50% Cu; 21% Ni; 29% Co)	0.34	53,000	8,000
Cunife (60% Cu; 20% Ni; 20% Fe)	0.54	44,000	15,000
Co_2Sm	0.95	760,000	
Co–Pt (23% Co; 77% Pt)	0.45	216,000	40,000
Ferrite $BaO.Fe_2O_3$	0.4	190,000	
Ferrite $SrO.Fe_2O_3$	0.34	256,000	

Table 13.3 *Properties of some magnetically soft materials.*

Material	B_r (T)	B_{sat} (T)	H_c (A/m)	Applications
Soft iron	1.3	2.16	7	electromagnets
Silicon steel (3% Si)	0.8	1.75	24	transformers
Silicon steel (4.25% Si)	0.7	1.90	12	transformers
Mumetal (74% Ni; 5% Cu; 1% Mn; 20% Fe)	0.6	0.8	4	magnetic shielding
Permalloy (55% Fe; 45% Ni)	1.1	1.6	24	magnetic shielding
Supermalloy (79% Ni; 16% Fe; 5% Mo)	0.5	0.8	160	magnetic shielding
Ferrite $NiO.Fe_2O_3$	0.11	0.27	950	h.f. transformers
Ferrite $MnO.Fe_2O_3/ZnO.Fe_2O_3$	0.14	0.36	50	h.f. transformers

materials are hard ferrites including barium iron oxide and strontium iron oxide materials made by powder compaction and sintering. The properties of some permanent magnet materials are given in Table 13.2.

Materials with low remanence, low coercivity and a very small hysteresis loop are required in components, such as transformer and electromagnet cores, which operate in a constantly alternating magnetic field. Soft iron is very useful for this purpose, as is silicon steel sheet containing 3–5 per cent silicon and virtually no carbon. This latter material, in the fully annealed condition, is used for laminated transformer cores. The silicon content gives the material a high electrical resistivity and thereby reduces eddy current losses. Another group of magnetically soft materials are those with low hysteresis but high permeability and used in communications equipment where a high permeability will utilise fully the small currents. These materials include Mumetal (a Ni–Cu–Mn–Fe alloy), Permalloy (78% Ni/22% Fe) and Supermalloy (79% Ni/16% Fe/5% Mo). Other magnetically soft materials include some ferrites which are suitable for use as cores in high-frequency transformers because of their low hysteresis losses and the fact that eddy currents cannot be generated within them. Some soft ferrites have square hysteresis loops and are used for information storage in computers. The properties of some soft magnetic materials are given in Table 13.3.

If a ferromagnetic material is heated beyond a certain temperature, the Curie temperature, its ferromagnetic properties will be lost and it becomes paramagnetic. The changes that occur at the Curie temperature are reversible, the material becoming ferromagnetic again on cooling through the Curie temperature. The Curie temperature varies from one material to another (Table 13.4), but for pure iron the value is 768°C.

Table 13.4 *The value of Curie temperature for some materials.*

Material	Curie temperature (°C)
Pure iron	768
1% C steel	695
Cobalt	1130
Gadolinium	16
Nickel	358
Alnico I	780
Alnico V	900
Cunico	855
Co_2Sm	725
BaO/Fe_2O_3 ferrite	450

13.10 | The Hall effect

If a current is passed through a thin slab of a conducting material and the conductor is situated in a magnetic field, such that the direction of the field is normal to both the surface of the slab and the current direction, an electric field (hence a potential difference) will be developed across the width of the conductor. This is known as the Hall effect (see Figure 13.8).

Figure 13.8 *The Hall effect – a potential difference is developed between X and Y.*

The magnitude of the electric field, E (V/m), that is developed is given by

$$E = R_H IB$$

where I is the current density in the conductor (A/m²), B is the magnetic field strength (T) and R_H is the Hall coefficient of the material. The Hall effect is observed in metals and in semiconductors. The polarity of the electric field produced in a semiconductor depends on whether the material is p-type or n-type, that is, whether the charge carriers are positive holes or electrons.

The magnitude of the Hall coefficient, R_H, is small in metals, but is very much greater in semiconductor materials and the Hall voltage developed across a crystal may be large enough to be put to practical use. The effect is used in a device for the measurement of magnetic-field strength. A Hall probe may be made small in size, for example 3 mm × 1 mm × 10 mm, and such an instrument may be used for the determination of magnetic-field strengths in confined spaces. One novel use of a Hall effect device is in the grading of ball- and roller-bearings. Bearings moving along an inspection line are magnetised before passing a scanning device consisting of a Hall probe, where the coercivity, H_c, is measured. There is

Figure 13.9 *Inspection line using a Hall device.*

a direct relationship between coercivity and the physical hardness of the bearing. The measuring device feeds information to an electronic sorting-gate so that all bearings with coercivities outside the acceptable limits for the product are separated from the rest. The satisfactory bearings are demagnetised before use. A schematic diagram of such an inspection line is shown in Figure 13.9.

SUMMARY

The electrical properties of materials are dependent upon their electronic structures. Throughout the range of materials, electrical conductivities range from very high, as in many metals, to extremely low, as in ceramics and many polymeric materials.

The electronic structures of metals are such that they possess either unfilled or overlapping valence bands, thus requiring very little energy to excite electrons. Any factor which disturbs the regularity of the crystal lattice will reduce electron mobility and, hence, reduce conductivity. In other words, the same factors that increase physical hardness and strength (see sections 8.6, 8.8 and 8.9) also increase electrical resistivity.

In some materials the energy gap between valence and conduction bands is relatively small and, when excited, some electrons will possess sufficient energy to cross the gap into the conduction band and giving some conductivity. These are the *semiconductors*. A rise in temperature will increase electron energies, permitting more to cross the energy gap and giving increased conductivity.

The energy gap in most ceramics and polymeric materials is large, making it extremely difficult for electrons to cross into the conduction band. These are classed as insulating materials. Some ceramics and glasses are, in part, bonded ionically. There will be a small, but measurable, conductivity if the charged ions can diffuse through the material. At high temperatures diffusion rates increase, thus increasing ionic conductivity.

Many insulating materials polarise to some extent when placed in an electric field and some show the *piezoelectric* effect, that is they polarise, creating an electric field or voltage, when physically strained and vice versa, will strain when subject to an electric field. These materials form the basis of transducers.

When placed in a magnetic field most materials show the characteristics of either diamagnetism or paramagnetism, that is either weakly repelled by or weakly attracted to a magnet. A few metals, including iron, and some metallic oxides with a spinel structure, are ferromagnetic and are strongly attracted by a magnet. When a ferromagnetic material is subject to a magnetic field, important properties are its *remanence* and its *coercivity*. Materials with high values of remanence and coercivity are termed magnetically hard and are suitable for use as permanent magnets.

Questions

13.1 Calculate the resistance of a 1 m length of constantan wire of 0.10 mm diameter, given that the resistivity of constantan at 20°C is $49 \times 10^{-8}\,\Omega\,\mathrm{m}$.

13.2 Determine the resistivity of a nickel silver alloy if a 1.5 m length of wire of 0.15 mm diameter has a measured resistance at 20°C of 18 Ω.

13.3 The resistivity of cadmium at 20°C is $7.5 \times 10^{-8}\,\Omega\mathrm{m}$. Given that α for cadmium is $40 \times 10^{-4}/°\mathrm{C}$, calculate the resistivity at −40°C and 120°C.

13.4 Why is pure silicon classed as a semiconductor? Describe the mechanism of electrical conduction in silicon and state the effect that a change in temperature has on the conductivity.

13.5 Distinguish clearly between a *p*-type and an *n*-type semiconductor. What happens at the interface if the two types of semiconductor are joined to one another? Describe what happens within the joined materials if connected to an external electrical supply, firstly with a forward-bias voltage and secondly with a reverse-bias voltage?

13.6 Explain what is meant by the term *transducer*. Give one industrial application for a transducer device.

13.7 The *B–H* loop for a ferromagnetic material is shown in Figure 13.10. Determine the saturation magnetisation, remanence and coercivity for the material. State, giving reasons, whether or not this material would be suitable for use as a permanent magnet.

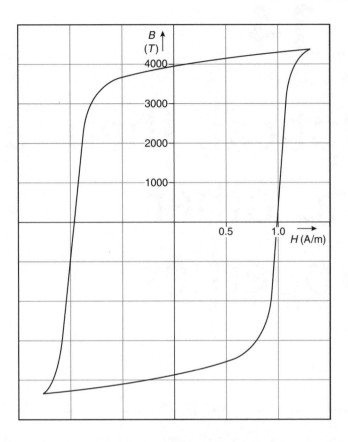

Figure 13.10

13.8 A ferromagnetic material reaches magnetic saturation with induced magnetisation of 1.80 T in a magnetising field of 39 A/m. The remanence of the material is 0.7 T and the coercivity is 11 A/m. Sketch the *B–H* curve for this material. State whether the material is magnetically hard or soft and suggest a possible application for a material with these magnetic characteristics.

13.9 Laminated sheet metal cores are used in small transformers. Why is a low carbon steel containing between 3 and 5 per cent of silicon preferred for this application?

13.10 Why are eddy current losses of lesser importance in ferrimagnetic materials than in ferromagnetic materials?

CHAPTER 14

Optical, thermal and other properties

Why are the optical and thermal properties of materials important?

The optical properties of materials are connected with the interactions between materials and incident electromagnetic waves. When such waves reach a material they are, in part absorbed, in part reflected and in some cases transmitted through the substance. The exact degree of response depends upon both the nature of the material and the frequency of the radiation. There are many manifestations of these interactions. The phenomenon of colour is the result of a selective absorption of some frequencies in the visible light portion of the spectrum. The understanding and utilisation of the interactions between specific types of material and certain ranges of wave energy frequencies has led to the development of lasers, photoconductive devices, solar cells and fibre-optic devices, while the ability of some materials to absorb radio waves in the frequency range used for radar, together with design features which aid the scattering of reflections, are used in the development of so-called stealth aircraft.

When a material is subject to heating, heat energy is absorbed and the temperature of the substance increases. The important thermal properties of materials are heat capacity, thermal expansion and thermal conductivity. Many engineering components are subjected to heating and cooling cycles during their operational life. It is important to know how the particular thermal properties affect such parameters as thermal shock resistance and development of thermal stresses. This information is vital for design engineers when considering materials for applications in which components are subject to major temperature variations.

LEARNING OBJECTIVES

1. Have knowledge of the total electromagnetic spectrum and know Planck's relationship between photon energy and radiation frequency.

2. Know that incident photons can excite electrons in a material and that the subsequent energy emission may occur at different frequencies.

3. Know that absorption, reflection and transmission of electromagnetic waves may occur when they are incident on a material.

4. Know what is meant by *refractive index* and how light can be refracted in passing from one transparent medium into another.

5. Understand the interactions that cause luminescence and photoconductivity and how these phenomena can be utilised in many devices.

6. Know the principle of operation of ruby and semiconductor lasers.

7. Have some knowledge of the development and use of optical fibres.

8. Know what is meant by *specific heat capacity*.

9. Understand the meaning of the term *coefficient of linear expansion*.

10. Know the mechanisms of thermal conduction and how conductivities vary between metals, non-metals and fluids.

11. Know how sound energy is transmitted through materials and what is meant by the term *damping capacity*.

14.1 Optical properties

The optical properties of materials are the results of interactions between incident electromagnetic energy and the material. The electromagnetic spectrum ranges from low-frequency radio waves, through infra-red, visible light and ultra-violet into the very high-frequency range of X- and γ-radiation. When electromagnetic energy is incident on a material the interaction may involve absorption, transmission, reflection, refraction or photoelectric effects such as luminescence. Photoelectric effects are utilised in the design and development of a number of devices of commercial importance including light emitting diodes and lasers. As a prelude to the discussion of optical properties, let us first examine the whole spectrum of electromagnetic energy.

14.2 The energy spectrum

Energy may be transmitted in the form of a wave-like radiation termed electromagnetic radiation. Electromagnetic waves travel through a vacuum at a velocity of $3 \times 10^8 \, \mathrm{m \, s^{-1}}$. This constant, termed the velocity of light is represented by the symbol c. The frequency, ν, and the wavelength, λ, of any radiation are related by

$$\nu\lambda = c$$

The frequencies of radiation range from about $10^4 \, \mathrm{Hz}$ for low-frequency waves up to about $10^{22} \, \mathrm{Hz}$ for some γ-rays. Within this total spectrum the section, termed light, that can be observed by the human eye occupies a small range of frequencies from about $4.3 \times 10^{14} \, \mathrm{Hz}$ for red to about $7.5 \times 10^{14} \, \mathrm{Hz}$ for violet. See Figure 14.1.

Electromagnetic waves can be regarded as a series of energy quanta, or *photons* but these photons do not all possess the same amount of energy. The energy of a photon, E, is related to the frequency of the radiation, ν, by Planck's quantum relationship:

$$E = h\nu$$

Figure 14.1 *The electromagnetic spectrum.*

where h is Planck's constant (6.625×10^{-34} J s). Photon energies, quoted in joules, are extremely small and it is customary instead to quote them using the energy unit, the electron volt, eV ($1\,\text{eV} = 1.602 \times 10^{-19}$ J). The energy of a photon for a low-frequency radio wave of 200 kHz is very small, having a value of about 8×10^{-10} eV (13×10^{-29} J). Photon energies in the visible light sector of the spectrum are about 2 eV (3.2×10^{-19} J) while for X- and γ-radiation with frequencies in excess of 10^{16}, photon energies will range from 100 to 10^6 eV (1.6×10^{-17} J to 1.6×10^{-13} J). The higher-energy photons of light, ultra-violet and X- and γ-radiation can stimulate chemical changes to occur in some materials. Examples of this are the effects on silver salts in photographic emulsions, the stimulation of photosynthesis in plants by light, the stimulation of branching and cross-linking reactions in polymers and the causing of physiological changes in living tissue by X- and γ-radiation.

14.3 | Electron excitation

The build-up of electrons in shells within atoms was discussed in Chapter 2 and it was stated that each electron shell and sub-shell represents a particular energy level. Normally the electrons within an atom occupy the lowest permissible energy levels, the 'ground' state, but it is possible for the energy of an electron to be raised to some higher level. This is termed the excited state and is an unstable state as excited electrons quickly revert to their ground state and in so doing cause energy to be emitted. The energy is emitted as a discrete packet, or *quantum*, also known as a photon. The electrons in atoms may be excited and raised from the ground state in several ways. For example, if crystals of common salt are placed in a flame, the flame colour changes to a characteristic yellow. This happens because the heat energy excites the electrons in the atoms making up the salt. When many of the excited electrons from sodium atoms revert to the ground state, photons with energies of about 2 eV are emitted. From Planck's equation this is equivalent to electromagnetic waves with a frequency of about 5×10^{14} Hz and this is the frequency of light in the yellow section of the spectrum. Atoms of other elements, when heated in this manner, will emit photons of different energies corresponding to waves of different frequencies. For example, strontium will give a red coloration and copper will give a green coloured flame. This effect is utilised and forms the basis of an analytical method termed flame photometry.

The electrons in atoms may be excited in other ways. When an electric arc is struck between two electrodes the electrons in atoms of the electrode materials are excited into photon emission. Similarly, when an electrical discharge is made through a vapour, as in a sodium or mercury vapour lamp, photon emission will occur.

During excitation the electrons of atoms will jump into any of the many unfilled energy levels. When they return to the ground state, with photon emission, they may fall back either in stages or in one jump (see Figure 14.2) and so there will be photon emission at many different discrete energy values, in other words radiation at many different frequencies. An excited element will, therefore, yield a spectrum

Figure 14.2 *Electron excitation and energy emission.*

containing many lines which are characteristic of the element. This fact is used as the basis of an analytical tool and spectroscopy is used for both qualitative and quantitative chemical analysis of substances.

A very high degree of excitation can be achieved if a material is bombarded by a stream of high-energy electrons. The emissions from such an excitation will consist of very-high-energy photons with energies of the order of 10^2 to 10^5 eV, in other words the emission of X-rays with frequencies in the range 10^{17} to 10^{20} Hz. This principle is used for the generation of X-rays.

If the energy input to an atom is large enough it is possible to excite one or more electrons sufficiently to allow them to be freed from the atom. If this happens and an electron moves from an atom completely then it leaves behind a positively charged ion. The minimum quantity of energy required to ionise an atom is termed the *ionisation energy* or *ionisation potential* of the element.

EXAMPLE 14.1

An electron beam is accelerated through a potential difference of 10,000 V to strike a metal target in an X-ray tube. What is the maximum frequency and minimum wavelength of the X-radiation generated?

SOLUTION

The photons emitted from the target metal after electron excitation by electron bombardment will have a range of frequencies from X-radiation, through visible light, to heat energy (infra-red), but the highest frequency emission cannot be greater than the energy of an incident electron, namely 10,000 eV.

Using Planck's expression: $E = h\nu$ then $\nu_{max} = (10,000 \times 1.602 \times 10^{-19})/6.625 \times 10^{-34}$
$$= 2.42 \times 10^{18} \text{ Hz}$$

and from $c = \lambda\nu$ $\lambda_{min} = 2.42 \times 10^{18}/3 \times 10^8 = 0.124 \times 10^{-9} \text{ m} = 0.124 \text{ nm}$

14.4 | Interactions of electromagnetic waves with materials

When a beam of electromagnetic energy reaches a material some of the energy may be absorbed, some may be reflected and some may be transmitted through the material. The exact mix of these interactions will be dependent upon both the frequency of the radiation and the nature of the material. The photons of incident energy when absorbed by the material cause excitation of some electrons into higher-energy levels. When the excited electrons fall back to the ground state energy will be emitted but the photons released may or may not be at the same frequency as those initially absorbed.

EXAMPLE 14.2

Ultra-violet light with photon energies of 4.2 eV strikes crystals of a non-metallic compound. Many photons are absorbed, causing electron excitation. The excited electrons return almost instantaneously to the ground state emitting photons with energies of 2.5 eV and 1.7 eV. Calculate the frequencies of the radiation emitted and comment on the significance of the results.

SOLUTION

Using the Planck relationship $E = h\nu$ the frequencies of the emitted radiation can be calculated.

For photons of 2.5 eV $\quad \nu = (2.5 \times 1.602 \times 10^{-19})/6.625 \times 10^{-34} = 6.05 \times 10^{14}\,\text{Hz}$

and for photons of 1.7 eV $\nu = (1.7 \times 1.602 \times 10^{-19})/6.625 \times 10^{-34} = 4.11 \times 10^{14}\,\text{Hz}$

The 2.5 eV photons are in the green sector of the visible-light portion of the spectrum while the 1.7 eV photons are in the infra-red sector. (The frequency range of green light is in the range from about 5.5×10^{14} to about 6.2×10^{14}. Frequencies lower than about 4.3×10^{14} are in the infra-red sector.)

The implication of this result is that the crystals show the phenomenon of *luminescence* when subjected to ultra-violet light and fluoresce with a green colour. In addition some infra-red energy is emitted. (Refer also to section 14.7.)

In metals the outer shell, or valency, electrons are in partially filled conduction bands (see section 13.2). When photons of comparatively low energy are incident upon a metal the photon energy can be transferred to valency electrons, raising the energies of these electrons within the conduction band. When this occurs the incident energy is absorbed by the material and not transmitted through it. Consequently, metals are opaque to radio waves, infra-red, visible and ultra-violet light, but the high-energy photons of X- and γ-rays can penetrate through them. There is some absorption of this high-energy radiation by the metal and so the amount of X- or γ-radiation that is transmitted through the metal will be inversely proportional to its thickness. The absorptive power of a metal for X- and γ-radiation is proportional to its density. This is why a metal of high density, such as lead, is used as a radiation shielding material. Both X- and γ-radiation are used for the radiography of metals and other materials. The penetrating power of the radiation increases as the radiation frequency is increased and the deeply penetrating very-high-frequency rays are referred to as 'hard' X-rays. There is some reflection of X-rays from atomic nuclei and diffraction of radiation by the regularly spaced planes of crystals. Because of this fact X-rays can be used for the investigation and analysis of crystal structures (see section 5.8).

Metal surfaces will largely reflect, rather than absorb, incident photons of visible light although there may be some selective absorption. For example, when visible white light strikes the surface of metals such as aluminium or silver most of it is reflected giving the metal a silvery white appearance but when light hits the surface of copper there is absorption of some of the higher frequencies and thus the copper shows a reddish appearance.

Dielectric materials possess filled electron energy bands and forbidden energy gaps. There cannot be transference of energy from photons to electrons in these materials in the same way as in metals and so, generally, these materials do not absorb electromagnetic radiation but transmit it. This is the reason that good radio reception can occur within a building. The construction materials, brick, stone and glass are largely transparent to radio waves. However, radio reception within a metal-bodied car tends to be difficult unless an external aerial is fitted. Almost all non-metallic materials are transparent to light in thin section. The fact that most are opaque in thick section is due largely to repeated reflection and refraction of light at internal surfaces within the bulk material.

EXAMPLE 14.3

What steps can be taken in the design of military aircraft to make them difficult to detect by radar?

SOLUTION

Radar transmissions are, generally, within the frequency range of 10^9 to 10^{11} Hz and detection of air-craft relies on a portion of the radio wave energy transmitted being reflected and picked up by a receiver situated close to the transmitter. Metals reflect radio waves and most aircraft produce a strong reflected signal which can be picked up by the radar receiver. A so-called stealth aircraft is invisible, or very nearly so, to radar. One major step is to replace, wherever possible, metals with materials which are transparent or absorbent to waves in the appropriate frequency range. Fibre-reinforced plastic com-posite materials fit this requirement. The plastic matrix will be transparent while the carbon reinforcing fibres in the composite will absorb some of the radiation. The wing structure, for example, can be strengthened by filling the space between the surfaces with a honeycomb structure. Repeated reflec-tions of wave energy within such a honeycomb structure will aid the dissipation of the incident radar energy. Many ceramic materials absorb radar energy but these materials tend to be dense and would give a serious weight penalty. They may be used, however, in small quantities in areas adjacent to the aircraft engines because of their good high-temperature properties. Finally, because the use of metals cannot be entirely eliminated, the aircraft design should be such that its contours would result in much of the incident radar energy being reflected at angles away from the transmitter rather than directly back to the radar site. By adopting these practices the radar signature of quite large aircraft can be reduced to very small values.

The explanation of absorption and transmission given above is a very simple one and does not cover the situation fully. Non-metallic materials are not fully transparent to all types of electromagnetic radi-ation and many materials show absorption peaks at particular frequencies. For example, window glass is opaque to the infra-red section of the spectrum. Also certain materials show definite colours because they strongly absorb certain frequencies within the visible part of the energy spectrum.

These colour effects are caused by the presence of certain types of crystal defects, termed *colour centres*. Impurity ions in certain materials may create donor or acceptor energy levels which cause selec-tive absorption of photons of particular wavelengths. It is this selective absorption which is responsible for the varied colour of materials. Many types of defect (colour centre) may occur, the best known example being the F-centre which is an electron trapped on an anion vacancy and this has a series of energy levels available to it. The energy required to transfer from one level to another falls within the visible part of the electromagnetic spectrum giving the characteristic colour, which depends on the host crystal. F-centres in sodium chloride, NaCl, give the characteristic greenish-yellow colour.

Semiconductor materials possess filled valence bands but with forbidden energy gaps of a much smaller size than those in insulators. Low-frequency radiation is not absorbed by semiconductors but if the frequency of the incident radiation is such that the photon energy is equal to the magnitude of the energy gap, energy may be transferred from photons to electrons. In other words, the incident energy is absorbed and electrons are excited into the conduction band. This gives rise to the effect termed *photoconduction* since the conductivity of the material increases in a similar manner to that caused by raising the temperature of the semiconductor.

14.5 Refraction and polarisation

A transparent material possesses a refractive index, μ, and this is defined as being the ratio of the velocity of light in a vacuum, c, to the velocity of light in the material, v:

$$\mu = \frac{c}{v}$$

There is a relationship between the refractive index of a material and the dielectric constant, K, for the material:

$$\mu^2 = K \quad \text{or} \quad \mu = \sqrt{K}$$

The velocity of light and the refractive index of a material are functions of frequency. An example of this is the separation of white light into the colour spectrum by a glass prism; this effect is called *dispersion*.

There are very many different types of glass available for optical purposes and they possess refractive indices in the range 1.5 to 1.7. A material with a refractive index of this order of magnitude is ideal for use in lenses and similar devices. Refractive index is not the only important property, and an optical material must also possess a high transmission value. The transmission ability of optical glass is about 98 per cent (that is, 98 per cent of incident light would be transmitted through a section of 10 mm thickness). A number of plastic materials are also suitable for optical applications, possessing similar refractive indices to glass and with transmission abilities of 90 per cent or more.

A glass is isotropic, but crystalline materials are not. If crystallisation takes place in a glass or glassy polymer the crystals will possess a different refractive index from the glass. Reflections at internal interfaces between glass and crystal regions will lead to the material becoming translucent.

Because of the anisotropy of crystals light will have a different velocity in the directions of the different crystal axes (except in cubic crystals where the velocity of light is the same for each major axis). Crystals in which the velocity, and hence the refractive index, differs from one crystal axis to another are termed *birefringent*. Birefringent crystals are doubly refracting. A beam of light incident upon the crystal is split into two beams. There is another interesting effect in that the two beams are polarised at right angles to each other. Some crystals have the effect of absorbing one polarised component of the split beam to a very much greater extent than the other resulting in the emergent beam being of polarised light. This is the effect when light passes through Polaroid.

The phenomenon of birefringence is used in the micro-examination of rocks and minerals under polarised light and it can also be used to study fibre orientation in fibre reinforcd composites.

An isotropic material, such as glassy polystyrene, becomes anisotropic to some extent when it is physically strained. When physically strained the material becomes birefringent. This stress-induced birefringence is termed the *photoelastic* effect. In photoelastic stress-analysis a model of a component, say a chain hook, is made from clear plastic. The model is loaded and, because of the birefringence, interference fringes can be observed within the model. The fringe pattern will indicate the stress distribution within the model and the number of fringes observed is in direct proportion to the magnitude of the stress induced.

14.6 Diffraction

Electromagnetic radiation can suffer diffraction. Consider a beam of monochromatic light passing through a narrow slit of width w (Figure 14.3(a)). The wavefront diverges as it emerges from the slit such that the light intensity on a plane surface set at some distance from the slit varies in the manner

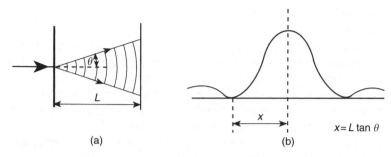

Figure 14.3 *(a) Diffracted beam from narrow slit. (b) Intensity of light on screen, first minimum at half-cone angle θ, sin θ = λ/w.*

shown in Figure 14.3(b). The first point of minimum intensity is at an angle θ from the incident beam where $\theta = \sin^{-1} \lambda/w$, λ being the wavelength. If, instead of passing through a single slit, the light beam is passed through a finely ruled grating on glass then diffraction at each of the narrow slits in the grating produces regions where emergent coherent beams overlap producing a series of fringes, narrow bright lines separated by dark zones. The angles, θ_n, at which these fringes occur is given by $n\lambda = d \sin \theta_n$ where n is an integer and d is the separation distance of the lines on the grating. The principle of optical diffraction fringe formation can be used as the basis of an extremely sensitive system of linear measurement. The regularly spaced planes of atoms in crystals act with respect to X-radiation or an electron beam in a similar manner to a diffraction grating and light. X-ray and electron beam diffraction patterns are used in the identification and measurement of crystal structures (see section 5.8).

EXAMPLE 14.4

A beam of monochromatic radiation is diffracted by a grating with 100 lines per millimetre and a series of diffraction fringes with an angular separation of 3.21° is observed. Determine the wavelength of the incident radiation.

SOLUTION

Use the relationship $n\lambda = d\sin \theta$ where d is the line spacing of the grating, namely 0.01 mm. So $\lambda = 0.01 \times \sin 3.21° = 0.560 \times 10^{-3}$ mm = 560 nm

14.7 | Luminescence

Luminescence is the general term covering the phenomenon of light emission from a material as a result of excitation of some type. During excitation, some of the valency electrons may be excited into a higher energy band, as mentioned in section 14.3. When the excited electrons drop back to the ground state, or into some intermediate energy level, energy will be emitted from the material. An emission which occurs only for the duration of the excitation is termed *fluorescence*. A delayed emission which persists for a period of time after excitation has ceased is termed *phosphorescence*. The intensity of a phosphorescent emission at some time t after excitation has ceased, I_t, decays with time according to the expression: $I_t = I_0 \exp(-t/\tau)$ where I_0 is the initial intensity and τ is a constant for the material termed the *relaxation time*. Excitation of a luminescent material may be caused in one of several ways, by incident photons, termed *photoluminescence*, by an incident electron beam, termed *cathodoluminescence*, by the effects of an alternating electric field, termed *electroluminescence*, or by certain chemical reactions, termed *chemiluminescence*.

Phosphorescent materials, or *phosphors*, are used to coat the inside surface of television and radar tubes, and excitation to cause luminescence is due to the moving electron beam within the tube which traverses the screen very rapidly. The phosphors used in a television tube have a short relaxation time otherwise the images produced would tend to overlap one another. Phosphors with longer relaxation times to give an afterglow of several seconds are selected for the coatings of search radar display tubes. The inside surface of fluorescent light tubes is coated with a layer of a phosphorescent material. Ultra-violet light is generated within the tube and this causes the coating to luminesce giving a white or soft rose-white light emission. Some phosphors will absorb high-energy photons of X- or γ-radiation and emit photons of lower energy within the visible light sector of the spectrum. This effect renders the X-radiation visible on a fluorescent screen and is used for viewing radiography in real-time rather than having to expose and develop radiographic film. Electroluminescence can occur in some semi-conductor devices and this is the principle utilised in light-emitting diodes (LEDs). A voltage applied to the diode in the forward-bias direction will bring about the recombination of electrons and holes at the junction with the emission of photons within the visible sector of the spectrum. Generally the light emitted is green or red, the colour being determined by the exact nature of the semiconducting material.

14.8 Lasers

The word *laser* is an acronym (**L**ight **A**mplification by **S**timulated **E**mission of **R**adiation). There are several types of laser including solid-state, gas and semiconductor lasers.

Let us look at the principle involved in the solid state ruby laser. Ruby is a crystalline form of alumina, Al_2O_3, which contains a very small percentage of Cr_2O_3. It is the presence of Cr^{3+} ions which gives ruby its characteristic red colour. The ruby laser is composed of a single-crystal ruby rod, the end faces of which are plane, parallel and polished. One end face of the rod is silvered to be fully reflective while the opposite end is silvered to make it partially reflective and to permit partial transmission of light. The stimulation is provided by a xenon flash lamp. (See Figure 14.4.)

Photons produced by the illuminating flash lamp excite electrons from the Cr^{3+} ions into higher energy states. When some of these excited electrons fall back to the ground state, photons are emitted at a frequency of 4.32×10^{-14} Hz (a red light) and this emission triggers a cascade effect causing the other excited electrons to decay to the ground state with photon emission. As these photons travel through the ruby crystal they cause further electron excitation and energy emission. The light beam travels backwards and forwards through the rod, repeatedly reflected from each end and so stimulating more emissions and increasing in intensity. Eventually, a very-high-intensity short-duration pulse of light is transmitted through the partially silvered end of the rod. This light beam is fully coherent, that is all the photon waves are completely in phase. Continued external stimulation from the flash

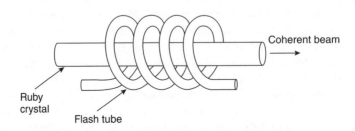

Figure 14.4 *Principle of a solid-state laser.*

lamp produces a high-intensity pulsed laser beam. Another crystal material used in solid-state lasers is yttrium aluminium garnet (YAG) doped with neodymium.

Some semiconductor materials can be used as lasers. The magnitude of the energy gap between the valence and conduction bands must be such that it corresponds to photon energies within the visible sector, that is energies within the range 1.77 to 3.10 eV. Gallium arsenide, GaAs, is a semiconductor which meets this criterion and is used in lasers. As with the ruby laser, one end surface of the GaAs is made fully reflective and the other end is made partially reflective. When a voltage is applied across the GaAs semiconductor electrons are given sufficient energy to move across the energy gap from the valence band into the conduction band, leaving behind an electron hole. Recombination of an electron–hole pair is accompanied by the emission of a photon with an energy equivalent to the energy gap value. This photon will stimulate the recombination of another electron–hole pair, and so on. The emitted photons will be reflected back and forth within the GaAs and all will be of the same wavelength and in phase, creating a coherent beam of light. The energy gap value of GaAs is such that the light emitted is red. Some of the beam will leave the semiconductor, being transmitted through the partially silvered surface, while the remainder will remain within the GaAs and stimulate further electron–hole recombination. The maintenance of a steady voltage across the semiconductor will ensure the continual 'freeing' of electrons into the conduction band, hence a continuing supply of electron–hole pairs. This type of semiconductor laser generates a continuous coherent laser beam, rather than a pulsed beam. The typical construction of a semiconductor laser is shown in Figure 14.5. The active GaAs element is sandwiched between other semiconductor materials which are selected to have higher energy gap values and a different refractive index so that photon emission is confined to the GaAs. The different refractive index value ensures that the generated laser light will be reflected at the interfaces and be confined within the gallium arsenide. A suitable material to use in conjunction with GaAs is gallium aluminium arsenide, GaAlAs.

Figure 14.5 *Schematic diagram showing layout of a semiconductor laser.*

Some of the newer semiconductor materials, such as GaN, have energy gap values of the order of 3 eV and emit higher frequency radiation at the blue end of the visible spectrum.

Lasers have many applications. High-power solid-state and gas lasers are used for welding, cutting and the piercing of holes in materials. Lower-power lasers are used for making accurate measurements and in surgery, while low-power semiconductor lasers are used in CD players, in bar-code readers and in communications applications. Current CD technology, using GaAs red light lasers, allows for the storage of approximately 650 Mb of information on a CD. This is the maximum density that can be read by a red light laser. A move to the use of higher-frequency blue light lasers would permit the pit size on a CD to be made smaller and thus increase the amount of information that can be stored. The next generation of CD equipment, now being developed, should allow the density of information stored to be approximately four times greater than that possible with current equipment using a red light laser.

14.9 | Optical fibres

In recent years there has been a revolution in the communications industry with the introduction of fibre optic technology. In modern telecommunications information, in digitised form, is transmitted by light pulses through an optical fibre rather than as electrical signals transmitted along copper wire. Signal transmission by photons is quicker than the electronic conduction of a signal along wires and, in addition, the amount of information that can be transmitted by means of light signals along a fibre optic guide is very much greater than is the case for electrical signals along wire. Consider the transmission of, say, a telephone conversation, which has already been converted from speech into an electronic signal. This electronic signal is first put into digital form and then this digitised signal is converted into an optical signal. Usually, the converter is a semiconductor laser, the output being a series of pulses of monochromatic coherent laser light fed into the optical fibre cable. At the receiving end the optical signal is converted back into an electrical signal which can then be decoded from the digitised form. For very long transmission distances it will be necessary to use repeaters, which amplify the signals, at intervals along the transmission line.

What are the characteristics required in an optical fibre material? Firstly, it must possess extremely good transparency and secondly, there should be negligible leakage of light from the fibre. The diameters of optical fibres are generally within the range from 5 to 100 μm and these fibres are produced by drawing. The carefully drawn fibres are virtually free from defects and have a very high degree of transparency and are also strong. The means by which light leakage from the fibre can be prevented is to ensure that the light beam is always totally reflected whenever it strikes the internal surface of the fibre. This can be achieved by cladding the light transmitting fibre core with a sheath which has a lower refractive index than the fibre core (Figure 14.6(a)). An alternative method is to modify the composition of the core fibre by infusing into its surface a small amount of an impurity oxide which will reduce the refractive index (Figure 14.6(b)). The refractive index of the fibre will vary across the section, being highest in the centre and reducing gradually towards the circumference. This type of fibre is capable of transmitting information with less signal distortion than the first-mentioned type. Like the other type, the fibre also possesses a cladding layer. In both cases the optical fibres will be given a protective coating to protect from damage.

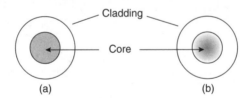

Figure 14.6 *Cross-sections of optical fibres: (a) normal type, (b) doped type.*

14.10 | Heat capacity

The thermal properties of a material are influenced by atomic vibration. At any temperature above absolute zero, the atoms that make up a material are in a state of continual movement in the form of a vibration. The amplitude and frequency of the vibration varies with temperature. The vibration of atoms produces elastic waves, termed *phonons*, and the energy of a phonon, E, can be related to temperature as $E = kT$, where k is Boltzmann's constant (1.38×10^{-23} J/K) and T is temperature (K). The material gains or loses heat by gaining or losing phonons. The *heat capacity* of a material is the amount of energy required to raise the temperature of one kilomole of a substance through one degree K.

Table 14.1 *Specific heats (at 15 °C) for some substances.*

Substance	Specific heat (kJ/kg K)	Substance	Specific heat (kJ/kg K)
Lead	0.13	Aluminium	0.88
Silver	0.23	Magnesium	1.03
Copper	0.38	Polystyrene	1.20
Nickel	0.44	Nylon (PA66)	1.70
Mild steel	0.44	Natural rubber	1.90
Plate glass	0.67	Polyethylene (LDPE)	1.90
Quartz	0.73	Polyethylene (HDPE)	2.31
Silica glass	0.84	Water	4.186

For most practical purposes the quantity used is the *specific heat capacity*, or *specific heat*, which is the amount of heat energy required to raise the temperature of unit mass of substance through one degree K. The specific heat of a substance is not constant and increases as the temperature rises. The values of specific heat for a number of substances is given in Table 14.1. There are two specific heats quoted for gases, the specific heat at constant pressure, C_p, and that at constant volume, C_v, C_p being the larger of the two. The ratio of the specific heats, C_p/C_v, for gases is denoted by the symbol γ.

14.11 | Thermal expansion

When a material is heated the vibrations of its constituent atoms increase. This is accompanied by an increase in the physical dimensions of the material, that is, expansion. The change in length, δl, that occurs when a material of length l_0 is heated from a temperature T_0 to T_1 is given by

$$\delta l = l_0 \alpha (T_1 - T_0)$$

where α is the linear expansion coefficient for the material. The value of α, the expansion coefficient, is small for many ceramic materials and approximately half of that for many metals. Polymer materials tend to have very high thermal expansion coefficients. The thermal expansion coefficients for some materials are given in Table 14.2.

Table 14.2 *Thermal expansion coefficients for some materials.*

Material	α (K^{-1})	Material	α (K^{-1})
Pyroceram	0.4×10^{-6}	Nickel	12.8×10^{-6}
Silica glass	0.54×10^{-6}	Copper	16.2×10^{-6}
Invar	0.9×10^{-6}	Silver	18.8×10^{-6}
Diamond	1.2×10^{-6}	Aluminium	21.6×10^{-6}
Graphite	5.4×10^{-6}	Magnesium	25.2×10^{-6}
Quartz (parallel to axis)	7.5×10^{-6}	Polystyrene	63×10^{-6}
Quartz (normal to axis)	13.7×10^{-6}	Phenol-formaldehyde	72×10^{-6}
Plate glass	9.0×10^{-6}	Natural rubber	80×10^{-6}
Building brick	9.0×10^{-6}	Nylon (PA66)	100×10^{-6}
Mild steel	11.0×10^{-6}	Polyethylene (HDPE)	120×10^{-6}
Concrete	12.6×10^{-6}	Polyethylene (LDPE)	200×10^{-6}

If the material is constrained during heating or cooling so that dimensional change is restricted, thermal stresses can be established (see section 7.4) and this could lead to distortion or cracking. The risk of cracking is pronounced during the rapid heating or cooling of non-ductile materials such as ceramics and glasses. The magnitude of the stress developed in a non-ductile material during rapid heating or cooling is proportional to α and, so, the susceptibility to crack formation through thermal shock is much less in materials with very low expansion coefficients. LAS (LiO–Al$_2$O$_3$–SiO$_2$) glass ceramics with extremely low expansion coefficients have been developed for applications in which a high resistance to thermal shock is required. The proprietary name for one such glass ceramic is Pyroceram. Some alloys with very low expansion coefficients have been developed, the best known being 'Invar', a steel containing 36 per cent Ni and 0.2 per cent C.

When a material melts, generally there is a marked increase in volume accompanying the transition from solid to liquid. In the case of metals the expansion at the melting point is of the order of 6 per cent. There are certain exceptions to this, a notable one being antimony, which contracts on melting. Use is made of this peculiarity in the manufacture of type metals. These contain some antimony to prevent solidification shrinkage thus ensuring that a good clean type-face is obtained. Ice is another solid that contracts on melting, with the reverse process, expansion during freezing, giving rise to burst water-pipes.

EXAMPLE 14.5

Of what diameter should a sand mould be made for the casting of a wheel in grey cast iron, if the diameter of the cast wheel is to be 1.0 m? Take the linear coefficient of expansion of cast iron as 1.2×10^{-5} and assume that the liquid iron solidifies at a temperature of 1100 °C.

SOLUTION

After the iron has solidified at 1100 °C it will contract within the mould as it cools to room temperature. The mould, therefore, should have a diameter of greater than 1.0 m.

Assuming room temperature to be 15 °C, the cast wheel will cool through (1100–15) = 1085 °C after it has solidified.

The dimensional change will be $\alpha \Delta T$, where α is the expansion coefficient and ΔT is the change in temperature:

$$\alpha \Delta T = (1.2 \times 10^{-5}) \times 1085 = 0.013$$

For a final casting diameter of 1.0 m the diameter of the mould would need to be 1.013 m.

14.12 | Thermal conductivity

If one end of a bar of material is heated the atoms in that portion will be stimulated into increased vibration. Energy will then be transmitted along the bar from the hot to the cool end in an attempt to reduce the energy at the heated end. For an electrically insulating material the energy transference along the bar will be a result of atomic vibrations only. The transmission of heat energy along the bar will be a function of the separation distance between atoms, the density of the material and the specific heat. The thermal conductivity of a glass or liquid will be less than the conductivity of the crystalline version of the same material, since the atoms in a glass or liquid are not in a fully ordered array and interatomic separation distances are variable.

In metals, with valence electrons existing as a free cloud or gas, energy may be transmitted along a bar through the medium of electron movement as well as by means of atomic vibration. Metals are, in general, good conductors of both electricity and heat. The class of materials known as electrical semiconductors are also reasonably good thermal conductors because heating the material can cause some electrons to be excited and cross the energy gap into the conduction band. The excited electrons may then take part in the transmission of heat energy from one part of the material to another.

Table 14.3 *Thermal conductivities of some materials at 20 °C.*

Material	Conductivity (W/m K)	Material	Conductivity (W/m K)
Silver	420	Building brick	0.63
Copper	395	Polyethylene	0.33
Aluminium	240	Nylon (PA66)	0.25
Magnesium	168	Phenol-formaldehyde	0.25
Graphite	125	Natural rubber	0.13
Nickel	88	Asbestos	0.12
Mild steel	51	Polystyrene	0.08
Lead	33.5	Cork	0.046
Quartz	5.1	Sand	0.042
Silica glass	1.25	Hydrogen*	0.17
Concrete	1.05	Oxygen*	0.025
Mica	0.84	Nitrogen*	0.024
Plate glass	0.75	Perfect vacuum	zero

* At atmospheric pressure.

The thermal conductivities of some materials are listed in Table 14.3. The units of thermal conductivity are watts per metre per degree Kelvin (W/m K).

It will be noticed that the thermal conductivities of gases, apart from hydrogen, are very much lower than the conductivities of other non-metallic materials. Hence the reason for most thermal insulating materials being porous. Cavity-wall construction and double glazing in buildings are other examples in which use is made of the insulating properties of gases. There is no thermal conduction at all in a vacuum, since there are no atoms present to vibrate and transmit energy. This principle is used in the vacuum flask in which there is an evacuated space between two layers of glass.

A quantity related to conductivity is *thermal diffusivity*. The thermal diffusivity, D_T, is given by: $D_T = k/(\rho C_p)$ where k is thermal conductivity, ρ is density and C_p is the specific heat capacity. The units of diffusivity are $m^2 s^{-1}$. Thermal diffusivity can be regarded as the diffusion coefficient for thermal energy. The resistance of refractory ceramic materials to spalling tends to increase as the diffusivity increases.

The rate of heat transmission through a material is given by the expression $kA(T_1 - T_2)/t$ where k is the thermal conductivity, A is the surface area of the material, T_1 and T_2 are the temperatures of each surface of the material and t is the thickness.

EXAMPLE 14.6

Determine the heat loss through a brick wall 4 m by 3 m of 0.25 m thickness if the inner surface is maintained at 20 °C and the outer surface temperature is 5 °C. The thermal conductivity of brick is 0.5 W/m K.

SOLUTION

$$\text{Heat loss} = \frac{kA(T_1 - T_2)}{t}$$
$$= \frac{0.5 \times 4 \times 3 \times 15}{0.25}$$
$$= 360\,\text{W}$$

14.13 | Thermal shock

The fracture of a material when it is subjected to rapid heating or cooling is termed *failure by thermal shock*. Rapid heating or cooling of a material may lead to severe temperature gradients being established. These, in turn, will give varying amounts of expansion or contraction and cause thermal stresses to be set up. If the magnitude of the thermal stress exceeds the fracture strength of the material failure will occur. The susceptibility of a material to failure by thermal shock is affected by a number of factors and these are given below:

1. The rate of heating or cooling. The more rapid the rate of temperature change the greater will be the chance of establishing major temperature gradients in the material.

2. The thermal conductivity. Materials with a high value of thermal conductivity will transfer heat energy quickly, reducing thermal gradients.

3. The expansion coefficient. The extent of dimensional changes will be less in materials possessing low thermal expansion coefficients and this will reduce the level of thermal stresses.

4. The fracture strength. Materials with low fracture strengths will show a greater tendency to failure by thermal shock. It should be remembered that the fracture strength is influenced by the presence of small cracks and other flaws. Many ceramics and glasses invariably contain small defects.

5. The elastic constants. Materials with low elastic moduli values may be strained by a relatively large extent before fracture occurs.

Thermal shock is not normally of great consequence in metals as they possess high conductivities, low expansion coefficients and also tend to be ductile. Brittle metals may fail through thermal shock and this is a possibility when some steels are quenched very rapidly from a high temperature to bring about a martensitic transformation. Many ceramics and glasses possess low resistance to thermal shock. They tend to have low thermal conductivities and low tensile fracture strengths. One measure of the resistance of a material to failure by this type of failure is the thermal shock parameter. This is

Thermal shock parameter $= \sigma_f k / E \alpha$

where σ_f is the fracture strength, k is the thermal conductivity, E is the modulus of elasticity and α is the coefficient of thermal expansion.

14.14 | Sound absorption and damping

Sound energy is transmitted through a medium both as a compression wave and as a shear wave. Sound cannot travel through a vacuum because there are no atoms available to vibrate and transmit energy. The velocity of sound within a material is related to both the density and the elastic moduli. The velocity, v_1, of a longitudinal compression wave within a material is given by

$$v_1 = \sqrt{\frac{K + (\frac{4}{3})G}{\rho}}$$

where K is the bulk modulus of elasticity, G is the modulus of rigidity and ρ is the density of the material. The shear wave velocity, v_s, is given by

$$v_s \sqrt{\frac{G}{\rho}}$$

The elastic constants of a material can be determined from a knowledge of the velocity of sound through it. All the elastic constants are related to one another (see section 7.2) and so values for the modulus of elasticity, E, and Poisson's ratio, v, can also be determined in this way.

When sound travels through a medium some of the energy is absorbed by the material. This absorption, or damping of vibrations, is due to the internal friction of the material. The internal friction of a material is a function of the degree of crystallinity and the nature of the interatomic bonds within the substance, but it is also greatly affected by the presence of discontinuities within the material, such as impurities, and cracks. These serve to increase the internal friction. The internal friction of metals, and a number of non-metallic crystalline solids, is generally low. The internal friction of brick and concrete is about 100 times greater than that of many metals, while that of timber is about 1000 times greater than that of many metals. Most thermoplastic materials have internal friction values similar to or greater than those for timber.

For sound-insulation purposes the most effective type of material is one of low density and high internal friction. Foamed plastics are an excellent example of this. When sound energy reaches an interface between two different media there will be some reflection and this reflection may approach 100 per cent at an air/dense-solid interface. It is for this reason that double glazing is such an effective sound insulator since there are four air/glass interfaces between the external noise source and the ear, with a low density medium, air, between the two glass layers.

A material of high internal friction will have great ability to attenuate vibrations, and vice versa. This ability is termed the damping capacity of the material. The damping capacity, ζ, is given by

$$\zeta = \frac{1}{n}$$

where n is the number of vibrations in free attenuation from an amplitude A to an amplitude A/e, that is, $0.368\,A$. For the manufacture of a bell a material of very low damping capacity is necessary, hence the use of the copper alloys, brass and bronze. Cast iron is a metal with a very high damping capacity because of the presence of graphite flakes and small flaws in its structure. It is therefore a most suitable material for components where vibrations should be absent, as in a lathe bed. For antivibration mountings, very high internal friction materials such as rubber and cork are used.

14.15 | Interface effects and surface tension

The properties at a surface are different from those within the heart of a material. An atom or molecule in the centre of a liquid is completely surrounded by other molecules and it is attracted equally by all its neighbours. A molecule at the surface, however, will be attracted inwards by the molecules within the bulk material. The net result of this inward pull is that the surface of the liquid tends to contract to the smallest possible area. It is for this reason that liquid droplets and gas bubbles within liquids tend to be spherical. The surface behaves as if it is in a state of tension, and this phenomenon is termed *surface tension*. The term signifies a separation between two substances, for example liquid–gas, and the value of surface tension is dependent on both substances present. The recorded values of surface tensions for liquids are generally for a liquid–air interface. The surface tension value may well be different for the interface between the liquid and some other gas. The interface surface tension, γ, has the units of newtons per metre (N/m) and the values of γ for some liquids are given in Table 14.4.

Table 14.4 *Surface tensions of some liquids in air (at 20 °C).*

Liquid	γ (N/m)	Liquid	γ (N/m)
Ethyl alcohol	0.022	Benzene	0.029
Acetone	0.024	Water	0.073
Carbon tetrachloride	0.027	Mercury	0.485

It will be noticed that the surface tension of water is high compared with organic liquids, even though it is very low compared with mercury.

Whether a liquid wets the surface of a solid or not depends on the values of interface surface tensions for the three interfaces involved, namely liquid–air (γ_{la}), liquid–solid (γ_{ls}) and air–solid (γ_{as}). Consider a liquid resting on the surface of a solid. At equilibrium there will be an angle of contact, θ. The angle θ is measured within the liquid (see Figure 14.7). At equilibrium, the interface forces will be in balance, and this can be represented by the statement

$$\gamma_{as} = \gamma_{ls} + \gamma_{la} \cos \theta$$

If $\gamma_{as} > \gamma_{ls}$ then $\cos \theta$ is positive and $\theta < 90°$; if $\gamma_{as} < \gamma_{ls}$ then $\cos \theta$ is negative and $\theta > 90°$. When the contact angle, θ, is less than 90° the liquid is said to wet the solid surface. This is the case with water on glass. If the contact angle, θ, is greater than 90°, as with mercury on glass, the liquid does not wet the solid surface.

Figure 14.7 *(a) Liquid–solid contact angle. (b) Liquid wetting solid. (c) Non-wetting liquid.*

The air–liquid interface for liquid in a tube will be curved. When the liquid wets the tube the surface will be concave upwards but with non-wetting liquids the surface will be convex upwards. The curved liquid surface is termed *meniscus* and readers are probably familiar with meniscus curves for both water and mercury in glass tubes.

The pressure of liquid close to a curved surface is different from that close to a plane surface by an amount, δP, given by

$$\delta P = \frac{2\gamma}{r}$$

where r is the radius of curvature. When the liquid surface is convex δP is positive and when the surface is concave δP is negative. When an open-ended fine-bore, or capillary, tube is partially immersed in water the meniscus within the tube will be concave with a small radius r, compared with the flat surface of water outside the tube. The pressure just below the meniscus in the tube will be less than the pressure just below the plane water-surface in the main vessel by an amount $2\gamma/r$ and so the water level within the tube will rise through some height h, where h is the height of a column of water having a pressure of $2\gamma/r$ at its base. (The level of a liquid with an angle of contact, θ, greater than 90°, for example mercury, will be depressed.) This phenomenon is termed *capillary attraction* or *capillarity*. In general, liquids of low surface tension will have low contact angles and show high capillarity.

This phenomenon is utilised in *liquid phase sintering* of ceramics. In this form of sintering of ceramic powder compacts one of the constituents melts and wets the surface of the solid particles. The resulting capillary pressure can cause movement of some particles to give a better densification and elimination of porosity in the sintered product.

The surface of a solid–gas interface is in a state of tension in a similar manner to a liquid–gas surface. In a strongly bonded crystalline solid the atoms in the surface layers have their bonds only partially satisfied and they can react or join with atoms from the other medium, providing an adsorbed film on the surface of the solid. In some cases the forces of attraction between atoms on the solid and the adsorbed layer are of the van der Waals' type but in other cases there may be ionic or covalent type bonding between the solid surface and atoms from the surrounding medium. This type of effect occurs between clean metal surfaces and oxygen and is an important first stage in the surface oxidation of metals. When there are ionic or covalent bonds formed at the surface the phenomenon is called chemisorption.

14.16 | Friction and lubrication

The interface between two solids is of considerable importance in engineering. The surface of a solid is normally far from smooth when examined on a microscale. A lathe-turned or flat-ground surface will possess innumerable 'hills and valleys'. The height difference between hills and valleys for a fine-ground surface is of the order of 1–3 μm. When two metal surfaces are put together the true contact area, because of surface irregularities, will be very much less than the apparent area of contact. In consequence the stresses involved at points of contact may be sufficiently high to cause localised deformation and possibly welding. If a shearing force is applied in such a way as to cause sliding between the contacting surfaces, sliding motion will be resisted. The resistance to sliding is termed friction and is due to both the roughness of the surfaces and localised welding that may have occurred at points of contact.

When dry solid surfaces are in sliding contact the work done in overcoming friction is converted to heat energy and there will be considerable abrasion between them. This leads to severe wear of the surfaces. The surface temperature build-up could be sufficient to cause the two surfaces to weld over an extensive area. This is termed *seizure*. The extent of the friction between surfaces is considerably reduced by the presence of a lubricant, that is, a fluid layer interposed between the two solid surfaces.

In normal fluid lubrication, also termed hydrodynamic lubrication, the two solid surfaces are kept apart by means of a fluid layer and provided that the separation distance between the surfaces does not fall below about 0.1 μm almost any fluid that does not react with the solids may be used. Most lubricating fluids are mineral oils but high-pressure air, as a fluid, is used in low-friction air bearings. Most lubricating oils contain boundary-layer additives. Such an additive is composed of polarised molecules which become strongly adsorbed on a metal surface. When conditions are such that hydrodynamic lubrication may break down, the adsorbed layers will still keep the solid surfaces apart. Organic fatty acids are used as boundary-layer additives in oils, as is also molybdenum disulphide.

SUMMARY

The optical properties of materials are based on the interaction between electromagnetic energy and the material. The incident energy may be absorbed, reflected or transmitted through the material, the exact effect depending upon the frequency of the radiation and the nature of the material, with metals being opaque to all electromagnetic energy except the very-high-energy photons of X- and γ-radiation. Photons of incident energy can excite electrons in the material into higher energy states and energy is emitted as photons when the excited electrons fall back to the ground state. The emitted energy may be at the same, or at a lower, frequency than the incident photons. When the emissions fall within the visible range the phenomenon is termed *luminescence*. The principle of luminescence finds applications in television tubes, LEDs and lasers.

In all materials at temperatures above absolute zero the atoms are vibrating and the amplitude of vibration and, hence, the vibrational energy, increases with temperature. The vibration of atoms produces elastic waves, termed *phonons*, and a material gains or loses heat by gaining or losing phonons. One important thermal property is *heat capacity*, or *specific heat*, this being the amount of energy required to raise the temperature of a given mass of substance through 1 degree K. Materials expand when heated, the amount of expansion being directly proportional to the temperature increase. The proportionality constant for each material is termed its *coefficient of linear expansion*. Ceramics and metals of high melting point, namely materials with high bond strengths, tend to have low expansion coefficients while metals of low melting point and polymer materials have high coefficients. Heat energy is conducted through materials by both electrons and phonons. Metals, with many free electrons, are good thermal conductors but in ceramics and polymer materials, with few free electrons, the mechanism of conduction is mainly through phonons and conductivities are low. Thermal stresses may be set up if there are thermal gradients within the material or the material is subject to constraint when heated or cooled. Materials may fracture through thermal shock if the rate of heating or cooling is high. The propensity for this type of fracture is greatest in materials of low conductivity and those which are inherently brittle.

Sound energy is transmitted through a material and the velocity of sound within the material is a function of both density and the elastic moduli. Some of the sound energy is absorbed by the material, this absorption being due to internal friction. The internal friction is a function of interatomic bond type and structure and it is also greatly increased by the presence of flaws or other discontinuities. Materials with high internal friction tend to attenuate vibrational energy rapidly. This ability to attenuate vibrations can be quantified and is termed the *damping capacity*.

Questions

14.1 Upon excitation an element emits radiation at a wavelength of 532 nm. Determine the energy transition involved.

14.2 The energy change when an excited electron within an element falls back to ground state is 18 eV. What type of radiation is emitted and at what frequency?

14.3 Thermionic electrons emitted from the filament of an X-ray tube are accelerated to the target by a tube voltage of 50 kV. Calculate the maximum frequency of the X-radiation produced when high-energy electrons strike the target material.

14.4 A beam of monochromatic light of wavelength 560 nm passes normally through a narrow slit of width 0.08 mm to a screen set 1.5 m from the slit. At what distance from the axis of the beam is the first point of minimum light intensity?

14.5 Red monochromatic light of wavelength 600 nm falls on a diffraction grating with 1×10^5 lines per metre. Calculate the angular separation between fringes.

14.6 Give a brief description of the principle of operation of (a) a ruby laser, and (b) a semiconductor laser.

14.7 Estimate the percentage reduction in dimensions which occurs when a magnesium casting cools from its melting temperature (649 °C) to 15 °C. What size should the mould cavity be for casting a wheel of 0.55 m diameter in magnesium? (α for magnesium is 25.2×10^{-6}.)

14.8 A room has two external walls measuring 3.3 m × 2.4 m and 5.4 m × 2.4 m. The walls are of brick of 0.3 m thickness and each contains a window of 3 mm thick glass. The window dimensions are 1.4 m × 2.0 m and 2.0 m × 3.0 m respectively. Calculate the amount of heat energy lost through these walls per day if the room is maintained at 21 °C and the outer surface temperature is 7 °C. Assume that the thermal conductivity of brick is 0.6 W/m K and that of window glass is 0.75 W/m K.

PART **III**

THE MATERIALS OF ENGINEERING

CHAPTER 15

Non-ferrous metals and alloys

What are the non-ferrous metals?

Usually metals and alloys are divided into two major categories, ferrous and non-ferrous. All the metallic elements other than iron (some 70 in number) are referred to as non-ferrous. This division is not as unbalanced as might at first appear because iron occupies a special position among metals owing to its availability, comparatively low cost and the very useful range of alloys, *steels* and *cast irons*, formed when iron is alloyed with carbon and other elements. Of the total world consumption of metallic materials more than 90 per cent is made up of steels and cast irons. Of all the non-ferrous metals only eight are produced in relatively large quantities, these being aluminium, copper, lead, magnesium, nickel, tin, titanium and zinc, and the main characteristics and applications of these are discussed in this chapter. Many other metals play an important role in engineering, both as metals in their own right and as alloying elements. The part played by some of these, including beryllium, cobalt, molybdenum and tungsten, as alloying elements in some non-ferrous alloys and in steels, is mentioned in this and in the following chapter, but detailed coverage of these metals is outside the scope of this book.

Some non-ferrous metals are used in the pure form but, in many cases, the metals are alloyed. The main reasons for alloying are:

(a) to strengthen by forming solid solutions,

(b) to form alloys which respond to strengthening heat treatments,

(c) to improve castability or workability, or

(d) to improve some other characteristic, for example, corrosion resistance.

15.1 | Introduction

Only a few non-ferrous metals are produced in relatively large quantity, but these form the basis of a wide range of useful engineering materials. In the latter half of the twentieth century the newer materials, polymers, industrial ceramics and composites, were used in many applications which, traditionally, had been the preserve of metals. This process of substitution in design is taking place still, for example, in the continued trend to include an ever increasing volume of polymer materials in cars. However, there are still many areas in which metals offer the only viable solution and metallic materials will continue to be used by engineers.

15.2 | Aluminium

Aluminium is one of the most abundant elements in the earth's crust, but, owing to its high affinity for oxygen, it cannot be reduced to the metallic state by reduction with carbon or carbon monoxide, as is the case with many metallic oxides. Aluminium was first produced in the early part of the nineteenth century by reduction with potassium. This was a very expensive process, and until 1850 the price of aluminium was about £250 ($437.5) per kilogramme and, because of its great cost at that time, some European royal houses used aluminium cutlery at state banquets! In 1886, Charles Hall devised a new method for the relatively cheap production of aluminium. The Hall method involved the electrolysis of a molten solution of alumina in the mineral cryolite (Na_3AlF_6) at a temperature of about 950°C.

The economic ore of aluminium is bauxite, a hydrated alumina, and this mineral normally has iron oxides and other minerals associated with it. The bauxite has to be purified before electrolytic reduction can take place. Pure alumina is obtained by the Bayer process, in which bauxite is digested with caustic soda. Aluminium hydroxide is then precipitated from solution and calcined to give a pure alumina, about 99.6 per cent Al_2O_3, which can be electrolysed by the Hall process. The aluminium produced by the Hall process is of a purity ranging from 99.5 to 99.8 per cent, with the major impurities being iron, silicon, and manganese. Aluminium of purities up to 99.9 per cent can be produced by refining molten aluminium from the reduction cell in another electrolytic cell, the Hoopes process, with a molten electrolyte.

The principal properties of pure aluminium (99.9 per cent) are given below:

Melting point	660°C
Crystal structure	face-centred cubic
Density	2.70×10^3 kg/m^3
Young's modulus, E	70.5 GPa
Tensile strength	45 MPa
Electrical resistivity	2.66×10^{-8} Ω m at 20°C
Corrosion resistance	very good

Aluminium possesses a number of properties that make it an extremely useful engineering material. Its good corrosion resistance and low density make it particularly suitable for applications in the field of transportation – land, sea, and air. The pure metal has a very low strength but, by alloying with other elements, the strength may be increased considerably to give alloys with very good strength to weight ratios. Young's modulus for aluminium is only one-third of the value for steels, however, and so the specific modulus (E/density) is almost identical for both materials. Like all metals with a face-centred cubic crystal structure, aluminium is highly ductile and can be shaped easily by a wide variety of methods. The good electrical conductivity of the metal makes it suitable for many applications in the electrical industry.

Aluminium has an extremely high affinity for oxygen and any fresh metal surface will oxidise rapidly. The surface oxide layer that forms is only a few atoms in thickness, but it is impermeable to oxygen and protects the surface from further attack. The corrosion resistance of aluminium is due to the presence of this oxide coating. The corrosion resistance of aluminium may be improved further by anodising. In an anodising operation the aluminium is made the anode of an electrolytic cell. Oxygen is liberated at the anode and the thickness of the protective aluminium oxide film is increased. The anodic coating is cellular in structure and immersion of the anodised material in boiling water will seal this cellular structure and give a hard, smooth surface. An anodised aluminium part with a cellular oxide layer may be dipped in a vat of dye, prior to the sealing operation. Dye will be absorbed into the pores to produce the type of colour-anodised finishes widely used for decorative purposes.

High-purity aluminium (purities of 99.5 per cent aluminium, and above) is too weak to be used for many purposes and its main applications are for use as corrosion-resistant linings for vats and other vessels in the food and chemical industries. The material commonly termed pure aluminium is, in reality, an aluminium–iron alloy. Commercial-purity aluminium alloy is made by adding up to 0.5 per cent of iron to the metal produced in the Hall process. This small iron addition gives a considerable increase in strength (see Table 15.1), although there is some reduction in ductility and corrosion resistance. Commercial-purity aluminium is used extensively, and accounts for about nine-tenths of aluminium product sales. Its principal applications are: thin foils for packaging, thermal insulation, and capacitor manufacture; cold-rolled sheet for panelling; and the production of kitchenware, rod and wire for electrical transmission cables, and armature windings.

15.3 | Aluminium alloys

Aluminium may be alloyed with a number of elements to produce a series of useful engineering materials. The alloys of aluminium may be subdivided into those that are not heat-treatable, and those that are heat-treatable. Alloys in the first group may be strengthened by cold working operations, and the only type of heat treatment that may be given is an annealing treatment to soften a work hardened sample. The heat-treatable alloys, on the other hand, may be strengthened by giving them a special type of heat treatment.

The non-heat-treatable alloys are those of the Al–Mn, Al–Mg and Al–Si systems. Manganese enters solid solution in aluminium to a limited extent, and the only commercial alloy in this system is that containing 1.25 per cent of manganese. This alloy is some 15 per cent stronger, and has a slightly better corrosion resistance, than commercial purity aluminium. Its principal uses are for the production of domestic utensils, including pressure cookers and, in the form of corrugated sheet, for the cladding of buildings.

Magnesium enters solid solution in aluminium to a limited extent and gives a considerable solid solution strengthening effect. Magnesium also improves the resistance to corrosion, particularly to corrosion in a marine environment. A range of aluminium–magnesium alloys is marketed, with magnesium contents of up to 10 per cent. The various alloys in this range are used for the production of pressed and deep-drawn articles, where high strength is required, for example, boxes, crates and milk churns, and as a constructional material for ship building. The majority of aluminium–magnesium alloys are manufactured using commercial-purity aluminium as a base, and so contain a range of impurity elements. Some alloys are made from very-high-purity ingredients and these can be polished to give a bright, attractive surface with a very high reflectivity. These highly pure alloys are used for decorative trim applications as a substitute for chromium-plated mild steel.

Silicon forms a eutectic with aluminium, and it also increases the fluidity of the molten aluminium. Alloys containing silicon are very suitable for the manufacture of both sand and die castings. The most widely used casting alloys are those containing 10–13 per cent of silicon, namely those compositions close to the eutectic composition. Normally, the eutectic mixture would be of fairly coarse structure, rendering the alloy brittle. The melt is treated immediately before casting by making a very small addition (about 0.01 per cent) of sodium. This is termed modification of the alloy and it produces a fine-grained eutectic structure. The modification treatment also causes a shift in the position of the eutectic point (Figure 15.1).

Figure 15.1 *Aluminium–silicon phase diagram. The addition of 0.01 per cent sodium modifies the diagram to position shown by broken lines.*

The principal heat-treatable alloys are those of the following systems: Al–Cu, Al–Cu–Ni, Al–Mg–Si, Al–Zn–Cu and Al–Li. These various alloy types respond to age hardening, or precipitation hardening, treatments (see also sections 8.9 and 12.6). This type of hardening is possible in alloy systems that show partial solid solubility, with a wide difference between the low-temperature and high-temperature solubility limits. The aluminium-rich end of the aluminium–copper phase diagram is of this type (Figure 15.2). Aluminium can hold 0.2 per cent, by weight, of copper in solid solution at room temperature, but it can hold up to 5.7 per cent, by weight, of copper in solution at 548°C. If an alloy containing, say, 5 per cent, by weight, of copper (2 atomic per cent of copper) is slowly cooled from high temperature, the microstructure will be as shown in Figure 15.2(b), with the compound $CuAl_2$ precipitated at the aluminium grain boundaries. If, alternatively, the alloy is heated to a high temperature and rapidly cooled, the structure at room temperature will consist of a supersaturated solid solution of copper in aluminium (Figure 15.2(c)). Such a solution is unstable, or metastable, and there will be a tendency for copper atoms to diffuse through the aluminium space lattice prior to precipitation of $CuAl_2$ particles. The concentration of copper in the compound $CuAl_2$ is 53.5 per cent, by weight, or 33.3 atomic per cent, so that considerable diffusion of copper must take place in order to produce copper-rich areas within the aluminium space lattice, as a prerequisite of the precipitation of $CuAl_2$

Figure 15.2 *(a) Aluminium-rich end of Al–Cu phase diagram. (b) 5% Cu alloy slow-cooled from 550°C showing CuAl₂ precipitate. (c) 5% Cu alloy rapidly cooled from 550°C – supersaturated solid solution.*

particles. As this pre-precipitation diffusion occurs there will be a considerable increase in the amount of lattice strain within the aluminium lattice, and, consequently, there will be an increase of hardness and strength with time following a solution heat treatment. With aluminium–copper alloys, this diffusion process and consequent hardening, will occur slowly at room temperature but in some alloy systems hardening does not take place spontaneously at room temperature. In these latter cases, the alloy has to be heated to some comparatively low temperature before the diffusion process will occur. Owing to there being insufficient thermal energy available at room temperature, the diffusion of copper in aluminium–copper alloys does not occur to the extent necessary for precipitation of $CuAl_2$ unless the temperature is raised. An increase in temperature will bring about an increase in copper diffusion rates, and if the temperature is high enough true precipitation of $CuAl_2$ particles will begin to take place. Once true precipitation occurs, the strain in the aluminium lattice will be relieved and softening of the alloy will take place. This softening is termed *overageing*. Figure 12.6 shows the relationship for hardness with ageing temperature and time for an alloy of aluminium and copper. When an alloy slowly hardens with time on being kept at ordinary temperatures after receiving a solution heat treatment, the alloy is termed an age-hardening alloy. If, after solution heat treatment, it is necessary to heat an alloy to some temperature to cause diffusion and hardening to occur, this second heat treatment is termed precipitation heat treatment, even though the treatment is stopped before true precipitation and over-ageing can occur.

The age-hardening alloys of aluminium and copper are used for structural applications, particularly in aircraft construction. The presence of copper reduces the corrosion resistance of the material, and sheet and plate forms of the alloy are often clad with a thin layer of high-purity aluminium to improve the corrosion resistance. This thin layer is applied at an early stage in the production sequence by sandwiching the cast ingot between two plates of high-purity aluminium prior to hot rolling. Pressure welding of the layers occurs during hot rolling.

In the ternary aluminium–copper–nickel alloys, both copper and nickel contribute to the age-hardening process. The presence of nickel also increases the temperatures at which overageing occurs. Consequently, these alloys may be used for service at elevated temperatures. The structure of the supersonic aircraft Concorde is made of this type of alloy. These alloys are also used for the manufacture of cylinder heads, cylinder blocks and pistons for internal combustion engines.

Magnesium and silicon form an intermetallic compound, Mg_2Si, and alloys of aluminium that contain magnesium and silicon in the correct proportions are effectively binary alloys of aluminium and the compound Mg_2Si. The aluminium-rich end of the Al–Mg_2Si phase diagram is similar in form to the Al–Cu diagram and heat-treatable alloys result. These alloys are relatively strong, and have good corrosion resistance. They are used in quantity for structural applications, and for panelling, in road and rail vehicles, and transit containers.

Alloys of the aluminium–zinc–copper system possess the highest strengths of any aluminium alloys. These alloys are difficult to work, and they are also subject to stress corrosion. They are mainly used in aircraft construction.

Considerable development work has been carried out in recent years on aluminium–lithium alloys. The density of lithium is extremely low (534 kg/m^3) and an addition of lithium will provide an alloy with a lower density than other aluminium alloys. The tensile strengths obtainable in heat-treated alloys containing 2–2.5 per cent lithium are comparable with those of the high-strength Al–Cu and Al–Zn alloys (see Table 15.1) and this is coupled with a modulus of elasticity, E, value which is about 10 per cent higher than other aluminium alloys. Weight saving in aircraft construction can offer significant reductions in operating costs and these alloys, with densities of around 2600 kg/m^3 compared with densities of about 3000 kg/m^3 for other high-strength aluminium alloys, show considerable potential for use in the aerospace industries.

An alloy of aluminium with beryllium is being developed. Beryllium has a very low density and a high modulus of elasticity. An Al–Be alloy with a density of 2100 kg/m^3 (compare with that of Al – 2700 kg/m^3) and a modulus of elasticity of 192 GPa (Al – 70 GPa) has been produced. This alloy with its very high stiffness/weight ratio offers significant potential for aerospace applications. The alloy is being used already in some space vehicles and in the braking systems of Formula 1 cars. Currently, its use for structural components in advanced military aircraft is being evaluated.

The alloys of the 2xxx series (Al–Cu), the 7xxx series (Al–Zn) and the 8xxx series (Al–Li) are all listed in Table 15.1 as materials used in aircraft construction. Because of their high cost the Al–Li alloys are not used greatly in civil aircraft construction and their main usage is in high performance military aircraft. In much civil aircraft construction both Al–Zn and Al–Cu alloys are used. The Al–Zn alloys, while possessing greater strength than the Al–Cu alloys, are more costly to produce and also have a lower fracture toughness. The design solution for the wings of some Airbus models utilises both types of alloy to produce a wing with the required strength characteristics, but at a minimum cost. The under surface of the wing, which is subjected to the highest tensile loading, is made of an Al–Cu alloy, usually alloy 2024. This alloy possesses good 'damage tolerance', that is its ability to resist crack propagation. The upper wing surface is made of a high strength Al–Zn alloy of the 7xxx series. Figure 15.3 shows a wing for the Airbus A340 constructed from both types of aluminium alloy. The wing dimensions are: length 37 m, maximum thickness 2.1 m.

In addition to the various alloys mentioned above, alloys of aluminium with tin form an important group of bearing materials (see section 15.12).

Table 15.1 *Properties of some aluminium alloys.*

Alloy number		Approximate composition	Condition*	Tensile strength (MPa)	Type of product	Uses
British	American					
		99.99% Al	O	45	Sheet, strip	Linings for vessels
1080	1060	99.8% Al	O	75	Sheet, strip	in food and chemical plant
1200	1200	99.0% Al	O	90	Sheet, strip, wire, extruded sections	Lightly stressed and decorative panelling, wire and bus-bars, foil for packaging, kitchen and other hollow-ware
			H14	120		
			H18	150		
3103	3103	Al + 1.75% Mn	O	110	Sheet, strip, extruded sections	Hollow-ware, roofing, panelling, scaffolding tubing
			H14	160		
			H18	210		
5251	5052	Al + 2% Mg	O	180 ⎫	Sheet, plate, tubes and extrusions	Stronger deep-drawn articles; ship and small boat construction and other marine applications
			H24	250 ⎪		
5154A	5454	Al + 3.5% Mg	O	240 ⎬		
			H24	300 ⎪		
5056A	5056A	Al + 5% Mg	O	280 ⎪		
			H24	335 ⎭		
LM6	S12C	Al + 12% Si	M	180†	Sand and die castings	Excellent casting alloy
			M	210‡		
6082	6082	Al + 0.9% Mg; 1% Si; 0.7% Mn	T4	220	Sheet, forgings, extrusions	Structural components for road and rail transport vehicles
			T6	320		
2014	2014	Al + 4.5% Cu; 0.5% Mg; 0.8% Mn	T4	440 ⎫	Sheet, forgings, extrusions, tubing	Highly stressed parts in aircraft construction and general engineering
			T6	480 ⎪		
2024	2024	Al + 4.5% Cu; 1.5% Mg; 0.6% Mn	T3	480 ⎭		
2L95		Al + 5.6% Zn; 1.6% Cu; 2.5% Mg	T6	500	Plate, rod and bar	Aircraft construction
L160						
7075	7075	Al + 7% Zn; 1.75% Cu; 2% Mg	T6	620	Sheet and extrusions	
2090	2090	Al + 2.2% Li; 2.7% Cu; 0.12% Zr	T6	580 ⎫	Sheet, plate	Aircraft construction
8090	8090	Al + 2.5% Li; 1.3% Cu; 0.7% Mg	T6	495 ⎭		

Key: * The symbols for condition are: O = annealed; M = as cast; H14 = partly work hardened; H18 = fully work hardened; H24 = work hardened and partially annealed; T3 = solution treated, cold worked and aged; T4 = solution treated and naturally aged; T6 = solution treated and precipitation hardened. † Sand cast. ‡ Chill or die cast.

Figure 15.3 *Wing for Airbus A340. (Courtesy of Airbus UK)*

EXAMPLE 15.1

What feature is required in the phase diagram for an alloy system to render the possibility of creating alloys that will respond to a precipitation strengthening process? Give one example of an alloy system in which precipitation hardening alloys occur.

SOLUTION

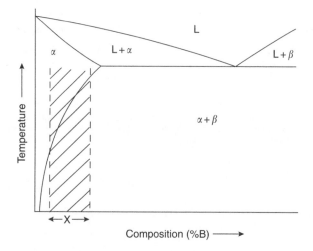

Figure 15.4

If the phase diagram for an alloy system shows partial solid solubility with a relatively wide difference between low and high temperature solubility limits, as in Figure 15.4, then there is the possibility that alloys in the composition range marked X in the diagram may respond to precipitation hardening. The aluminium–copper system is one example of such a system and alloys containing 4–4.5 per cent of copper will respond to precipitation strengthening.

EXAMPLE 15.2

State, giving reasons, the type of aluminium alloy that would be suitable for the manufacture of each of the following articles:

(a) a saucepan and lid,

(b) fuselage skin panels for a civil aircraft,

(c) the body and impeller for a small water pump, and

(d) structural members for the body frame of a bus.

SOLUTION

(a) An aluminium saucepan would be manufactured from aluminium sheet by either deep drawing or spinning, while the lid would be made by spinning. A requirement for these processes is that the material be soft and highly ductile. These conditions would be met by either alloy 1200 (commercial purity aluminium) or by 3103 (Al–Mn).

(b) Aircraft fuselage skin panels are highly stressed. The material used needs to have a high strength. Alloys from the 2xxx (Al–Cu–Mg) or 7xxx (Al–Zn–Mg) series would be suitable for this application. The latter alloys are the stronger but also the more expensive. They would be fully strengthened by precipitation heat treatment after they have been formed to shape.

(c) Because of their shape the pump body and impeller would be best made by a die casting process. Alloy LM6 (Al–Si) would be ideal for this purpose.

(d) Weight saving, while important in road transport vehicles, is not as critical as in aircraft. An Al–Mg–Si alloy, such as 6082, would be suitable for this application. It is less strong than the 2xxx series alloys but is also considerably less costly.

15.4 | Copper

Copper is one of the oldest metals known to man, and, together with tin in the alloy bronze, has been worked for over 5000 years. Copper and bronze were used for both utilitarian and decorative purposes in the early civilisations. Today, there are very many extremely useful copper alloys but, owing to the present high price of the metal, copper and its alloys are being replaced by cheaper materials, such as aluminium and plastics, in many applications.

The principal ores of copper are generally complex mixtures of copper and iron sulphides, and copper production from the ore may either be carried out by a smelting process, or by electrolytic extraction. Smelting operations are carried out in a blast furnace, or reverberatory furnace, and roasted sulphide ore, coke, and flux, make up the charge. A *matte*, which is a high-density double sulphide of iron and copper, collects at the base of the furnace and the gangue, or waste material, forms a lower-density slag, which separates easily from the matte. The matte is reduced to copper in a converter, similar to a Bessemer converter. The copper produced in this way is highly oxidised, and is termed blister copper. It is of the order of 99 per cent pure and it may be refined to higher purities by either electrolytic, or fire refining, techniques.

For electrolytic extraction, the finely divided ore is digested with dilute sulphuric acid. Copper compounds are dissolved, but most other minerals in the ore are not attacked. The solution obtained is then purified and electrolysed. The copper produced in this way is termed cathode copper, and may be of purity ranging from 99.2 to 99.7 per cent copper. Cathode copper may be refined either electrolytically, or in a reverberatory furnace.

The fire refining of blister or cathode copper is an interesting process. The copper is melted under oxidising conditions in a reverberatory furnace. Most impurities are oxidised and form a slag which floats on the surface of the copper. The bath of highly oxidised copper is then subjected to reducing conditions to reduce the oxygen content, the resulting product being termed *tough pitch* copper with a residual oxygen content of about 0.03 per cent. Formerly the necessary reducing conditions were obtained by *poling* the liquid bath. Poles, or trunks, of green wood were immersed in the bath of copper and the stream of reducing gases evolved removed the bulk of the oxygen from the copper. Poling is stopped when the oxygen content of the bath is about 0.03 per cent. Over-poling would cause the copper to become contaminated with excess hydrogen. The purity of tough pitch copper is not less than 99.85 per cent. Tough pitch copper is unsuitable for welding owing to the oxygen content of the material. It is also unsuitable for applications requiring high electrical conductivity. Copper is commonly deoxidised by adding phosphorus, in the form of a copper–phosphorus alloy, but this is not a suitable method for the production of electrical grades of copper. For high-conductivity purposes tough pitch, or electrolytically refined copper is remelted under controlled conditions and deoxidised, if necessary, with lithium. Some grades of copper contain up to 0.5 per cent of arsenic. Arsenical copper, which has improved tensile properties and an improved resistance to oxidation at high temperatures, is made by adding up to 0.5 per cent of arsenic to tough pitch or deoxidised copper.

The principal properties of pure copper are given below:

Melting point	1083°C
Crystal structure	face-centred cubic
Density	$8.93 \times 10^3 \, \text{kg/m}^3$
Young's modulus, E	122.5 GPa
Tensile strength	220 MPa
Electrical resistivity	$1.67 \times 10^{-8} \, \Omega \, \text{m}$ at 20°C
Corrosion resistance	very good

The microstructure of deoxidised copper is shown in Figure 15.5. This material has been annealed after having been cold worked and the structure shows an equiaxial grain structure. Some of the crystals show annealing twins, a characteristic of face-centred cubic metals.

High-conductivity copper is virtually indispensable for many applications in electrical and electronic systems and it is used for the manufacture of power cables, message cables for telephones and computer networks, printed circuit boards, connectors, etc. Figure 15.6 is a section of a submarine power cable linking the grid systems of Britain and France, while Figure 15.7 shows the interior of a modern telephone with many copper components. Other grades of pure copper are used in the form of sheet for architectural cladding and for shaping into articles such as domestic water tanks and vessels used in the food and chemical industries, and in the form of tubing for heat exchangers and in domestic water and heating installations.

Copper may be alloyed with a number of elements to provide a range of useful alloys. The important alloy systems are

copper–zinc (brasses)
copper–tin (zinc) (bronzes and gun-metals)

Figure 15.5 *Photomicrograph of cold worked and annealed deoxidised copper (etched) ×200. Some of the equiaxial crystals show pronounced annealing twins. α brass and other α-phase copper alloys have a similar microstructure. (Courtesy of The Copper Development Association)*

Figure 15.6 *Section of a submarine power cable. (Courtesy of The Copper Development Association)*

copper–aluminium (aluminium bronzes)
copper–nickel (cupronickels)

Small additions of beryllium or chromium to copper give high-strength alloys, a small addition of cadmium gives a significant increase in strength with little loss of electrical conductivity, while an addition of tellurium to copper gives an alloy with very good machineability.

Figure 15.7 *Interior of a modern telephone containing copper coils, cables, connectors and printed circuit boards. (Courtesy of The Copper Development Association)*

15.5 | Copper alloys

The phase diagrams for the binary systems copper–zinc, copper–tin, and copper–aluminium are complex and a number of intermediate phases are formed in each system (Figure 15.8). In each case, the α (copper-rich) phase possesses a face-centred cubic structure, and all the α alloys are ductile and suitable for cold working. The β phase in each system is body-centred cubic in structure, and alloys containing this phase cannot be readily cold worked. They are used in the cast form, and for making hot worked products. (In the Cu–Sn and Cu–Al systems it will be noticed that β phase exists only at elevated temperatures.) The γ phase (and δ phase in bronzes) possesses a complex cubic crystal structure, and is a very hard and brittle phase.

Brasses

The range of alloys known as the brasses are, primarily, alloys of copper and zinc. The addition of zinc to copper brings about an increase in strength because the zinc enters into solid solution in the copper. An unusual feature is that the ductility of the alloys also increases with dissolved zinc content, reaching a maximum value at a zinc content of 30 per cent. The ductility of the material then decreases with further zinc content until at a composition of 37 per cent of zinc, the limit of the α phase, the ductility of the alloy is of similar value to that of pure copper. The α brasses are widely used for deep drawing, spinning, and tube manufacture, because of their high ductilities. The alloy containing 30 per cent of zinc, 70/30 brass, is known as *cartridge brass* and is used for the manufacture of cartridge and shell cases because of its high ductility. For general purpose cold working applications, an α brass containing 35 per cent of zinc is normally used. The increased zinc content causes a reduction in ductility, but the alloy is somewhat cheaper than 70/30 brass. The properties of α brasses may be modified by the addition of small amounts of other elements. Small additions of tin or aluminium improve resistance to marine corrosion, while an addition of 1 or 2 per cent of lead improves the machineability of the alloy. Cold worked α brasses are prone to a form of stress corrosion termed *season cracking*, unless they are properly stress relieved before being put into service.

Figure 15.8 *Binary phase diagrams for copper alloys: (a) copper–zinc; (b) copper–tin; (c) copper–aluminium.*

The ($\alpha + \beta$) Alloys are generally formed into shapes by hot working or by direct casting. The most common composition for an ($\alpha + \beta$) brass is 60/40, that is, 40 per cent of zinc, and this alloy is termed *Muntz metal* (refer to Table 15.2 for properties of brasses). Small additions of other elements may be made to ($\alpha + \beta$) brasses in order to improve properties. A small addition of lead will impart good free-machining characteristics to the alloy, while the addition of elements such as manganese, tin, aluminium, iron, or nickel will considerably increase the strength of the material. The alloy containing an addition of manganese was formerly known as *manganese bronze*, but the improved strength alloys are more properly termed *high-tensile brasses*. The structures of high-tensile brasses are generally ($\alpha + \beta$), but may sometimes be wholly β. They are used in the form of castings and hot worked products. A typical application is for the manufacture of ships' screws and marine fittings. β brass is used as the basis for some brazing alloys. No commercial alloys are made containing more than 50 per cent of zinc, as above this value the presence of γ phase in the structure would embrittle the alloy. (Photomicrographs of some brass structures are shown in Figures 15.9, 15.10 and 15.11.)

Molten brasses do not require deoxidation before casting. This is because zinc possesses a very low boiling point (907°C) and the volatilisation of some zinc from the melt expels oxygen, and also the heavy zinc vapour forms a protective blanket over the liquid metal surface.

Table 15.2   *Properties of copper and some copper alloys.*

Alloy	Approximate composition	Condition*	Tensile strength (MPa)	Type of product	Uses
Pure copper	99.95% Cu	O	220	Sheet, strip, wire	High-conductivity electrical applications
	99.85% Cu	H	350	All wrought forms	Chemical plant, deep drawn and spun articles
Arsenical copper	99.25% Cu; 0.5% As	O	220	All wrought forms	Retains strength at elevated temperatures. Heat exchangers, steam pipes
		H	360		
Brasses					
Gilding metal	90% Cu; 10% Zn	O	280	Sheet, strip, wire	Imitation jewellery and decorative work
		H	510		
Cartridge brass	70% Cu; 30% Zn	O	325	Sheet, strip	High-ductility brass for deep drawing
		H	700		
General cold working brass	65% Cu; 35% Zn	O	340	Sheet, strip, extrusions	General-purpose cold working alloy
		H	700		
Muntz metal	60% Cu; 40% Zn	M	375	Hot rolled plate and extrusions	Condenser and heat exchanger plates
High tensile brass	35% Zn; 2% Mn; 2% Al; 2% Fe; balance Cu	M	600	Cast and hot worked forms	Ships' screws, rudders and high-tensile applications
Bronzes	95.5% Cu; 3% Sn; 1.5% Zn	O	325	Strip	British 'copper' coinage
		H	725		
	5.5% Sn; 0.1% P; balance Cu	O	360	Sheet, strip, wire	Springs and steam turbine blades
		H	700		
	10% Sn; 0.5% P; balance Cu	M	280	Castings	General-purpose castings and bearings

Alloy	Composition	Condition*		Form	Applications
Gunmetal	10% Sn; 2% Zn; balance Cu	M	300	Castings	Pressure-tight castings, pump and valve bodies
Aluminium bronze	95% Cu; 5% Al	O	400	Strip, tubing	Imitation jewellery and condenser tubes
		H	770		
	10% Al; 2.5% Fe; 2.5–5% Ni; bal. Cu	M	700	Hot worked and cast products	High-strength castings and forgings
Cupronickel	75% Cu; 25% Ni	O	360	Strip	British 'silver' coinage
		H	600		
	70% Cu; 30% Ni	O	375	Sheet, tubing	Condenser tubing, excellent corrosion resistance
		H	650		
Monel	29% Cu; 68% Ni; 1.25% Fe; 1.25% Mn	O	550	All forms	Excellent corrosion resistance, used in chemical plant
		H	725		
Beryllium–copper	$1\frac{3}{4}$–$2\frac{1}{2}$% Be; $\frac{1}{2}$% Co; balance Cu	WP	1300	Sheet, strip,	Springs, non-spark tools
Cadmium–copper	99% Cu; 1% Cd	O	285	Wire, rod	Overhead electrical wire, spot-welding electrodes
		H	500		
Chromium–copper	0.4–0.8% Cr; bal. Cu	WP	450	Wrought forms and castings	Welding electrodes, commutator segments
Tellurium–copper	0.3–0.7% Te; bal. Cu	O	225	Wrought forms	Free-machining properties
		H	300		

* Key: O = annealed; H = work hardened; M = as manufactured (cast or hot worked); WP = solution heat treated and precipitation hardened.

Figure 15.9 *Photomicrograph of 70/30 brass, sand cast (etched) ×100 showing cored α crystals. The dendritic growth pattern is visible. (Courtesy of The Copper Development Association)*

Figure 15.10 *Photomicrograph of 60/40 brass, sand cast (etched) ×100. A two-phase structure with α precipitate in a β matrix. (Courtesy of The Copper Development Association)*

Bronzes

The bronzes are high-strength alloys with a good resistance to corrosion. Although, according to the phase diagram, copper can hold up to 14 per cent of tin in solid solution, cast alloys containing more than 5 per cent of tin will show some δ phase in the microstructure. This is because pronounced

Figure 15.11 *Photomicrograph of high tensile brass (etched) ×200, β crystals with precipitated α at grain boundaries and within the grains. (Courtesy of The Copper Development Association)*

Figure 15.12 *Photomicrograph of 95/5 bronze, cold worked (etched) × 200. The α crystals are distorted and show strain bands. (Courtesy of The Copper Development Association)*

dendritic coring occurs owing to the wide freezing-temperature range of these alloys. Coring may be removed by a lengthy annealing treatment. The α phase cold working alloys do not normally contain more than 7 per cent of tin. These α alloys are ductile and are suitable for cold working, although they work harden rapidly. The uses of these alloys include the manufacture of non-ferrous springs, coinage, and sheet metal for ornamental work. Figure 15.12 shows the microstructure of a bronze containing 5 per cent of tin which has been cold worked after annealing. The α crystals are distorted. Alloys containing the brittle δ phase are unsuitable for working, and are used in the 'as cast' condition.

Figure 15.13 *90/10 phosphor bronze, sand cast × 500, showing the very hard Cu$_3$P phase (dark in colour) and the α + δ eutectoid. The light background is α phase. (Courtesy of The Copper Development Association)*

Bronzes need to be deoxidised before casting and phosphorus is used for this purpose. Sufficient phosphorus is normally added to give a certain residual phosphorus content in the deoxidised alloy. The materials are then termed *phosphor bronzes*. The effect of the residual phosphorus is to strengthen the alloy. α bronzes contain about 0.1 per cent of phosphorus, and the residual phosphorus content in an (α + δ) bronze may be as high as 1.0 per cent.

A major application for (α + δ) bronze is as a bearing material (see also section 15.12). The duplex structure of the material is most suitable for this purpose as the α phase matrix will be highly resistant to shock, and the very hard δ phase particles will sustain some of the load and reduce wear. Some of the residual phosphorus in (α + δ) bronze will form copper phosphide, Cu$_3$P, and this extremely hard constituent is associated with δ phase in the microstructure, resulting in an improvement in the bearing characteristics (see Figure 15.13).

Zinc and lead may be added to bronzes. Zinc can be used to cheapen the alloy, and it also improves the casting performance of the material. Alloys containing zinc are termed *gunmetals*, as they were used in former days for the casting of heavy ordnance. Small lead additions improve the machining characteristics of bronzes and gunmetals, while large additions, up to 25 per cent of lead, are added to some bearing metals (see section 15.12). A range of gunmetal water fittings is shown in Figure 15.14.

Aluminium bronzes

The aluminium bronzes possess similar strengths to bronzes. Again, the α alloys are cold working materials while the (α + γ) alloys are used for the manufacture of castings and hot worked products. The main characteristic of aluminium bronze is its excellent corrosion resistance (Figure 15.15), particularly in marine environments. The α alloys are used in the form of sheet and tubing for applications including condensers and heat exchangers. With aluminium contents of greater than 9.5 per cent, the alloy possesses an (α + γ$_2$) structure at ordinary temperatures, although the γ$_2$ phase will transform into β phase on heating to temperatures above 565°C. On cooling through 565°C, the β phase undergoes

Figure 15.14 *Gunmetal water fitting. (Courtesy of The Copper Development Association)*

Figure 15.15 *Aluminium bronze impellers for use in a power station system. These impellers, associated castings, valves and pipework are favoured because their strength and corrosion resistance give long-term reliability. (Courtesy of The Copper Development Association)*

a eutectoid transformation into α and γ phases. The $(\alpha + \gamma_2)$ eutectoid consists of alternate layers of ductile α and brittle γ_2 (compare with pearlite eutectoid in steels, see Chapter 16). Slow cooling through the eutectoid temperature will give a coarse lamellar structure, which will tend to be brittle, while more rapid cooling will give a fine structure which will be tough. Consequently, sand castings should be heat treated followed by cooling in air, or by quenching, in order to produce a fine-grained structure. If the β phase is rapidly quenched, the formation of γ_2 phase may be suppressed, and a structure similar to martensite in steels produced. This martensite structure may be tempered by heating the quenched alloy at about 500°C to produce a very fine-grained $(\alpha + \gamma_2)$ microstructure. Aluminium bronzes containing about 10 percent of aluminium are used for ships' screws, valves and pumps for marine use.

Cupronickels

Nickel is soluble in solid copper in all proportions (see phase diagram Figure 11.7(a)), and the cupronickel alloys are strong, ductile alloys with a face-centred cubic crystal structure. Cupronickels possess an excellent resistance to corrosion. British 'silver' coinage is a cupronickel containing 25 per cent of nickel. Other applications include use for condenser tubing and heat exchangers.

Nickel silvers

The addition of zinc to copper and nickel produces the alloys termed *nickel silvers*. These are also face-centred cubic in structure, and ductile. A major use for nickel silver alloys is the manufacture of cutlery and decorative articles. Generally these articles are given a thin coating of electrodeposited silver to enhance their appearance. The term EPNS (electro-plated nickel silver) characterises products made in this way.

Beryllium–copper

The copper–beryllium materials, containing up to 2.7 per cent of beryllium, are the highest strength copper alloys available. The alloys may be strengthened by solution heat treatment and precipitation hardening (compare with Al–Cu) and tensile strengths of up to $1400\,MN/m^2$ can be obtained. Normal beryllium–copper contains 2 per cent of beryllium but other alloys with beryllium contents ranging from 0.4–2.7 per cent beryllium and with small additions of cobalt or nickel have been developed. The alloys are used for the manufacture of springs, pressure diaphragms and capsules, and non-sparking tools.

Cadmium–copper

The addition of 1 per cent cadmium to copper gives an alloy with a tensile strength approximately 50 per cent higher than that of high-conductivity copper but with little loss of electrical conductivity. In the annealed condition the conductivity is only about 7 per cent less than that of annealed high-conductivity copper while the fully work hardened material has a conductivity which is about 85 per cent of that of annealed pure copper. The main applications of this material are for overhead contact wires for electric railways, telephone wires and electrodes and electrode holders for electric resistance welding equipment.

Chromium–copper

Chromium–copper is a precipitation hardening alloy containing, typically, 0.5 per cent of chromium, which has a high strength coupled with high electrical and thermal conductivities. A tensile strength of about 400 MPa is achieved by precipitation hardening and, in this condition, the alloy has a conductivity of about 80 per cent of that of annealed high-conductivity copper. Chromium–copper is used mainly for the manufacture of spot and seam welding electrodes.

Tellurium–copper

This material contains 0.3–0.7 per cent tellurium and is a free-machining alloy of high electrical conductivity. Tellurium is virtually insoluble in copper and appears in the microstructure as small particles of copper telluride. These second phase particles act as internal chip-breakers giving discontinuous chip formation during machining.

EXAMPLE 15.3

(a) Commercial brasses may be termed either *cold working* or *hot working*. Explain why these terms are used and give the basic compositional and metallurgical differences between these two types.
(b) For what purposes are small additions of the following metals made to brasses: aluminium, lead, manganese and tin?

SOLUTION

(a) Commercial brasses are alloys of copper with up to 50 per cent zinc. Reference to the Cu–Zn phase diagram shows that alloys with up to 37 per cent of zinc are single α-phase alloys. This is a highly ductile face-centred cubic solid solution which can be cold worked readily. With zinc contents between 37 and 50 per cent the alloy structure is either two-phase a $\alpha + \beta'$, or single-phase β'. The β' phase is an ordered body-centred cubic structure which is hard and strong and very difficult to work cold. These alloys can be formed into shapes either by casting or hot working processes such as rolling or forging.

(b) Aluminium enters solid solution giving additional strength and an improvement in corrosion resistance, particularly in marine environments. It may be added to either α or $\alpha + \beta'$ brasses.

Lead is insoluble in copper–zinc alloys and appears in the microstructure as discrete globular particles. Up to 3 per cent of lead may be added to either α or $\alpha + \beta'$ brasses. The discrete particles of lead act as internal chip-breakers, thus giving the brass free-machining characteristics.

Manganese enters solid solution and strengthens. It is used as an addition to $\alpha + \beta'$ materials to produce high-tensile brasses. An early name for high-tensile brasses containing manganese was *manganese bronze*.

Tin improves the resistance to marine corrosion and 1 per cent may be added to either cold working or hot working brasses. For example, *Admiralty brass* (70% Cu, 29% Zn, 1% Sn) is an α brass while *Naval brass* (62% Cu, 37% Zn, 1% Sn) is an $\alpha + \beta'$ brass.

EXAMPLE 15.4

(a) A small quantity of phosphorus is added to bronzes before casting. For what reasons is this addition made? Explain why the phosphorus addition is made before the addition of tin.

(b) Phosphor bronzes containing 10–13 per cent of tin make good bearing materials. Explain the part played by each of the constituent phases in the alloy when the material is used as a bearing.

SOLUTION

(a) Phosphorus is used as a deoxidiser and, often, sufficient is added to leave some residual phosphorus in the alloy as this strengthens the alloy. If tin is added to the liquid copper before it has been deoxidised some of the tin may oxidise forming tin oxide, which is very difficult to remove from the alloy. This oxide could seriously reduce the strength of the bronze.

(b) Cast phosphor bronzes containing 10–13 per cent tin have a duplex microstructure with a very hard $\alpha + \delta$ eutectoid in a matrix of more ductile α phase. If the amount of residual phosphorus exceeds 0.3 per cent, the limit of solubility of phosphorus in copper, the excess will appear in the structure as the very hard constituent Cu_3P, and this is associated in the microstructure with the $\alpha + \delta$ eutectoid. When in use as a bearing the very hard eutectoid and Cu_3P sustain some of the load and have excellent resistance to abrasion and wear, while the tough and more ductile a matrix will withstand mechanical shock.

15.6 | Lead

The principal properties of lead are:

Melting point	327°C
Crystal structure	face-centred cubic
Density	11.34×10^3 kg/m^3
Young's modulus, E	16.5 GPa
Electrical resistivity	2.1×10^{-7} Ω m at 20°C
Corrosion resistance	very good

Lead is a metal that has been used by man for over 4,000 years. It is soft and malleable, and possesses an excellent resistance to corrosion. Because of these properties it has been used for water pipework, and weather flashing on buildings. Because of its high density it is very suitable as a radiation-shielding material. Unfortunately, lead is a highly toxic metal. The use of lead for water pipework and waste disposal has ceased and the metal has been replaced by other materials for these applications. However, the metal is used as the basis of a number of useful alloys. The principal alloys of lead are those with tin and antimony and some of the important alloys of lead are listed in Table 15.3.

A major application for lead is in the manufacture of lead–acid storage batteries. The battery plates are grids of a hard lead–antimony alloy, into which a paste of lead compounds is pressed. Storage batteries account for almost 30 per cent of the annual world consumption of lead.

The soft solders are basically lead–tin alloys. The lead–tin phase diagram is shown in Figure 15.16. The low melting temperatures of these alloys make them very suitable for use as solders. The soft solders often contain a little antimony for added strength. The eutectic composition alloy, known as tinman's solder, freezes at 183°C and is used for electrical jointing and the sealing of tinplate cans.

Table 15.3 *Some lead alloys.*

Alloy	Composition (per cent)					Applications
	Pb	Sb	Sn	Bi	Hg	
Antimonial lead	99	1	—	—	—	Cable sheathing
Hard lead	94	6	—	—	—	Lead–acid batteries, lead shot
Plumber's solder	60	2.5	37.5	—	—	
Common solder	50	—	50	—	—	} Soft solders
Tinman's solder	38	—	62	—	—	
Type metal	62	24	14	—	—	For casting into printing type
Linotype metal	81	14	5	—	—	
Wood's alloy	24	—	14	50	—	(+ 12% Cd) Alloy with M.Pt. of 71°C used for fusible plugs in sprinkler systems
Rose's alloy	28	—	22	50	—	Low melting point alloy with M.Pt. of 100°C
Dental alloy	17.5	—	19	53	10.5	Dental cavity filling (M.Pt. 60°C)

Figure 15.16 *Phase diagram for lead–tin system.*

Plumber's solder contains a much higher lead content, 60 per cent of lead, and solidifies over a range of temperature. This enables it to be used for making wiped joints in lead pipework. Because of the toxicity of lead, soft solders containing lead are being phased out. The aim is that all soldering in the manufacture of electrical circuitry will be using lead-free solders by the year 2010. Refer also to section 24.2.

Alloys of lead, tin and antimony are used for casting printer's type. Tin, usually in amounts of up to 25 per cent, increases the fluidity of the melt, and increases the toughness of the alloy. Antimony, unlike most metals, expands as it solidifies from liquid. An alloy containing between 20 and 30 per cent of antimony shows a negligible volume change on solidification, and this ensures the production of a good, clean type face.

Alloys with very low melting points have been developed. These are based on lead, bismuth and tin and are used for the manufacture of fusible plugs for use in automatic fire alarm water-sprinkler systems and also for making patterns for the production of precision castings.

Lead is also a major constituent of many bearing alloys, and these will be discussed in the section 15.12.

15.7 | Tin

The principal properties of tin are as follows:

Melting point	232°C
Crystal structure	body-centred tetragonal (β or white tin)
Density	$7.29 \times 10^3 \, kg/m^3$
Young's modulus, E	40.8 GPa
Electrical resistivity	$1.15 \times 10^{-7} \, \Omega \, m$ at 20°C
Corrosion resistance	very good

Tin can exist in two allotropic forms, α (grey tin) possessing a diamond-type crystal structure and β (white tin) with a body-centred tetragonal structure. The equilibrium phase transition temperature is 18°C but, in practice, white tin does not transform into grey tin until it is cooled to sub-zero temperatures. When the change begins areas of grey powder begin to appear on the bright white tin surface. This was referred to as *tin disease*.

The attractive appearance of tin, its low melting point, good corrosion resistance and its lack of toxicity made tin an ideal metal for coating mild steel to make tin-plate for the manufacture of cans. When first produced, tin-plate was made by dipping sheets of mild steel into baths containing molten tin. A small amount of tin-plate is still made in this way, but the vast majority is made by the continuous electrodeposition of a small thickness of tin of the surface of mild steel strip as the strip passes through a bath of electrolyte at speeds of 12–30 m/s. The electro-deposited layer of tin is matt grey in appearance and after the strip has passed through a drying zone this tin layer is flash-melted to give the bright finish associated with tin-plate.

Another major use of tin is in the manufacture of white metal bearings (see section 15.12), and this probably accounts for about 40 per cent of the total world tin production. The bulk of the remaining tin production is used for alloying purposes with other elements.

Tin can be rolled into very thin foils, and tin foil was used for packaging before World War II. The high price of tin has led to its replacement by aluminium in this field.

Pewter is an alloy of tin that is widely used for making decorative articles. Earlier forms of pewter contained lead but in recent years non-toxic lead-free pewters have been developed. Pewter or Britannia metal, is tin alloyed with 6 per cent of antimony and 2 per cent of copper. Antimony greatly reduces the $\beta \rightarrow \alpha$ transition temperature and tin disease does not occur with this alloy.

15.8 | Magnesium

The principal properties of magnesium are given below:

Melting point	649°C
Crystal structure	close-packed hexagonal
Density	$1.74 \times 10^3 \, \text{kg/m}^3$
Young's modulus, E	44 GPa
Electrical resistivity	$3.9 \times 10^{-8} \, \Omega \, \text{m}$ at 20°C
Corrosion resistance	very good

Perhaps the most important property of magnesium is its very low density of $1.74 \times 10^3 \, \text{kg/m}^3$, compared with $2.71 \times 10^3 \, \text{kg/m}^3$ for aluminium. Pure magnesium is comparatively weak and is not used unalloyed. Magnesium can be strengthened considerably by suitable alloying. The principal alloying element is aluminium, in amounts up to 10 per cent. Additions of aluminium, thorium, zirconium or zinc will all give alloys that can be strengthened by precipitation hardening heat treatments (see section 12.6). Small amounts of other elements may also be added to magnesium alloys. Manganese will improve the corrosion resistance while both thorium and cerium give increased strength and improve creep resistance at elevated temperatures.

Because of its hexagonal crystal structure magnesium is not as ductile as aluminium and is not as easy to cold work as that metal. Magnesium alloys are readily hot worked and can be easily cast into shapes by both sand and die casting methods.

The corrosion resistance of magnesium is good and this can be further improved by anodising.

Some of the major uses of magnesium alloys are in aircraft construction, in the form of castings, forgings and sheet material for such components as landing wheels, airscrew blades, fuel tanks and engine components. The alloys are also used for cast car and motorcycle wheels, petrol engine crankcases and as a canning material for nuclear fuel elements.

The composition and properties of some magnesium alloys are given in Table 15.4.

Pure magnesium is used as sacrificial anode material for the cathodic protection of steel structures and ship's hulls against corrosion. The magnesium anodes are corroded and have to be renewed periodically while the steelwork is protected against corrosion. Zinc may also be used for this application.

Table 15.4 *Some magnesium alloys.*

Composition (per cent)						Condition	Tensile strength (MPa)
Al	Mn	Zn	Zr	Th	Ce		
3	0.3	1	—	—	—	Annealed	250
						Cold rolled	280
6	0.3	1	—	—	—	Forged	280
8	0.3	0.7	—	—	—	Cast	140
						Heat treated	210
10	0.3	0.7	—	—	—	Cast	200
—	—	3	0.7	—	—	Rolled	260
—	—	4	0.7	—	1	Cast	170
						Heat treated	220
—	—	—	0.7	3	—	Heat treated	210
—	1	—	—	—	3	Rolled	280

15.9 | Nickel

The principal properties of nickel are summarised below:

Melting point	1455°C
Crystal structure	face-centred cubic
Density	8.88×10^3 kg/m^3
Young's modulus, E	210 GPa
Electrical resistivity	6.84×10^{-8} Ω m at 20°C
Corrosion resistance	very good

Nickel is one of the most important metals in this technological age. It is used to a large extent as an alloying addition in many steels and cast irons, and in a number of non-ferrous alloys. Nickel is also used as a pure metal in its own right, and as the base for a number of useful engineering materials. The use of nickel as an alloying element in steels and irons will be dealt with in Chapter 16.

Pure nickel possesses an excellent resistance to corrosion by alkalis and many acids and, consequently, is used in chemical engineering plant, and in the food industry. For cheapness, nickel is frequently used as a cladding of thin sheet on a mild steel base. Nickel may also be electroplated on to a number of materials, and an intermediate layer of electrodeposited nickel is essential in the production of chromium-plated mild steel.

The principal nickel base alloys used industrially are *Monel, Inconel,* and the *Nimonic* series of alloys. Monel is an alloy containing 68 per cent of nickel, 30 per cent of copper, and 2 per cent of iron. It is a single-phase alloy with an exceptionally good corrosion resistance, and it is used for steam turbine blades, and in chemical engineering plant. Inconel contains 80 per cent of nickel, 14 per cent of chromium and 6 per cent of iron. Inconel combines good corrosion resistance with good properties at elevated temperatures, and it is used in chemical engineering plant, and as heater sheaths for electric cooker elements.

The Nimonic series of alloys are basically nickel–chromium alloys containing between 55 and 80 per cent of nickel, and around 20 per cent of chromium. In addition, the various alloys in the range contain small amounts of titanium, cobalt, iron and aluminium, together with a very small amount of carbon. These alloys are high-temperature creep resistant, and have been very fully developed to meet the requirements of gas turbine designers. The alloys are used for the manufacture of gas turbine discs and blades, and for flame tubes. They owe their high-temperature strengths to finely dispersed precipitates of intermetallic compounds of nickel with titanium and iron, and also some metallic carbides. Other nickel–chromium and nickel–chromium–iron alloys that have been developed include the proprietary alloys Brightray and Hastelloy (see Table 15.5). These materials, together with the Nimonic alloys, all developed to meet the very stringent demands of service at elevated temperatures, are sometimes referred to as *superalloys*. (See also section 15.13.)

15.10 | Titanium

The principal properties of titanium are summarised below:

Melting point	1660°C
Crystal structure	α close-packed hexagonal up to 880°C
	β body-centred cubic above 880°C
Density	4.54×10^3 kg/m^3
Young's modulus, E	106 GPa
Tensile strength	30 MPa
Corrosion resistance	excellent

Table 15.5 *Composition and uses of some nickel alloys.*

Name	Composition (per cent)										Uses
	Ni	Cu	Cr	Fe	Mo	W	Ti	Al	Co	C	
Monel	68	30	—	2	—	—	—	—	—	—	Chemical engineering plant
Inconel	80	—	14	6	—	—	—	—	—	—	Steam turbine blades Chemical engineering plant Electric cooker heating elements. Exhaust manifolds
Brightray	80	—	20	—	—	—	—	—	—	—	Heating elements for kettles, toasters, electric furnaces
Hastelloy C	55	—	15	5	17	5	—	—	—	—	Chemical engineering plant
Hastelloy X	47	—	22	18	9	1	—	—	—	—	Furnace and jet engine components
Nimonic 75	77	—	20	2.5	—	—	0.4	—	—	0.1	Thermocouple sheaths, furnace components, nitriding boxes
Nimonic 90	56.5	—	20	1.5	—	—	2.4	1.4	18	0.06	Gas turbine discs and blades
Nimonic 115	56.5	—	15	0.5	4	—	4	5	15	0.1	Gas turbine discs and blades

The useful properties of titanium are its relatively high strength coupled with a low density, and its excellent corrosion resistance with respect to most acids, alkalis and chlorides. It does possess some characteristics, though, that make processing both difficult and costly. At very high temperatures, titanium will readily dissolve oxygen, nitrogen and carbon, all of which will cause embrittlement. Consequently, titanium is produced by reducing titanium chloride with magnesium in a closed reaction vessel followed by vacuum distillation to remove any unreacted magnesium. The pure titanium sponge cannot be melted and cast in the same manner as other metals. Instead, the sponge is made into an electrode and this is arc-melted into a water-cooled copper crucible either under an inert gas atmosphere or in vacuo. The liquid titanium is then poured into a graphite-faced ingot mould, again either in vacuo or in an inert atmosphere. Ingots of titanium alloys may be made by compacting the required alloying additions with the sponge when making the electrode. This process for the production of titanium is expensive and is a major factor in determining the high cost of titanium products. In recent years a number of patents have been taken out covering an electrolytic process for the production of pure titanium. When such a process has been developed sufficiently for large-scale production we should see a significant reduction in the cost of titanium and a consequent increase in usage of the metal.

Titanium oxidises rapidly in air at all temperatures in excess of 700°C and oxygen will dissolve in the metal. Small amounts of oxygen will cause strengthening, with loss of ductility, while larger amounts will cause almost total brittleness. The α phase (hexagonal) is difficult to plastically deform and the amount of cold working that can be accomplished is limited. Hot forging and hot rolling of titanium alloys can be carried out at temperatures between 700°C and 1100°C. After hot working operations, surface contamination is removed by grinding and machining followed by pickling in hydrofluoric acid and nitric acid solutions. Castings in titanium alloys can be made by melting and casting in vacuo into graphite-lined moulds.

The principal alloying element in titanium alloys is aluminium and up to 6 per cent will enter solid solution in α titanium giving a strengthening effect. Some alloying elements including iron, chromium vanadium and molybdenum tend to stabilise the β phase and give alloys with $\alpha + \beta$ or β structures. Other alloying elements, including copper, will produce alloy structures that respond to precipitation hardening. Tensile strengths of up to 1200 MPa can be achieved with some of the complex alloys.

The range of titanium alloys available can be divided into four categories, commercially pure titanium, near α alloys, $\alpha + \beta$ alloys and metastable β alloys. The properties of commercially pure titanium materials are affected by small amounts of oxygen and nitrogen. By careful control of these, and other trace elements, various grades of commercially pure titanium can be produced with tensile strengths ranging from 350 to 700 MPa. The near α alloys contain 5–6 per cent of aluminium with a low content of β stabilising elements. They can be hot worked but have very limited cold forming ability. These alloys retain considerable strength at temperatures of up to 600°C and possess good creep resistance. Because of this they find application in the compressor stages of aero gas turbines. Figure 15.17 shows the large titanium alloy compressor fan of the Rolls-Royce Trent 500 aero engine. The $\alpha + \beta$ alloys can be hot worked but the possibility for cold forming is, at best, extremely limited. The alloys find

Figure 15.17 *Rolls-Royce Trent 500 aero engine showing titanium alloy compressor fan. (Courtesy of Rolls-Royce)*

Figure 15.18 *EADS Typhoon, showing canard foreplanes. (Courtesy of BAE Systems)*

application for aero engine compressor fans, as bio-medical implants and in general engineering. The β alloys contain up to 15 per cent of vanadium or molybdenum to stabilise the β phase. These alloys are cold formable and high tensile strength values of up to 1300 MPa can be developed by a solution heat treatment with rapid cooling to produce homogeneous β phase, followed by an ageing heat treatment to form a finely dispersed precipitate of α particles. The β alloys find many applications, including components for aircraft landing gear, jet engine nozzle assemblies, pressure vessels and fabricated sheet metal structures. The canard foreplanes of the EADS Typhoon (Eurofighter) aircraft (see Figure 15.18) are manufactured from a β alloy. These are formed superplastically and diffusion bonded to give high strength with minimum weight.

At present the use of titanium is restricted by its high cost (about £25/kg, as compared with £1–5/kg for stainless steel or aluminium). About one-half of the titanium produced is used in chemical engineering and general industrial applications while about 40 per cent is used in the aerospace industries. New and emerging applications account for the remainder, including jewellery, sports equipment, some automotive components and architecturally. Titanium has been used for building cladding and roofing. The cost becomes reasonable when it is considered against the high strength and excellent corrosion resistance of the metal. By good design, with the cladding supporting some of the load, lighter supporting structures are required. Lower maintenance costs are incurred because of the superior corrosion resistance of titanium. There has been considerable use of titanium architecturally in Japan because, traditionally, marine corrosion has been a major cause of building deterioration in that country. One example of the use of titanium architecturally in Europe is seen in the Guggenheim Museum at Bilbao in Spain, where the structure of the building is clad entirely with commercial purity titanium sheet (Figure 15.19).

Figure 15.19 *Guggenheim museum clad with titanium sheet. (© TIMET)*

15.11 | Zinc

The principal properties of zinc are summarised below:

Melting point	419°C
Boiling point	907°C
Crystal structure	close-packed hexagonal
Density	7.14×10^3 kg/m³
Young's modulus, E	90 GPa
Electrical resistivity	5.92×10^{-8} Ω m at 20°C
Corrosion resistance	good

   Pure zinc, having a melting point only slightly in excess of 400°C, makes an excellent material for die-casting purposes. The die-casting alloys contain 4 per cent of aluminium and, in some cases, 1 or 2 per cent of copper. These alloying additions increase the tensile strength of the zinc, in the die-cast state, to about 250 MPa. The zinc used for die casting purposes must be of extremely high purity, as traces of cadmium, tin, or lead, in the material would lead to the incidence of intercrystalline corrosion failure in service. Zinc alloy die castings are used extensively, and their applications include automotive components, such as carburettor bodies, fuel pump bodies and door handles, and the production of accurately scaled model toys.

Metals possessing hexagonal crystal structures are normally difficult to work cold. Zinc is quite brittle at ordinary temperatures, but it can be plastically deformed with ease at temperatures within the range of 100°C to 150°C. Sheet zinc is produced by rolling the metal at these temperatures. Sheet zinc is used in the manufacture of battery cases for dry cells, the zinc being the negative electrode of the battery.

There is a eutectoid transformation in the zinc–aluminium alloy system at a composition of 22 per cent aluminium. The eutectoid alloy can be produced with an extremely fine grain size and is super-plastic at the eutectoid temperature of 275°C (see section 8.12). This material is used to manufacture deep-drawn products such as instrument cases.

In addition to the above, a major use for zinc is for galvanising, that is the coating with zinc of steel sheet, structural sections, and nails, as a protection against corrosion. Like magnesium, zinc may also be used as a sacrificial anode to give cathodic protection to steel structures.

15.12　Bearing materials

There are several requirements which should be possessed by a successful bearing material. It should possess sufficient hardness and wear resistance so that it does not wear away during service but its hardness should be low relative to the shaft or journal in order to avoid wear or damage to the shaft, particularly during a start-up when, because of low oil pressure, there may be metal-to-metal contact. The strength of the bearing material should be sufficient to sustain the load without deformation and yet possess considerable toughness to resist shock loading. Many of these apparently conflicting requirements can be met by metallic alloys possessing a duplex structure with hard constituents, to sustain the load, embedded in a softer and tough matrix.

There are very many bearing materials, each suited to particular types of loading, running speed and service conditions. The main types are: white metals, copper-base alloys, aluminium-base alloys, cast iron, plastics materials and ceramics.

White metals

These bearing metals are either tin-base or lead-base, of which the former are the more widely used. The tin-base bearing metal alloys are of better quality than the cheaper lead-base variety and are suitable for use in many low- and medium-duty bearing applications, particularly in the automotive industry. The lead-base alloys tend to be used for low-pressure/low speed bearing applications. Both types contain about 10 per cent of antimony and usually 1.5–4.0 per cent of copper. The tin-base bearing metals are known as Babbit metals after the original patentee, Isaac Babbit. Tin and antimony combine to form a very hard intermetallic compound, with the formula SbSn. The compound SbSn appears in the microstructure as small hard cubic crystals, termed cuboids. Within the tin–lead–antimony alloys these hard SbSn cuboids are dispersed within a soft matrix of ternary tin–lead–antimony eutectic. During the solidification of bearing metals, the hard SbSn cuboids solidify from the liquid first and, being of low density, they tend to rise to the surface of the liquid metal. This segregation of the hard constituent may be overcome by introducing about 3 per cent of copper to the alloy composition. In the alloys containing some copper, it is a hard intermetallic compound of copper and tin, Cu_6Sn_5, that solidifies first. This compound forms as a series of needle-shaped crystals, creating a network throughout the melt. This network largely prevents the SbSn cuboids from segregating to the surface of the melt when they begin to solidify (see Figure 15.20).

Bearing bronze

There are several types of bronze used in bearing applications; straight Cu–Sn alloys, phosphor bronzes or leaded bronzes. Generally, a bearing bronze contains 10–15 per cent tin and has an $\alpha + \delta$ structure.

Figure 15.20 *Photomicrograph of white bearing metal (80%Sn, 11% Sb,6% Pb, 3% Cu) (etched) × 60, showing cuboids of SbSn and network of Cu$_6$Sn$_5$ needles.*

The hard δ phase, present as part of an ($\alpha + \delta$) eutectoid within a softer α matrix, is capable of sustaining heavy loads. In a phosphor bronze, the very hard phase Cu$_3$P is associated with the $\alpha + \delta$ eutectoid and provides improved load bearing capacity. A typical composition for a leaded bronze is 2% Sn, 24% Pb, 74% Cu. Lead is insoluble in copper and the presence of lead globules in the microstructure reduces the dry coefficient of friction of the material. There is a tendency for major segregation of lead during solidification and the addition of a small amount of nickel to the alloy reduces this tendency. The leaded alloys are less strong than other bearing bronzes but will sustain higher loads at higher speeds than white metals. They find application in aero, diesel and automobile engines.

Some bronze bearings are made by compacting and sintering a powder. The powder particle size and compaction pressures are controlled to produce a porous bearing which can be saturated with lubricant before installation. This type is used widely in applications where regular lubrication is difficult, as for example in automotive starter motors and alternators.

Aluminium bearings

Aluminium-base alloys containing tin, copper and nickel are used widely as bearing materials and have replaced the more expensive tin-base bearing metals in many applications, particularly in the automotive industry. Typical compositions lie within the ranges 5.5–7.0 per cent tin, 0.7–1.3 per cent copper and 0.7–1.3 per cent nickel. The basis alloys tend to be of low strength and frequently are bonded to a steel backing shell.

Some polymeric materials, notably PTFE and the polyamides (nylons), have very low coefficients of friction and are used for low load applications and where the presence of a lubricating oil is undesirable.

Ceramic materials have been used as bearings in small precision instruments for many years, for example jewel bearings in watch movements. Modern industrial ceramics such as alumina are used as bearings in high speed precision movements.

15.13 | Superalloys

The development of the aero gas turbine, with its requirements for materials capable of operating at high levels of stress at high temperatures for extended periods, has been the stimulus for much research into the development of creep resistant alloys. Superalloy is a general term used to describe the nickel-base and cobalt-base alloys which have been developed for use at elevated temperatures. These alloys tend to contain many alloying elements and are complex in structure. An alloy that will resist high-temperature creep successfully must possess structural elements that will greatly impede the movement of dislocations. Some alloying elements will enter solid solution and strengthen the matrix while others, together with other elements, will form intermetallic compounds and carbides. When carbides and other precipitate particles are present in a highly dispersed state in high density it will be extremely difficult for dislocation movement to occur within the material, even at very high temperatures. Examples of some of the stable precipitate particles that are used to good effect in superalloys are Ni_3Al, Cr_7C_3, TiC and Ni_3Ti. An alloying addition that will promote a large grain size or that will cause grain boundary strengthening will also be beneficial in resisting creep because the viscous flow of grain boundaries contributes to the creep process at temperatures close to the melting point. The functions of the various alloying additions that are made to nickel in the creation of superalloys is given in Table 15.6.

Table 15.6 *Functions of elements added to superalloys.*

Functions	Al	B	C	Cr	Co	Hf	Mo	Nb	Ta	Ti	W	Zr
To strengthen the matrix				x	x		x				x	
To strengthen the grain boundaries		x	x			x						x
To form carbides				x		x	x	x	x	x	x	
To create protective oxide scale	x			x								

Development of processing technology has taken place alongside alloy composition developments. The introduction many years ago of vacuum melting and casting gave clean gas-free materials and resulted in improved alloy performance. In the late 1970s DS casting (directional solidification) in vacuo was introduced for the production of turbine blades. In a DS casting the crystal grains are aligned along the length of the blade, that is in the direction of maximum direct stress, and there are very few grain boundaries normal to this direction. The next logical development was the production of single-crystal turbine blades by a directional solidification process (see section 20.6) and this is the method used today.

SUMMARY

This chapter surveys the eight most commonly used non-ferrous metals and their principal alloys. Aluminium is a soft, ductile metal of low density and with a high resistance to corrosion. It can be alloyed to make a wide range of alloys, some of which will respond to precipitation strengthening heat treatments to give alloys with high strength/weight and stiffness/weight values. Copper is a highly ductile metal with high electrical conductivity and the pure metal is used in many electrical and electronics applications. Copper can be alloyed to give a wide range of alloys, many of which possess good corrosion resistance. Lead is a soft metal of high density which has been used in plumbing, electrical storage batteries, bearings, soft solders and as a radiation shielding material. Because of its high toxicity it is being phased out from a number of traditional applications, being replaced by other, less toxic materials. Tin is a white metal with a low melting point, a high resistance to corrosion and is non-toxic. Its main uses are for coating mild steel sheet (tin plate), as the basis for bearing materials and solders, and as an alloying addition in other metals. Magnesium has a very low density and excellent corrosion resistance. The pure metal can be strengthened considerably by alloying and magnesium alloys find many applications in the automotive and aircraft industries. Nickel has an excellent resistance to corrosion and is used extensively in the chemical and food industries. Nickel is the basis metal in a number of alloys developed to have excellent corrosion resistance or high-temperature creep resistance. It is also an important alloying addition in many steels. Titanium has a relatively high strength, coupled with a medium density, about half that of iron, and an excellent resistance to corrosion. With suitable alloying, alloys with tensile strengths of up to 1200 MPa can be achieved. Because of its relatively high cost, the major usage at present is in aerospace and chemical engineering applications. Zinc, with its low melting point, is an ideal metal for the production of die castings. Also, because of the good corrosion resistance of the metal, much zinc is used for giving galvanised coatings to steels. A range of metallic alloys capable of operating under stress for long periods at high temperatures has been developed. These alloys, also termed *superalloys*, are based on nickel and cobalt.

Questions

15.1 Portions of the phase diagrams for the Al–Cu and Al–Mg$_2$Si phase diagrams are shown in Figure 15.21.

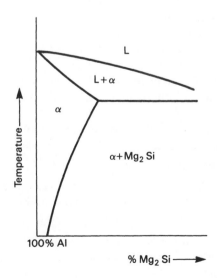

Figure 15.21

(a) Explain how maximum strength can be developed in alloys in each of these two alloy systems.

(b) Outline some of the advantages and disadvantages of each type of alloy and give examples of the engineering applications of these materials.

15.2 You have small samples of four materials, all similar in appearance. The identification marks are missing from the samples but it is known that the materials are

(a) commercial purity aluminium,

(b) Al–13% Si alloy,

(c) Al–4% Cu alloy, and

(d) Al–1% Mg–1% Si alloy.

How may each of the four materials be identified if the only equipment available is

(i) a hardness testing machine,

(ii) a furnace with a working temperature range from 100°C to 600°C,

(iii) temperature measuring equipment accurate to 5°C, and

(iv) a water supply?

15.3 (a) Alloys of aluminium with silicon are used for the manufacture of castings. Why are Al–Si alloys so useful for this purpose?

(b) What is meant by the term *modification*, in respect of Al–Si alloys? How is modification achieved and what effects does it have on the structure and properties of the alloys?

15.4 Refer to the Cu–Zn phase diagram (see Figure 15.8(a)).

(a) State what phases exist, and the relative proportions of the phases present, in (i) a Cu–Zn alloy containing 30 per cent zinc at 950°C and (ii) an alloy containing 40 per cent zinc at 25°C.

(b) Sketch the expected room temperature microstructures for the following two materials:

(i) a brass containing 30 per cent Zn which has been rapidly solidified, and

(ii) a brass containing 40 per cent zinc which has been slowly cooled from 870°C.

(c) For the rapidly cooled 30 per cent Zn alloy, what treatment is needed to give a fully homogeneous microstructure?

(d) Of the two alloys quoted above, state, giving reasons, which would be chosen for

(i) manufacture of cold rolled sheet,

(ii) manufacture of hot pressings.

15.5 Refer to the Cu–Al phase diagram (see Figure 15.8(c)).

(a) Describe the room temperature microstructures for

(i) an alloy containing 5 per cent aluminium,

(ii) an alloy containing 10 per cent aluminium treated in the following ways:

(A) cooled slowly from 900°C,

(B) cooled rapidly, but not quenched, from 900°C,

(C) quenched in water from 900°C.

(b) State, giving reasons, which of the above two alloys would be suitable for fabrication as:

(i) condenser tubing,

(ii) a cast high-pressure pump body.

15.6 What are the necessary metallurgical characteristics required in a good bearing metal? Compare and contrast lead-base, tin-base, copper-base and aluminium-base bearing alloys.

15.7 Discuss the statement 'Nickel is vital for the maintenance of a manufacturing economy'. List some of the major applications of nickel and its alloys in engineering.

15.8 Give an account of the main applications of magnesium and zinc as engineering materials.

15.9 Give the main characteristics of the metal titanium. What are the main advantages and disadvantages of titanium and its alloys as engineering materials?

15.10 State, giving reasons, which non-ferrous metals or alloys could be selected for each of the following applications. Give two possible alternative materials in each case.

(a) Wire for the armature winding for small, low-power electric motors.

(b) Waterproof flashing material in buildings.

(c) 'Big-end' and 'little-end' bearings in engines for saloon cars.

(d) Structural panels for supersonic aircraft.

CHAPTER 16

Iron and steel

Why are irons and steels of importance?

Iron has been used by man for many thousands of years. The economic ores of iron are distributed widely throughout the world and they can be reduced to the metal with relative ease. Essentially, steels are alloys of iron and carbon but many steels also contain other alloying elements, including chromium, molybdenum and nickel. Because the metal iron can exist in two allotropic crystalline forms it becomes possible to modify the crystal structures of steels by various heat treatments and, thus, produce a wide range of mechanical properties. The many different steel alloys, coupled with the property development brought about by heat treatment, gives the engineer a very wide range of extremely useful materials. The principal features of the metallurgy of iron and the main categories of plain and alloy steels and irons, together with some typical engineering applications, are discussed in this chapter.

LEARNING OBJECTIVES

1. Gain a basic knowledge of the structures, properties and applications of the more common types of steels.

2. Understand the iron–carbon phase diagram, know the names of the major microstructural constituents and how they are formed.

3. Understand T–T–T and C–C–T diagrams and be able to show how non-equilibrium phases can be formed in steel structures.

4. Know the range of heat treatments which may be given to steels and understand how these treatments affect the structure and properties of the material.

5. Know the principal effects on both structure and properties of the major alloying elements used in steels.

6. Know the basic structures and properties of grey and white cast irons and understand how improved cast irons can be created by alloying and/or heat treatment.

16.1 | Introduction

Iron was first used by man for the manufacture of tools and weapons more than 3000 years ago. Iron ores are relatively abundant and are widely distributed throughout the world. The availability, comparatively low cost and the wide range of properties that can be achieved through alloying and heat treatments have made irons and steels materials of great importance. In the first industrial revolution in the mid-18th century, the main constructional materials were cast and wrought iron. These materials held sway until the latter part of the 19th century, with many road and rail bridges built of these materials still in service today. The invention of the Bessemer converter in 1856, and other developments later in that century, enabled steels of consistent quality to be mass produced. Chromium and nickel alloy steels were first produced just over 100 years ago and since then the pace of development has been rapid so that today a very large range of steels with enhanced properties are available for use by engineers. Irons and steels account for more than 90 per cent of the total tonnage of all metals used today.

16.2 Irons and steels

Iron ores are smelted in large blast furnaces and the iron produced contains a number of impurities. The principal impurities in blast furnace iron are carbon (about 4 per cent), silicon (about 2 per cent), manganese (about 1 per cent), sulphur (about 0.5 per cent) and, in many cases, phosphorus (up to 2 per cent). The exact level of these impurities present in a batch of iron will depend upon the quality of the raw materials used and the furnace operating conditions. When the liquid iron from a blast furnace is cast into ingots it is referred to as *pig* iron. Selected grades of pig iron are remelted and used to make cast products (*cast iron*) but the bulk of blast furnace production is converted into *steel*. The steel-making process is an oxidation process in which the impurities present in the blast furnace iron are oxidised away. This oxidation may be carried out in an open-hearth furnace under an oxidising slag layer rich in iron oxide, in a large converter using oxygen as the oxidising agent, or in an electric arc furnace. During the latter half of the 20th century there was a major increase in the use of oxygen-blown converters and electric arc processes at the expense of the older open-hearth process so that nowadays less than 10 per cent of total steel production comes from open-hearth furnaces and more than 60 per cent of steel production is by the oxygen converter process. At the end of the refining process the amounts of carbon, silicon, manganese, sulphur and phosphorus in the iron will be very small, usually less than 0.05 per cent of each, but the iron will be in a highly oxidised state. For the production of steels the liquid metal must be deoxidised. Silicon and manganese are generally used as deoxidisers and these elements are added, in the form of ferro-alloys, in sufficient quantity to leave some residual silicon and manganese in the steel. Another element often used as a deoxidiser is aluminium. After deoxidation is complete, anthracite is added to bring the carbon content up to the desired level.

Essentially, steels are alloys of iron and carbon but they will contain some silicon and manganese, traces of sulphur and phosphorus, and also they may contain other elements such as nickel or chromium as alloying additions. We can now give the following definitions:

Plain carbon or *non-alloy steels* are steels containing up to 1.5 per cent of carbon, together with not more than 0.5 per cent of silicon and not more than 1.5 per cent of manganese, and with only traces of other elements.

Alloy steels are steels that contain either silicon or manganese in amounts in excess of the values quoted in the previous definition, or that contain any other element, or elements, as the result of deliberate alloying additions. Alloy steels can be divided into two main groups – *low alloy* and *high alloy* steels.

Low alloy steels, which contain up to 3 or 4 per cent of one or more alloying elements, have similar microstructures to, and require similar heat treatments to plain carbon steels.

High alloy steels are steels which, by virtue of the relatively large alloy content (usually more than 5 per cent), possess structures, and require heat treatments, that differ considerably from those of plain carbon steels.

16.3 | Constituents in steels

As mentioned in section 11.11, iron is allotropic and exists in two crystalline forms – body-centred cubic and face-centred cubic. These forms are stable over certain ranges of temperature. For pure iron the ranges are:

α iron – body-centred cubic – stable at temperatures up to 908°C

γ iron – face-centred cubic – stable between 908°C and 1388°C

δ iron – body-centred cubic – stable between 1388°C and 1535°C (the melting point)

There is another change at a temperature of 768°C, the *Curie temperature*. Iron loses its ferromagnetic characteristics when heated above the Curie temperature. Originally, iron at temperatures between 768°C and 908°C was termed β iron. The term β iron was discontinued when it was discovered that there was no change in crystal structure at the Curie temperature.

Before we can understand properly how various microstructures occur in steels and the effects of these structures in determining the properties, it is necessary to examine the iron–carbon phase diagram and to know the special names that are used to denote the various phases and types of structure that occur. It is appropriate to list and explain these terms at this stage as they will be used in the text in the remainder of the chapter.

Ferrite. Ferrite is the name given to the body-centred cubic allotropes of iron, α and δ iron, and to body-centred cubic solid solutions.

Austenite. Austenite is the name given to the face-centred cubic, or γ, variety of iron, and to face-centred cubic solid solutions.

Cementite. Cementite is the name given to the carbide of iron, formula Fe_3C. This is an extremely hard and brittle constituent.

Pearlite. Pearlite is the eutectoid mixture of ferrite and cementite, and is formed when austenite decomposes during cooling. It consists of alternate thin layers, or lamellae, of ferrite and cementite.

Martensite. This is the name given to the very hard and brittle constituent that is formed when a steel is very rapidly cooled from the austenitic state. It is a ferrite, highly supersaturated with dissolved carbon.

Tempered martensite. Martensitic and bainitic structures undergo changes when the material is tempered, that is heated to some temperature not exceeding 700°C for the purpose of reducing hardness and brittleness. The appearance of the tempered structure will depend upon the temperature and the time of treatment. The names *sorbite* and *troostite* were used formerly to describe these tempered structures.

Bainite. This is the term that is given to the decomposition product that is formed when austenite decomposes by either isothermal transformation, or at a cooling rate intermediate between the very rapid cooling necessary for martensite formation and the slower rate of cooling at which pearlite is formed.

16.4 | The iron–carbon phase diagram

The phase diagram for the iron–carbon system is shown in Figure 16.1. This diagram may appear at first sight to be extremely complex, but it can be divided into sections that, in themselves, are straightforward. For the consideration of steels it is convenient to consider only that portion of the diagram up to a carbon content of 1.5 per cent, and up to a temperature of 1000°C (see Figure 16.2).

It will be seen that ferrite cannot hold carbon in solid solution to any great extent, the limits being 0.04 per cent of carbon at 723°C and 0.006 per cent of carbon at 200°C. Austenite, however, can hold a considerable amount of carbon in solid solution, ranging from 0.87 per cent at 723°C to 1.7 per cent at 1130°C. Carbon is held in interstitial solid solution in both ferrite and austenite. The eutectoid

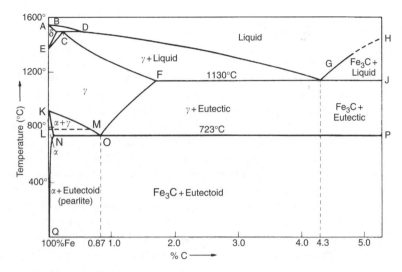

Figure 16.1 *Fe–Fe₃C phase diagram.*

point occurs at a temperature of 723°C and at a carbon content of 0.87 per cent. The terms *hypoeutectoid* and *hypereutectoid* are used to denote steels that contain less carbon than, and more carbon than, the eutectoid composition, respectively.

The presence of carbon depresses the α–γ transformation temperature of iron. Line KMO in the figures denotes this transformation temperature and its dependence on composition. Lines OF and QN are solvus lines and denote the maximum solubility limits of carbon in γ and α iron respectively. Point O is the eutectoid, or pearlite, point. The line LMOP indicates the Curie temperature.

If a sample of a steel is heated or cooled, and accurate measurements are taken, thermal arrest points will be noted corresponding to the phase transformation lines (and Curie temperature) on the phase diagram. The phase line NOP is known as the A_1 transformation, the Curie temperature, LM as the A_2 transformation, the phase line KMO as the A_3 transformation, and the line OF as the A_{cm} transformation. These arrest points, or transformation temperatures, are also known as the *critical points*, or *critical temperatures*, for the steel. The eutectoid temperature, A_1, is known as the *lower critical temperature*, and the α to γ transformation, line KMO, is known as the *upper critical temperature*. If a steel is heated or cooled very slowly, so that equilibrium conditions are approached, the measured arrest temperatures will agree with the values shown on the iron–carbon phase diagram. With more rapid heating or cooling rates, the measured arrest points will differ from equilibrium values. They will be higher than the equilibrium values when determined during heating, and lower when determined during cooling. Values measured during heating are written as Ac_1, Ac_2 and Ac_3, while values determined during cooling are written as Ar_1, Ar_2 and Ar_3.*

The temperatures and composition values quoted in Figures 16.1, 16.2 and in the above text are the values for alloys of pure iron and carbon. Steels, as has been stated, contain other elements as well. Of what relevance, then, is Figure 16.2? The iron–carbon phase diagram still has great relevance. The presence of small amounts of other elements will affect the position of the eutectoid but will not affect the general shape of the phase diagram. In all cases the effect of additional elements will be to move the eutectoid point towards the left, in other words to reduce the carbon content of the eutectoid, pearlite. The eutectoid temperature will also be affected, being either raised or lowered by the added elements. Manganese and nickel, for example, will lower the critical temperatures while chromium and silicon will raise them.

---

* From the French: chauffage – heating; refroidissement – cooling.

16.5 | Structures of plain carbon steels

Let us now consider the changes that occur during the cooling of steels of various compositions. Refer to Figure 16.2 and consider first the cooling of a hypoeutectoid steel of composition (1). At a high temperature the steel structure will be composed of homogeneous crystals of austenite solid solution. On cooling to the upper critical temperature, point l on the diagram, austenite will begin to transform into ferrite. The ferrite can hold very little carbon in solid solution and so the remaining austenite becomes enriched in dissolved carbon. As the temperature falls, more ferrite is formed, and the composition of the remaining austenite increases in carbon content, following the line KMO. When the lower critical temperature is reached, the austenite, which is now of eutectoid composition, transforms into the eutectoid mixture pearlite, a mixture composed of alternate layers of ferrite and cementite.

For a hypereutectoid steel of composition (2) the homogeneous austenite structure that exists at high temperatures will begin to change when the temperature has fallen to a point m, on the solvus line OF. This is the saturation limit for dissolved carbon in austenite and on cooling below the temperature of point m, excess carbon precipitates from solid solution in the form of cementite. The cementite appears in the microstructure as a network around the austenite crystals, and also in the form of needles within the austenite crystal grains. This latter type is termed *Widman–Statten* type precipitation. The carbon

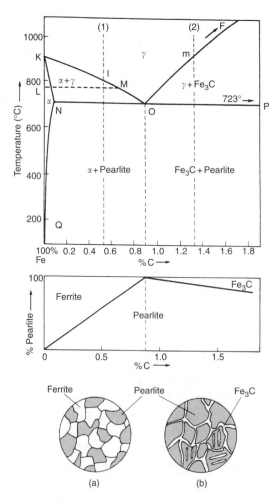

(a)   (b)

Figure 16.2   *Steel portion of the Fe–Fe$_3$C diagram: (a) microstructure of hypoeutectoid steel (1); (b) microstructure of hypereutectoid steel (2).*

content of the austenite reduces with further cooling and when the lower critical temperature is reached all remaining austenite, which is now of eutectoid composition, transforms into pearlite. The presence of cementite in the form of needles, or as a boundary network, renders the steel brittle, and heat treatment is necessary to put the steel into a suitable condition for many applications.

Ferrite is a comparatively soft and ductile constituent possessing an ultimate tensile strength of about 280 MPa. The tensile strength of pearlite formed by slow cooling from the austenitic range is about 700 MPa, but its ductility is very much less than that of ferrite. The variation of strength, hardness and ductility with carbon content, for slowly cooled steels, is given in Figure 16.3. It must be emphasised that the properties quoted apply only to slowly cooled steels. An increase in the rate of cooling through the critical temperature range will alter the structure, and hence the properties, of any steel. When the cooling rate is increased, there is some undercooling of austenite to below the equilibrium transformation temperatures. Once the phase change from undercooled austenite commences it occurs very rapidly. In the case of a hypoeutectoid steel, the ferrite grain size will be very small, as a result of many nucleation sites generated in the undercooled austenite, and also, once the transformation into pearlite begins, a pearlite with very fine lamellae of ferrite and cementite is formed. As stated in section 8.5, a reduction in grain size brings about an increase in strength and hardness. Also, the hardness and strength of pearlite is dependent on its interlamellar spacing and the very fine pearlite formed by rapid cooling may have tensile strengths of up to 1300 MPa.

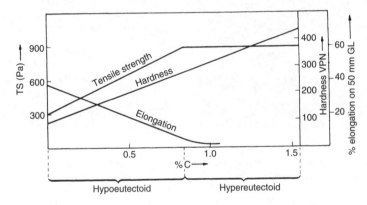

Figure 16.3 *Properties of normalised plain carbon steels, and their variation with carbon content.*

Figure 16.4(a) shows the microstructure of a commercially pure iron, ingot iron. This is composed of equiaxial grains of ferrite. A low carbon steel, in the normalised condition, will have a duplex structure of ferrite and pearlite and Figure 16.4(b) shows the microstructure of such a steel. The pearlite has etched dark. Pearlite itself is duplex in structure and the alternate layers of ferrite and cementite can be seen clearly in Figure 16.4(c).

If a steel is cooled extremely rapidly there will be insufficient time allowed for austenite to decompose into pearlite and, instead, the austenite changes into a body-centred lattice with all the carbon trapped in interstitial solid solution. Ferrite can, theoretically, hold virtually no carbon in solution. The rapidly cooled structure that is formed is highly strained and distorted by the large amount of dissolved carbon into a body-centred tetragonal lattice. This constituent is termed *martensite*, and it is extremely hard and brittle. The hardness of martensite depends on the carbon content, and is greatest in high carbon steels, that is, the greatest degree of lattice strain. Martensite, under the microscope, appears as a series of fine needle-like (*acicular*) crystals. A martensite structure can be formed by rapidly quenching a heated steel, from the austenitic state, into water or oil. This is the treatment termed *hardening*.

It will be seen from Figure 16.3 that while an increase in carbon content will give increased strength and hardness, this is accompanied by a major reduction in ductility. A major advantage given by alloying elements, such as manganese, is that they will bring about an increase in strength with comparatively

Figure 16.4 *(a) Ingot iron (etched) ×500 showing ferrite grains; (b) 0.3 per cent carbon steel (etched) ×300 showing ferrite and pearlite (dark); (c) 0.8 per cent carbon steel (etched) ×2000 showing lamellar structure of pearlite; (d) 1.2 per cent carbon steel (slowly cooled from above the A$_{cm}$) (etched) ×500 showing pro-eutectoid cementite (white) and pearlite (dark).*

little loss of ductility. However, the presence of most alloying elements in steels will effectively increase the carbon content as the major alloy elements have a carbon equivalent value. The carbon equivalent value (CEV) for a steel is given by

$$CEV = C + \frac{Mn}{6} + \frac{Cr + Mo + V}{5} + \frac{Ni + Cu}{15}$$

So that, for example, a steel containing 0.2 per cent carbon and 0.9 per cent manganese will have a CEV of 0.35 per cent. In other words, it will have the same relative proportions of ferrite and pearlite in its structure as a 0.35 per cent carbon steel.

EXAMPLE 16.1

For two plain carbon steels containing 0.4 per cent C and 1.0 per cent C respectively, and with reference to the Fe–Fe$_3$C phase diagram:

(a) Calculate the amount of austenite present in each steel at a temperature of 724°C.

(b) Calculate the amount of pearlite present in each steel at 25°C.

(c) Estimate the proportion of cementite existing at the grain boundaries as free cementite and the proportion present with ferrite in the pearlite in the hypereutectoid steel at room temperature (Fe_3C contains 6.67 per cent C).

(d) Sketch the microstructures of each steel after the following treatments:

    (i) for the 0.4 per cent C steel, water quenching from a temperature of $(A_3 + 30)°C$.

    (ii) for the 1.0 per cent C steel, water quenching from a temperature of $(A_1 + 30)°C$.

(e) Suggest a typical application for each steel in its fully hardened and tempered condition.

SOLUTION

The relevant section of the phase diagram is shown in Figure 16.5. All phase boundary lines are assumed to be straight and the lever rule (see section 11.4) is used.

(a) (i) For 0.4 per cent C steel, the percentage of γ in structure is given by

    $(AB/AC) \times 100 = (0.36/0.83) \times 100 = 43$ per cent

    (ii) For 1.0 per cent C steel, the percentage of γ in structure is given by

    $(DE/CE) \times 100 = (5.67/5.80) \times 100 = 98$ per cent

(b) (i) For 0.4 per cent C steel, the percentage of pearlite in structure is given by

    $(FG/GH) \times 100 = (0.40/0.87) \times 100 = 46$ per cent

    (ii) For 1.0 per cent C steel, the percentage of pearlite in structure is given by

    $(JK/HK) \times 100 = (5.67/5.80) \times 100 = 98$ per cent

(c) It is valid to use the tie-line FK for room temperature proportions as F is, to all extents, at 0 per cent C.

From (b)(ii), 98 per cent of the structure is pearlite and 2 per cent is grain boundary cementite. Within pearlite the percentage is given by

$(FH/FK) \times 100 = (0.87/6.67) \times 100 = 13$ per cent

So in a 1.0 per cent C steel, 13 per cent of 98 per cent of the structure, namely 12.8 per cent, is the cementite within pearlite. The amount of cementite existing as grain boundary cementite is, therefore

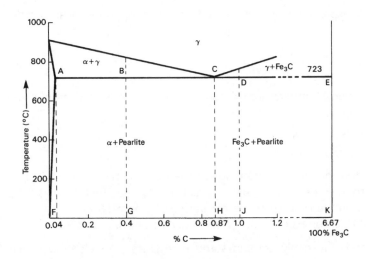

Figure 16.5

(2/14.8) × 100 or 13.5 per cent and the remaining 86.5 per cent of the cementite is present with ferrite in the pearlite areas.

(d) The type of microstructures expected after water quenching are shown in Figure 16.6. For the 0.4 per cent C steel, the structure is martensite. For the 1.0 per cent C steel, about 2 per cent of the structure is spheroidised cementite and the remainder is martensite.
The hardness values which would be achieved are H_D = 700–800 for the 0.4 per cent C steel and H_D = 1000–1100 for the 1.0 per cent C steel.

0.4% C   1.0% C

Figure 16.6

(e) 0.4 per cent C steel is suitable for the manufacture of shafts. Correctly heat treated it will have a high surface hardness to withstand wear and possess good strength and toughness to transmit power.
1.0 per cent C is a tool steel. The high surface hardness obtainable would make it suitable for use as, for example, a milling cutter.

EXAMPLE 16.2

A thermal analysis of an unknown steel to determine the critical temperatures gave the following results: A_{c1} = 720°C, A_{c3} = 840°C, A_{r1} = 660°C, A_{r3} = 780°C.

(a) Assuming that the rates of heating and cooling were the same throughout the test, calculate the critical temperatures for the steel and estimate the carbon content.

(b) Explain why the experimentally determined critical temperature values differ from those in the phase diagram.

SOLUTION

(a) The equilibrium critical temperature values will be the arithmetic mean of those determined during heating and cooling.

So A_1 = (720 + 660)/2 = 690°C,
and A_3 = (840 + 780)/2 = 810°C.

The A_1 temperature of 690°C is 33°C lower than the A_1 temperature in the phase diagram. To estimate the carbon content of the steel correct the A_3 value by adding 33°C and mark these critical temperatures on the Fe–Fe$_3$C phase diagram (see Figure 16.7).
An A_3 value of 843°C corresponds to a carbon content of 0.3 per cent.

(b) Plain carbon steels invariably contain some manganese, an element which lowers the critical temperatures.

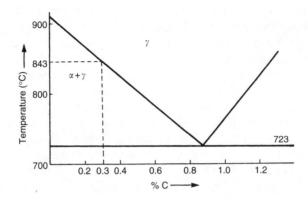

Figure 16.7

16.6 T–T–T and C–C–T diagrams

Martensite and bainite are non-equilibrium phases that do not appear in the Fe–Fe$_3$C phase diagram. In order to show the influence of varying cooling rates, that is time, on the transformation of austenite other types of diagram are necessary. These are the time–temperature–transformation, or T–T–T, and the continuous cooling–transformation, or C–C–T, diagrams. These diagrams are sometimes referred to as 'S' or 'C' curves because of the general shape (see also section 12.7).

When a steel is cooled slowly through the A_3 temperature there will be a small amount of undercooling of the austenite before nucleation and growth of pearlite occurs. The pearlite formed will be coarse in structure. A more rapid rate of cooling will give a greater amount of undercooling before nucleation begins, but the rate of nucleation will be more rapid leading to the formation of a finer pearlite structure. If the rate of cooling exceeds a particular value, known as the critical cooling velocity, austenite will not transform into pearlite but, at a much lower temperature, will transform into the non-equilibrium phase martensite. A typical T–T–T diagram for a plain carbon steel is shown in Figure 16.8.

Figure 16.8 *T–T–T diagram (S curve) for a plain carbon steel.*

The transformation product bainite, which is a finely divided dispersion of carbide particles in ferrite, is formed by the isothermal transformation of undercooled austenite, namely by the rapid quenching of the steel to a position below the nose of the T–T–T curve and then maintaining constant temperature until the transformation is complete.

The position of the nose of the curve, and hence the value of the critical cooling velocity, is not constant for all steels. An increase in the carbon content of the steel, or an increase in the content of other alloying elements, will reduce the value of the critical cooling velocity, that is, move the T–T–T curve to the right. The rate of cooling possible by quenching a steel from a temperature above the critical temperature range in water is roughly equal to the critical cooling velocity of a plain carbon steel containing 0.3 per cent of carbon. Consequently, it is impracticable to harden, by quenching, plain carbon steels with a lesser carbon content than this. However, many low alloy steels with carbon contents of less than 0.3 per cent are quench hardenable. The T–T–T diagrams for a few different steels are given in Figure 16.9 and these curves show that the critical cooling velocity is reduced as either carbon or alloy element content is increased. The curves for some alloy steels show a double nose and there are two critical cooling velocities, a high rate for martensite formation and a lower rate for bainite formation. The curve for a Ni–Cr–Mo steel (Figure 16.9(d)) shows this feature.

The data for the construction of T–T–T diagrams are obtained from the isothermal transformation of austenite at differing temperatures. Most practical heat treatments involve continuous cooling

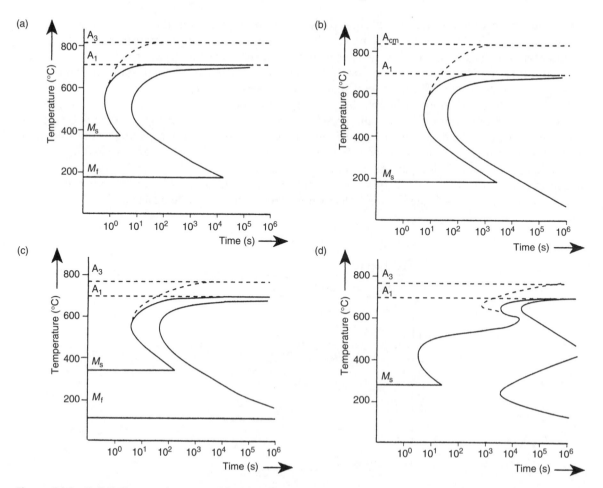

Figure 16.9 *T–T–T diagrams for some steels: (a) carbon steel 0.4 per cent C, 0.6 per cent Mn; (b) carbon steel 1.2 per cent C, 0.7 per cent Mn; (c) low alloy steel 0.4 per cent C, 1.25 per cent Ni, 0.6 per cent Cr; (d) low alloy steel 0.3 per cent C, 1.8 per cent Ni, 0.8 Cr, 0.25 per cent Mo.*

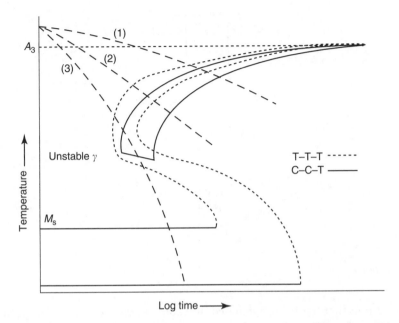

Figure 16.10 *Relationship between a T–T–T diagram and a C–C–T diagram for a plain carbon steel.*

from a high temperature and a T–T–T diagram may not give a fully accurate representation of the temperatures and times of the transformations occurring. The continuous cooling diagram (C–C–T) gives a more accurate picture. Very approximately the C–C–T diagram may be considered as the T–T–T diagram moved slightly to the right and down, in other words showing the transformations occurring at lower temperatures and requiring longer times than those indicated by the T–T–T curve. The relationship between the two types of diagram is shown in Figure 16.10. The cooling rates shown as (1) and (2) show that austenite transforms into pearlite while cooling rate (3) shows that some austenite would transform into fine pearlite and the remainder transforming into martensite at a much lower temperature.

16.7 Hardenability

As stated earlier, the hardness of a martensite is related to the carbon content of a steel. The term *hardenability*, however, refers not to the maximum hardness obtainable, but to the depth of hardening that may be obtained. When a heated bar of a steel is quenched in water from a high temperature, the surface is cooled extremely rapidly, but the interior of the bar is cooled at a slower rate. This means that, while the surface layers may be converted into martensite, the rate of cooling at some little distance below the surface of the bar may be less than the critical cooling velocity of the steel, and pearlite is formed. The depth of hardening obtained in a plain carbon steel may be only 2 or 3 mm. Alloy steels, because of their lower critical cooling velocities, will possess much greater hardenabilities than plain carbon steels and, in some cases, may show martensite at distances of 50 mm, or more, below the surface of the material. A term connected with hardenability is *ruling section*. The ruling section of a steel is the maximum diameter of bar which can be hardened throughout by a quenching treatment. Alloy steels of high hardenability may be air hardening, that is, martensite is formed at the surface of the steel even when it has been cooled in air from a high temperature. A test that is frequently used for assessing the hardenability of a steel is the *Jominy* test. In this test a bar of steel is heated within the austenite range and it is then quenched by directing water on to one end only (see Figure 16.11). The rate of cooling at points along the length of the bar will approximate to the rate of cooling at points below the

Figure 16.11 *Jominy test.*

surface of a large section that has been quenched in water. When the test bar has fully cooled, surface flats are carefully ground, and a series of hardness measurements are then made along the length of the bar, working from the quenched end. Standard conditions are laid down for the dimensions of the test-piece, and for the quenching procedure, in the Jominy test (see BS 4437 (1969) and ASTM A 255–89).

EXAMPLE 16.3

Jominy hardenability curves for three different steels are shown in Figure 16.12.

(a) Which curve relates to the steel of greatest hardenability?

(b) Which curve relates to the steel of highest carbon content?

(c) Estimate the ruling section of steel B.

(d) What would be the likely microstructure of a 15 mm diameter bar of steel A after water quenching from the austenitising temperature?

Figure 16.12

SOLUTION

(a) Steel C is the steel of greatest hardenability.

(b) The hardness of martensite is related to the carbon content, so steel A has the highest carbon content.

(c) Steel B is fully hard for a distance of 20 mm from the quenched end, so its ruling section is 40 mm (the maximum diameter bar that can be hardened throughout).

(d) For steel A martensite is formed for a distance of about 4 mm from the surface. The central core of a 15 mm diameter bar would have a structure of very fine pearlite with, possibly, some bainite.

EXAMPLE 16.4

(a) The T–T–T diagram for a 0.6 per cent carbon steel is shown in Figure 16.13. Draw on the diagram the cooling rates that would be expected if this steel was (i) annealed, (ii) normalised, (iii) oil quenched and (iv) water quenched.

Sketch the microstructures that would be obtained at 25°C for each of (i), (ii) and (iv).

(b) Using the diagram, calculate (i) the critical cooling velocity and (ii) estimate the holding time at 400°C for a 50 per cent transformation into bainite, if the steel is rapidly quenched to 400°C and held at that temperature.

(c) What effect does (i) the addition of carbon and (ii) the addition of alloying elements have on the hardenability of a steel? Why is hardenability an important property in the industrial application of a steel?

Figure 16.13

SOLUTION

Refer to Figures 16.13 and 16.14.

(a) The rates of cooling which are achieved are approximately as follows: (i) slow furnace cool, 0.01–0.05°C/s; (ii) air cool, 5–10°C/s; (iii) oil quench, 100–200°C/s; (iv) water quench, 500–1000°C/s. See the solution diagram (Figure 16.14). It will be seen that during annealing transformation of austenite occurs at a high temperature, only just below the critical temperatures, giving coarse pearlite, but during normalising the transformation into pearlite occurs between about 670°C and 600°C and a finer pearlite will be formed. With oil quenching, transformation into a very fine pearlite will commence at about 570°C and some 70 per cent of the austenite will have transformed by the time the temperature has fallen to 450°C. The remaining austenite will transform into martensite when the M_s temperature is reached. Water quenching will keep to the left of the nose of the curve and austenite will transform into martensite.

   The microstructures expected at 25°C for (i), (ii) and (iv) are shown in Figure 16.15. In (i) and (ii) transformation will have been completed giving coarse pearlite + ferrite and fine pearlite + ferrite respectively. In (iv) transformation into martensite begins at the M_s temperature of about 220°C. At 25°C the structure is of martensite with some retained austenite.

(b) The cooling rate curve which just grazes the nose of the T–T–T diagram is the critical cooling rate and it is about 900–1000°C/s.

   For a sample quenched rapidly to 400°C and maintained at that temperature, 50 per cent transformation is achieved in about 1.5 s.

(c) Both the addition of carbon and the addition of other alloying elements have the effect of moving the T–T–T diagram toward the right, in other words, reducing the critical cooling velocity and increasing hardenability. The effect of alloying elements, such as manganese, nickel and chromium, is much greater than that of carbon in this respect.

Figure 16.14

(i)　　　　　　　(ii)　　　　　　　(iv)

Figure 16.15

Hardenability is an important property of a steel. A steel of high hardenability will harden to a greater depth below the surface than one of lesser hardenability for the same rate of cooling. Also, a martensitic structure can be achieved with a slower rate of cooling in a steel of high hardenability than is the case for a low hardenability steel, in some alloy steels even cooling in air is sufficiently rapid. The less drastic quenching techniques that are needed lessen the risk of cracking and distortion occurring in heat-treated components.

16.8 | Tempering

Martensite, although very hard, is also extremely brittle and a hardened steel requires a further heat treatment, known as *tempering*, before it can be put into service. When a martensite structure is heated it becomes possible for the carbon trapped in supersaturated solid solution to diffuse through the lattice and precipitate from solution in the form of iron carbide particles. This precipitation will relieve the strain within the lattice and cause the hardness and brittleness of the material to be reduced. This diffusion process can commence at temperatures of about 200°C, but the rate of diffusion is extremely slow at this temperature. An increase in the temperature will cause an increase in diffusion and precipitation rates and, therefore, increase the extent of the softening. At temperatures up to 450°C the carbide precipitate particles are much too fine to be resolved under the optical microscope, although their presence may be detected by using more sophisticated techniques. At higher temperatures the carbide particles increase in size and, at 700°C, the cementite coalesces into a series of fairly large, and roughly spheroidal particles (700°C is just below the lower critical temperature). This gives rise to a soft, but incredibly tough, material. Originally, the microstructures observed in tempered steels were given the names *troostite* and *sorbite*, but these names have largely gone out of use today, and structures are generally described as *tempered martensite*. See Table 16.1.

Table 16.1 *Tempering temperatures.*

Tempering temperature (°C)	Suitable for
220	Hacksaw blades
230	Planing and slotting tools, hammers
240	Milling cutters, drills, reamers
250	Taps, shear blades, punches, dies
260	Stone-cutting tools, drills for wood
270	Axes, press tools
280	Cold chisels, wood chisels, plane blades
290	Screwdrivers
300	Wood saws, springs
450–700	To produce great toughness, but at the expense of hardness

Tempering temperatures and times have to be fairly accurately controlled in order to produce the desired properties in the material. In small workshops this is accomplished by observing the colour that develops when a clean steel surface is heated. The colour developed is an interference colour, owing to the thickness of the oxide film formed on the metal surface, and this colour ranges from a pale-yellow, at 200°C, through brown, at 250°C, to dark-blue at 300°C.

16.9 | Heat treatments for steels

Some of the effects of heat treatments on steels have been mentioned in earlier paragraphs, but it is now convenient to give fuller definitions of the various steel heat treatments.

Full annealing is the treatment given to produce the softest possible condition in a hypoeutectoid steel. It involves heating the steel to a temperature within the range 30–50°C above the upper critical temperature and then allowing the steel to cool slowly within the furnace. This produces a structure containing coarse pearlite.

Full annealing is an expensive treatment and when it is not absolutely essential for the steel to be in a very soft condition, but a reasonably soft and ductile material is required, the process known as *normalising* is used instead. Normalising involves heating the steel to 30–50°C above the upper critical temperature, but followed by cooling in still air. The structure of the pearlite produced is finer than in the fully annealed state, giving a slightly harder and stronger material.

A different process has to be used for softening hypereutectoid steels. If a hypereutectoid steel is slowly cooled from above the A_{cm} temperature, a brittle grain boundary network of cementite will be formed in the structure. Hypereutectoid steels are softened by giving them a *spheroidising* treatment, sometimes known as *spheroidising annealing*. This involves heating the steel to a temperature just below the lower critical temperature. During this treatment cementite forms as spheroidal particles in a ferrite matrix, putting the steel into a soft, but very tough, condition.

Process annealing, or *sub-critical annealing*, is often used for softening cold worked low carbon, or *mild*, steels. To fully anneal such a steel would involve heating to a temperature of more than 900°C, with subsequent high cost. In a mild steel ferrite makes up about 90 per cent of the structure, and the recrystallisation temperature of cold worked ferrite is only about 500°C. Annealing a cold worked mild steel in the temperature range 550–600°C will result in complete recrystallisation of ferrite, although the cold worked pearlite will be largely unaffected.

Hardening. This involves rapidly quenching the steel, from a high temperature, into oil or water, in order to bring about the formation of martensite. Hypoeutectoid steels are heated to 30–50°C above the upper critical temperature prior to quenching. If hypereutectoid steels are quenched from above the A_{cm} temperature, it is possible that some cementite would be precipitated at austenite grain boundaries. This would have an embrittling effect, which would still be evident after subsequent tempering. Consequently, hypereutectoid steels are hardened by quenching from 30–50°C above the lower critical temperature. At this temperature the structure becomes one of spheroidal cementite particles in an austenite matrix. Quenching transforms the austenite into martensite. (Note that it is impracticable to harden, by quenching, plain carbon steels containing less than 0.3 per cent of carbon.)

Tempering. Hardened steels may be tempered by heating them within the temperature range 200–700°C. This treatment will remove internal stresses set up during quenching, remove some, or all, of the hardness, and increase the toughness of the material (see also section 16.8). For most steels, cooling from the tempering temperature may be either cooling in air, or quenching in oil or water. Some alloy steels, however, may become embrittled if slowly cooled from the tempering temperature, and these steels have to be quenched.

The various heat treatments quoted above apply to both plain carbon and low-alloy steels. See Figure 16.16.

Figure 16.16 *Heat treatment temperatures for plain carbon steels.*

16.10 | Types of steels and their uses

As has already been stated, the hardness and strength of steels vary very considerably with both carbon content and type of heat treatment. Certain names, which relate to the carbon content, are used in connection with steels. *Mild* or *low carbon steels* are those containing up to 0.3 per cent of carbon. Steels containing between 0.3 and 0.6 per cent of carbon are termed *medium carbon steels,* and these may be hardened and tempered. Steels containing more than 0.6 per cent of carbon are always used in the hardened and tempered condition, and these are known as *high carbon steels,* or *tool steels.*

Table 16.2 gives some typical uses of steels of various carbon contents.

Table 16.2 *Compositions and typical applications of steels.*

C (per cent)	Name	Applications
0.05	Dead mild steel	Sheet and strip for presswork, car bodies, tin-plate; wire, rod, and tubing
0.08–0.15	Mild steel	Sheet and strip for presswork; wire and rod for nails, screws, concrete reinforcement bar
0.15	Mild steel	Case carburising quality
0.1–0.3	Mild steel	Steel plate and sections, for structural work
0.25–0.4	Medium carbon steel	Bright drawn bar
0.3–0.45	Medium carbon steel	Shafts and high-tensile tubing
0.4–0.5	Medium carbon steel	Shafts, gears, railway tyres
0.55–0.65	High carbon steel	Forging dies, railway rails, springs
0.65–0.75	High carbon steel	Hammers, saws, cylinder linings
0.75–0.85	High carbon steel	Cold chisels, forging die blocks
0.85–0.95	High carbon steel	Punches, shear blades, high-tensile wire
0.95–1.1	High carbon steel	Knives, axes, picks, screwing dies and taps, milling cutters
1.1–1.4	High carbon steel	Ball bearings, drills, wood-cutting and metal-cutting tools, razors

16.11 Surface hardening

Generally, the toughness of a material decreases as the hardness and strength increase. There are very many service conditions where the requirement is for a tough material of very high surface hardness. This is the case with shafts and gears, in particular. An ideal solution to the material problem would be to create a very hard surface layer on a comparatively soft and tough material. The surface hardening of steels is readily accomplished by a variety of methods.

Case carburising is one of the methods used for producing a hard surface on a ductile steel. It involves the introduction of additional carbon into the surface of a mild steel, effectively producing a composite material consisting of a low carbon steel with a thin case, about 0.5–0.7 mm in thickness, of high carbon steel. The principal methods used are *pack carburising* and *gas carburising*. In pack carburising the parts to be treated are heated above the upper critical temperature in contact with wood or bone charcoal and barium carbonate, within a cast iron container. In gas carburising the parts are heated above the upper critical temperature in a furnace with an atmosphere of methane, or mixed hydrocarbon gases.

Considerable grain growth occurs in the material during a carburising treatment, and a three-stage heat treatment must be given to the carburised parts to produce the desired final properties. This heat treatment involves:

(a) refining the core structure by heating to a value above the upper critical temperature for the core composition, followed by rapid quenching;

(b) hardening the case by heating to just above the lower critical temperature, followed by quenching;

(c) tempering the case.

Another commonly used surface hardening process is *cyaniding*. In this process the steel components are heated in a bath of molten sodium cyanide and sodium carbonate at a temperature of about 950°C. During this treatment both carbon and nitrogen diffuse into the surface of the steel. The formation of hard iron nitrides contributes to the surface hardening of the material. After cyaniding, the parts require the same type of three-stage heat treatment as is given to case-carburised articles.

Carburised and cyanided articles have to be ground to final dimensions after all heat treatments have been given, as some dimensional changes occur during the processes.

Nitriding is another process for the surface hardening of a steel. Nitriding is not suitable for the hardening of plain carbon steels, as iron nitrides could be formed to a considerable depth below the surface of the steel, embrittling the material. Steels for nitriding are low-alloy steels containing chromium and molybdenum, together with, in some cases, nickel and vanadium. Some nitriding steels contain about 1 per cent of aluminium. Aluminium nitride is extremely hard, and does not diffuse very far below the surface (about 0.8 mm). Nitriding is carried out by heating the steel parts at about 500°C in a gas-tight chamber, in an atmosphere of ammonia. The ammonia dissociates at the steel surface into nitrogen and hydrogen and some of the former is absorbed by the steel. The major advantages of the nitriding process are:

(a) an extremely hard surface is formed,

(b) the treatment is conducted at comparatively low temperatures, minimising cracking and distortion,

(c) no subsequent heat treatment is necessary.

Parts for nitriding are first heat treated, to produce the best core properties. Machining to final dimensions is then carried out, while the material is in a soft condition, allowing for the small growth of 0.02 mm that occurs during nitriding.

Flame hardening and *induction hardening* are very rapid processes for the production of hard surfaces on steels with carbon contents in excess of 0.35 per cent. The steel surface is rapidly raised to the critical temperature range by means of a flame, or by being enclosed within a high-frequency induction coil. The surface is then rapidly quenched to promote martensite formation. After quenching, the out-flow of heat from the core to the surface is sufficient to temper the surface layers. These processes are widely used, and they are particularly suitable for hardening the journals of shafts.

The above-mentioned and other methods for surface hardening are summarised in Table 16.3.

16.12 | The effects of alloying elements in steels

It is possible to improve the properties of steels greatly by alloying with other elements. The general effects of alloying elements in steels are as follows:

(a) An alloying element entering into solid solution will increase the strength of the steel. An increase in carbon content will increase the strength of the steel, but with a considerable loss of ductility. Many of the commonly used alloying elements increase the strength of steels with little or no reduction in ductility.

(b) Almost without exception, alloying elements reduce critical cooling velocities and, so, increase the hardenability of a steel.

(c) The alloying element will have an effect on the α to γ transformation temperature of iron, and may either increase or decrease this (Figure 16.17). Chromium, tungsten and silicon are some of the elements that raise the transformation temperature, while nickel and manganese are elements that lower it. There will be similar alterations in the critical temperatures of steels that contain alloying elements in addition to carbon. It will be noted that if a large amount of alloying element is present, a transformation could be eliminated and the steel becomes either wholly ferritic, or wholly austenitic.

(d) The alloying element could form stable carbides. Manganese, chromium and tungsten are some of the elements that have a strong carbide-forming tendency, and these carbides are extremely hard.

(e) The alloying element could cause the breakdown of cementite, resulting in the appearance of graphite in the structure of the steel. Nickel and silicon have this effect and, for this reason, are not added to high carbon steels.

(f) The alloying element may confer some property, characteristic of itself, on the steel. For example, chromium, which is corrosion resistant, will make a steel corrosion resistant, or stainless, if it is present in the steel in amounts in excess of 12 per cent.

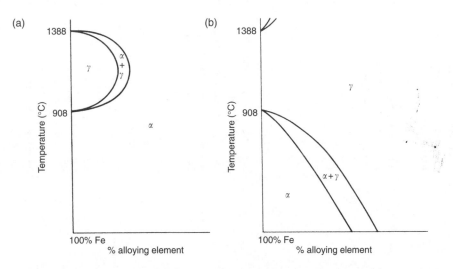

Figure 16.17 *Effect of alloying addition on allotropic transformation temperatures of iron: (a) ferrite-stabilising (γ loop) element – (Cr, W, Si); (b) austenite-stabilising element – (Ni, Mn).*

Table 16.3 *Surface hardening methods.*

	Carburising	Nitriding	Cyaniding	Carbonitriding
Effect	A high carbon surface is produced on a low carbon steel and is hardened by quenching	A very hard nitride-containing surface is produced on the surface of a strong tough steel	A carbon and nitride-containing surface is produced on a low carbon steel and is hardened by quenching	Carbon and nitrogen are added to the surface of a low carbon steel and permit hardening by an oil quench
Suitability	Suitable for plain carbon or alloy steels containing about 0.15 per cent C	Nitralloy steels containing aluminium; a typical nitriding steel contains 0.3 per cent C, 1.6 per cent Cr, 0.2 per cent Mo, 1.1 per cent Al. This steel is hardened by oil quenching from 900°C and tempered at 600–700°C before being nitrided	Suitable for plain carbon or alloy steels containing about 0.15 per cent C	Suitable for plain carbon steels containing about 0.15 per cent C
Method	Low carbon steel is heated at 850–930°C in contact with gaseous, liquid, or solid carbon-containing substances for several hours. The high carbon steel surface produced is then hardened by quenching	The steel is heated at 500–540°C in an atmosphere of ammonia gas for 50–100 hours. No further heat treatment is necessary	Low carbon steel is heated at 870°C in a molten 30 per cent sodium cyanide bath for about one hour. Quenching in oil or water from this bath hardens the surface of the steel	Low carbon steel is heated at 700–870°C for several hours in a gaseous ammonia and hydrocarbon atmosphere. Nitrogen in the surface layer increases hardenability and permits hardening by an oil quench
Result	Case depth is about 1.25 mm. Hardness after heat treatment is H_{RC} 65 (H_D 870). Negligible dimension change caused by carburising. Distortion may occur during heat treatment	Case depth is about 0.38 mm. Extreme hardness (H_D 1100). Growth of 0.025–0.05 mm occurs during nitriding. Case is not softened by heating for long times up to 420°C. Case has improved corrosion resistance	Case depth is about 0.25 mm. Hardness is about H_{RC} 65. Negligible dimension change is caused by cyaniding. Distortion may occur during heat treatment	Case depth is about 0.5 mm. Hardness after heat treatment is H_{RC} 65 (H_D 870). Negligible dimension change occurs. Distortion is less than in carburising or cyaniding
Applications	Typical uses are for gears, camshafts and bearings	Typical uses are for valve guides and seatings, and for gears	Typical uses are for small gears, chain links, nuts, bolts and screws	Typical uses are for gears, nuts and bolts

	Flame hardening	Induction hardening	Siliconising (Ihrigising)	Hard chromium plating
Effect	The surface of a hardenable steel or iron is heated by a gas torch and quenched	The surface of a hardenable steel or iron is heated by a high-frequency electro-magnetic field and quenched	A moderately hard corrosion-resistant surface containing 14 per cent silicon is produced on low carbon steels	A hard chromium plate is applied directly to the metal surface
Suitability	Steel containing 0.4–0.5 per cent carbon or cast iron containing 0.4–0.8 per cent combined carbon may be hardened by this method	Steel containing 0.4–0.5 per cent carbon or cast iron containing 0.4–0.8 per cent combined carbon may be hardened by this method	Suitable for plain carbon steels containing 0.1–0.2 per cent carbon	Generally used on steels, low or high carbon, soft or hardened
Method	A gas flame quickly heats the surface layer of the steel and a water spray or other type of quench hardens the surface	The section of steel to be hardened is placed inside an induction coil. A heavy induced current heats the steel surface in a few seconds. A water spray or other type of quench hardens the surface	The steel parts are heated at 930–1000°C in contact with silicon carbide and chlorine gas for two hours. No further heat treatment is required	The steel parts are plated in the usual plating bath but without the usual undercoat of nickel. The plating is a thousand times thicker than decorative chromium plating
Result	The hardened layer is about 3 mm thick. Hardness is H_{RC} 50–60 (H_D 500–700). Distortion can often be minimised	The hardened layer is about 3 mm thick. Hardness is H_{RC} 50–60 (H_D 500–700). Distortion can often be minimised. Surface remains clean	Case depth is about 0.63 mm. Hardness is about H_D 200. Case has good corrosion resistance. Growth of 0.025–0.05 mm occurs during siliconising	Plating thickness is about 0.125 mm. Extreme hardness H_D 900. Plating has good corrosion resistance and a low coefficient of friction
Applications	Used for gear teeth, sliding ways, bearing surfaces, axles and shafts	Use for gear teeth, sliding ways, bearing surfaces, axles and shafts	Typical uses are for valves, tubing and shafts	Typical uses are for dies, gauges, tools and cylinder bores

I6.13 | Alloy steels

Alloy steels are generally classified into two major categories, *low-alloy* steels and *high-alloy* steels. The main effects conferred by specific alloying elements are given in Table 16.4.

Low-alloy steels. A low-alloy steel generally contains up to 3 or 4 per cent of one or more alloying elements and is characterised by possessing similar microstructures to, and requiring similar heat treatments to, plain carbon steels. Frequently they are referred to as pearlitic alloy steels as the normalised structure contains the eutectoid pearlite. The presence of alloying elements provides enhanced properties such as increased strength without loss of toughness and increased hardenability. The applications of low-alloy steels are similar to those quoted in Table 16.2 for plain carbon steels of equivalent carbon content, for example, 0.4 per cent carbon is quoted as typical for the manufacture of shafts – an alloy steel containing 1 per cent nickel and 0.4 per cent carbon is an ideal shaft material with improved strength and toughness over a plain carbon steel and also an increase in the resistance to fatigue conferred by the presence of nickel.

A major category of low-alloy steel is the group known as *micro-alloyed steels* or *high-strength low-alloy steels* (HSLA). The grain size of the ferrite formed by the transformation of austenite during cooling is a key factor in the control of strength and toughness of low carbon steels. The strongest and toughest steels of this type have been developed by making micro-alloying additions of the elements aluminium, niobium and vanadium, either singly or in combination, to give a major grain refinement. The amount of alloying addition is very carefully controlled and for each element is in the range 0.003 to 0.08 or 0.10 per cent. These elements form carbonitrides which inhibit austenite grain growth during hot working operations, give a highly refined ferrite grain size, and can also introduce a capacity for the steel to respond to a precipitation hardening treatment. In order to achieve the finest grain sizes the hot rolling temperatures must be closely controlled with a low finishing temperature, just above the A_{c1} temperature, and with close control of the cooling rate after hot working. The micro-alloyed steels form a very useful range of low-carbon high-strength structural steels with tensile strengths of up to 700 MPa and tensile yield strengths in the range 350–500 MPa.

Another important range of structural steels are those which are weather resistant. These are low carbon structural steels with additions of about 0.3 per cent copper, 0.5–1.5 per cent chromium and, in one case, about 0.5 per cent nickel. Unlike other steels, after exposure to wind and rain, these do not corrode with the formation of the usual porous and flaky rust deposit but, rather, develop an adherent protective oxide film which, after about two years' exposure to the weather, has a pleasing purplish-copper colour.

High-alloy steels. High-alloy steels are those that possess structures, and require heat treatments, that differ considerably from those of plain carbon steels. Their room temperature structures after normalising may be austenitic, martensitic or contain precipitated carbides. Generally, they contain more than 5 per cent of alloying element. A few examples of some high-alloy steels are given below.

(a) *High-speed tool steels.* Tungsten and chromium form very hard and stable carbides. Both elements also raise the critical temperatures and, also, cause an increase in softening temperatures. High-carbon steels rich in these elements provide hard wearing metal-cutting tools, which retain their high hardness at temperatures up to 600°C. A widely used high-speed tool steel composition is 18/4/1 steel, containing 18 per cent of tungsten, 4 per cent of chromium, 1 per cent of vanadium and 0.8 per cent of carbon. In a steel of such a composition very hard and stable chromium carbides, vanadium carbide and complex iron–tungsten carbides are formed instead of the normal cementite. The annealed material, produced by very slow cooling from a 'soaking' temperature of about 900°C, will possess a microstructure of ferrite and spheroidal carbide particles, with most of the alloy element content combined as carbide. The heat treatment that is needed to put the material into a hard and tough condition is heating at a very high temperature (1200–1300°C), to take a proportion of the alloying elements and carbides into solution, followed by oil quenching. The oil-quenched structure will comprise carbides, martensite and a high proportion of retained austenite. Quenching is followed by a double tempering

Table 16.4 *Effects of alloying elements in steels.*

Alloying element	General effects	Typical steels
Manganese	Increases the strength and hardness and forms a carbide; increases hardenability; lowers the critical temperature range, and when in sufficient quantity, produces an austenitic steel; always present in a steel to some extent because it is used as a deoxidiser	Pearlitic steels (up to 2 per cent Mn) with high hardenability used for shafts, gears and connecting rods; 13 per cent Mn in Hadfield's steel, a tough austenitic steel
Silicon	Strengthens ferrite, raises the critical temperatures; has a strong graphitising tendency; always present to some extent because it is used, with manganese, as a deoxidiser	Silicon steel (0.07 per cent C; 4 per cent Si) used for transformer cores; used with chromium (3.5 per cent Si; 8 per cent Cr) for its high-temperature oxidation resistance in internal combustion engine valves
Chromium	Increases strength and hardness, forms hard and stable carbides; raises the critical temperatures; increases hardenability; amounts in excess of 12 per cent render steel stainless	1.0–1.5 per cent Cr in medium and high carbon steels for gears, axles, shafts and springs, ball bearings and metal-working rolls; 12–30 per cent Cr in martensitic and ferritic stainless steels; also used in conjunction with nickel (see below)
Nickel	Marked strengthening effect, lowers the critical temperature range; increases hardenability; improves resistance to fatigue; strong graphite-forming tendency; stabilises austenite when in sufficient quantity	0.3–0.4 per cent C with up to 5 per cent Ni used for crankshafts and axles, and other parts subject to fatigue
Nickel and chromium	Frequently used together in the ratio Ni/Cr = 3/1 in pearlitic steels; good effects of each element are additive, each element counteracts disadvantages of the other; also used together for austenitic stainless steels	0.15 per cent C with Ni and Cr used for case carburising; 0.3 per cent C with Ni and Cr used for gears, shafts, axles and connecting rods; 18 per cent, or more, of chromium and 8 per cent, or more, of nickel give austenitic stainless steels
Tungsten	Forms hard and stable carbides; raises the critical temperature range, and tempering temperatures; hardened tungsten steels resist tempering up to 600°C	Major constituent in high-speed tool steels; also used in some permanent magnet steels
Molybdenum	Strong carbide-forming element, and also improves high-temperature creep resistance; reduces temper-brittleness in Ni–Cr steels	Not normally used alone; a constituent of high-speed tool steels, creep-resistant steels and up to 0.5 per cent Mo often added to pearlitic Ni–Cr steels to reduce temper-brittleness
Vanadium	Strong carbide-forming element; has a scavenging action and produces clean, inclusion-free steels	Not used on its own, but is added to high-speed steels, and to some pearlitic chromium steels
Titanium	Strong carbide-forming element	Not used on its own, but added as a carbide stabiliser to some austenitic stainless steels
Aluminium	Soluble in ferrite, also forms nitrides	Added to nitriding steels to restrict nitride formation to surface layers
Cobalt	Strengthens, but decreases hardenability	Used in 'Stellite' type alloys, magnet steels, and as a binder in cemented carbides
Niobium	Strong carbide former, increases creep resistance	Added for improved creep resistance and as a stabiliser in some austenitic stainless steels
Copper	Increases strength and corrosion resistance. >0.7% Cu permits precipitation hardening	Added to cast steels to improve fluidity, castability and strength. Used in corrosion resistant architectural steels
Lead	Insoluble in iron	Added to low-carbon steels to give free-machining properties

process. The first-stage tempering at 300–400°C will allow most of the retained austenite to transform into martensite while the second tempering at 550–600°C relieves internal stresses and forms a structure of carbide particles in a toughened martensite matrix. This high-alloy content martensite does not soften appreciably until it is heated at temperatures in excess of 600°C making them usable as cutting tools at high cutting speeds.

(b) *Stainless steels.* When chromium is present in amounts in excess of 12 per cent, the steel becomes highly resistant to corrosion, owing to the protective film of chromium oxide that forms on the metal surface. Chromium also raises the α to γ transformation temperature of iron, and tends to stabilise ferrite in the structure. There are several types of stainless steel, and these are summarised below.

(i) Ferritic stainless steels contain between 12 and 25 per cent of chromium and less than 0.1 per cent of carbon. These steels are ferritic in structure at all temperatures up to their melting points. As austenite cannot be formed, it is impossible to form hard martensitic structures by quenching the steels from high temperature. This type of stainless steel cannot be heat treated, but may be strengthened by work hardening.

(ii) Martensitic stainless steels contain between 12 and 18 per cent of chromium, together with carbon contents ranging from 0.1 to 1.5 per cent. The presence of carbon restores the α to γ transition, and these steels can be hardened by quenching from the austenite range.

(iii) Austenitic steels contain both chromium and nickel. When nickel is present, the tendency of nickel to lower the critical temperatures overrides the opposite effect of chromium, and the structure may become wholly austenitic. The austenitic stainless steels, which are also non-magnetic, are widely used in chemical engineering plants, for cutlery and for architectural work. A notable alloy is 18/8 stainless steel, which contains 18 per cent of chromium, and 8 per cent of nickel. Carbon contents are kept below 0.15 per cent in the austenitic steels in order to minimise the formation of chromium carbides in the structure, as this would cause a reduction in corrosion resistance. Carbides may form in these steels if they are allowed to cool slowly from high temperature, or if they are reheated in the range 500–700°C. As this latter condition may apply in the heat-affected zones adjacent to welds, the type of corrosion failure that can occur, owing to the presence of carbide particles, is known as *weld decay*. Austenitic stainless steels that are required for welding contain small stabilising additions of titanium, or niobium, and these stabilisers prevent the intercrystalline corrosion, weld decay. An austenitic stainless steel with improved strength, hardness and wear resistance can be achieved by introducing a fine dispersion of titanium nitride particles throughout the material. The method is applicable to parts made by the powder metallurgy route. Prior to compaction and sintering, the fine metal powder is nitrided in an ammonia atmosphere at 700°C, during which some of the titanium content of the steel is converted to the nitride.

(c) *Maraging steels.* These are very high-strength materials that can be hardened to give tensile strengths of up to 1900 MPa. They contain 18 per cent nickel, 7 per cent cobalt and small amounts of other elements such as titanium. The carbon content is low, generally less than 0.05 per cent.

The heat treatments that are necessary to develop maximum properties are a solution treatment at 800–850°C, in order to produce a uniform austenite structure, followed by rapid quench. A martensitic structure is formed after quenching but this is relatively soft as the carbon content of the steel is very low. This is followed by precipitation treatment at about 450°C. A very finely dispersed precipitate of complex intermetallic phases occurs which causes great strengthening.

A major advantage of maraging steels is that after the solution treatment they are soft enough to be worked and machined with comparative ease. The precipitation treatment is at a fairly low temperature when distortion of machined parts is negligible. Although the basic material cost of a maraging steel is very high the final cost of a complex component made from a maraging steel may be less than if made from a more conventional and cheaper high-strength steel because of the much lower machining costs. One major use of maraging steels is for the manufacture of aircraft undercarriage components.

(d) *Hadfields manganese steel.* This is a high-alloy steel that contains 12–14 per cent of manganese and 1 per cent of carbon. The high manganese content renders this steel composition austenitic at all temperatures. It is also, therefore, non-magnetic. This type of steel is unique in that any attempt to deform, or abrade, the material greatly increases the surface hardness. Because of the exceptionally high resistance of this type of steel to abrasion, coupled with the fact that the core of the material remains comparatively soft and tough, it is used for pneumatic drill bits, rock crusher jaws, excavator bucket teeth, and railway points and switches.

16.14 Cast irons

Cast irons are selected grades of pig iron, which are remelted and cast in sand moulds. The carbon content of cast irons is generally between 2 and 4 per cent. Although normal grades of cast irons are not particularly strong, and are quite brittle, they are widely used as engineering materials, because of their cheapness, ease of melting and casting, very good machineability and a high damping capacity. According to the phase diagram (Figure 16.1) the phases that should exist at ordinary temperatures are ferrite, cementite and pearlite. Cementite, however, is only metastable, and most cast irons contain graphite in their structures. Cast irons may be classified as either *white* or *grey*. These terms arise from the appearance of a freshly fractured surface. An iron that contains graphite in its structure shows a fracture surface that is grey and dull in colour. If, however, the whole of the carbon content of an iron is in the combined form, as cementite, a fresh fracture surface appears bright and silvery, and this is termed a white iron structure. White irons are exceptionally hard, but brittle. White irons are not often used, except as the basis for the production of malleable irons, although portions of many sand castings are made to solidify white, by the insertion of chills in the sand mould, in order to produce hard and wear resistant surfaces in parts of the castings.

According to the phase diagram (Figure 16.1), the eutectic occurs at 4.3 per cent carbon, but the presence of other elements, notably silicon and phosphorus, will affect the position of the eutectic. The effects of these elements can be assessed by assigning carbon replacement values to them. The carbon equivalent value, CE, is given by

$$CE = \%C + 0.33\,(\%Si + \%P)$$

For example, a cast iron with a composition of 3.0% C, 2.8% Si and 1.1% P would have a carbon equivalent value of: CE = 3.0 + 0.33 (2.8 + 1.1) = 4.3%, the eutectic composition.

The structure of cast irons is affected by the following factors:

(a) The rate of solidification. Slow rates of solidification allow for graphite formation and castings made in sand moulds tend to solidify grey. More rapid solidification will tend to give white iron structures. Metal chills are sometimes inserted into parts of sand moulds in those areas where a high surface hardness is required.

(b) Carbon content. The higher the carbon content of the iron, the greater will be the tendency for it to solidify grey.

(c) The presence of other elements. Some elements promote the formation of graphite in an iron structure. Silicon and nickel have strong graphitising tendencies. Figure 16.18 illustrates the effect of silicon on the structure of a cast iron.

(d) The effect of heat treatment. The prolonged heating of a white iron will cause graphitisation to occur. This phenomenon is used as the basis for the production of malleable irons (refer to section 16.15). Graphite is less dense than cementite and, if cementite decomposes into ferrite and graphite during service, this change will be accompanied by a reduction in the density of the iron, and a corresponding increase in dimensions. This phenomenon is termed the *growth* of cast irons, and irons for high-temperature service must be in a fully graphitised state before being put into use.

When an iron solidifies from the molten state, the crystals of graphite usually form as irregularly shaped flakes, many of which may be quite large. The presence of graphite in this form gives the material a low value of fracture toughness. Cast iron containing flake graphite of this type is perfectly suitable for many low- and medium-duty applications but it is possible, by using special melting and/or melt treatment procedures, to produce high strength irons. These treatments are given below.

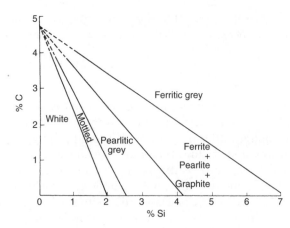

Figure 16.18 *Effect of silicon and carbon content on the structure of sand-cast irons.*

Lanz–Perlit iron. This is a high-strength iron with a microstructure of fine graphite flakes and pearlite produced by casting an iron with controlled low carbon and silicon contents into a mould which has been preheated to about 400°C. This composition iron, if cast in the normal way, would solidify to give either a white or mottled structure but, with the slower solidification rate within a heated mould, it solidifies giving a structure of fine graphite flakes and pearlite, even in thin sections. The growth of graphite in thicker sections of the casting is restricted because of the low silicon content of the iron. This means that castings of variable section will have fairly uniform micro-structures. This type of cast iron is used for high quality castings, including automotive cylinder blocks.

Meehanite. This is the proprietary name for a high-strength cast iron. It is an iron composition which, ordinarily, would solidify white but is treated by superheating the melt to about 1500°C and inoculating it with calcium silicide immediately before casting. This produces a grey iron structure with very small graphite flakes, resulting in improved mechanical properties.

SG iron. SG (spheroidal graphite) or nodular cast irons possess structures in which graphite is present in the form of roughly spherical-shaped nodules rather than as flakes. This type of structure is achieved by inoculating the melt, prior to casting, with small additions of magnesium and/or cerium. The amounts required are about 0.05 per cent Mg or 0.01 per cent Ce. The presence of spheroidal, as opposed to flake, graphite improves the strength and fracture toughness of the material. SG irons may be made in a wide range of compositions and the properties may be improved further by means of alloying additions.

The two photomicrographs in Figure 16.19 show the appearance of large irregular graphite flakes in a grey cast iron and graphite nodules in a SG iron.

16.15 Malleable irons

Malleable irons are produced by the heat treatment of certain white irons. There are two types of process used giving rise to *blackheart* and *whiteheart* irons, respectively. These names arise from the appearance of the fracture surface of a treated iron.

In the blackheart process, white iron castings are heated at a temperature of about 900°C for between 2 and 3 days, and then cooled very slowly, at about 3°C per hour. A neutral atmosphere is maintained in the furnace during the treatment. Cementite in the white iron structure breaks down into ferrite and roughly spherical aggregates of graphite. If the iron is cooled more rapidly from the treatment temperature, the resulting structure will consist of graphite in a matrix of ferrite and pearlite.

Figure 16.19 *(a) Large flakes of graphite in a grey cast iron (×200). (b) Graphite nodules in a spheroidal graphite cast iron (×100).*

The whiteheart process involves packing the white iron castings into boxes with haematite ore, and heating to a temperature of 900°C for a period of 2–5 days. Because the casting is in contact with haematite, carbon is oxidised away from the surface, and the resulting structure is composed of ferrite at the edge of the casting and ferrite, pearlite, and some graphite nodules at the centre.

16.16 | Alloy cast irons

Alloying elements may be added to cast irons in order to improve their properties, and alloy cast irons have been developed to give high-strength materials, hard and abrasion-resistant materials, corrosion resistant irons, and irons for high-temperature service. For the production of high-strength irons, additions of chromium, nickel and molybdenum are commonly made. The addition of about 5 per cent of nickel causes the formation of a martensitic structure at the surface of a sand casting. This very hard constituent makes this type of iron very suitable for the manufacture of metal working rolls.

The best irons for corrosion resistance are those containing 15–25 per cent of nickel. These irons are austenitic in structure, and non-magnetic. They usually contain some chromium and copper, in addition to the nickel content, to give some further improvement in corrosion resistance.

For prolonged service at elevated temperature it is important that the iron does not grow. In order to ensure that the structure of the iron is completely graphitic when cast, the carbon content is kept down to about 2 per cent, and the alloy contains about 5 per cent of silicon, or silicon and nickel. The presence of silicon also reduces the rate of oxide scale formation at high temperatures.

Table 16.5 *Composition and properties of some cast irons.*

Approximate composition	Tensile strength (MPa)	Type and uses
3.2% C, 1.9% Si	250	Pearlite and graphite. Motor brake drums
3.25% C, 2.25% Si	220	Pearlite and graphite. Engine cylinder blocks
3.25% C, 2.25% Si, 0.35% P	185	Ferrite, pearlite and graphite. Light machine castings
3.25% C, 1.75% Si, 0.35% P	200	Ferrite, pearlite and graphite. Medium machine castings
3.25% C, 1.25% Si, 0.35% P	250	Pearlite and graphite. Heavy machine castings
3.6% C, 2.8% Si, 0.5% P	370	Wear resistant. Piston rings
3.6% C, 1.7% Si	540	Pearlitic SG
3.6% C, 2.2% Si	415	Ferritic SG
2.8% C, 0.9% Si	310	Blackheart malleable
3.3% C, 0.6% Si	340	Whiteheart malleable
2.9% C, 2.1% Si, 1.75% Ni, 0.8% Mo	450	Shock resistant. Crankshafts for petrol and diesel engines
2.9% C, 2.1% Si, 15% Ni, 2% Cr, 6% Cu	220	Ni-resist. Corrosion-resistant austenitic iron. Used in chemical plant
2.5% C, 5% Si	170	Silal. A growth-resistant iron for high-temperature service

SUMMARY

Steels and cast irons are, essentially, alloys of iron and carbon and, together, account for more than 90 per cent of the tonnage of metals used This is because of their availability, relatively low cost and the wide range of properties that can be developed by heat treatment and alloying.

Low-carbon steels are relatively strong, highly ductile and can be processed readily by cold working.

Medium- and high-carbon steels can be hardened and tempered by heat treatment to obtain the optimum properties of hardness and toughness required for a wide range of engineering applications.

Low-alloy steels have been developed to possess enhanced properties as compared with plain carbon steels.

High-alloy steels have structures, and require heat treatments, that differ from those of plain-carbon and low-alloy steels and have been developed to possess one or more special characteristics, for example, as stainless steels or with a high proportion of hard carbides in the structure for high-speed metal cutting.

Cast irons contain between 2.5 and 4.5 per cent carbon and, depending on composition and solidification rates, solidify as either *white* or *grey*. All the carbon is present in the form of Fe_3C in a white iron but in a grey iron some or all of the carbon exists as graphite.

Grey irons containing flake graphite tend to be brittle but nodular irons, with the graphite present as spheroidal nodules has increased toughness. Malleable irons may be produced by heat treating certain white iron compositions.

Alloy cast irons have been developed to have improved or special properties, such as high corrosion resistance.

Questions

16.1 Microexamination of two plain carbon steels shows that steel 1 contains about 60 per cent ferrite and 40 per cent pearlite and steel 2 contains about 95 per cent pearlite and 5 per cent free cementite. Estimate the percentage carbon in each steel.

16.2 The graphs showing dimensional change with temperature for a hypoeutectoid steel during heating and cooling, both at 15°C per minute, are shown in Figure 16.20.
  (a) Determine the equilibrium values of the critical temperatures for this steel and explain any differences from values expected from consideration of the iron–carbon phase diagram.
  (b) Estimate the percentage carbon in the steel.

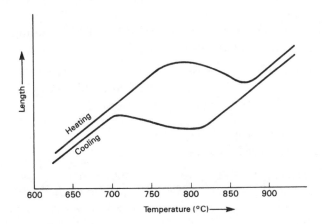

Figure 16.20

16.3 The T–T–T diagram for a steel is shown in Figure 16.21. Use the diagram to estimate:
  (a) the critical cooling velocity,
  (b) a cooling rate to give a coarse pearlitic structure,
  (c) a cooling rate to give very fine pearlite with no bainite or martensite,
  (d) the minimum holding time, after rapid quenching to 425°C, to give a fully bainitic structure.

Figure 16.21

16.4 How would the following heat treatments be conducted, quoting approximate temperatures and cooling rates?
(a) Normalising a plain carbon steel containing 0.4 per cent carbon.
(b) Hardening a plain carbon steel containing 0.8 per cent carbon.
(c) Process annealing a cold rolled low carbon steel.

16.5 (a) A plain carbon steel containing 0.6 per cent carbon is heat treated as follows:
    (i) heated to 800°C and quenched in cold water,
    (ii) heated to 800°C and slowly cooled in the furnace,
    (iii) heated to 800°C, quenched in cold water and reheated at 250°C,
    (iv) heated to 800°C, quenched in cold water and reheated at 600°C.
Describe, with the aid of diagrams, the structures obtained by these treatments.
(b) Indicate, in a general way, the type of physical properties you would expect the different treatments to produce.
(c) Which heat treatment would you consider best for (i) cutting tools and (ii) dynamically stressed engineering components?

16.6 State, giving reasons, the type of steel which should be selected for each of the following applications, and the nature of the heat treatment, if any, which should be given:
(a) steel sheet for making domestic refrigerator casings,
(b) a hacksaw blade,
(c) a crankshaft for a four-cylinder petrol engine,
(d) plain reinforcement bar for concrete.

16.7 (a) Explain the concept of T–T–T curves. How are they obtained and how do they help in understanding the various transformation products observed in steels?
(b) What is meant by the term *hardenability*? Use this concept to account for the difference in properties you would expect from air quenched and tempered samples of (i) steel 080M40 – 0.4 per cent C, 0.8 per cent Mn, and (ii) steel 817M40 – 0.4 per cent C, 0.8 per cent Mn, 1.5 per cent Ni, 1.0 per cent Cr, 0.2 per cent Mo. Give examples of the use of both of these steels in engineering applications.

16.8 (a) Describe how the hardenability of a steel may be determined experimentally.
(b) Sketch the types of hardenability curves which would be obtained from tests on the following steels:
    (i) steel 080M40 – 0.4 per cent C, 0.8 per cent Mn,
    (ii) steel 817M40 – 0.4 per cent C, 0.8 per cent Mn, 1.5 per cent Ni, 1.0 per cent Cr, 0.2 per cent Mo,
    (iii) steel 835M30 – 0.3 per cent C, 0.5 per cent Mn, 4.0 per cent Ni, 1.25 per cent Cr.

16.9 Three of the methods available for surface hardening steel components are:
(i) carburising, (ii) induction hardening and (iii) nitriding.
Compare and contrast these three processes, quoting conditions, approximate temperatures, necessary heat treatments and applicability.

Select the appropriate method for case hardening an intricate component made from a 0.4 per cent C plain carbon steel. Justify the choice and explain why the remaining two methods would be unsuitable.

16.10 Summarise the effects of the following elements as alloying additions to steels: nickel, chromium, manganese, molybdenum and vanadium.
Indicate some types of commercial alloy steels which make use of the effects you have referred to in the answer above.

16.11 What are the main classifications of stainless steels?
Discuss the different types of stainless steel, making reference to approximate compositions, structures, heat treatments. Give typical applications for each of the main categories of stainless steel.

16.12 (a) Describe the structures of the main types of cast irons and account for their continued use as engineering materials.
(b) In what way does the silicon content and sectional thickness affect the structure of a cast iron?
(c) What is meant by the term *carbon equivalent?*
(d) Why was the discovery of a method of manufacturing nodular cast iron important and what are the advantages of nodular irons over malleable irons?

16.13 (a) What are the factors that influence the form that carbon takes up in a cast iron?
(b) Alloying additions may be made to irons to produce corrosion-resistant irons, heat-resistant irons and high-strength irons. What are the principal additions made for these purposes and in what ways are the microstructures and properties affected?

16.14 The compositions of several cast irons are given in the table below:

	C %	Si %	Mn %	P %
1	3.5	1.2	0.8	0.1
2	3.3	2.2	0.5	0.3
3	3.6	1.8	0.5	0.8
4	3.0	5.0	0.3	0.1
5	3.3	0.6	0.5	0.1

(a) Determine the carbon equivalents of these irons.
(b) State, giving reasons, which iron would:
(i) be most likely to solidify white when sand cast,
(ii) be most suitable for prolonged service at high temperatures.

CHAPTER 17

Polymer materials

Why study polymer materials?

There are very many polymer materials and they can be divided into three main groupings: *thermoplastics, elastomers* and *thermosets*. The polymer, or plastics, industry had it beginnings in the early 20th century. The rate of growth of the industry was high, at about 7 per cent per annum, for most of the latter half of that century and the industry continues to develop and introduce new products. Some of the attractive characteristics of polymer materials are low density, good thermal and electrical insulation properties, high resistance to chemical attack, ease of fabrication into simple or complex shapes and relatively low cost. Generally, they have low strength and elastic modulus values, as compared with metals, and also relatively low softening temperatures. Polymer materials are used widely in industrial applications and the engineer (and designer) needs to have knowledge of the availability and general properties of these and how properties can be modified by the use of fillers and other additives.

LEARNING OBJECTIVES

1. Understand the differences between thermoplastic, elastomeric and thermosetting materials.
2. Know that polymer moulding compounds contain many ingredients in addition to the polymer.
3. Gain some knowledge of the properties and applications of some of the more common thermoplastic materials.
4. Gain some knowledge of the properties and applications of the main members of the group of materials termed elastomers.
5. Have some knowledge of the main thermoset materials.

17.1 | Introduction

Plastics or polymer materials can be divided into three main groupings: *thermoplastics, elastomers* and *thermosets*. The plastics, industry, which had its beginnings in the early 20th century, is based almost entirely on petrochemicals derived from petroleum and natural gas. The rate of growth of the industry was high, at about 7 per cent per annum, for most of the latter half of the last century and the industry continues to develop and introduce new products.

In many instances plastics materials have replaced metals but it should not be thought that plastics materials will eventually replace all metals. There are many areas of application in which the selection of a metal or alloy is the only feasible solution to a design requirement. However, in many cases a plastic material can provide a better solution than a metallic material. A simple example of this is the domestic washing-up bowl. In the middle of the last century bowls made of injection moulded polyethylene were introduced, replacing bowls made from low carbon steel sheet and finished with a vitreous enamel finish. Polyethylene bowls are low cost, light, durable and can be produced in a wide range of attractive colours, whereas steel bowls, although long lasting with careful use, could easily have the enamelled surface chipped through accidental damage.

The large increase in the use of polymer materials during recent decades has been due to their many useful properties, coupled with comparatively low cost, which more than offset their disadvantages. Their useful properties include low density, good thermal and electrical insulation properties, high resistance to chemical attack, and ease of fabrication into simple or complex shapes. The main disadvantages of polymer materials are low strength and elastic modulus values, as compared with metals, and also relatively low softening temperatures. However, their low strength and stiffness may often be improved by using certain fillers as reinforcement. Polymer materials are used widely in industrial applications and the engineer (and designer) needs to have knowledge of the availability and general properties of these and how properties can be modified by the use of fillers and other additives.

Plastics materials, as supplied to a component manufacturer for the production of mouldings or extrusions, are not pure polymers. The moulding compound of, say, polypropylene is supplied in the form of small granules or as a powder and the actual polymer content may be as little as 50 per cent of the total mass. The remainder is made up of inert fillers and reinforcing materials, dyes and pigments, stabilisers, antistatic agents, flame retardants, plasticisers, and so on. The exact nature of the additives and their proportion may vary from one supplier to another. This means that the polypropylene from supplier A may have properties which differ markedly from those shown by polypropylene from supplier B. This is reflected in the wide variation in some of the properties quoted for plastics in this chapter. Information on some of the more important polymer materials is given in the subsequent sections of this chapter. Composite materials based on polymer materials are discussed in Chapter 19.

EXAMPLE 17.1

List the main classes of polymer material.
   Which general type of material would be suitable for the manufacture of the following: (a) a luncheon box, (b) the casing for a domestic light switch, and (c) disposable surgical gloves?

SOLUTION

The main classes of polymers are *thermoplastics, thermosets* and *elastomers.*
(a) A luncheon box needs to be non-toxic, fairly rigid, resistant to damage if dropped accidentally, yet have sufficient flexibility so that the lid may be fitted to give an airtight closure. These requirements will be met by a thermoplastic material, several of which are suitable.
(b) A light switch housing needs to be strong, hard and rigid. A thermoset moulding will be ideal for this application.
(c) Surgical gloves must have great flexibility and be capable of considerable elastic strain under low deformation forces so that they may be pulled over the hands to give a close fit. This need is fulfilled by an elastomeric material.

17.2 | Thermoplastic materials

Many thermoplastic polymer materials have been developed during the last 70 years. Some, such as polyethylene and polyvinyl chloride, are produced in great quantity while others, for example PEEK and polysulphones, have been developed for use at relatively high temperatures and these latter are costly, made in small quantity and used for certain specific applications where their special characteristics are needed.

Table 17.1 gives the generic names of the main thermoplastic materials together with some of their main applications, while the range of mechanical properties of the thermoplastics is given in Table 17.2 and the structural formulae of some common polymers are given in Appendix B.

17.3 | Hydrocarbon polymers

The major hydrocarbon thermoplastic polymers are polyethylene, polypropylene and polystyrene.

Polyethylene (PE)

Polyethylene, or as it is more commonly known, polythene, is made by the polymerisation of ethene (ethylene) ($CH_2=CH_2$). The properties and applications of this polymer vary over wide ranges depending on the molecular weight, the method of manufacture, and differences in structure and density. There are several varieties of polyethylene made, these being *high density* (HDPE), *low density* (LDPE), *linear low density* (LLDPE) and *ultra-high-molecular weight* (UHMWPE).

The raw polymer is compounded with various additives to produce a material suitable for moulding. The principal additions that may be made are: (1) up to 2 per cent of carbon black to improve the stability of the material when exposed to sunlight; (2) up to 10 per cent butyl rubber to prevent in-service cracking in blow moulded containers for oils and detergents; (3) pigments to produce the desired colour.

The various types of polyethylene have excellent resistance to most solvents and chemicals and are tough and flexible over a wide range of temperature. With the exception of UHMWPE, they have low softening temperatures and can be readily moulded and formed into shapes. They are also non-toxic and possess good electrical insulation properties.

UHMWPE has an extremely good chemical resistance to both acids and alkalis, is very tough and has a very good wear resistance. It finds many applications, particularly in machine construction and in the chemical industry. Figure 17.1 shows a large pump housing moulded from this material.

Polyethylene can be cross-linked to give higher strength and stiffness at elevated temperatures. The cross-linking is developed after the polyethylene has been moulded or formed into the desired shape and is brought about either by exposure to irradiation (see section 4.5) or by a chemical reaction with peroxides. The former method requires the use of expensive equipment while the latter method may cause a deterioration in electrical properties. Polyethylene can also be modified to give materials of increased strength and high stiffness by using glass fibre as a filler.

Polyethylene can also be formed as a foam. Polyethylene foam can be moulded and shaped and it has a very good chemical resistance. It is more expensive than polystyrene foam but has a good resilience and finds use for acoustic insulation and for the packaging of delicate equipment.

A number of useful copolymers based on ethylene have been developed. Ethylene and propylene can be copolymerised in various proportions to produce a range of different products. One of these is termed *polyallomer*. Other ethylene–propylene copolymers are ethylene–propylene thermoplastic elastomers, the elastomers EPM and EPDM, the latter being a terpolymer between ethylene, propylene and a diene (see section 17.10).

Ethylene–vinyl acetate copolymers (EVA) are block copolymers which have a better low-temperature flexibility and tear resistance than LDPE but their chemical resistance is poorer than that material.

Figure 17.1 *Pump housing made from UHMWPE. (Courtesy of Ticona)*

They are alternative materials to PPVC and do not have to be compounded with additional plasticising agents, as is the case with PVC. Major applications for EVA copolymers are for the manufacture of ice-cube trays, flexible toys, disposable gloves, vacuum cleaner hose and flexible pipe connectors. EVA can be produced as film and this, which has a high clarity, is used for shrink-wrapping.

Ethylene may be copolymerised with some fluorocompounds (see section 17.4).

Ionomer. Ionomer is the name given to a copolymer of ethylene with an organic acid, for example methacrylic acid, together with metallic ions such as NA^+, K^+, Mg^{2+} or Zn^{2+}. Ionomer is a soft flexible transparent material with a better resistance to stress-cracking than LDPE. Ionic cross-links exist between polymer molecules at ordinary temperatures and this makes the material more rigid than LDPE. These ionic links break when the material is heated, thus allowing the polymer to be moulded with the same degree of ease as LDPE. The ionic cross-links reform as the material cools.

Polypropylene (PP)

The monomer propene (propylene), $CH_2{=}CH{-}CH_3$, is related to ethene and, like ethane, possesses a double covalent bond that may split to allow addition polymerisation to occur.

Polyproplyene is a highly crystalline olefinic polymer and the presence of CH_3 groups attached to the linear chain molecules gives polypropylene a greater strength and stiffness than HDPE although its density is similar to that of LDPE. It has a higher softening point than HDPE and can be used to make bottles that may be boiled or steam sterilised without softening. It has a good resistance to most forms of chemical attack and excellent fatigue resistance. One very useful property of polypropylene is its excellent fatigue resistance and its ability to be bent repeatedly without fear of cracking. An example of how this property has been utilised by designers is the integrally moulded hinge. Many small boxes and components such as car accelerator pedals are produced as one-piece mouldings in polypropylene with an integrally moulded hinge.

Polypropylene is brittle at low temperatures and, to overcome this, high impact grades are available. High impact polypropylene may be made by mechanically blending the polymer with small amounts of EPM or EPDM elastomers. A toughened grade of polypropylene, which is a block copolymer of propene with between 2 and 10 per cent of ethene, is also available. This block copolymer is also known as *pollyallomer*. The brittleness temperature of polyallomer is reduced to about −40 °C.

Polystyrene (PS)

Polystyrene is a low cost, brittle, glassy (T_g = 100 °C), transparent polymer. It is easy to process and has good dimensional stability. It has a poor chemical resistance and is susceptible to degradation if exposed to ultra-violet light. Unmodified polystyrene is used for toys, food boxes, light diffusers and components for domestic appliances.

High impact polystyrene is a hybrid material composed of small particles of an elastomer (BR or SBR) dispersed throughout a polystyrene matrix with some graft polymerisation having taken place at the boundaries between the two phases. This is a much tougher material than unmodified polystyrene and again is used for toys, mouldings for domestic and office appliances, boxes and cases.

Much polystyrene is used in expanded form as a solid foam. The foam is available in a rigid massive form, which can be cut to size and shape or as 'sandwich' sheets for decorative shapes and mouldings. Expandable beads and shapes are available also, and these may be expanded in place or pre-expanded and moulded. Styrene polymer compounded with 6 per cent *n*-pentane, a gas-releasing chemical, is one of the most common expandable foam polymer grades for in-place expansion. The polymer beads are heated to about 120 °C in a jacketed mould, at which temperature the polymer softens and the expanding agent decomposes. Major applications of moulded, expanded foam is for food trays, cups and food packaging. In pre-expansion the beads are heated in a steam bath. This causes the beads to expand to about 30 times their previous size. These beads are broken apart and can be used for insulation and packaging.

Styrene is also used as the basis of some addition copolymers. The copolymer of styrene and butadiene (SBR) is an elastomer and is discussed in section 17.9. Other copolymers are ABS, a terpolymer of acrylonitrile, butadiene and styrene, and SAN (styrene–acrylonitrile).

ABS has good mechanical properties, outstanding impact strength, and is dimensionally stable over a variety of conditions. It is also resistant to many acids, alkalis and many petroleum-based solvents. It is readily mouldable and can be calendered into sheet for vacuum forming.

SAN, which generally contains some 20–30 per cent of acrylonitrile, possesses a good combination of strength, toughness, rigidity and transparency, with a better chemical resistance than polystyrene.

All the styrene based materials may be glass-filled for added strength and rigidity.

17.4 | Chlorocarbon and fluorocarbon polymers

Polyvinyl chloride (PVC)

The monomer vinyl chloride, CH_2=CHCl, is a derivative of ethylene and can be readily polymerised by an addition process to give the linear polymer PVC, $(-CH_2-CHCl-)_n$, a relatively cheap thermoplastic material with a good resistance to acids and alkalis. The presence of chlorine atoms causes the molecular chains to be polarised and there are, in consequence, polar attractive forces between adjacent molecules. This gives a material which is quite hard and rigid at ordinary temperatures.

About one-quarter of the PVC produced is used in the unplasticised form (UPVC), while the remainder is produced as plasticised PVC (PPVC).

UPVC. UPVC is a strong, rigid and tough material with good resistance to ultra-violet light. The main applications for the material include many products for the building industry.

PPVC. The raw polymer is compounded with a semi-solvent or an oil as a plasticising agent to make PPVC, a soft, flexible, rubbery material. The amount of plasticiser may range from 5 to 50 per cent of the final weight of the moulding material, the amount and type determining the final properties. Also, the properties of the material may vary with time due to migration of plasticiser. The chemical resistance of PPVC is less than that for UPVC. Inert fillers may be added to PPVC, both to cheapen and also to improve the properties for some applications, for example in flexible floor tiles and floor coverings.

PVC has a tendency to decompose with the liberation of hydrochloric acid gas when it is heated or exposed to sunlight. To prevent this occurring a small amount of a lead salt, lead silicate or lead stearate, is blended into the polymer at the compounding stage. The lead salt acts as a *stabiliser*. Other additives that may be compounded with PVC are pigments to produce the desired colour, plasticisers and fillers. PPVC may suffer from microbial attack unless fungicides and bactericides are included in the formulation.

Copolymers. Vinyl chloride may be copolymerised with other monomers, including vinyl acetate and acrylonitrile. In copolymers with either of these two monomers, no external plasticising agent need be compounded with the material as both vinyl acetate and acrylonitrile act as *internal plasticisers*.

There are two other polymers that are related to PVC. These are CPVC (chlorinated PVC), a polymer of dichloroethylene, $(-CHCl-CHCl-)_n$, and PVDC (polyvinylidene chloride), with the formula $(-CCl_2-CH_2-)_n$.

Polytetrafluoroethylene (PTFE)

PTFE (polytetrafluoroethylene), formula $(-CF_2-CF_2-)_n$, is the most widely used fluorocarbon polymer. It is a highly crystalline material that does not soften appreciably on heating. It loses its crystallinity above 327 °C but does not flow significantly under stress even at these high temperatures. Generally, components in PTFE are made by hot pressing and sintering the polymer powder, in a manner similar to that used for ceramic and powder metal parts.

The carbon–fluorine covalent bond is extremely strong and the presence of these bonds gives a material of extreme stability, PTFE is almost completely chemically inert, will not stick to most materials and possesses an extremely low coefficient of friction. Its main uses are as chemically resistant coatings, seals and gaskets in chemical plant, non-stick coatings and low friction bearing bushes. Certain properties can be improved by using a filler, for example, chopped glass fibre filled PTFE has greatly increased creep resistance, while the use of a bronze powder or bronze–graphite mix as a filler gives material suitable for bearing applications.

There are other fluorocarbon polymers, these being PVDF, PVF and PFA. PVDF (polyvinylidene fluoride), $(-CH_2-CF_2-)_n$, has the highest tensile strength of all the melt processable fluorocarbon materials, together with good resistance to abrasion and chemical attack. Its main applications are for pipes, pipe fittings, bearings and cable insulation for the chemical processing industry, and for chemical laboratory apparatus. PVF (polyvinyl fluoride), $(-CH_2-CHF-)_n$, is a crystalline polymer with an excellent resistance to weathering, staining and chemical attack. It is very slow burning and has low permeability to vapours. Its low thermal stability normally precludes injection moulding and the material is usually only available as film. It is used mainly as a coating material for metals, glazing panels and as a film covering in aircraft interiors to reduce flammability. PFA (perfluoroalkoxyethylene) is a melt processable thermoplastic material with the highest temperature resistance of any of the melt processable fluoroplastics and a resistance to chemical attack similar to that of PTFE. It is a material with a high toughness which possesses a similar stiffness and strength to PTFE at room temperatures and retains its stiffness and strength at elevated temperatures. It is, however, more expensive than PTFE. Its main applications include sheathing for heater cables and chemically resistant linings for pumps and pipes that also require a high temperature resistence.

Tetrafluoroethylene can be copolymerised with ethylene to produce the copolymer ETFE. This material possess a very high impact strength and, for a flurocarbon polymer, a high stiffness. It has a poorer chemical resistance than PTFE and is very expensive. The main applications for it include chemical apparatus, components for pumps and valves and as resistant lining. ECTFE is another fluoro-copolymer, being a copolymer of ethylene and chlorotrifluoroethylene, $(CFCl=CF_2)$. The material is very expensive and its main applications include chemically resistant linings, pump and valve components, and tubing for optical fibres.

17.5 | Acrylic materials

Acrylic materials are thermoplastic polymers based on the polymerisation of esters of acrylic acid and/or methacrylic acid. The most widely used acrylic polymers are PMMA (polymethyl methacrylate) and PAN (Polyacrylonitrile). Other polymers in the group are based on ethyl, propyl, or butyl acrylate or methacrylate.

PMMA is produced by the addition polymerisation of methyl methacrylate and is commonly known as *perspex* or *plexiglass*.

The polymer is in the amorphous state at ordinary temperature and is well below its T_g of 110 °C. PMMA is a hard, rigid and high impact strength thermoplastic that is highly transparent to light and can be readily formed by most of the forming techniques used for thermoplastics. It is resistant to most household chemicals but is attacked by petrol, acetone and cleaning fluids. It is one of the most widely employed optical materials, because, in cast form, it has a higher transparency than many types of optical glass. The material can be readily coloured and its colour and decorative properties are excellent. Moulded components are often decorated by hot stamping, spray painting and vacuum metallisation. Although the material is classed as slow-burning, acrylic smoke is a hazard. The applications are numerous and include aircraft glazing.

Polyacrylonitrile (PAN) is an acrylic polymer made by the addition polymerization of acrylonitrile. Polyacrylonitrile is remarkably stable and is resistant to acids and greases. It is used mainly for synthetic fibre manufacture. As its density is considerably lower than that of cotton or wool, it is used as a fibre in the manufacture of attractive, but lightweight, garments. Some proprietary names for acrylonitrile fibre are *Acrilan, Courtelle* and *Orlon*. It is also a precursor for carbon fibre manufacture, which gives a higher yield than rayon. Filaments of the polymer are highly oriented and decompose without melting when heated to the carbonisation temperature.

Some polyacrylates are elastomers (see section 17.10).

17.6 | Polyamides (nylons) (PA)

The polyamides, or *nylons*, are the products of condensation reactions between an amine and an organic acid (or the self-condensation of an amino acid) and they possess amide groups as an integral part of a linear chain. This family of polymers was developed as a synthetic replacement for natural silk in the late 1930s. The polyamides resemble linear polyesters in that they are long carbon chains with recurring functional groups in the chain.

The orginal reactants were hexamethylene diamine and adipic acid. As both of these reactants contain six carbon atoms, the product is termed PA 6.6.

$$\underset{\text{Hexamethylene diamine}}{\overset{H}{\underset{H}{>}}N-(CH_2)_6-N\overset{H}{\underset{H}{<}}} \quad + \quad HO\ OC-(CH_2)_4-COOH \atop \text{Adipic acid}$$

The nomenclature of the polyamides is based on the number of carbon atoms in the monomer molecules. In the example above, both monomers have six carbon atoms. PA 6 is made from *caprolactam*, or its parent compound *ω-amino caproic acid*, both compounds with 6 carbon atoms. These compounds contain an amine group at one end of the molecule and an acid group at the other permitting self-condensation. PA 6.10 is made from hexamethylene diamine (6 carbon atoms) and sebacic acid (10 carbon atoms), PA 11 from 11 amino-undecanoic acid (11 carbon atoms), and so on.

The polyamides, as a group, are very strong and tough and possess good abrasion resistance. They are flexible and have high impact strengths. They are thermoplastic but possess high softening temperatures, in excess of 200 °C and so moulding is difficult. Much PA material is produced as fine fibre for conversion into fine fabrics and rope but they are also good moulding compounds. In general they possess a good resistance to most solvents and chemicals but are affected by phenols. They tend to absorb water, with a consequent reduction in strength, but the amount of absorption is less with some types than others. Fillers and property modifiers may be used and very many grades are available with mineral fillers, chopped carbon or glass fibre reinforcement, with solid-phase lubricants such as molybdenum disulphide or PTFE, and many others. Some of the characteristics of the main PA polymers are given below.

PA 6.6. PA 6.6 has a good abrasion resistance, better than PA 6, and is the strongest and stiffest aliphatic polyamide. It possesses a good resistance to fatigue and has a better low temperature toughness than either PA 6, PBT or acetal. It is relatively difficult to process as it has an exceptionally low melt viscosity. It also shows high mould and post-moulding shrinkage. It has a high water absorption (8 per cent saturated). The main applications are as gears, cams, bearings, nuts and bolts, wheels, power tool casings and many automotive components, including rotationally moulded fuel tanks, radiator tops, timing chain covers and fan blades.

PA 6. This polymer has the highest rate of water absorption and the highest equilibrium water content of all the polyamides, but it is easier to process than PA 6.6 and is also castable and may be reaction injection moulded (RIM). The material has a good fatigue resistance but a lower strength and stiffness than PA 6.6. The stiffness may be improved by addition of glass fibre. The range of applications for PA 6 is generally similar to that for PA 6.6. The choice between PA 6 and PA 6.6 is often made on the basis of availability and cost or a familiarity with the product rather than on analysis of all the properties.

PA 6.10. This material has a lower brittle temperature and a lower water absorption (4 per cent saturated) than either PA 6 or PA 6.6 and is stronger than PA 11, PA 12 or PA 6.12. This material shows a high mould shrinkage and is more expensive than PA 6.6.

PA 6.12. PA 6.12 has a similar water absorption (4 per cent saturated) and brittle temperature to PA 6.10 but has a better heat resistance than that material. It possesses a lower mould shrinkage than PA 6, PA 6.6 or PA 6.10. It is more expensive than PA 6 or PA 6.6.

PA 11. This material has a low water absorption (2.5 per cent saturated) and a better resistance to ultra-violet light than other polyamides. It is expensive and its applications include compressed air, hydraulic and petrol hosing, gears, cams and tool handles.

PA 12. This material is very expensive but has the lowest water absorption of any polyamide (2 per cent saturated). Its applications include precision engineering components and components requiring good low temperature toughness. It is used as a sports shoe sole material also.

PA (RIM). RIM (reaction injection moulded) polyamide, a proprietary name for which is Nyrim, shows a better stiffness than RIM polyurethanes. Although more difficult to mould because of the higher temperatures required and having some dimensional instability due to moisture absorption, this material is used for large structural parts, such as automotive panels and housings for business machines, because of its combined properties of stiffness and toughness.

17.7 Heterochain polymers

Many polymer materials possess linear molecular chains with some of the atoms in the chain being of elements other than carbon. These are termed *heterochain molecules* and they may contain oxygen, nitrogen or sulphur. These heterochain compounds give rise to a large range of materials which are very useful for many engineering applications.

Acetal (POM)

Acetal, also known as *polyacetal* and *polyoxymethylene*, is based on the polymerisation of formaldehyde and has the structural formula $(-CH_2-O-)_n$. Acetals possess good strength and toughness coupled with a low coefficient of friction. They also retain their properties at temperatures of up to 120 °C. Known also by the proprietary name of *Delrin*, the materials' main applications are as a substitute for metals for the manufacture of water taps, pump parts and other engineering components.

Polycarbonate (PC)

Polycarbonate is a linear heterochain polymer made from the condensation of bisphenol A and carbonic acid. It is a tough, transparent material with a high impact resistance at temperatures down to −40 °C and can also be used continuously up to 115 °C. It does, however, have low fatigue and wear resistance, is attacked by some organic solvents and is susceptible to stress cracking. It is, though, the second engineering thermoplastic, after polyamides, in terms of tonnage produced, being used for such applications as safety helmets, shields and goggles, lenses, glazing, lighting fittings, instrument casings and machine housings, and in electrical switchgear. It is used also for sterilisable medical components, kitchen ware and, in the form of laminated sheet, for bullet-proof glazing. Other grades of polycarbonate available include glass and carbon fibre reinforced materials, PTFE lubricated polymer, structural foams and ultra-violet stabilised and fire retardant materials. Polycarbonate is also a component in hybrid polymers, the main example being PC/ABS and PC/PBT.

Saturated polyesters

An *ester* is the product of the condensation reaction between an organic acid and an alcohol. PET (polyethylene teraphthalate), also known as PETP, is a linear polyester made by the condensation polymerization of ethylene glycol, and teraphthalic acid.

PET is a high strength, high stiffness thermoplastic produced as a fibre, as transparent film and as a moulding material. Moulding grades may be reinforced with glass fibre. Injection moulded PET has sufficient strength, toughness and dimensional stability to be processed into large structural parts. The Chrysler Composite Concept Vehicle (CCV) (Figure 17.2) has been designed with a body consisting of only six large PET mouldings which is then directly bonded to the steel chassis. (See also Chapter 30.)

Fibre is extruded from the liquid state in a similar manner to polyamide fibre and is drawn at a temperature above its T_g to give maximum molecular orientation. One proprietary name for PET fibre is *terylene*. PET suffers from a high mould shrinkage and a limited resistance to hydrolysis but possess good

Figure 17.2 *The Chrysler Composite Concept vehicle (CCV). (Courtesy of Ticona)*

Figure 17.3 *Gearbox for an automotive electric window lifter injection moulded in PBT. (Courtesy of Ticona)*

electrical and thermal properties and can operate continuously at temperatures of up to 120 °C. It is used for electrical parts such as transformer coil bobbins and components for tuners and fire alarms, and the manufacture of blow-moulded bottles for carbonated and alcoholic drinks, perfumes and drugs. In film form it is used for packaging, photographic and radiographic film, audio/video tapes, capacitor film and drawing office transparencies.

PBT (polybutylene teraphthalate) is a thermoplastic with a good combination of stiffness and toughness. It can withstand continuous service at 120 °C and possesses good arcing and tracking resistance. However, its high mould shrinkage and poor resistance to hydrolysis are problems. The main applications for the material are similar to those for PET. Glass fibre reinforced and fire retardant grades of PBT are also available. One example of a complex injection moulded part in PBT is illustrated in Figure 17.3.

PCDT (poly (1.4) cyclohexylene dimethylene teraphthalate) is more expensive than PET but possesses higher temperature stability and it has a superior resistance to strong alkalis. It is used for similar applications to PET.

The only thermoplastic polyesters are of the saturated type. Unsaturated polyesters form the basis of some useful thermosetting materials and are dealt with in section 17.15.

Cellulosics

Cellulose is a naturally occurring high polymer, formed by the photosynthesis of glucose, and it is the chief constituent of most forms of plant life, being one of the most widely distributed chemical materials available. It forms the framework of cell walls in trees and most green plants. The composition of wood comprises about 50 per cent cellulose although the exact amount varies from species to species. The natural fibre cotton is about 98 per cent cellulose. The most important sources for chemical processing are cotton, linen and wood pulp. The structure of cellulose is shown in section 19.2.

Cellulose, generally in the form of cotton linters, can be treated to produce a range of materials, cellulose nitrate (CN), cellulose acetate (CA), cellulose acetate butyrate (CAB), cellulose propionate (CP), ethyl cellulose (EC), viscose and viscose rayon.

The first synthetic resin derivative of cellulose was *celluloid* or cellulose nitrate (CN). Cellulose fibres, in the form of cotton linters or purified wood cellulose, is nitrated in an aqueous solution of 55 per cent nitric and 25 per cent sulphuric acids to give cellulose nitrate, often referred to as nitro-cellulose. After separation and washing, the cellulose nitrate is mixed with camphor, as a plasticiser, and stabilisers and compacted into a block. This block can be sliced into celluloid sheets in thicknesses ranging from 0.125 to 25 mm. A major disadvantage of celluloid is its very high flammability. Celluloid is used for brush backs, knife handles and table tennis balls.

Cellulose acetate (CA) is made by reacting cellulose with acetic anhydride and acetic acid, using sulphuric acid as a catalyst. Another process is the reaction between cellulose, acetic anhydride and methylene chloride with perchloric acid as a catalyst. It is mainly the diacetate which is formed but by modifying the process cellulose triacetate can be produced. Cellulose acetate is a transparent, glossy thermoplastic which can be injection moulded and formed into film and sheet. Acetate fibre can be formed by spinning from a solution of cellulose acetate in acetone, the fibre hardening as the solvent evaporates. The diacetate suffers from high moisture absorption, with associated dimensional instability, and is flammable, though much less so than celluloid. The triacetate has a lower moisture resistance than the diacetate and also has a lower flammability. Applications for CA include brush backs, cosmetic containers, spectacle frames, tool handles and toys, as sheet and film for packaging and, the triacetate, as the base for photographic film, and as fibres. Acetate fibre is weaker than rayon but is more elastic and resistant to wrinkling.

Cellulose acetate butyrate (CAB) is made by reacting cellulose with a mixture of acetic and butyric anhydrides. It is a transparent, glossy polymer which is tougher and has a lower moisture absorption than cellulose acetate and has a better resistance to weathering than either cellulose acetate or cellulose propionate. Applications include bathroom fittings, decorative automobile trim, gun stocks, pens, set squares, stencils and other drawing aids, tool handles and for packaging, especially blister packaging.

Cellulose propionate (CP) is made by reacting cellulose with propionic anhydride. It is a transparent, glossy polymer which is stiffer, has a better low temperature impact strength, but poorer resistance to weathering than both cellulose acetate and cellulose acetate butyrate. It is more expensive than both CA and CAB but is used for a similar range of applications.

Ethyl cellulose (EC) is a polymer with good toughness, thermoplasticity, low flammability, and low water absorption. The unplasticised grades remain flexible down to temperatures as low as about $-40\,°C$. However, the material is readily attacked by organic liquids. An important use for ethyl cellulose is in the production of protective hot dips by mixing with oils, plasticisers and stabilisers. Metal components such as tools, gauges and valves are dipped in the hot molten mix to give a tough, transparent, strippable coating. The material is readily injection moulded and extruded and applications include helmets, furniture trim, luggage and tubing. Ethyl cellulose is also used as an additive in laquers, varnishes and adhesives to improve toughness.

Cellulose reacts with sodium hydroxide and carbon disulphide to form a compound called cellulose xanthate. Viscose is a solution of cellulose xanthate in dilute sodium hydroxide, from which regenerated cellulose can be produced. The degree of polymerisation of regenerated cellulose is much lower than that of natural cellulose. Viscose is made into two main products, a fibre termed viscose rayon and a sheet material, *cellophane*. Cellophane is a transparent and tough material used widely in packaging.

PEEK

PEEK (polyether etherketone) is a linear crystalline heterochain polymer developed in the 1970s for service at high temperatures. It has a melt temperature of about $330\,°C$ and, though melt processable, is difficult to process, but can be used continuously at temperatures up to $250\,°C$.

PEEK has good fatigue and chemical resistance and is an inherent fire retardant material. Glass fibre and carbon fibre reinforced grades are available. The main uses are in aerospace applications and in radiation environments as injection moulded components, wire covering and as a resin in fibre prepregs.

Polyphenylenes

PPO (polyphenylene oxide), also known by the proprietary name *Noryl*, is an amorphous linear heterochain polymer developed in the 1960s with good mechanical properties and possessing a high impact strength over a wide range of temperature ($-40°$ to $+150\,°C$), and suitable for continuous use at temperatures up to $80\,°C$.

PPO has a low mould shrinkage and shows very good dimensional stability over a wide range of temperature. It has a good resistance to hydrolysis but not to many solvents. PPO is compatible with polystyrene and the two are miscible in all proportions. Modified PPO is a single phase hybrid polymer of PPO and polystyrene and is a more easily processable material. It is relatively expensive but is widely used for such applications as electrical fittings, components for dishwashers and washing machines. It has found widespread use as a replacement for zinc-based die-castings in business machines, pumps and impellers and is particularly suitable for pumping fluids at high temperatures.

PPS (polyphenylene sulphide) is a related material and is also a linear heterochain polymer which was developed in the 1970s. It has a good temperature stability and is suitable for continuous use at temperatures up to 190 °C.

PPS has good resistance to chemicals and solvents, and good flame resistance, but is difficult to pigment and, hence, is only available in dark colours. PPS, although a thermoplastic, has some of the characteristics of a thermoset and can be cured by heating at about 200 °C. It is thought that chain length extensions and some cross-linking are brought about by heat treatment. The main uses of the polymer are similar to those for PPO.

An example of mouldings in PPS is illustrated in Figure 17.4.

Figure 17.4 *Pump casing mouldings for use in a soft drinks dispensing machine made in PPS. (Courtesy of Ticona)*

Polysulphones and polyarylates

Polysulphones and polyarylates are tough, glassy polymers with a value of T_g in the range 200–250 °C. They are melt processable at temperatures of the order of 350 °C and can be extruded and injection moulded. They possess a very good resistance to oxidation and creep and can be used at elevated temperatures. They are both heterochain aromatic polymers. The main members of these groups are PSU (polysulphone), PES (polyethersulphone), PPSU (polyphenylenesulphone), PAS (polyarylsulphone) and PAE (polyarylether). Another heterochain aromatic polymer with good high temperature stability is PPS (polyphenylene sulphide).

PSU (polysulphone) is a heterochain linear thermoplastic polymer containing $-SO_2$ groups in the chain. PSU, which has been available since the mid 1960s, has a good high temperature resistance, up to 170 °C, is non-toxic, possesses an excellent resistance to acids, alkalis and oils, and has good electrical properties.

PSU is melt processable but processing temperatures of about 350 °C are required. The main applications are for integrated circuit boards, coil bobbins, high temperature and flame resistant cable insulation, parts for milking machines and for food dishes and trays.

Table 17.1 *Names and applications of some thermoplastics.*

Name	Notes	Applications
HDPE High-density polyethylene	A linear polymer with a high degree of crystallinity. It is easily moulded but is susceptible to stress cracking and has low resistance to ultra-violet light	Used for packaging film, bowls, buckets, bottles, toys, refrigerator parts, pipes and pipe fittings for water
LDPE Low-density polyethylene	The molecules are not completely linear, showing some branching. It softens in boiling water but remains flexible down to −75 °C	Used for packaging film, bowls and other kitchenware, squeeze bottles, toys, seals and gaskets, and wire and cable insulation
LLDPE Linear low-density polyethylene	It has higher tensile and impact strengths than LDPE and can be blown into thinner gauge films	Used for similar applications to those for LDPE
UHMWPE Ultra-high mol. wt. PE	High toughness, excellent chemical resistance and good resistance to cutting and abrasive wear	Used for chopping boards, gears, bearings and other machine parts
XLPE Cross-linked polyethylene	Cross-linked after moulding	Steam-sterilisable bottles and other articles, piping for domestic water and heating systems
PP Polypropylene	Higher strength and stiffness than PE, very good fatigue resistance but poor resistance to ultra-violet light	Used for buckets, bowls, battery cases, bottle crates, toys, bottles and bottle caps, automotive parts, vacuum cleaner bodies, etc., in fibre form for ropes and carpeting
Polyallomer	A block copolymer of propene with 2–10 per cent ethene	Containers and other parts with integrally moulded hinges
EVA copolymer	Very good low temperature flexibility and tear resistance	Ice-cube trays, flexible toys, disposable gloves, vacuum cleaner hose, flexible pipe connectors and shrink wrap film
Ionomer	A soft, flexible material with better resistance to stress cracking than has LDPE	Blow moulded bottles for vegetable oils and cosmetics, tubing to carry beer and liquid foods, disposable hypodermic syringes
PS Polystyrene	Easily moulded material with good resistance to most foods and household acids	Food boxes, light diffusers, toys
HIPS High impact polystyrene	Easily mouldable, tough general-purpose material	Toys and modelling kits, mouldings for office and domestic appliances, boxes and cases
ABS acrylonitrile–butadiene–styrene	Readily mouldable, resistant to many acids and solvents	Automotive parts, mouldings for telephones, business machines, domestic cleaners and mixers, bathroom fittings
SAN styrene–acrylonitrile copolymer	A strong, rigid, tough, transparent material	Containers, trays, toothbrush handles, dials, knobs, lenses
UPVC Unplasticised polyvinyl chloride	A strong, rigid, tough material very resistant to ultra-violet light	Building industry products, guttering, cladding panels, pipes and pipe fittings, doors and window frames
PPVC Plasticised polyvinyl chloride	A soft, flexible, rubbery material	Used for seals and gaskets, simulated leather cloth for clothing, luggage and furniture, floor and wall coverings, wire and cable insulation, hoses, sachets and bottles

Table 17.1 (Continued)

Name	Notes	Applications
CPVC Chlorinated PVC	Softening point of CPVC is 10–15 °C higher than PPVC	Used for piping to carry hot water and hot aqueous liquids
PVDC Polyvinylidene chloride	Good resistance to organic solvents and almost impermeable to water vapour	Used for pipes, seals and gaskets in chemical engineering plant
PTFE Polytetrafluoroethylene	An almost completely inert material with very low friction coefficient	Chemically resistant coatings, non-stick coatings, low-friction bearing bushes
PMMA Polymethyl methacrylate	A rigid, tough, highly transparent material	Aircraft glazing, automotive lenses, light covers and instrument panels, display signs, baths and shower units
PAN Polyacrylonitrile	A highly stable material, resistant to oils and greases	Used mainly as fibres for the clothing trade
PA Polyamides (nylons)	A range of strong, tough materials, but strength reduced by water absorption	Used as both fibres and as moulding compounds
PA 6.6	The strongest, stiffest polyamide. Low friction	Gears, cams, bearings, fastenings, many automotive parts
PA 6	Lower water absorption than PA 6 or PA 6.6 but more costly	Applications similar to those for PA 6.6
PA 6.10	Better heat resistance than PA 6.10	Zip fasteners, electrical insulators, filaments for brushes
PA 6.12		Zip fasteners, mechanical components, abrasion-resistant coatings
PA 11	Low water absorption, good resistance to ultra-violet light, but more expensive	Gears, cams, hosing for compressed air, hydraulic fluids and petrol
PA 12	Lowest water absorption and most costly of all polyamides	Precision engineering components and sports shoe soles
PC Polycarbonate	A tough, transparent material with a high impact strength	Safety helmets, shields and goggles, glazing, light fittings
POM Polyoxymethylene (Acetal)	Good strength and toughness, and can be used up to 120 °C.	Gears, cams, bearing bushes, levers, fans, pump parts and water taps
PET Polyethylene teraphthalate	High-strength, high-stiffness thermoplastics produced as a moulding compound, as transparent	Electrical connectors, coil bobbins, parts for domestic goods such as domestic iron handles, power tool casings and automotive components.
PBT Polybutylene teraphthalate	film and as fibre. They can operate continuously at temperatures up to 120 °C	In film form for packaging, photographic and radiographic film and audio/video tapes
PEEK Polyether etherketone	Polymer developed for high temperature service (up to 250 °C)	Used mainly in aerospace and radiation environments
PPO Polyphenylene oxide	A strong, high impact strength material usable up to 80 °C	Electrical fittings, components for dishwashers, washing machines, VDU cases, car fascia panels, pump impellers
PPS Polyphenylene sulphide	Related to PPO but usable up to 190 °C	Electrical connectors, pumps, pipes and gaskets in the chemical industry
PSU Polysulphone	Non-toxic materials with good resistance to	Electrical components and cable insulation
PES Polyether sulphone	acids, alkalis and oils. Suitable for service at	for use at high temperatures. Components used in
PPSU Polyphenylene sulphone	temperatures up to 180 °C	the medical, food and agricultural industries

313

Table 17.2 *Properties of some thermoplastic materials.*

Material	Density (kg/m³)	E (GPa)	Tensile strength (MPa)	% elongation	Impact strength (J/m²)
LDPE	915–930	0.12–0.3	1–16	90–650	no break
HDPE	960–970	0.42–1.4	22–38	20–1300	80–100
HDPE (30% glass filled)	1170	7	70	2–3	5.5
EVA (5% vinyl acetate)	920	0.16	16	700	
EVA (15% vinyl acetate)	970	0.045	17	700	
Ionomer	930–960	0.07–0.38	20–30	320–520	
PP	900–910	0.5–1.9	27–40	30–200+	2–12
HIPP	900–910	0.8–1.3	19–35	30–200+	8–64
PP (30% glass filled)	1120–1130	4–7	47–103	2–4	3–16
Polyallomer	900–910	0.5–0.8	21–28	400–500	30–180
PS	1040–1100	2.8–3.5	35–84	1–4.5	1.3–3.4
HIPS	980–1100	0.9–3.5	12–62	7–60	2.7–40
PS (25% glass filled)	1200–1330	5.9–9	63–105	0.75–1.3	2–22
ABS	1040–1700	0.9–2.8	17–62	10–140	17.5
SAN	1070–1100	2.8–3.8	63–84	1.5–3.7	1.7–2.5
UPVC	1400–1540	2.4–4.1	24–62	2–40	2–100
UPVC (20% glass filled)	1580	7.6	96	3	8
PPVC (low % plasticiser)	1300–1350	0.35–0.4	28–42	200–450	
PPVC (high % plasticiser)	1200–1550	0.003–0.02	7–56	200–450	
CPVC	1380–1580	2.5–3.2	35–62	4.5–65	5–30
PVDC	1650–1750	0.3–0.5	21–34	up to 250	2–5
PTFE	2100–2250	0.35–0.62	17–28		
PMMA	1180	2.7–3.5	50–75	5–8	
PA 6 (dry)	1120–1140	2.6–3.0	65–86	20	5.3
PA 6 (50% rel. humidity)	1120–1140	1.4	24–53	180–300	16
PA 6.6 (dry)	1130–1150	3.2	70	60	5
PA 6.6 (50% rel. humidty)	1130–1150	2.6	58	300–540	10
PA 6.6 (30% glass filled)	1340–1420	9.6	117–200	3	4.5–11
PA 6.10 (dry)	1070–1090	1.9	50–70	85	3.2
PA 6.10 (50% rel. hum.)	1070–1090	1.1	48	300	8.5
PA 6.12 (dry)	1060–1080	2.0	60	150	5.3
PA 6.12 (50% rel. hum.)	1060–1080	1.2	51	340	7.5
PA 11 (dry)	1040	1.3	55	100–300	7–11
PA 12 (dry)	1010	1.1–1.4	49–65	120–350	6.4–27.4
PC	1200	2.2–2.4	59–70	80–120	65–85
POM	1370–1430	3.5	62–70	25–60	7–12
PET	1320–1340	2.2–2.5	55	300	2
PBT	1300	2.6–2.7	52	200–250	2.5
PEEK	1300	3.8	240	4	
PPO	1060	2.5	61	60	21
PPS	1340	3.3	75	1.6	
PSU	1245	2.5	60–74	50–100	
PES	1370	2.45	84	40–80	
PPSU	1290	2.14	72	60	

PES (polyethersulphone) is a linear heterochain thermoplastic developed in the 1970s and suitable for continuous service at temperatures up to 180 °C.

PES is transparent and yellow and is expensive and difficult to process, requiring processing temperatures of the order of 350 °C. Its uses include electrical components to operate at elevated temperatures, such as coil bobbins, circuit boards and lamp holders, and in medical, agricultural and food industry components which require repeated sterilisation in service. Glass and carbon fibre reinforced grades of this material are also available.

PPSU (polyphenylene sulphone) is one of the group of polysulphone thermoplastics. Like other polysulphones, PPSU has good electrical properties, excellent chemical and thermal resistance and can be used for continuous service at temperatures up to 180 °C. PPSU is not used on a wide scale but it does find application for electrical parts, such as connectors, where high temperatures are involved, and in the construction of fuel cells.

Polyarylates are completely aromatic thermoplastics developed in Japan in the 1970s and they possess a linear heterochain structure. They are tough, glassy, transparent materials with a high T_g value of about 200 °C. The materials have a similar impact resistance to that of ABS and are resistant to ultra-violet radiation. However, the polymers are susceptible to stress cracking in contact with hydrocarbon solvents and require high processing temperatures, of the order of 350 °C. Applications for the materials include electrical connectors, glazing, components in hot drinks vending machines and housings for appliances.

Polyimides (PI)

The polyimides are a group of linear aromatic polymers, some of which are thermoplastic and some are thermosets. They possess good mechanical properties with a low coefficient of friction and they possess excellent thermal resistance up to about 250 °C and with inherent fire retardance. They are produced by a condensation reaction between pyromellitic dianhydride and a diamine. Several diamines, mainly aromatic amines have been used for the production of polyimides.

These materials, though expensive and difficult to process, have good resistance to organic solvents but are attacked by alkalis and concentrated acids. They are transparent to microwaves and are not affected by radiation. PTFE, graphite and molybdenum disulphide can be used as fillers to create bearing materials that can operate at temperatures up to 250 °C. The main uses are as bearings and seals for high-load/high-speed applications, printed circuit boards, high temperature electrical cable insulation, mechanical components exposed to irradiation and wet or dry spun textile fibres. The resins may also be used as composite matrices. They have similar properties to epoxies at room temperature, but are stable up to higher temperatures than them.

Polyetherimide (PEI). This is a thermoplastic polymer related to polyimide but with a greater chain flexibility and, hence, a better processing performance than polyimide.

This polymer can be used continuously at temperatures up to about 170 °C (compared with 180 °C for PES). The material possesses a higher strength than PES at ambient temperature and is less expensive than either PES or PEEK. It does have a low value of notched impact strength and is susceptible to stress cracking in contact with chlorinated solvents. The main applications include electrical switchgear and connectors for high temperature service, cookware for microwave ovens and automotive components.

Polyamideimide (PAI). Related to polyetherimide and the polyimides, this is a linear aromatic polymer for use at high temperatures. This material has properties approaching those of polyimide and can be used at temperatures up to about 230 °C, yet it is easily melt processable. Unfortunately, the material is expensive and is attacked by alkalis. Applications include valves, bearings, electrical connectors, printed circuit boards and components for gas turbine and spark-ignition engines. Grades with glass or graphite reinforcement are available.

EXAMPLE 17.2

Polymer moulding compounds are not pure polymers but contain also numerous additives. List *four* types of additive, giving the reasons for their use.

SOLUTION

The types of additive used to make polymer moulding compounds include fillers, dyes and pigments, stabilisers, plasticisers, flame retardants and antistatic agents.

Fillers. Fillers are used either to increase bulk, thus reducing the amount of costly pure polymer resulting in a cheaper material, or to improve certain properties. As examples, a wood flour filler will give a less costly moulding compound while the use of chopped glass fibre will increase strength and modulus values, but at the expense of ductility and impact resistance.

Dyes and pigments. These are used to produce the desired colour.

Stabilisers. Some polymers may deteriorate when exposed to light and/or ultra-violet radiation, or react with oxygen or ozone. Stabiliser additives tend to inhibit these effects.

Plasticisers. Plasticisers, which are often liquids of low molecular weight, tend to act as a lubricant between the large polymer molecules, thus permitting easier molecular movement, or plastic flow. The presence of a plasticiser will reduce the strength and rigidity of a polymer but cause an increase in ductility and toughness. A good example of the effect of a plasticiser can be seen by comparing the properties of unplasticised and plasticised PVC (refer to Table 17.2).

Flame retardants. Many pure polymers are flammable. The flammability of the materials can be reduced significantly by the addition of a flame retardant.

17.8 | Elastomers

Elastomers are materials that have low elastic moduli which show great extensibility and flexibility when stressed but which return to their original dimensions, or almost so, when the deforming stress is removed. In essence, the molecules act like a series of small helical springs. There is some cross-linking between molecules (see section 4.6), the flexible cross-links causing the material to return to its original shape. An increase in the number of cross-links formed will increase the hardness and stiffness of the material.

There are several classes of elastomers, these being:

R class – unsaturated carbon chains (this group includes natural rubber)
M class – saturated carbon chains
O class – heterochain polymers with O in the chain
U class – heterochain polymers with C, O and N in the chain
Q class – silicones – heterochain polymers with Si in the chain

Tables 17.3 and 17.4 summarise the properties and applications of some elastomers.

17.9 | The R class elastomers

Natural rubber (NR)

Natural rubber is *cis*-polyisoprene and this is an unsaturated linear molecule, that is, there is double covalent bonding between some atoms in the molecular chain.

The source of natural rubber is the milky liquid known as *latex*, which is obtained from the cultivated tree *Hevea Brasiliensis*. The major rubber plantations are in Malaysia. To obtain the latex, a cut is made in the bark of the tree and the milky liquid slowly flows out into a collecting cup. Latex is an emulsion containing tiny particles of rubber. The latex is treated with an organic acid, such as acetic acid, to coagulate the rubber particles and precipitate them from the emulsion. The separated rubber is then dried thoroughly. It is necessary to mill the raw rubber between heavy rolls in order to make it plastic and suitable for forming. During this milling operation, the mechanical shearing effect breaks up some of the polymer chain molecules, reducing the average molecular weight of the rubber. The material is now in a plastic state and can be compounded with fillers and other additives, including plasticisers and antioxidants. In order to restore rubber-like elasticity to this plastic mass, an addition of a controlled amount of sulphur is made to allow for some cross-linking, or vulcanisation, to take place. The plastic rubber mass is then moulded or extruded to the desired shape and, finally, the article is *cured* by heating at some temperature in the range 100–200 °C to vulcanise. During vulcanisation, some of the double covalent bonds in the rubber molecules split allowing sulphur atoms to form cross-links between adjacent molecules (see section 4.6). For ordinary soft rubber, about 4 per cent of sulphur is added to the mix and this will cause cross-linking between approximately 10 per cent of the double bonds in the polymer. If the sulphur content is increased, the amount of cross-linking is increased giving a harder and less flexible product. Full vulcanisation, requiring about 45 per cent of sulphur, will give rise to a fully rigid network structure. Fully vulcanised natural rubber is the hard elastic solid termed *ebonite*.

Oxygen or ozone will react with the double bonds in natural rubber molecules forming cross-links in a similar manner to the vulcanisation reaction. Oxygen cross-links are not flexible and they embrittle the material. The rate of oxidation, or *perishing*, of the rubber can be retarded by compounding anti-oxidants and anti-ozonants into the material.

The role of fillers in rubber is two-fold. The use of a cheap inert filler will cheapen the material, but a filler may also be used to improve certain properties. Carbon black and silica are two fillers commonly used in rubbers and both of these give reinforcement and increase the resistance to abrasion and tearing.

Cis-polyisoprene can be synthesised and the nomenclature for synthetic polyisoprene rubber is IR. IR is less expensive to produce than natural rubber and it is often blended with natural rubber for reasons of economy.

Natural rubber accounts for some 30–35 per cent of the elastomeric materials used and synthetic isoprene makes up about 10 per cent of world output of synthetic elastomers.

Synthetic R class elastomers

BR. There are a number of synthetic elastomers with unsaturated carbon chains which permit vulcanisation. An important member of this group is polybutadiene (BR), made by the addition polymerisation of butadiene.

BR is a cheap and resilient elastomer with a good abrasion and tear resistance. However, it is difficult to process and has a poor resistance to ozone and oxidizing chemicals. Its sensitivity to heat and light results in poor ageing characteristics and the poor stability of the carbon bonds often results in low pigmentation stability. BR is mainly used for making physical blends with NR and SBR, for example a blend of 25 per cent, BR with 75 per cent SBR is the most common material for rubber tyres. This is used also for footwear, seals, domestic floor tiles and sponges, a considerable quantity of BR production is used as the dispersed rubber phase in high impact polystyrene.

SBR. SBR is a random copolymer of styrene and butadiene and accounts for over half of the world production of synthetic rubbers. This material is cheaper than natural rubber and is used, sometimes alone, and sometimes in physical blends with NR or BR.

NR, SBR and BR absorb organic solvents, petrol and oils and swell considerably. Two types of synthetic rubber which may be used in contact with oils and other solvents are the types CR and NBR.

CR. CR is polychloroprene. It is also known as *chloroprene* and *neoprene*. This polymer can be vulcanised by heat alone but its properties are enhanced by compounding with zinc oxide and magnesium oxide, both of which act as vulcanising agents. It has an outstanding resistance to oils and many other chemicals.

NBR. NBR is a copolymer of butadiene with between 10 and 40 per cent of acrylonitrile. It is also known as *nitrile rubber*. The nitrile rubbers possess excellent resistance to fuels and oils, even at temperatures up to 160 °C.

IIR. This rubber, also known as *butyl*, is a copolymer of iso-butylene with a small amount of isoprene.

IIR is fully saturated when polymerised. The presence of a small amount of isoprene provides some double bonds, and hence cross-linking sites, within the molecules. The rubber is fully saturated after vulcanisation and this makes the material more resistant to oxidation than NR or SBR. IIR has an extremely low permeability to air and other gases and its main applications are for hoses, tubing and inner tubes. It is also possible to react the material with halogens and substitute chlorine or bromine for some of the hydrogen atoms in the molecular chains. These derivatives, chlorobutyl and bromobutyl rubbers have even greater resistance to oxidation and improved properties at elevated temperatures.

17.10 | Other elastomer classes

M class elastomers

The M class of elastomers are saturated carbon chain elastomers. EPM and EPDM are, respectively, a copolymer of ethylene and propylene and a terpolymer of ethylene, propylene and diene. EPM is fully saturated but cross-links (*vulcanisation*) can be brought about by reaction with peroxides. In EPDM, the presence of a small amount of diene introduces a small number of double covalent bonds into the structure, thus permitting vulcanisation by sulphur. Both of these elastomers are highly resistant to oxidative degradation.

Polyethylene can be chlorinated by direct exposure to gaseous chlorine or to a solution of chlorine. There is a simple substitution of chlorine for some hydrogen atoms in the linear chain molecules. Similarly, polyethylene can be reacted with both chlorine and sulphur dioxide to give chloro-sulphonated polyethylene. Both of these derivatives can be cross-linked to produce elastomers. They are identified by the nomenclatures CM and CSM respectively. Both of these elastomers have excellent resistance to oxygen, ozone and many chemicals and good high temperature characteristics. FKM is the term for a fluorinated hydrocarbon elastomer. This type of material is suitable for use at temperatures up to 230 °C and possesses an excellent resistance to oils and chemicals.

Some polyacrylates are elastomers. These are a group of synthetic elastomers based on ethyl acrylate copolymerised with a small amount of an active monomer which will permit some subsequent cross-linking. Acrylate elastomers possess excellent resistance to oxidative degradation. ACM is polyethylacrylate and ANM is a copolymer of ethyl acrylate with acrylonitrile.

O class elastomers

A polyether is a heterochain polymer containing the –C–O–C– ether element in the chain. Epichlorhydrin can be polymerised to give a polyether with elastomeric properties, with the nomenclature CO or ECO. This is an elastomer produced in relatively small quantities for special purposes. It possesses an excellent resistance to petrols, oils and greases but with superior thermal and oxidation resistance to NBR.

Polysulphide rubbers. Polysulphides are heterochain polymers which are sulphur analogues to the polyethers with some carbon–sulphur–sulphur–carbon links within the molecular chain. Polyethylene polysulphide is an elastomer which has an extremely high resistance to swelling in the presence of organic solvents. It is, however, weak but finds application as an adhesive and sealant.

U class elastomers

The U class of elastomer is characterised by having a carbon–oxygen–nitrogen heterochain structure. These are the materials termed *polyurethanes* and there are two types, AU and EU, these being polyester- and polyether-urethanes respectively. The term polyurethane is given to the series of polymers produced by the reaction between aromatic di-isocyanates and low molecular weight polymer molecules which may be polyesters, polyethers or a mixture of a polyester and a polyamide. They can be tailored chemically to give a range of products in several forms, some of the polyurethanes being elastomers. These materials have excellent abrasion and tear resistance but have a major disadvantage in that isocyanates are highly toxic and there is a major evolution of poisonous smoke when these materials degrade and burn.

Q class elastomers

The Q class are heterochain polymers with silicon in the chain, in other words, *silicones* or *siloxanes*. The backbone chain in a silicone is not composed of silicon alone but are repeating units of silicon and oxygen, $-Si-O-$. As silicon is tetravalent and oxygen is divalent, the $-Si-O-$ unit has four valency bonds. Two of these are used up in linking with other $-Si-O-$ units and the other two bonds link with hydrogen or organic groups, such as the methyl group, $-CH_3$, or the phenyl group, $-C_6H_5$. The nomenclature code for these materials lists the types of organic groups attached to the siloxane chain, for example, MQ is methyl silicone, PMQ is phenyl-methyl silicone, and VMQ is vinyl-methyl silicone.

The major advantage of silicone rubbers is that they possess rubber-like elasticity over a very wide temperature range and some of them retain their flexibility and elasticity down to temperatures as low as $-80\,°C$. They are, however, very expensive. Applications of silicone rubbers include cable insulation, seals for hydraulic systems, and door and canopy seals for aircraft.

17.11 Thermoplastic elastomers

The large group of materials known as thermoplastic elastomers are materials which have elastomeric properties at ordinary temperatures but can be processed in the same way as thermoplastics. The thermoplastic elastomers do not have the type of cross-linking possessed by other elastomers, for example the sulphur links in a vulcanised rubber. A typical thermoplastic elastomer structure is one in which the very large polymer molecules consist of a block copolymer with three segments, a large central segment of a 'soft' flexible polymer with two somewhat smaller blocks of a harder more rigid polymer at either end. In styrene–butadiene thermoplastic elastomers, for example, the central flexible polybutadiene block may show a degree of polymerisation of between 600 and 2000, while the outer polystyrene blocks may be about one-tenth of that size. A similar structure occurs in ethylene–propylene thermoplastic elastomers, in this case with the polyethylene forming the central flexible segment. The hard end segments in this type of copolymer tend to come together into small groups giving, in effect, a series of small hard blocks, or domains, within a softer elastomeric matrix. These domains act as cross-links. When the temperature is increased and the hard block domains 'melt' the material becomes truly thermoplastic and can be processed using any of the usual techniques for thermoplastics.

Thermoplastic elastomers can also be made by blending an elastomer with a thermoplastic material to make an 'alloy'. Examples of this are blends between natural rubber, SBR or NBR, with polypropylene.

17.12 Thermosetting materials

Unlike a thermoplastic material which will soften when heated, a thermosetting material undergoes chemical changes when first heated and is converted from a plastic mass into a hard and rigid

Table 17.3 *Properties of some elastomers.*

Type	Temperature range (°C)	Maximum % elongation	Hardness (Shore A)
NR	−50 to +100	700	30–90
BR	−60 to +100	600	40–80
SBR	−60 to + 70	600	40–90
CR	−20 to +130	600	30–90
NBR	−20 to +120	600	40–90
IIR	−50 to +125	800	40–80
EPM	−40 to +150	600	30–90
ACM	−40 to +130	400	40–85
CSM	−40 to +120	500	50–90
FKM	−40 to +230	300	60–95
AU	−50 to +100	700	40–100
EU	−50 to +100	700	40–100
MQ	−60 to +230	700	30–85
PMQ	−60 to +230	700	30–85

Table 17.4 *Applications of some elastomers.*

Type	Applications
NR	Low cost, good properties. Tyres, engine mountings, flexible hoses
BR	General-purpose rubber with better abrasion resistance than NR. Used for hose, shoe soles, tyre treads
SBR	Low cost, similar properties to NR. Tyres, transmission belts
CR	Good heat resistance and resistance to oils. Car radiator hoses, gaskets and seals, conveyor belts
NBR	Best resistance to oils and heat. Seals, hoses and linings
IIR	Good resistance to ozone, low permeability. Inner tubes
EPM	Good mechanical properties. Resistant to hot water and superheated steam. Steam hoses, conveyor belts
ACM	Good heat resistance. Excellent resistance to hot oils. High temperature gaskets and seals
CSM	Excellent resistance to chemicals. Acid hoses
FKM	High temperature resistant, no swell in oils. 'O'-rings and gaskets in automotive applications
MQ/PMQ	High cost. Mainly used for seals and gaskets. Aircraft applications

material. During the heating, or *curing*, process further polymerisation occurs and a full network molecular structure is formed. Figure 17.5 is a schematic representation of a three-dimensional network structure. There are also a number of materials that will set hard and rigid at ordinary temperatures. These *cold-setting* materials are of the same general type as the thermosetting plastics since the curing process is fundamentally the same in both cases.

17.13 Phenolic materials

The first thermosetting materials to be commercially exploited were polymers based on phenol and formaldehyde. These two compounds polymerise by a series of condensation reactions. During the reaction the phenol molecules become linked by $-CH_2-$ groups. The $-CH_2-$ linkages may be

| | Phenol group |
| | CH$_2$ bridge from formaldehyde |

Figure 17.5 *Schematic representation of a network structure.*

formed at three positions in each phenol molecule, namely at the 2, 4 and 6 sites, giving a full network polymer.

Within a cyclic carbon compound the sites may be numbered thus:

The resultant pure polymer is opaque and milky-white in colour, but it darkens with time. Consequently, the phenolic moulding powders are always mixed with dark pigments to give constant dark-coloured materials. One trade name for phenolic materials is 'Bakelite'.

In the production of moulding powder polymerisation is allowed to proceed only to a limited extent to give low molecular weight resin that can be readily melted. This material is then ground into powder and compounded with suitable fillers and other ingredients. When the part-polymerised powder is next heated it melts and can flow under pressure into a mould. Curing is accompanied by a volume contraction of the resin and an inert filler must be mixed in with the resin powder in order to reduce the effect of this shrinkage. The use of a filler will also tend to reduce the cost of the moulding material and many fillers also give an improvement in some properties. Other additives used in the compounding of thermosetting moulding powders are catalysts to accelerate the curing process, pigments to colour the resin, and lubricants to prevent the material from sticking to the moulds. A variety of materials may be used as fillers and the filler generally accounts for between 50 and 80 per cent of the total weight of moulding powder. Coconut shell flour and wood flour, a fine soft-wood sawdust, are two widely used fillers. Wood flour increases the impact strength of the material. Coconut shell flour, a very cheap filler, would, if used alone, give rise to mouldings with comparatively poor properties, and so it is generally used in conjunction with wood flour.

Cotton flock and chopped fabric are often added as fillers where a high resistance to impact is required. Paper pulp and nylon fibre may also be used for this purpose. Other fillers that may be used to improve specific properties are asbestos, for improved resistance to heat, mica, for improved electrical resistance, and graphite, to reduce the coefficient of friction and give good sliding properties.

The compounded moulding powder may then be processed into a wide range of shapes by compression or transfer moulding. The properties of some phenolic resin moulding materials are given in Table 17.5.

Some of the principal applications of phenol–formaldehyde (PF) mouldings are given in Table 17.6.

Some phenolic resins are cold-setting. These have a syrupy consistency and the addition of a catalyst causes curing to occur at ordinary temperatures. These frequently form the basis for glues and adhesives, and may be used for the bonding of plywood and in the manufacture of hardboard. Powdered resin may be used as a binder in the production of shell moulds and cores for metal casting (refer to section 20.4).

17.14 | Amino-formaldehyde materials

Compounds containing an amino group, $-NH_2$, can condense with aldehydes to form rigid polymers. Of these, the most important are urea and melamine.

It is possible for both of the hydrogen atoms in each $-NH_2$ group to take part in a condensation reaction with an aldehyde molecule to give a highly rigid network polymer structure. In the same way as in the preparation of phenolic moulding powders, urea and formaldehyde are first part-polymerised and the low molecular weight polymer that is produced is then ground into powder and compounded with fillers, pigments and a catalyst to make the finished moulding powder. The amino-formaldehyde resins are light in colour and pigments can be added to give a wide range of coloured moulding materials.

Melamine, a cyclic amino compound, can also be condensed with formaldehyde to give a strong and rigid material. The general properties of melamine–formaldehyde (MF) materials are similar to those of UF materials, but the MF resins have a better resistance to heat and can be used at temperatures of up to 95 °C. UF mouldings have good stability at temperatures up to 80 °C but they tend to darken and decompose at higher temperatures than this.

In addition to their use as moulding powders, both UF and MF materials, like the PF resins, are used in the manufacture of laminates, as adhesives, and as binding agents in shell moulding sands.

The properties of amino-formaldehyde (UF and MF) materials are given in Table 17.5 and some of the principal applications of these materials are given in Table 17.6. These materials do not impart a taste to foodstuffs and are used for the manufacture of cups, saucers and plates.

17.15 | Polyester materials

As mentioned in section 17.7, an ester is formed when an acid condenses with an alcohol. If an unsaturated polyester is made, that is, the molecules contain reactive double covalent bonds, it becomes possible to cause cross-links to form between molecules, creating a rigid network structure. Unsaturated polyesters are principally used as the basis of fibre reinforced plastics, and glass fibre is the most usual reinforcing material.

There are many polyester formulations possible but the ester produced by the condensation

of maleic acid,
$$\begin{array}{l} CH\!-\!COOH \\ \| \\ CH\!-\!COOH \end{array}$$
, and propylene glycol,
$$\begin{array}{l} CH_2OH \\ | \\ CH.OH \\ | \\ CH_3 \end{array}$$
, is typical of the type.

The unsaturated polyester is blended with an unsaturated monomer, such as styrene. Subsequently, on mixing in an *initiator*, an addition polymerisation will occur between the styrene and the polyester, giving a hard and rigid molecular network. Benzoyl peroxide is generally used as an initiator and curing will occur rapidly at elevated temperatures, up to 100 °C. When certain amino compounds are present with benzoyl peroxide the curing process will take place at ordinary temperatures and the application of heat is not necessary. It is customary to mix the initiator thoroughly into the polyester resin immediately before use, and then make up a resin/fibre laminate over a former, as outlined in section 21.11. The final properties of the laminate will depend, not only on the nature of the resin, but also on the type and form of reinforcing fibre used.

Polyester laminates are produced to meet a number of specific requirements and separate grades are available for general purposes, heat resistance, chemical resistance, water resistance and for good impact resistance. Some typical applications of the various grades are

General purpose   – boats, baths and bathroom fittings, trays
Heat resistant   – aircraft components and electronic parts
Chemical resistant – tanks, pipes and ducts
Water resistant   – structural panels
High impact   – safety helmets, machine guards

17.16 Epoxide materials

These are a group of materials that, although two or three times as expensive as polyesters, are finding a growing use as engineering materials. The uncured resin consists of polymer molecules with a highly reactive epoxy group at both ends of each molecular chain:

$$\underset{CH_2-CH}{\overset{O}{\triangle}} - - - - - - \underset{CH-CH_2}{\overset{O}{\triangle}}$$

Most of the polymers used are based on the polymerisation of bisphenol A and epichlorhydrin. By controlling the degree of polymerisation, the physical state of the resin produced can vary from solid to fully liquid. In order to convert the epoxy resin into a fully rigid material it is necessary to mix a hardener into it. There are several types of hardener that may be used but some of the more commonly used types are reagents possessing amino groups, $-NH_2$. The amine hardener will form cross-links between the epoxy polymer molecules. It is necessary that the hardener possesses two or more amino groups. The equation showing cross-linking is given below:

$$NH_2-R-NH_2 + \underset{CH_2-CH}{\overset{O}{\triangle}} - - - - \rightarrow NH_2-R-\underset{\overset{|}{H}}{N}-CH_2-\underset{\overset{|}{OH}}{CH} - - - -$$

Both hydrogen atoms in every amino group can react with an epoxy group in this manner. Consequently, considerable cross-linking can occur and an extremely hard and rigid molecular structure can result. It should be noted that this is not a condensation reaction, and that there is no by-product evolved. There is virtually no shrinkage associated with this cross-linking reaction.

Epoxide materials are generally used with reinforcing fibres, such as carbon and glass fibre, to produce hard and strong structural materials. Carbon reinforced epoxy resins are extremely strong materials with a very high value for Young's modulus, compared with other plastics materials, and they are finding applications in the aerospace field as compressor blades in gas turbine engines, wing leading edges and flaps, and rocket motor cases.

Table 17.7 gives the properties of some polyester and epoxide laminates.

Another property of the epoxide materials is that they will adhere to virtually all materials. Consequently, they find a use as high-strength adhesives, particularly in the aerospace industries.

Other important applications of epoxides are for the encapsulation of miniature electronic components and for road surfacing. The extreme resistance to wear of these materials makes them suitable for use at busy road junctions. The high cost of the material is offset by the greatly reduced need for repair operations and the consequent disruptive effect on traffic.

17.17 | Other thermosets

Polyurethanes

Polyurethane is a family name given to a series of polymers that are produced by the reaction between aromatic di-isocyanates and low molecular weight polymer molecules, which can be polyesters, poly-ethers or a mixture of a polyester and a polyamide.

There are a number of different polyurethanes with differing characteristics. They all possess a very good wear resistance and are resistant to oils, greases and petrol but, depending on the exact formulation, they may be thermosets, thermoplastics, which can be processed by injection moulding or extrusion, elastomeric materials, or foams, and such foams may be either flexible or rigid.

The principal uses for polyurethane moulded components are for hoses, car bumpers, shoe heel tips, hammer heads, gears, including gears for quiet running in film cameras and projectors.

Allyl resins

Allyl phthalates can be polymerised in the presence of an alkaline catalyst to give a thermoset product. Diallyl phthalate (DAP) is a thermoset resin used in the manufacture of laminates, as a coating resin and as a moulding compound.

DAP possesses a higher strength than polyester or epoxy materials and has good electrical properties, and a better resistance to tracking than phenolic resins. The electrical properties are maintained at high temperatures (up to 160 °C) and high humidities. The resin is non-toxic and, unlike epoxies, does not cause dermatitic reactions. It is used as a laminating resin for the manufacture of radomes, trunking, printed circuit boards and high temperature applications. Typical moulded components include electrical connectors, insulators, tool handles and pump impellers. As a coating resin it is used as a wood sealant. Glass fibre reinforced and mineral filled grades are available.

Diallyl isophthalate (DIAP) is a material similar to DAP but has a poorer resistance to alkaline chemicals. It has a better temperature resistance than DAP (up to 180 °C) and a higher strength. It is used for applications similar to DAP but where a better temperature capability is required.

Table 17.5 *Properties of some thermosets.*

Material	Density (kg/m³)	E (GPa)	Tensile strength (MPa)	Compressive strength (MPa)	% elongation
PF unfilled	1.25–1.3	5.2–7.0	35–55	165	1–1.5
PF wood flour filled	1.32–1.45	5.5–8.0	45–60	180	0.4–0.8
PF glass filled	1.5–2.0	7.0–14.0	40–70	260–280	0.5
UF cellulose filled	1.5–1.6	7.0–10.0	40–90	175–315	0.5–1.0
UF wood flour filled	1.5–1.6	7.0–10.0	40–55	140–190	0.5–1.0
MF cellulose filled	1.5–1.6	7.0–11.0	41–70	280–330	0.5–1.0
MF glass filled	2.0	12–17	35–138	140–245	0.5

Table 17.6 *Names and applications of some thermosets.*

Name	Notes	Applications
PF Phenol–formaldehyde	The phenolics are hard, rigid thermosets stable up to about 150 °C. They have low thermal conductivity, good electrical insulation properties and good resistance to oils, greases and many solvents	Electrical plugs, sockets, switches, door knobs and handles. Also used laminated with paper or fabric for electrical insulation, circuit boards, small gears and water-lubricated bearings
UF Urea–formaldehyde MF Melamine–formaldehyde	These amino-formaldehyde materials are hard, rigid thermosets and they have good electrical insulation properties and good resistance to most chemicals	Electrical fittings, plugs, sockets, switches, etc. Bottle caps, cups, saucers, plates, etc., as they do impart a taste to foodstuffs. Like the phenolics, can be used in laminates
Polyesters	Unsaturated polyesters form the basis of many thermosetting and cold setting resins	Used mainly as the matrix for fibre reinforced composites. FRPs used for boats, road and rail vehicles and some aircraft components, tanks, ducting and pipes, safety helmets, machine covers and structural panels
Epoxides	Very hard, rigid materials available as both thermosetting and cold setting resins	Used for moulded components, as the matrix for fibre reinforced, particularly carbon fibre, composites for many applications in aerospace – structural panels, helicopter rotor blades, rocket motor casings, etc., and in general engineering structures. Also the basis for high strength adhesives

Table 17.7 *Properties of some polyester and epoxide laminates.*

Material	Reinforcement	Reinforcement content (%)	Density (kg/m^3)	E (GPa)	Tensile strength (MPa)
Polyester	Chopped glass mat	25–35	1.5–1.6	5.5–12.5	70–105
Polyester	Plain weave glass fabric	60–65	1.8	7–20	140–300
Polyester	Carbon fibre	40	1.54	150	720
Epoxide	Plain weave glass fabric	60–65	1.8	21–25	200–420
Epoxide	Carbon fibre	60	1.5	125–150	1100–1350

SUMMARY

The range of polymer, or plastics, materials is large but polymer materials can be put into three main classifications: thermoplastics, elastomers and thermosets. Polymer moulding materials are not pure polymers but contain a number of additives. Generally, these additives are present to enhance certain properties or to improve the resistance of the material to exposure to light or ultra-violet light, or oxidation effects.

The properties of thermoplastics range from those of LDPE, with a tensile strength value of the order of 10 MPa, an E value of about 0.2 GPa, and which softens in boiling water, to those of PEEK (tensile strength 240 MPa, E 3.8 GPa, and which can be used at temperatures of up to 240 °C).

Elastomeric materials are characterised by possessing great elastic extensibility and, mainly, are usable over a wide temperature range from sub-zero temperatures up to more than 100 °C. They can be vulcanised and, through careful control of the vulcanising process, be produced in a range of hardnesses.

Thermoset materials tend to be hard, strong and rigid materials which, once they have been first moulded, cannot be reshaped. The polyester and epoxide materials will generally cure at ambient temperature after the active ingredients have been mixed together. These latter two materials can be used on their own to make moulded products but, more often they are used as the matrix material in fibre reinforced composites. They are used also as the basis of high strength adhesives.

Questions

17.1 Explain the main differences in properties between thermoplastics, thermosets and elastomers on the basis of polymer structure.

17.2 Polypropylene scrap may be recycled but scrap Bakelite is worthless. Why is this so?

17.3 Some polyester materials are thermoplastic while others form thermosets. What are the reasons for this?

17.4 State, giving reasons for the choice, polymer materials that would be suitable for each of the following applications:
    (a) cups for a machine dispensing cold drinks,
    (b) cups for a machine dispensing hot drinks,
    (c) bottles to contain still water,
    (d) bottles to contain carbonated drinks.

17.5 Select suitable materials for the manufacture of the following, giving reasons for the choice:
    (a) bottle crates,
    (b) pipes and pipe fittings for the conveyance of domestic waste water,
    (c) the casing for a domestic iron,
    (d) the rear light lens cluster for an automobile,
    (e) the casing for an automobile door mirror.

17.6 Natural rubber (NR) is widely used for the manufacture of tyres for road vehicles. Would it be a suitable material for the tyres of vehicles to be used in Arctic conditions where the temperature may fall to −55 °C? If not, what material could be used for tyres for use in these conditions?

17.7 A relatively large shrinkage occurs when a PF material is moulded but very little shrinkage takes place when an epoxide moulding is made. Give an explanation of this difference between the two types of material.

CHAPTER 18

Ceramics and glasses

Why study ceramics and glasses?

The term *ceramic* embraces not only earthenware, pottery and porcelain, all materials used by man from earliest times, but also the newer industrial ceramics which are mainly the oxides, carbides and nitrides of metals. Ceramics, generally, are hard and brittle materials which are electrically insulating and are resistant to high temperatures and to chemical attack. They tend to have high elastic moduli, be strong in compression but be relatively weak in tension. Glasses are amorphous materials, often based on silica and other metal oxides. Again they tend to be brittle materials, fairly weak in tension. There is also a group of newer materials, known as *glass ceramics*. These are fine-grained polycrystalline materials produced by the controlled crystallisation of glasses. Ceramic materials are used in varied engineering applications, including tools for grinding and cutting, seals, bearings and other components for engines and pumps, and for many applications in the electrical and electronic industries. It is therefore necessary for the engineer to possess knowledge of these important classes of materials.

LEARNING OBJECTIVES

1. Gain some knowledge of the properties and applications of some of the various categories of ceramic and glass materials.
2. Understand the main characteristics and engineering applications of some of the newer industrial ceramics.

18.1 | Introduction

The word *ceramic* conjures up different images with different people. For many, the word signifies fine porcelain and pottery, but the engineer tends to think of the newer industrial ceramics, namely the oxides, carbides and nitrides of metals. In this chapter we will give brief consideration to a wide range

Table 18.1 *Types of ceramic.*

Class	Material types	Applications
Whitewares	Porcelain, china, stoneware	Tableware, wall tiles, sanitary ware, insulators
Heavy clayware	Clay-based material	Bricks, roof tiles, pipes
Refractories	Acid, intermediate, basic	Furnace linings
Stone	Granite, sandstone, limestone	Concrete aggregate, building stone
Glass	Various glasses, glass-ceramics	Containers, flat glass
Abrasives	Diamond, carborundum, corundum	Grinding wheels, abrasive cloth and paper
Cements	Portland cement, alumina	Concrete, structural products
Engineering ceramics	Oxides, carbides, nitrides, cermets	Dies, seals, bearings, spark-plugs, diesel and turbine engine components
Electrical and magnetic ceramics	Many types	Various, including capacitors, thermistors, varistors, magnets
Nuclear ceramics	Oxides and carbides of uranium	Nuclear fuel elements

of inorganic materials including not only the materials normally termed ceramics, but also stone, cement, concrete, bricks and other clay products, and a range of non-crystalline materials, the commercial glasses. A summary of some of the main groupings of ceramic and glass materials is given in Table 18.1.

18.2 | Building stone

The rocks that make up the earth's crust vary considerably, both in composition and structure. Rocks may be classified as *igneous, sedimentary,* or *metamorphic* according to the method of their formation. Igneous rocks are those that have formed by solidification from hot liquid rock material – *magma*. The structure of igneous material is one of interlocking crystal grains. The various constituent minerals crystallise by a process of nucleation and growth from liquid to give a compact mass of irregular shaped crystals. Igneous rocks, which include *granite* and *dolerite*, are hard, weather-resistant materials and possess compressive strengths in the range 110–180 MPa. Quartz (silica), which is a very hard mineral, is a major constituent in granites.

Because of the interlocked grain structure, igneous rocks fracture in an irregular manner. There are not normally major lines or planes of weakness in an igneous rock mass, and quarrying and dressing this type of stone into regular shapes for building purposes is both difficult and costly. These days comparatively little stone is used for building, but granites and other igneous material are widely used in the form of crushed chips for road surfacing, as railway track ballast, and as aggregate for concrete. The irregular intercrystalline fracture surfaces allow for extremely good bonding with cement.

Much of the earth's crust is made up of sedimentary rocks. These are derived from sediments laid down hundreds of millions of years ago, which have become consolidated into rock with the passage of time. Sedimentary rocks are often well stratified and show well-defined bedding planes. Frequently, they are also well jointed, joints being normal to the bedding planes and formed by lateral contraction of the sediment during consolidation. Many sedimentary rocks can be split and cleaved easily, both in the bedding direction and normal to the bedding planes, to make building blocks of fairly regular shape. The properties of sedimentary rocks vary considerably, depending on the nature of the sediment and the type of bonding that exists between adjacent sediment grains. *Sandstones*, formed by the consolidation of sand deposits, consist principally of small rounded grains of silica. In some sandstones the silica grains are bonded together with a clay or other soft mineral. Sandstones of this type will be comparatively soft and will weather rapidly. In other types of sandstones the silica grains may be bonded with hard minerals. In the *millstone grits* of the carboniferous period a very hard type of rock is found composed of silica grains bonded with a silica cement.

Other types of sedimentary rocks are *limestones* and *dolomites*, formed by the precipitation of calcium carbonate, or calcium and magnesium carbonates, from solution. Some limestones and dolomites are very hard, with compressive strengths of the order of 150 MPa, while others, such as chalk, are very soft. Portland stone is a well-bedded and jointed limestone which can be easily dressed into regular shapes for building purposes.

Some sedimentary deposits have been caused to recrystallise, during geological time, by the action of heat or pressure to give new structures, termed *metamorphic* rocks. *Slate* is a metamorphic material formed by the recrystallisation of a clay or shale. In the recrystallisation to form slate, very small plate-like crystals are formed, all with a similar orientation. The hard, weather-resistant slate that results may be readily cleaved into thin sheets.

18.3 Cement and concrete

Portland cement powder is made by firing a mixture of limestone with shale or clay in a rotating kiln. The maximum temperature in the cement kiln is about 950°C, and at this temperature the lime and clay partially fuse together as a hard clinker. Cement clinker is ground into powder and mixed with a small amount of *gypsum* (calcium sulphate) to produce dry cement powder. The function of the gypsum is to control the setting characteristics of the cement.

When cement powder is mixed with water a series of complex chemical reactions occurs, forming hydrated silicates and aluminates of calcium. It is these chemical reactions that cause the wet cement to harden and set as a rigid material. Some of the hydration reactions take place very slowly and, although the cement will set fairly rapidly, great strength and hardness will not be developed for several days, or weeks, depending on the composition of the cement, the amount of water added, and the temperature. The amount of water added to dry cement powder is important. If insufficient water is added, full hydration of the cement particles will not occur. For full hydration and the development of maximum strength, a water/cement ratio of about 0.4/1 is necessary. If the water/cement ratio is much in excess of this value the strength of the hardened cement will be reduced.

Concrete consists of a mixture of cement, sand, and an aggregate of small stones. When water is added to a dry concrete mix, the cement paste formed should fully coat all sand and aggregate particles, and fill in the void spaces between aggregate particles. The cement paste hardens, owing to the hydration reactions, and bonds the inert sand and aggregate together.

The final properties of a concrete will be dependent on a number of factors, including the relative proportions of water, cement, sand and aggregate in the material, the average size of aggregate particles, the type of aggregate stone used, and the surface texture of the aggregate. The compressive strengths of plain concrete may be up to 65 Mpa, but the strength in tension is only about one-tenth of the value of compressive strength. (This compares with a compressive strength of about 100 MPa for hardened cement.)

When a concrete is subjected to stress, failure probably commences at the interface between aggregate and cement. Aggregate particles with rough surfaces will give concretes of higher strengths than will smooth-surfaced aggregate. The tensile strength of concrete is low (up to 5 MPa) and, to overcome this disadvantage, concrete fabrications are very often reinforced with steel. In plain reinforced concrete, a network of steel rods or bars is assembled and the concrete is allowed to set around this framework. The steel reinforcement is positioned in the portion of the concrete member that will be subjected to tensile stresses. For example, in a simply supported beam (Figure 18.1(a)) the steel lies along the lower portion of the beam. There is a purely mechanical bonding between concrete and steel, and the reinforcement bars are often twisted, or possess surface projections (these may be formed by rolling the bars through patterned rolls) in order to increase the adhesion between steel and concrete.

Figure 18.1 *(a) Plain reinforced concrete beam. The upper layers of the beam are in compression and the lower layers are in tension. Steel bars help sustain the tensile stress. (b) Pre-stressed concrete beam. Steel wires in tension exert a compressive force on the concrete.*

EXAMPLE 18.1

Usually the reinforcing bars used in plain reinforced concrete are made from a low-carbon steel. In some circumstances the conditions may be such as to cause the reinforcing bars to corrode. If this occurs it will have adverse effect on the long-term strength of the concrete structure. Suggest possible alternative reinforcement materials that could be used, giving reasons for the choice.

SOLUTION

One solution to the problem of corrosion of reinforcing bars would be to coat the bars with some impervious layer which would keep the steel insulated from the corrosive environment. Steel bars coated with a layer of an epoxide have been used successfully in practice.

Two suitable alternative reinforcing materials are austenitic stainless steel and fibre-reinforced plastic composites with glass, carbon or aramid fibres in a resin matrix. Both of these materials possess sufficient strength to act as good reinforcement materials for concrete and neither would corrode during long-term service. These materials are much more expensive than plain carbon steel but, although the initial cost of reinforced concrete structures would be high the overall cost over the lifetime of the structure could be lower as there would be less need for repair and maintenance, as the costs of rectification work to structures in which the reinforcement bar has corroded can be very great.

Another form of reinforced concrete is known as *pre-stressed* concrete. The concrete is put into a state of compression by means of highly stressed steel wires (Figure 18.1(b)).

Glass fibre reinforced plastic (GFRP) cables have been considered as a viable alternative to high tensile steel wires. Although possessing a lower tensile yield strength than the high tensile steels normally used, larger diameter pre-stressing cables can be used yet still giving an overall reduction in weight. The first successful use of GFRP pre-stressing cables in a major structure in Europe was in a road bridge in the German city of Dusseldorf in 1987.

When a pre-stressed concrete beam is in service, the initial compressive stresses must be overcome before tensile stresses can be developed within the material. Concrete may be pre-stressed by pre-tensioning, or by post-tensioning. In the former method steel wires are placed in tension before being surrounded by concrete. The externally acting stress on the steel is removed when the concrete has set. In post-tensioning, the concrete is allowed to set and harden around a tube, or tubes. Steel wires are then put through the tubes and these wires are stretched and anchored to the concrete.

18.4 Clay and clay products

Muds and *clays* have been used as the basis for artefacts, including structural products, throughout human history. Clay is a general name for minerals containing small plate-like crystals which have been formed from the decomposition of felspar in igneous rocks, such as granite, by the action of air and water over long periods of time. A major product of this decomposition is the clay mineral *kaolinite*, a hydrated aluminium silicate. Clay formed from the parent rock may be deposited at its place of origin, and this is known as residual or primary clay, or it may be transported by water and redeposited at some distance and this is known as sedimentary or secondary clay. The latter is usually of much finer particle size and contains more impurities than residual clay deposits. Many types of clay occur in nature and they are mixtures of the clay minerals with non-clay minerals such as mica and quartz. The kaolinitic clays include *china clay, ball clay* and *fireclay*. Ball clay is a fine textured clay with a high plasticity which consists of kaolinite, quartz and, in some cases, mica. China clay is almost pure kaolinite. It is coarse and has a low plasticity but fires to give a pure white product. The various types of clay products can be broadly classified into *whiteware*, which includes china, porcelain and stoneware, *heavy clayware*, which includes brick, roofing tiles, earthenware pipes and insulators, and *clay refractories*. This last group is dealt with in section 18.5.

Wet clays are plastic and can be moulded or extruded to form clay products such as bricks, tiles and earthenware articles. After the plastic clay has been formed into the required shapes, the products are allowed to dry out. This drying process is followed by firing the clay in a kiln within the temperature range 800–1500°C. During the firing process, water of hydration is driven off with resulting crystal structural changes within the material. A partial melting of some phases occurs and, on cooling, these molten areas vitrify and the glass phase cements the fired clay particles together. The amount of glass phase formed depends on both the clay composition and the firing temperature, a higher firing temperature giving a higher glass content. After firing the clay product is both porous and brittle. Many clay products are glazed after the first firing to produce a non-porous surface layer that will render the product impermeable. The glaze, which is a slurry of mixed oxides – mainly silica, soda and lime – is sprayed or painted onto the surface and the article kiln-fired a second time. During this second, or *glost*, firing the surface coating is converted into a hard glass.

Whiteware and *porcelain* are general terms for fine-textured clay ceramics including china, stoneware and fine earthenware. They are mixtures of ball and china clays compounded with silica as a filler and a flux, usually felspar, to form a glass phase. Firing temperatures, and the flux content, for porcelains are higher than for clay products such as bricks and tiles so as to give a high glass content with low levels of porosity. Both china and stoneware are highly vitrified, which gives high strength. The compressive strength of a low-porosity porcelain may be as high as 600 MPa.

As stated above, most clay products are porous. There are two types of porosity, open and closed. Open porosity means that many of the pores are interconnected and reach the surface of the body, so that the body becomes permeable to fluids. Closed porosity means that the pores are discontinuous and that the body, while containing some porosity, is impermeable to fluids. Glazing a fired clay body will render the surface impermeable. China cups, for example, would be useless without being surface glazed, as they would be permeable to fluids.

The extent of open porosity in a body can be determined by weighing the body dry, when it is fully saturated with a fluid, and when it is suspended in a fluid, as shown in Example 18.2.

EXAMPLE 18.2

A fired clay body is weighed under several conditions, the results being:

weight dry – 873 g.
weight saturated with kerosene – 898 g.
weight suspended in kerosene – 587 g.

The density of kerosene is 829 kg m^{-3} (0.829 g cm^{-3})

Calculate: (a) the apparent porosity,
  (b) the apparent density,
  (c) the bulk density,
  (d) the apparent volume, and
  (e) the bulk volume.

SOLUTION

The dry body absorbs $(898 - 873) = 25$ g of kerosene into its open pores. (There may be some closed porosity into which the kerosene cannot penetrate.) So the volume of open porosity is $25/0.829 = 30.16 \text{ cm}^3$.

The difference between the saturated weight and the weight suspended in kerosene is $898 - 587 = 311$ g.

Therefore, the volume of kerosene displaced $= 311/0.829 = 375.15 \text{ cm}^3$. This is the bulk volume of the body, including both open and any closed porosity.

The difference between the dry weight and the weight in kerosene is $873 - 587 = 286$ g and the volume of kerosene displaced $= 286/0.829 = 345.00 \text{ cm}^3$. This is equal to the apparent volume of the body which excludes any open porosity but includes any closed porosity.

So (a) the apparent porosity
  $= $ (vol. of open porosity/bulk volume) $\times 100$
  $= (30.16/375.15) \times 100 = 8.04$ per cent
  (b) the apparent density
  $= $ dry weight/apparent volume
  $= 873/345 = 2.53 \text{ g cm}^{-3} = 2530 \text{ kg m}^{-3}$
  (c) the bulk density
  $= $ dry weight/bulk volume
  $= 873/375.15 = 2.327 \text{ g cm}^{-3} = 2327 \text{ kg m}^{-3}$
  (d) the apparent volume
  $= 345.00 \text{ cm}^3$
  (e) the bulk volume
  $= 375.15 \text{ cm}^3$

18.5 | Refractories

Refractories are mineral products that are stable at high temperatures and that are used in the construction of furnaces and allied equipment. Refractory materials must also resist the action of molten metals, slags, glasses, abrasive particles and hot gases. Refractory materials are classified as acidic, basic and neutral. Acidic refractories are rich in silica, SiO_2, an acidic oxide, while basic refractories are composed mainly of magnesite, dolomite and lime. There are many neutral refractories and the group includes alumina, fireclays, chrome, carbon and other pure metal oxides such as zirconia.

Acid refractories. Silica, SiO_2, is an excellent refractory as its melting point is in excess of 1700°C, and materials with a high silica content (over 93 per cent) are used in metallurgical furnaces to resist high temperatures and attack by acid slags, namely those with high silica and a low basic oxide content. Silica refractories are also used as regenerator bricks and for the roofs of furnaces in the steel and glass industries. The raw material for silica refractories is *quartzite* and *ganister*. These are both naturally occuring siliceous rocks containing 96–98 per cent of silica. The rock is crushed and broken down into a fine state of subdivision. Pressings made from dry silica grains would possess virtually no strength, so a small amount of an organic binder is added to the powder to aid the production of moulded silica shapes. A small amount of lime is also added to the silica before moulding. The moulded shapes are then dried and fired in a kiln. The lime content forms calcium silicate during firing and this acts as a bonding agent for the silica grains. Silica refractories can be used successfully at temperatures up to 1700°C.

Basic refractories. These are refractory materials consisting mainly of basic oxides such as magnesia, MgO, and lime, CaO, used in metallurgical and other furnaces to resist high temperatures and attack by basic oxides and slags. Naturally occurring magnesite, $MgCO_3$, calcined to the oxide has been used as a refractory material but most magnesite deposits contain some silica and this adversely affects the quality of the refractory. Much of the magnesia, MgO, used for refractories is derived from seawater. The most widely used raw material for basic refractories is *dolomite*, a natural rock which is a mixed carbonate of magnesium and calcium. When calcined at 800–900°C, dissociation occurs to give calcium oxide, magnesium oxide and carbon dioxide. The mixture of the two oxides is susceptible to hydration and is referred to as *caustic dolomite*. In refractory bricks made from dolomite, the calcined grains are usually chemically bonded. A bonding agent, frequently tar, is used to coat the grains which are then pressed into bricks at pressures up to 70 MPa and fired or 'burnt' at 1500–1700°C. The resulting product is more resistant to hydration owing to partial sintering, and is referred to as *dead-burnt dolomite* and used as a refractory in the steelmaking and glass industries. Lime, calcium oxide, CaO, produced from limestone has been used as a basic refractory. It needs to be stabilised by tar impregnation or surface treatment to prevent hydration, but has been used for lining rotary furnaces and converters in steelmaking.

Neutral refractories. A major neutral refractory is *chrome-magnesite*. The term covers a range of refractory materials made from chrome ore and calcined magnesite. When chrome predominates the resulting refractories are *neutral* whereas when magnesite predominates, the materials are slightly *basic* and the term *magnesite-chrome* is used. Chrome-magnesite refractories are more resistant to thermal shock than magnesite. Courses of chrome-magnesite brick are used as a separating layer between basic and acid refractories used in the linings of some types of high temperature furnace. Chrome-magnesite is used also as a structural material in furnaces for handling lead, copper and aluminium. Magnesite-chrome refractories are neutral or slightly basic refractories and they have a higher degree of refractoriness under load than the chrome-rich materials. They are used as structural materials in furnaces for copper, aluminium and other metals.

Fireclays are naturally occurring sedimentary clays, often found underlying coal seams, consisting predominantly of kaolinite ($Al_2O_3.2SiO_2.2H_2O$) and quartz. They are of varied composition, depending on location, and may contain between 25 and 40 per cent of alumina, between 50 and 70 per cent of silica, and be low in Fe_2O_3, MgO and alkalis. They are refractory clays with high softening temperatures, depending on composition, in the range from 1400°C to more than 1600°C. Fireclays are used in the manufacture of refractories for use as linings for blast-furnaces, lime and cement kilns, pottery kilns, furnaces for aluminium melting, and for gas heaters and domestic fire-backs. Some fireclays possess a high silica content and these siliceous clays, containing 75 per cent or more of silica, are used for the manufacture of semi-silica brick, which may be used at higher temperatures than normal fireclays.

Another range of successful refractories is the high-alumina group of materials. These range from aluminous fireclays containing >45 per cent alumina up to almost pure alumina. They are made by a variety of processes from bauxite, sillimanite, mullite, fused and calcined aluminas and fireclays. They are suitable for use at temperatures up to 1600°C and find application in steelmaking furnaces, ceramic kilns, cement kilns, glass tanks and as crucibles for melting metals. Castable refractories can be made from a mixture of high alumina cement and a refractory grog blended with water. They are used for repairing furnace linings or building up monolithic structures.

Carbon is used successfully as a refractory material in applications where reducing conditions exist. A good example is the use of carbon as a blast furnace refractory. Carbon refractories made from coke or anthracite, bonded with pitch or tar are used for blast-furnace hearth linings. Clay-bonded graphite is also used as a refractory for some purposes.

EXAMPLE 18.3

The hearth area and lower portion of the walls of a steelmaking furnace are lined with dolomite refractory brick while the upper walls and arched roof of the furnace are built of silica brick. Why is it necessary to use several courses of chrome-magnesite refractory brick in the walls separating the other two refractories?

SOLUTION

Dolomite is a basic refractory while silica is an acid refractory material. At ordinary temperatures the two materials can rest against each other indefinitely with no reaction occurring. However, at the operating temperatures of a steelmaking furnace (1500–1700 °C) the two materials would react together rapidly and to avoid such a reaction taking place they need to be separated by a neutral material such as chrome-magnesite.

18.6 | Industrial ceramics

Considerable attention has been focused in recent years on the oxides, carbides, borides and nitrides of metals. These compounds are generally of high hardness, and possess high elastic moduli. They are also of high stability and possess high melting points. The principal disadvantage of all these materials is their brittle nature. Their use is, in general, only considered when the conditions to be met in service preclude the use of other materials. Industrial ceramics can be considered for use when the requirements are one or more of the following:

(a) high resistance to abrasion and wear,

(b) high strength at high temperature,

(c) good chemical stability,

(d) good electrical insulation characteristics.

Some of the areas in which ceramic materials are used is as abrasives, cutting tools, bearings, insulators, magnetic materials, piezoelectric materials, superconductors, crucibles for metals of high melting point, and pumping equipment for liquid metals.

The properties of a number of oxides, carbides and nitrides are given in Table 18.2.

Many ceramic components are made by the process of pressing a powder into a compact, followed by sintering. The powder compact will have a high degree of porosity but a densification of the material will take place during sintering and the amount of porosity will tend to be very small. Considerable shrinkage takes place, therefore, during the sintering of most products but this is predictable. The dies in which the powder compacts are pressed are made larger than the required final product dimensions.

Table 18.2 *Properties of some pure oxides, carbides and nitrides.*

Compound	Melting point (°C)	Density (kg/m³ × 10⁻³)	Hardness (Mohs' scale)
Al_2O_3	2015	3.97	9
CaO	2600	3.32	4.5
Cr_2O_3	2265	5.21	
MgO	2800	3.38	6
SiO_2	1723	2.32*	6–7
ThO_2	3300	6.5	6.5
TiO_2	1840	4.24†	5.5–6
Cr_3C_2	1890	6.68	7+
NbC	3560	7.82	9
SiC	2830	3.21	9
TaC	3870	14.48	8+
TiC	3250	4.25	8
WC	2630	15.5	9
ZrC	3530	6.7	9
AlN	2230	3.05	7+
Si_3N_4	1950	3.44	9
TiN	2930	5.43	8+
ZrN	2980	7.32	8+

* Value for quartz. † Value for rutile.

EXAMPLE 18.4

Components in an industrial ceramic are formed by compaction of powder and sintering. If the total porosity of the 'green' powder compact is 22 per cent and the total porosity of the sintered product is 2 per cent, what amount of linear shrinkage should be allowed for?

SOLUTION

In the 'green' state 22 per cent of the total volume of the compact is porosity, so 78 per cent of the volume is ceramic powder. In the sintered state, with 2 per cent porosity, 98 per cent of the sintered volume is ceramic and therefore the volume of the sintered compact as a percentage of the original 'green' volume is 78/0.98 = 79.6. So the volume shrinkage during sintering is 100 − 79.6 = 20.4 per cent. Assuming that the shrinkage is uniform in all directions, the linear shrinkage is the cube root of 20.4 = 2.73 per cent, or a shrinkage of 27.3 mm/m.

18.7 Alumina

High alumina ceramics, containing 75–99.8 per cent Al_2O_3, have been extensively developed. Most properties are improved as the alumina content increases (see Table 18.3). Precise control of crystal size, degree of calcination and levels of impurities is required in the manufacture of commercial aluminas and a variety of ceramics based on alumina are available including fusion-cast alumina and various grades of sintered alumina. High density, gas impermeable ceramics can be obtained by sintering alumina with small quantities, about 0.25 per cent of magnesium oxide, MgO, to prevent exaggerated grain growth. High alumina ceramics are used for the manufacture of spark-plug insulators, ceramic/metal assemblies in vacuum tubes, substrates for the deposition of electronic microcircuits, and

Table 18.3 *Properties of alumina ceramics.*

	Alumina (per cent)			
	75	86–94	94–98	> 98
Density (kg/m^3 × 10^{-3})	3.2	3.3	3.5	3.7
Hardness (Mohs' scale)	8.5	9	9	9
Compressive strength (MPa)	1250	1750	1750	
Flexural strength (MPa)	270	290	350	380
Maximum working temperature (°C)	800	1100	1500	

metal-cutting tool tips. Also, because of the low neutron absorption cross-section of alumina, the material finds application in nuclear equipment.

18.8 | Silicon nitride

Silicon nitride, Si_3N_4, is a very useful engineering ceramic which is fully resistant to most strong acids and to molten aluminium and other low melting point metals. It exists in two hexagonal crystalline structural modifications, α-$Si_{12}N_{16}$ and β-Si_6N_8 and silicon nitride powder, consisting mainly of α, is made by a number of routes including the nitriding of silicon, the carbothermal reduction of silica in nitrogen, and the reaction of silicon tetrachloride with ammonia. There are several different types of silicon nitride ceramics:

reaction bonded silicon nitride (RBSN),
hot pressed silicon nitride (HPSN),
sintered silicon nitride (SSN)

and variants of those including hot isostatically pressed or pressure sintered silicon nitride, and sintered RBSN. Silicon nitride is used for a range of engineering applications at high temperatures and as wear parts.

Reaction bonded silicon nitride (RBSN). Silicon nitride ceramic made by reaction bonding in which a silicon powder compact, normally isostatically pressed, is converted into silicon nitride by heating in a nitrogen atmosphere at temperatures in the range 1100–1450°C according to the reaction: $3Si + 2N_2 \rightarrow Si_3N_4$ which occurs by a gas/solid, solid/liquid/gas or gas vapour reaction depending on temperature and the levels of impurity in the silicon and nitrogen. The resulting microstructure consists of sub-micron sized grains of silicon nitride with 15–30 per cent porosity, including some large pores of up to 50 µm size, as a result of the melting out of silicon/iron impurity particles during nitriding, and small interconnected channels which may be < 0.1 µm in diameter. No shrinkage occurs during the process as the silicon nitride which is formed fills up the pore space of the original silicon compact. Thus, after partial nitriding to give the compact sufficient strength, the process may be stopped and the compact machined, using conventional tools, to its final shape and dimensions. This material is used as pouring tubes for aluminium, and for high temperature engineering components.

Hot pressed silicon nitride (HPSN). This is a dense silicon nitride engineering ceramic produced by uniaxially pressing α-silicon nitride containing sintering additives such as MgO and Y_2O_3 in an induction heated graphite die at temperatures in the range 1700–1850°C, during which an oxynitride liquid is formed which allows densification and the transformation of α to β silicon nitride. A residual glassy or crystalline grain-boundary phase remains on cooling. The process is restricted to fairly simple shapes.

Table 18.4 *Properties of silicon nitride ceramics.*

	RBSN	HPSN
Density (kg/m$^3 \times 10^{-3}$)	2.3–2.6	3.12–3.18
Open porosity (per cent)	18–28	0.1
Hardness (Mohs' scale)	9	9
Young's modulus (GPa)	160	290
Flexural strength at 20°C (MPa)	110–175	550–680
Flexural strength at 1200°C (MPa)	210	350–480

Sintered silicon nitride (SSN). This is a dense silicon nitride engineering ceramic produced by mixing α-silicon nitride, with its associated surface silica, and one or more metal oxides, such as MgO or Y_2O_3, as sintering additives. The powder mixture is formed into a shaped body and fired under a nitrogen atmosphere, protected by a powder bed of boron nitride. During the process an oxynitride liquid is formed which aids densification. On cooling, the oxynitride remains as a grain boundary phase, either amorphous or crystalline, depending on the type and amount of oxide used and this controls the high temperature properties of the ceramic.

Pressureless sintered silicon nitride (PSSN). This is a silicon nitride ceramic fabricated by the addition of an oxide or combination of two oxides to α-silicon nitride powder, consolidation to the required shape and firing in a nitrogen atmosphere at a pressure of one atmosphere at 1700–1800°C. During the firing, the oxide or oxides react with surface silica, SiO_2, and some nitride to form a M–Si–O–N liquid phase which aids densification and converts the α to β-silicon nitride. The M–Si–O–N phase, on cooling, is retained as a secondary phase at the grain boundaries which controls the high-temperature properties of the ceramic. This material is under development for engine components.

The properties of reaction-bonded and hot-pressed silicon nitride are given in Table 18.4.

18.9 Sialons

The sialons are derivatives of silicon nitride and the name sialon is an acronym derived from the elements involved, namely Si–Al–O–N. Much research and development work is being done on these materials and, at present, two major groups of sialon ceramics are manufactured commercially. These are known as low-substitution and high-substitution sialons. The former, and principal, material, is a solid solution of β-silicon nitride, Si_6N_8, in which Si is partially substituted by Al and nitrogen is partially substituted by oxygen, to give an expanded structure designated β'-sialon with a range of composition given by $Si_{(6-z)}Al_zO_zN_{(8-z)}$. β'-sialons sinter by a liquid-phase mechanism which aids densification and, on cooling, a glassy phase exists at grain boundaries. This glass phase, which softens at high temperatures, limits the operating temperature for the material to about 1000°C. In the highly substituted type of sialon, an yttrium aluminium garnet (YAG) crystalline phase forms at grain boundaries and these materials can operate at temperatures up to 1400°C. The sialons possess higher strength and hardness than silicon nitride and are used as cutting tool materials and for engine components and bearings.

18.10 Silicon carbide

Silicon carbide has been used as an abrasive for many years but is now also available as an engineering ceramic material.

Silicon carbide, SiC, is a hard, semiconducting ceramic existing in two crystalline structural modifications, α-SiC (hexagonal), made by the reduction of silica sand with carbon in an arc furnace, and

β-SiC (cubic), produced by vapour phase reactions. The α form converts to the β form at temperatures above 2000°C. There are a number of different types of silicon carbide ceramics available, these being:

reaction bonded silicon carbide,

clay-bonded silicon carbide containing up to 50 per cent fireclay; this material is coarse grained and used as a refractory for thermal shock applications,

hot-pressed silicon carbide containing 2 per cent alumina as a densification aid and to give a fine grained product with zero porosity,

sintered silicon carbide containing boron and carbon as dopants to aid densification,

recrystallised silicon carbide, sintered to a less dense, fairly coarse-grained product for use as furnace elements, refractory worktubes etc., and

nitride-bonded silicon carbide produced by the reaction bonding of a material containing coarse α-SiC and silicon powders in a nitrogen atmosphere at 1350–1450°C used for refractory tubes and containers.

Reaction bonded silicon carbide (RBSC) is made by reaction bonding in which a porous compact, consisting of silicon carbide, carbon (graphite) with a binder, is first formed and the binder is carbonised in the pores. The body is then heated in contact with molten silicon, during which process the silicon moves through the material under the influence of capillary forces, reacting with the graphite and converting it to silicon carbide which deposits epitaxially on the surfaces of the original silicon carbide particles, thus bonding them together. All the carbon reacts and excess silicon remains. The final microstructure consists of the original silicon carbide grains overcoated with the silicon carbide formed *in situ*, and interconnected silicon-filled pores. It was originally developed as a cladding material for high-temperature nuclear-reactor fuel elements and is now used for mechanical seals, bearings and engine components. It is also known as reaction sintered silicon carbide or self-bonded silicon carbide.

Silicon carbide is also formed as fibres and whiskers for use as reinforcement in composite materials.

18.11 | Boron nitride and other ceramics

Boron nitride is a synthetic nitride of boron, BN, which is isoelectronic with carbon and exists in two structural modifications similar to graphite (hexagonal) and diamond (cubic). The hexagonal form, with a density of 2270 kg/m³, is soft and platelike and is manufactured as a ceramic by hot-pressing. This produces a material exhibiting anisotropy because the platelets line up in planes perpendicular to the direction of pressing. Thin walled shapes, such as crucibles for special glasses, may be produced by a chemical vapour deposition process. The cubic form, with a density of 3480 kg/m³, is the hardest substance known, next to diamond. It is produced from the hexagonal form at high temperature and pressure. The conditions for the formation of the cubic form are about 1500°C at a pressure of about 60 GPa. It is used as grinding material in grit form, or as a solid ceramic for the manufacture of metal cutting tools, and is referred to as CBN (cubic boron nitride). It is possible to sinter together small crystals of CBN to produce large polycrystalline masses. The resulting material, PCBN (polycrystalline cubic boron nitride) has a structure of randomly orientated crystals and is isotropic.

Aluminium nitride, AlN, is a ceramic manufactured by the method of *tape casting* (see Chapter 22) with suitable densification additives, such as CaO or Y_2O_3, mixed with the nitride. Suitable shapes for forming substrates are blanked out from the cast product and sintered at 1700–1800°C. Aluminium nitride is used for *chip-carriers* where its high thermal conductivity allows good heat dissipation. It is also used as a raw material for sialons.

Aluminium titanate is a mixed oxide of alumina and titania with the formula Al_2TiO_5. The ceramic is formed at high temperature but it slowly dissociates below 1150°C into alumina and titania and this gives a microstructure with a network of fine microcracks. This structure gives the material an effective thermal expansion coefficient of zero up to 700°C. It is used, therefore, in applications where a high resistance to thermal shock is required.

Titania is the ceramic dioxide of titanium, TiO_2, occurring naturally as the mineral rutile and crystallising as a body-centred tetragonal structure. Titania was formerly used as a dielectric material (dielectric constant = 100) to replace mica and electric porcelains (dielectric constants, 5–10) in applications requiring high permittivity but it has now been replaced for these applications by barium titanate and its derivatives (dielectric constants, 10^4–10^5). Titania is still used as a silicate-bonded ceramic for mechanical applications such as thread guides, the material being manufactured from titania with a small amount of clay and a calcium based flux. It is fired to 1350–1400°C in a reducing atmosphere to give a black non-stoichiometric ceramic which is semiconducting, thus allowing the discharge of static electricity from synthetic textile fibres.

Titanium carbide, TiC, is a hard ceramic material with a density of 4920 kg/m³, usually used in combination with alumina, Al_2O_3, to give added strength and a higher elastic modulus. Compositions typically contain 20–40 per cent TiC and are hot-pressed to give a fine-grained two-phase material which is used as tool inserts for metal cutting.

Zirconia. The oxide of zirconium, ZrO_2, existing in three different crystalline modifications. The high temperature form is cubic and, during cooling, this transforms first into a tetragonal form and then, at a lower temperature, into a monoclinic phase. Transformation on cooling causes cracking because of a 3 per cent volume change and so it is difficult to fabricate pure ZrO_2 ceramics. For this reason, 'stabilising' oxides, for example, MgO, CaO or Y_2O_3, are added and these form solid solutions of a cubic structure at room temperature.

The different types of alloyed zirconia ceramics are as follows:

fully stabilised ZrO_2 (FSZ), containing CaO, and having a cubic structure,
partially stabilised ZrO_2 (PSZ), containing MgO, having a duplex structure of cubic crystals with precipitated tetragonal phase,
tetragonal zirconia polycrystals (TZP), containing Y_2O_3 and having a fully tetragonal microstructure,
zirconia toughened alumina (ZTA) which is a composite polycrystalline alumina ceramic containing up to 20 volume per cent ZrO_2 as a dispersed phase.

18.12 | Glass ceramics

Glass ceramics are complex silicates that can be readily fabricated by the conventional forming techniques for glasses but which are then converted into fine-grained polycrystalline materials by controlled crystallisation brought about by heat treatment. The resulting materials have considerably higher strengths than most glasses, and retain their strengths to much higher temperatures than glasses, being capable of use in stressed conditions at temperatures up to 900°C. In addition, ceramic glasses possess excellent resistance to thermal shock.

The controlled crystallisation or devitrification is brought about by a two-stage heat treatment. The first stage is to promote the formation of a large number of nuclei for crystal growth. The glass compositions generally contain small additions of oxides such as TiO_2 and P_2O_5 to act as nucleating agents. The material is then heated at a higher temperature to give controlled grain growth until the desired grain size structure is achieved. There is very little dimensional change during the crystallisation process. The degree of crystallinity achieved is in the range 70–100 per cent and the final structure is dense with no porosity. A range of glass–ceramic compositions exists in both the Li_2O–Al_2O_3–SiO_2 and Li_2O–MgO–ZnO–SiO_2 systems. The former type, LAS glass ceramics, have very low thermal expansion coefficients together with a very high thermal shock resistance, while the latter type possess very high thermal expansion coefficients.

The applications for glass ceramics include missile nose cones, domestic ovenware, bearings and ceramic-to-glass seals. The proprietary name for one commercial LAS glass ceramic is *Pyroceram*.

18.13 | Glasses

Glass is an amorphous, brittle material obtained by melting together a mixture of various compounds and supercooling the liquid to a temperature where its viscosity exceeds 10^{14} poise (10^{13} Pa s). There is no discontinuity in the cooling curve and the viscosity gradually increases throughout cooling. At low temperatures the material retains the internal structure of the liquid but with a viscosity so high that, for all practical purposes, it behaves as an elastic solid.

The viscosity of a glass varies with temperature according to an Arrhenius type relationship:

$\eta = A\exp(B/T)$

where η is viscosity, T is temperature (K), and A and B are constants.

Because they behave as highly viscous fluids over a relatively wide range of temperatures, glasses may creep slowly with time when subjected to a constant stress, even at fairly low temperatures. The creep rate, $d\varepsilon/dt$, at any particular temperature is given by the expression: $d\varepsilon/dt = B\,\sigma^n$ where n is the time exponent. $n = 10$ for glasses.

EXAMPLE 18.5

The tensile strength of a glass, as measured in a short-term test of 1-minute duration, is 50 MPa. What is the maximum continuous tensile stress that a component made from this glass should be subjected to if the component life is to be 10,000 hours? Assume that the time exponent for the glass is 10.

SOLUTION

Use the expression $d\varepsilon/dt = B\,\sigma^n$.
The maximum strain, ε_{max}, is the strain at break in the short-term tensile test, when t is 1 minute (60 s).
So: $d\varepsilon/dt = \dfrac{1}{60} = B\,50^{10}$ (1)

and, for a time of 10,000 hours ($10{,}000 \times 3600$) s

$d\varepsilon/dt = \dfrac{1}{10{,}000 \times 3600} = B\,\sigma^{10}$ (2)

Dividing equation (2) by equation (1) gives

$\dfrac{60}{10{,}000 \times 3600} = \dfrac{\sigma^{10}}{50^{10}}$

from which $\sigma = 13.2$ MPa.

There are several points or temperatures of importance with respect to glasses. The *glass transition temperature, T_g,* or *fictive temperature*, is the temperature at which viscous flow ceases. This is taken as the temperature at which the viscosity is 10^{17} poise (10^{16} Pa s) (see also section 6.1). The temperature at which the viscosity is 10^{14} poise (10^{13} Pa s) is known as the *strain point*. At this level of viscosity, there is little atom movement and a glass can be rapidly cooled from this temperature without causing the development of internal stresses. The *annealing point* for a glass is that temperature at which the viscosity is 10^{13} poise (10^{12} Pa s) and at this temperature any internal stresses are rapidly relieved. Glasses are usually worked and formed into shapes at temperatures where the viscosity is in the range 10^3–10^6 Pa s.

Silica is the principal glass forming oxide but it has a high melting point (>1700°C). Other oxides, particularly alkalis, lower the softening point so that lower melting temperatures can be employed.

The major commercial glasses are:

Silica glass (>99.5 per cent SiO$_2$),
96 per cent silica glass,
Soda–lime–silica glass,
Lead glasses,
Borosilicate glasses, and
Aluminosilicate glasses.

Silica glass, also referred to as vitreous silica or fused silica, is a glass consisting entirely of silica, SiO$_2$, with good chemical stability, a very low thermal expansion coefficient (0.5×10^{-6} K^{-1}), and usable up to 900°C. It is made in two forms: (1) translucent, requiring temperatures >2000°C for 'melting' but still containing gas bubbles, and (2) transparent, produced according to the following high temperature reaction:

$$SiCl_4 + 2H_2O \rightarrow SiO_2 + 4HCl$$

the silica condensing as an optical disc. There is also a silica glass containing 96 per cent SiO$_2$ and 4 per cent B$_2$O$_3$. This has good thermal properties and is used for space and missile applications.

Soda–lime–silica glass is the most common type of glass with an approximate composition of 15 per cent Na$_2$O (soda), 10 per cent CaO (lime) and 75 per cent SiO$_2$ (silica). Potash, K$_2$O, may be used as partial replacement of soda and magnesia, MgO, as partial replacement for lime, particularly in flat glass, to avoid premature devitrification during drawing. This type of glass is used also in containers and light bulbs. The electrical properties of the glass vary with composition.

Lead glasses have compositions within the range 30–70 per cent SiO$_2$, 18–65 per cent PbO, 5–20 per cent Na$_2$O or K$_2$O. The original name for lead crystal glass was *flint glass*. It was developed for optical purposes using powdered flint as the source of high-purity silica needed to maintain a colourless product. Now the term is used to refer to any colourless glass other than flat glass. Lead-crystal glass refers to a range of lead–alkali–silicate glasses with 18–32 per cent PbO, 5–10 per cent each of K$_2$O and Na$_2$O, 50–70 per cent SiO$_2$, used for high-quality tableware. It is also called crystal ware because of its especially lustrous quality as a result of its high refractive index. It is softer than other glass, thus making it easy to cut and grind the material to achieve brilliant decorative effects.

Borosilicate glasses are glasses in which the major constituents are silica (60–80 per cent) and boric oxide (10–25 per cent). They also contain 1–4 per cent Al$_2$O$_3$ and 1–4 per cent Na$_2$O. Borosilicate glasses have low thermal expansion, high chemical resistance, high dielectric strength and a higher softening point than soda glass. They are used for heat-resistant ovenware, laboratory ware, industrial piping and in electrical applications.

Aluminosilicate glasses contain 10–60 per cent SiO$_2$, 20–40 per cent Al$_2$O$_3$, 5–50 per cent CaO, 0–18 per cent B$_2$O$_3$ and possess a high chemical resistance and low thermal expansion.

Halide glasses. These are single halide ion special glasses, the most stable of which are based on fluorides of zirconium and barium (50–60 per cent ZrF$_4$, 30–40 per cent BaF$_2$ with LaF$_3$ and AlF$_3$ to improve stability). They possess a broad transparency range from near ultra-violet (300 nm) to the mid–far infrared (6–9 μm) and are under development for fibre-optic communications.

SUMMARY
....

Ceramics are crystalline inorganic compounds of metals which are hard, non-ductile and also electrically non-conductive. They can be classified, broadly, into clay products, refractories and industrial ceramics.

Clays are complex aluminosilicates which can be moulded readily into shapes. The firing process at high temperatures causes chemical change and the formation of a hard crystalline material with small areas of a glassy phase helping to cement the crystals together. The two main classifications are whiteware, which includes china and porcelain, and heavy clayware, or earthenware, used for products such as bricks, roofing tiles and pipes.

Refractories are materials which have very high melting temperatures and are used mainly in the construction of furnaces and allied equipment. Some of the main refractory materials are the oxides of silica, calcium, magnesium, chromium and aluminium.

Modern industrial ceramics are the oxides, carbides and nitrides of metals, including aluminium, silicon, boron, titanium and zirconium. Generally, they have high hardness and elastic moduli, and are highly stable materials with high melting points.

Most ceramic materials are non-ductile, are strong in compression but possess low tensile strengths. They contain, almost invariably, defects and flaws, for example, porosity, from which non-ductile fracture can initiate at relatively low values of tensile stress.

Glasses are non-crystalline, brittle materials. At low temperatures the glass has the same amorphous structure as the liquid phase but with such a high viscosity that it behaves as an elastic solid. Most glasses are complex silicates but pure silica can also be produced in the glass state.

Glass ceramics are complex silicate glasses which can be converted into a fine-grained polycrystalline state by controlled heat treatment. They possess higher strengths than the glass from which they are formed, and retain this strength to higher temperatures than the glass form.

Questions

18.1 Distinguish between a glass and a ceramic material.

18.2 What are the main classifications of ceramic materials?
Explain the main differences, in terms of composition and processing methods, between a traditional porcelain product and a product, for example a spark-plug insulator, made from a modern industrial ceramic.

18.3 Briefly explain the changes which occur when:
(a) a wet clay is processed to make a solid ceramic article,
(b) a dry oxide powder is pressed and sintered to form a solid ceramic component, and
(c) a molten silicate is cooled to form a glass.

18.4 Silicon carbide powder, which has a density of $3200\ kg\ m^{-3}$, is compacted and sintered to form a ceramic component. The finished component has a dry weight of 360 g and a weight of 385 g when saturated with water. The weight of the component when suspended in water is 224 g. Calculate both the apparent porosity and the true total porosity of the sintered compact. Why do these values differ?

18.5 The density of alumina powder is $3970\ kg\ m^{-3}$. An alumina component, made by pressing and sintering the powder has a dry weight of 120 g. When fully saturated with water the part weighs 138 g. Its weight, when suspended in water is 87 g. Determine the amounts of both open and closed porosity in the component.

18.6 The bulk density of a pressed powder compact of alumina is $2985\,kg\,m^{-3}$. After sintering the bulk density of the component is $3800\,kg\,m^{-3}$. Estimate the linear shrinkage which takes place during the sintering, assuming that the shrinkage is uniform in all directions.

18.7 The viscosities of a glass at 1450°C and 780°C are 10 and $10^5\,Pa\,s$ respectively. Determine the temperature below which glass would not creep. Assume that there would be no creep at viscosities of $10^{16}\,Pa\,s$ or greater.

18.8 Explain why window glass shows viscous behaviour at high temperatures, for example 850 K, yet behaves as a brittle material at room temperature.

18.9 The window in a vacuum chamber is to be made of a glass. Tests on a sample of the glass show that the material has an equivalent tensile strength of 53 MPa after 6 minutes. Determine the maximum stress that can be applied to the window to satisfy a design life of 1500 hours. Assume that the time exponent for the glass is 10.

18.10 The tensile strength of a glass, as measured in a short-term tensile test of 1-minute duration, is 55 MPa. What would be the effective life of a component made from this glass and subjected in service to a continuous stress of 16 MPa? Assume the time exponent for the glass is 10.

CHAPTER 19

Composite materials

Why study composite materials?

There are situations for which a single material is not fully suitable. For example, a very hard and strong material may be very brittle while a tough material may possess a low yield strength. It is possible, often, to produce a composite material in which the good characteristics of the components are present without many of the disadvantages of either. Most composites are composed of a matrix material reinforced with either fibres or particles of one or more other constituents. Many interesting and useful composites have been developed in recent decades and development work on composites is continuing. Fibre reinforced composites, in particular, give materials of high strength and stiffness, coupled with low density, and are used to great effect in all forms of transportation, road and rail vehicles, boats and aircraft. Composite structures are used increasingly in engineering design, one example of such a structure being the sandwich structure in which a light-weight core is contained between high-strength skins. This method is used to fabricate structural members with a high stiffness-to-weight ratio. It is important that engineers appreciate the general characteristics and properties of composite materials and structures so that they can utilise this important class of materials to the full.

LEARNING OBJECTIVES

1. Understand the principles involved in both fibre and particulate reinforced composite materials.
2. Know the characteristics of the major reinforcing fibres and matrix materials used in the manufacture of composites.
3. Gain some knowledge of metal matrix composites and cermets.
4. Discover some engineering applications in which composite materials can be used to advantage.

19.1 | Introduction

There are very many situations in engineering where no single material will be suitable to meet a particular design requirement. However, two materials in combination may possess the desired properties, and provide a feasible solution to the materials-selection problem. In this chapter some of the composites in current use, and under development, will be mentioned.

The term *composite* can refer to any multi-phase material. However, it is usually restricted to 'tailor made' materials in which two or more phases have been combined to yield properties not provided by the constituents alone. The continuous phase in a composite is referred to as the *matrix*, other phases provide reinforcement. The term includes fibre, whisker and platelet reinforced materials, particulate composites such as dispersion strengthened alloys and cermets, laminates and sandwich materials.

The principle of composite materials is not new. The use of straw in the manufacture of dried mud bricks, and the use of hair and other fibres to strengthen plasters, dates back to ancient civilisations. Also, many of the materials that have been discussed in earlier chapters could be regarded as composite in character, for example, multi-phased metals, case hardened metals, filled plastics and concrete.

The first material that will be discussed, timber, may seem out of place in this chapter, but woods are complex in structure, and are, effectively, composite materials.

19.2 Timber and plywood

Wood has been used as a structural material for the whole of man's history. Timbers, naturally occurring composites, may be classified into two broad types, *softwood* and *hardwood*. Softwood is obtained from trees of the coniferous type. This group includes pine, fir, cedar, spruce and redwood. The term hardwood is used for wood obtained from deciduous trees, and this includes oak, ash, elm, walnut, beech, hickory and mahogany.

Wood contains the natural polymer cellulose. This polymer is also the basis of other natural materials such as cotton, hemp and flax.

The cross-section of a tree shows a characteristic ring structure. In growth, a ring is added each year. In a mature tree, the inner rings, which were the first to form, have ceased to carry sap. These inner rings are termed *heartwood*. The outer rings still carry sap and are, thus, known as *sapwood*. The widths of individual rings vary, not only from one type of tree to another but also within the same species, as the rate of growth of a tree is governed by climate and soil condition. The rate of growth of a tree is greatest in spring, and is almost zero during late autumn and winter.

In softwoods the cellulose molecules form fibrous tubular cells, termed *tracheids*, and these lie parallel to the trunk and limbs of the tree. In addition, there are also some cells set at right angles to the fibrous tracheids, and lying in a radial direction. These radial cells are termed *rays* and it is through these that plant food passes across the section of a tree. The spaces between cells are mainly composed of *lignin*, a high molecular weight polymer of carbon, oxygen and hydrogen. The structure of hardwoods is more complex, and one major difference is that there is a higher proportion of radial cells than in softwood. The majority of the cells in wood are tracheids, and these give the wood its characteristic *grain*.

Green timber contains a high moisture content and this moisture content is reduced during seasoning. During seasoning, the timber is exposed to air for a period of several months. During this period the moisture content of the timber is reduced from about 33 per cent to about 12 per cent. After it has been seasoned the timber is kiln dried within the temperature range 70–85°C, and this process reduces the moisture content to about 3 per cent. As the moisture content of the wood decreases so the strength is increased.

Because of the fibrous nature of wood, it is highly anisotropic and the tensile and compressive strengths are very much greater along the grain than in a direction at right angles to the grain. The shearing strength along the grain is, however, considerably lower than the shearing strength across the grain. The low shear strength along the grain makes it difficult to use wood as tension members of a structure, but it is a good material for compression members.

One way in which the anisotropy of wood may be overcome is by bonding thin layers together as *plywood*. Plywood is built up of thin layers of wood bonded with a water-resistant glue or a thermosetting resin, with the grain of successive layers at right angles to each other. Plywoods contain an odd number of layers, in order that shrinkage stresses shall be symmetrical about the centre with a consequent minimum of warping.

19.3 | Fibre-reinforced materials

Very hard and strong materials are generally brittle, while the soft, but very tough materials, tend to have low yield strengths. In a fibre-reinforced material, high-strength fibres are encased within a tough matrix. The functions of the matrix are to bond the fibres together, to protect them from damage, and to transmit the load from one fibre to another. The greatest reinforcing effect is obtained when fibres are continuous and parallel to one another, and maximum strength is obtained when the composite is stressed in tension in a direction parallel to the line of fibres. When such a composite is stressed within the elastic range, the strain developed in both fibres and matrix will be the same and so the stresses induced in fibres and matrix will be proportional to the values of Young's modulus for each component. (Refer to section 7.3.)

The expressions for density, ρ, tensile strength, σ, and modulus of elasticity, E, all have the same form and are, respectively:

$$\rho_c = \rho_f V_f + \rho_m V_m$$

$$\sigma_c = \sigma_f V_f + \sigma_m V_m$$

$$E_c = E_f V_f + E_m V_m$$

where V is the volume fraction and the suffixes c, f and m refer to composite, fibre and matrix respectively.

For a composite with a ductile metal matrix, the matrix will begin to yield in a plastic manner when the stress in the matrix exceeds its yield stress. Beyond this point, the matrix will deform plastically but the fibres will continue to strain in an elastic manner. At this stage E_c will be approximately equal to $E_f V_f$. If fibres are discontinuous, their strengthening effect will be less than that of continuous fibres, and short discontinuous fibres will be considerably less effective than long fibres.

A material with all fibres aligned parallel will be highly anisotropic. In order to obtain some degree of isotropy in, say, flat sheet material, fibres would have to be arranged in the form of a two-dimensional mat. In this case the maximum strength in any direction would be about one-third of the maximum strength of a material possessing only one-directional fibres. In a composite containing a random three-dimensional fibre arrangement, the maximum strength would only be about one-sixth of that of a fully aligned material.

If very few fibres are added to a matrix the material is weakened rather than strengthened and there will be a critical value of volume fraction of fibres, V_f, below which strengthening will not occur. In the case of carbon fibre filled epoxides the critical value of V_f is 0.02 when all the fibres are fully aligned but $V_{f\,crit}$ is 0.052 for random distribution of fibres in one plane. Both of these values are considerably exceeded in the manufacture of carbon fibre filled epoxides. However, in the case of fibres distributed randomly in three dimensions the value of $V_{f\,crit}$ is 0.1. Because of the difficulty of packing a large proportion of fibre with a wholly random orientation into a matrix it is not always possible to produce strengthening because it may be impracticable to exceed the value of $V_{f\,crit}$.

Many different matrix/fibre combinations have been developed over the years. Glass, carbon, polymer and ceramic fibres and very fine metal wire filaments have been used in conjunction with thermoplastic and thermosetting resin, glass, ceramic and metal matrices. Not all of these combinations have yielded viable commercial materials but much research and development work is being conducted into new composite materials.

A major stimulus for this effort is the requirement for materials of high specific strength and stiffness. Such materials are needed for applications in all forms of transport where a saving in weight is important. A reduction in the weight of a car or truck will mean that less power will be required to move the vehicle or, in the case of racing cars, a higher performance can be achieved for the same power. The saving of even 1 kg of structural weight can have a significant effect on the operating costs of a civil aircraft or the performance limits of a military aircraft. Many modern aircraft incorporate comparatively large amounts of composite materials in their structures. In the latest version of the Harrier V/STOL aircraft carbon/epoxide composites account for more than 25 per cent of the structural weight, and there is a considerable utilisation of composites in the Typhoon (Eurofighter) aircraft. In the civil aircraft field the later models of Airbus and Boeing aircraft have many composite components, thus achieving a considerable weight saving, as compared with earlier designs. Also, much use is made of high performance composite materials in Formula 1 racing cars.

19.4 Fibres

A fibre is a thread-like form with length-to-diameter ratios of the order of 10^3 or greater. The term embraces thin metal wires as well as natural and synthetic fibres such as wool, cotton, polyacrylonitrile and nylon. Many materials exhibiting low or moderate strengths and elastic moduli in bulk form demonstrate greatly enhanced properties when in the form of small diameter fibres. Strong, stiff fibres are produced from cold-drawn metals (normally referred to as wires or filaments), ceramics and highly orientated polymers. They may be incorporated in polymer, metal, or ceramic matrices to form fibre-reinforced composites. Glass and carbon fibres are the principal materials used in fibre-reinforced composites although other polymeric fibres are being used increasingly.

Glass in fibre form with diameters in the range 6–10 μm exhibits very high tensile strength when compared with the bulk material. This results from the much reduced size and number of flaws as the diameter of the fibre is less than the critical Griffith crack length. Two main types of glass are used for the production of glass fibres. These are known as e-glass, a non-alkaline borosilicate, and s-glass, a high strength magnesium aluminosilicate. Although it has a lower strength and stiffness, e-glass is more widely used than s-glass. The availability and relatively low price of glass fibre compared with carbon and other high modulus fibres makes glass an attractive reinforcement where high stiffness composites are not required. Typical properties of glass fibres are given in Table 19.1.

Carbon fibres, with diameters in the range 6–10 μm, possess high elastic moduli and strengths and are used as a reinforcing material in epoxy and polyester resins for the manufacture of high strength and stiffness composites. Carbon fibre reinforcement is used also in some thermoplastics, metals and in carbon itself. There are several grades of carbon fibre, the main types being Type I (high modulus) and Type II (high strength). Two other grades are Type A and fibre made from mesophase pitch. Commercial production is by the carbonising of polyacrylonitrile fibre at 900°C followed by a heat treatment, with the fibres under tension, to give the final structure which is a graphitic crystalline structure with the basal planes orientated approximately parallel to the fibre axes. Heat treatment at 2000–2500°C will give Type I and heat treatment at 1300–1500°C gives Type II fibre. Type A, with intermediate properties, is treated at an intermediate temperature. Typical properties for the various grades of carbon fibre are given in Table 19.1.

Table 19.1 *Properties of some fibres and wires.*

Material	Young's modulus (E) (GPa)	Tensile strength (GPa)	Density (kg/m$^3 \times 10^{-3}$)
Natural cellulose	48–100	0.4–1.0	1.5
Terylene	1.2	0.6	1.4
Polyamide (nylon)	2.9	0.8	1.1
Aramid (Kevlar 49)	130	3.6	1.4
e-glass	70	2.0	2.5
s-glass	84	4.6	2.6
Boron	410	2.9	2.5
Carbon (Type I)	400–410	2.0–2.8	2.0
Carbon (Type II)	200–240	3.0–3.5	1.7
Carbon (Type A)	220–300	2.4–3.6	1.85
Silicon carbide	400	6.0–8.0	3.2
Alumina	470	2.0	4.0
Beryllium	300	1.5	1.8
Tungsten	350	3.5	19.0
Steel	210	2.8	7.8

An aramid is an aromatic polyamide and a polymer of this group which has been used successfully for the production of high strength, high modulus fibres is poly p-phenylene diamine teraphthalate. This is marketed under the proprietary name *Kevlar*. The high strength and stiffness result from the rigid nature of the benzene rings and rod-like molecules of the polymer. Aramid fibres are used in high-performance composites. The best known examples of aramid fibres are Kevlar 49, a competitor with carbon fibre, and Kevlar 29, a similar fibre but with a lower elastic modulus. The properties of Kevlar 49 are given in Table 19.1.

Thermoplastic polymers, including polyethylene, polypropylene and polyamides, can be produced as high-strength, high-modulus fibres. The development of high strength and stiffness is conditional upon creating highly aligned and orientated molecular structures. Polyethylene, usually regarded as a low strength and stiffness material, can be produced in fibre form with a value of E up to 100 GPa and a tensile strength as high as 3–4 GPa. Polyethylene and polypropylene fibres are only suitable for low temperature use as the highly orientated structure is lost when the material is heated.

Boron fibres, made by the pyrolytic deposition of boron on a core fibre of carbon or tungsten from the reaction between boron chloride and hydrogen at 1200°C, have been used in conjunction with epoxy and aluminium matrices to give high strength fibre-reinforced composites. Boron fibres made in this way are much thicker than carbon and glass fibres and have diameters of between 0.1 and 0.2 mm. E for the fibres is 400 GPa and the tensile strength is in excess of 3 MPa. There has been some use of boron fibre/epoxy composites in the US space program.

Silicon carbide can be produced as light, stiff and strong fibres with a typical diameter of 0.1 mm by chemical vapour deposition onto a tungsten substrate, using chlorinated silanes. Reaction with the tungsten substrate makes the fibres unsuitable for use above 900°C. Finer fibres (0.001 mm diameter) may be produced from carbon/silicon polymers in a process similar to that used for the manufacture of carbon fibres. Considerable research into the use of silicon carbide as a reinforcing fibre in aluminium and titanium matrices is being carried out.

19.5 Matrix materials

The main requirements of a good matrix material are that it can fully infiltrate between fibres and form a good interfacial bond, that there should be no chemical interaction between fibre and matrix, other than that involved in bonding, and the matrix should not cause physical damage to the reinforcing fibres.

Thermosetting resins

Unsaturated polyester resins (see section 17.15) were the first type of matrix material to be used in conjunction with glass fibre reinforcement and polyester GRP materials have been used as constructional materials for more than 40 years. They are relatively cheap, are easy to work and cure at room temperature, although cure rate can be accelerated by heating. Polyester resins are the most widely used matrix materials, the reinforcing fibres used with them being, mainly, glass and carbon. Many components, both large and small, are fabricated using polyester/glass composites, including machine housings, body assemblies for car and other motor vehicles, canoe and boat hulls. Some of the largest GRP mouldings have been the hulls for naval minesweepers.

Epoxide resins. Epoxides (see section 17.16) possess several significant advantages over polyester resins: they are stronger and have greater stiffness, show very little cure shrinkage, have much better adherence to reinforcing fibres, and can be used at higher temperatures than polyesters. Epoxides are the main matrix materials for use with carbon and aramid fibres. Other thermoset matrix materials have been investigated and polyimides offer the possibility of fibre composites with a higher maximum service temperature than is possible with epoxides.

Thermoplastics

The use of thermoplastic matrices for fibre reinforcement can give materials with a greater fracture toughness than is possible with thermosets. Virtually all thermoplastics can be strengthened and stiffened by incorporating glass fibre but in many cases these are in the form of glass-filled polymers, that is polymer with chopped glass fibre as a filler, rather than as FRPs with continuous fibre reinforcement. One disadvantage of thermoplastics as matrix materials for continuous fibre reinforcement is that infiltration around all fibres is difficult owing to the high viscosities of polymer melts. Many materials have been tried as matrices, including PA 6.6, PAI, PBT, PET, PES, PPS, PSU and PEEK. Considerable effort is being placed into development of thermoplastic matrices, particularly PEEK, with carbon and aramid reinforcement, for high-performance composites. The fracture toughness and fatigue strength of a PEEK-based composite is considerably higher than for epoxide-based materials and these offer potential for use in aerospace applications.

Metal matrices

The use of a metal matrix offers the possibility of obtaining composites of high stiffness and strength and with the ability to operate at higher temperatures than polymer-based composites (see also section 19.7). There are, however, difficulties in fabrication which result in these becoming high-cost materials. Focus has been mainly centred on the use of aluminium, titanium and magnesium, together with their alloys, as possible matrix materials with silicon carbide and alumina as the main reinforcing fibres. The manufacturing processes used to achieve these composites include powder metallurgy techniques, vapour deposition, diffusion bonding and infiltration of liquid metal into fibre bundles under pressure.

19.6 | Some applications of FRPs

Polyester and epoxide resins with glass, carbon or aramid fibre reinforcement have been in use for many years. They offer a number of very useful properties as engineering materials. They are of low density yet possess good strength and stiffness. The specific strengths, that is strength/density, and specific elastic moduli of glass fibre-reinforced plastics are roughly comparable with those for steels, while the specific properties of carbon fibre-reinforced plastics are considerably greater than those properties for many steels. These characteristics give the fibre-reinforced materials great potential for applications in many sectors. They are particularly suitable for use in all forms of transportation, whether by land, sea or air, as they give savings in structural weight with a consequent reduction in the power required to move the vehicle. Other advantages of the materials are that they possess a good fracture toughness and are resistant to many forms of chemical attack. A disadvantage of a composite reinforced with continuous parallel fibres is that it is anisotropic, being very much stronger and stiffer in the fibre direction than in a direction transverse to the fibres. This can be countered by using a woven fibre fabric as the reinforcement or by making laminates with the reinforcing fibres in the various lamellae lying in different directions. Another disadvantage is that, under certain conditions, delamination and debonding of fibres from matrix may occur with consequent adverse effects on properties.

Land vehicles

Glass reinforced plastics (GRPs) have been used for many years to make cab body shells for trucks, for parts of the body shells of buses and coaches, particularly the front and rear ends, and for portions of the body shells of rail vehicles. The use of composites in the construction of mass-produced cars has been limited but the former Reliant Motor Company used GRPs for the complete body shells of its three-wheel small cars and vans and its Scimitar sports car. Other small production sports car makers have also used GRP structures extensively. In recent years the Chrysler Motor Company has been developing a material suitable for car body shells which can be fabricated by injection moulding. This is polyethylene teraphthalate (PET) reinforced with chopped glass fibre. The experimental vehicle produced has a kerb weight of 545 kg compared with a kerb weight of 1090 kg for a steel vehicle of similar size. Advanced carbon fibre materials (CFRPs) are used for the main body shell components in Formula 1 racing cars.

In the military field there is development work taking place to produce an armoured fighting vehicle (AFV) constructed from GRP. The world political situation today often calls for the rapid deployment of armed forces to local areas of conflict. The main battletanks in use with most armies are too heavy to be deployed by air. DERA in the United Kingdom has produced an experimental AFV in which the hull is a GRP moulding with a thickness of 60 mm. The body shell has to be protected by a layer of metallic armour but, even so, there is a considerable reduction in weight as compared with that of the conventional AFV.

Marine craft

Today most small pleasure craft have hulls made from GRP. However, the majority of ships are still constructed with steel hulls, an exception being naval patrol craft and minesweepers. GRP is an ideal material for the hull of a minesweeper as the vessel may have to deal with magnetic mines. The hull is made as a one-piece monolithic moulding and, typically, a minesweeper hull is about 60 m in length, with beam dimensions of about 10 m. The experience gained in building vessels of this size over many years shows that FRPs have the potential to move into other areas of ship building. The cost of making a mould for the construction of a GRP hull is high but ship construction costs would become competitive with steel if more than 10 identical hulls are needed.

Aerospace

CFRP materials offer considerable weight reduction possibilities for aerospace applications. Weight reduction in aircraft construction yields significant advantages, in civil aircraft it equates with improved efficiency and, hence, increased profitability, in military aircraft it results in improved performance. In airliner design and construction FRPs were used first for cabin flooring and for what are termed secondary structures, namely wing slats, flaps and rudders but, with the experience gained FRPs are now being increasingly used for primary structural parts. In a number of the Airbus variants (Figure 19.1) the complete tailfin and tailplane structures are in CFRP material. The trend for the future is that there will be increasing use of FRP materials in aircraft and the Airbus A380, scheduled to enter commercial service in 2006, will have the centre wing box and pressure bulkheads constructed from CFRP also. (See also Chapter 30.)

Similarly for military aircraft. In the 1980s the successful BAe Harrier aircraft underwent major redesign and development (Figure 19.2). This was done jointly by BAe Systems and McDonnell and resulted in production of the Harrier GR Mk 5 (in the USA the AV-8B). In this aircraft composite materials are used for the main wing and tailplane structures, and the forward section of the fuselage, accounting

Figure 19.1   *Airbus A300. (Courtesy of Airbus UK)*

CFRP

GFRP

Figure 19.2   *Usage of carbon fibre and glass fibre composites in the BAe Harrier GR Mk. 5/McDonnell AV-8B aircraft. (From information supplied by BAe Systems)*

Figure 19.3 *EADS Typhoon (Eurofighter). (Courtesy of BAe Systems)*

for about 25 per cent of the total airframe weight. This resulted in a weight saving of 240 kg. From its initial conception, the design team decided that much of the structure of the multi-national EADS Typhoon, or Eurofighter, would be made using composites (Figure 19.3). In this aircraft more than 80 per cent of the surface area is made from carbon fibre and glass fibre composites. (See also Chapter 30.)

For many years helicopter rotors have been made using FRPs. One problem in this application is erosion of the leading edge. The leading edges can be protected against erosion by shields of titanium and nickel. Many components in space vehicles can also be made in FRPs, one example being the nose cone of the Ariane rocket.

The examples mentioned above are all cases where the design teams decided at an early stage to incorporate composite materials in the aircraft structure. Now for an example of how the incorporation of a composite material is being used to prolong the useful life of an aircraft. In 2000 a Concorde aircraft crashed shortly after taking off from Paris. The cause of this fatal crash was identified as being due to rupture of the wing fuel tanks by pieces of burst tyre and runway debris, followed by ignition of the escaping fuel. An important modification made to the remaining Concorde aircraft, in order to regain their airworthiness certification, was to fit new fuel tank liners made from a rubber compound reinforced with Kevlar. These Kevlar fuel tank liners should remain unbroken even if the wing skin is punctured.

FRPs in bridge building

GFRP materials have been used recently in the construction of several major bridges in Britain. Generally, they are not used on their own but as structural skin panels to protect the main steel or reinforced concrete structure from corrosion and erosion. The design life of a major structure, such as a bridge, will be well in excess of 100 years and the use of a GFRP structural skin will greatly reduce maintenance and repair costs. The cost of rectifying corrosion damage in a bridge could be extremely high, as well as causing major disruptions in traffic while repairs are taking place. A GFRP structural skin was fitted to a steel plate girder bridge carrying a major road over the river Tees near Middlesbrough. The corrosion rates of the bare steel behind the GFRP panels have been monitored by

Figure 19.4 *Glass fibre-reinforced composite panels used as permanent formwork for the reinforced concrete beams of a bridge on the M4 motorway.*

Figure 19.5 *A new concept for bridge building – a space frame structure covered with a fibre composite stressed skin.*

the Road Research Laboratory since construction and found to be about 5 per cent only of the corrosion rates of control samples outside the structure. Another example of the use of GFRP panels in bridge design was used in the construction of bridges forming part of the M4 motorway approach roads to the second Severn crossing at Caldicot. In this, specially designed GFRP structural panels were used as permanent formwork for the casting of the reinforced concrete road-deck cantilever sections (Figure 19.4). For the future, the availability of strong, advanced fibre composite materials has suggested a new bridge design approach, in which, typically, the structure would be a tubular space frame to carry the road deck, the space frame being covered with a stressed skin made of a fibre composite (Figure 19.5). In this type of design the skin would carry a proportion of the service stress and, in addition, give corrosion protection. Also, the skin panels could be shaped to give good aerodynamic features and be aesthetically pleasing.

Several bridges entirely constructed in GFRP materials have been built in China, but the first one in Britain was a suspension footbridge with a central span of 63 m and a deck width of 2.23 m. The GFRP

Figure 19.6 *Aberfeldy footbridge constructed of GFRP. (Courtesy of Maunsell & Partners)*

towers are 18 m high and the cables that support the GFRP decking are of Kevlar fibres in a polyethylene sheath. This bridge of novel design spans the river Tay at Aberfeldy (Figure 19.6). Refer also to Chapter 30.

EXAMPLE 19.1

(a) From the data below determine the densities and maximum values of Young's modulus and tensile strength for uniaxially aligned fibre/epoxy composites based on each of the three fibre types. (Assume a fibre volume fraction of 0.45 in each case.)
(b) Discuss the relative merits of the three composites for use as lightweight floor panels in civil aircraft.

Material	Density (kg/m³)	Tensile strength (MPa)	E (GPa)
epoxy resin	1300	65	4
e-glass fibre	2500	2000	70
Type I carbon fibre	2000	2200	400
Type II carbon fibre	1700	3300	200

SOLUTION

(a) The expressions
$$\rho_c = \rho_f V_f + \rho_m (1 - V_f)$$
$$\sigma_c = \sigma_f V_f + \sigma_m (1 - V_f)$$
$$E_c = E_f V_f + E_m (1 - V_f) \text{ are used}$$

For epoxy/e-glass
$$\rho_c = (2500 \times 0.45) + (1300 \times 0.55) = 1840 \text{ kg/m}^3$$
$$\sigma_c = (2000 \times 0.45) + (65 \times 0.55) = 935.75 \text{ MPa}$$
$$E_c = (70 \times 0.45) + (4 \times 0.55) = 33.7 \text{ GPa}$$

For epoxy/Type I C $\rho_c = (2000 \times 0.45) + (1300 \times 0.55) = 1615 \, \text{kg/m}^3$
$\sigma_c = (2200 \times 0.45) + (65 \times 0.55) = 1025.75 \, \text{MPa}$
$E_c = (400 \times 0.45) + (4 \times 0.55) = 182.2 \, \text{GPa}$

For epoxy/Type II C $\rho_c = (1700 \times 0.45) + (1300 \times 0.55) = 1480 \, \text{kg/m}^3$
$\sigma_c = (3300 \times 0.45) + (65 \times 0.55) = 1520.75 \, \text{MPa}$
$E_c = (200 \times 0.45) + (4 \times 0.55) = 92.2 \, \text{GPa}$

(b) The requirements for aircraft floor panels are low density and high stiffness. The epoxy/glass composite has the highest density and the lowest modulus. If this were to be used, a larger thickness section would be needed to give the necessary stiffness than for the other materials. This would impose a greater weight penalty.

   Considering the epoxy/carbon materials, the Type I fibre gives a higher modulus, but lower strength, than the Type II fibre. A floor panel is a plate subject to a bending mode, and the ratios $E^{0.33}/\rho$ and $\sigma^{0.5}/\rho$ are important parameters for this situation. These ratios for the two carbon fibre composites are

	epoxy/Type I C	epoxy/Type II C
$E^{0.33}/\rho$	3.51	3.05
$\sigma^{0.5}/\rho$	19.8	26.3

From these figures it is seen that the strength parameter for the Type II carbon composite is about 33 per cent greater than that for the Type I carbon composite while the stiffness parameter of the Type II composite is only about 15 per cent less than that of the Type II. As weight saving is extremely important in aircraft design, coupled with the fact that Type I carbon fibre is more expensive than Type II, it is likely that a Type II carbon fibre composite would be preferred for this application. The panel stiffness can be increased by incorporating stiffening ribs within the design with little added weight penalty.

19.7 | Metal matrix composites (MMCs)

The term *metal matrix composite* (MMC) covers a range of materials and not merely composites with continuous fibre reinforcement. Early materials, which today could be regarded as MMCs, were tungsten filament wire strengthened with a dispersion of thoria and sintered aluminium powder (SAP), effectively a fine dispersion of alumina in aluminium. SAP containing 10 per cent of dispersed alumina has a room temperature tensile strength of 390 MPa, compared with a strength of 90 MPa for annealed aluminium of commercial purity. At 400°C the tensile strength of SAP is as high as 180 MPa although at this temperature the conventional high strength aluminium alloys possess little, if any, strength. Other materials, strengthened in similar manner, developed later were TD (thoria dispersion) nickel alloys for high temperature service. Other examples have been the development of DS (directionally solidified) eutectic nickel-based superalloys for high temperature service. These can be regarded as *in situ* composites in which rod or plate shaped precipitate particles, produced during a controlled solidification process, provide the reinforcement.

   There has been much interest in the development of MMCs in recent years, with the main emphasis being placed on aluminium and titanium matrices with either continuous fibre reinforcement or discontinuous particle reinforcement. The use of MMCs offers the potential for weight saving in the manufacture of components. For example, a fibre reinforced aluminium matrix will have a relatively low density coupled with a higher strength and stiffness than conventionally processed aluminium. Because of this several of the major car manufacturers have played a leading part in the developments. Aluminium with particulate reinforcement by alumina or silicon carbide can be used for the manufacture of cylinder liners, pistons and pulleys. These MMCs may be made by thoroughly stirring fine ceramic powders into the molten metal or by blending the component powders and processing by powder metallurgy. Continuous fibre composite components are made either by infilling fibre preforms with liquid metal or by the hot isostatic pressing of alternate layers of fibres and metal foils. In this latter forming process there is diffusion bonding of the constituents.

The Toyota company has used an alumina fibre reinforced aluminium in its diesel engines for some years and Honda has produced a cast aluminium cylinder block with selective reinforcement by both alumina and carbon fibres, the latter material aiding lubricity in the cylinder wall. This redesign did away with the need for cast iron cylinder liners. Another example of the potential for MMCs in the automotive industry is the use of a cast Al–Si alloy with about 15–20 per cent silicon carbide, as a fine particulate reinforcement, for brake discs. The surfaces of the disc are coated with a thin coating of a ceramic, which is applied using a plasma spray process, to give good wear-resistant characteristics. The use of this material offers a weight saving, as compared with steel brake discs.

Discontinuous particle reinforcement using whiskers or platelets as the reinforcing component may have potential in the future development of MMCs.

19.8 Ceramic matrix composites (CMCs)

Many pure ceramic materials are hard, wear-resistant and can withstand prolonged service at high temperatures. However, a major limitation of these materials is their inherent brittleness. The development of ceramic matrix composite materials with reinforcing fibres offers the possibility of materials which are more damage tolerant. The demand for greater efficiency and higher operating temperatures in aero gas turbine engines has provided the stimulus for research and development into CMCs as they offer the potential for use in future generations of jet power units.

There are many ceramic materials which possess the necessary high temperature properties but current research into CMCs tends to concentrate on two systems, namely silicon carbide (SiC) fibre reinforcement of silicon carbide and glass–ceramic composites reinforced with silicon carbide. Silicon carbide fibre, like carbon fibre, can be processed readily into complex shapes using braiding and weaving techniques. Component manufacture in SiC/SiC composites involves making a fibre preform to net or near net shape and then using a chemical vapour infiltration (similar to chemical vapour deposition) process to infill the spaces between the fibres and thus form the silicon carbide matrix. This involves treating the preform in a furnace at a temperature of 1000–1200°C and under reduced pressure (10–100 millibar) in an atmosphere containing a gas such as methyl trichlorosilane ($CH_3.SiCl_3$). The methyl trichlorosilane decomposes into SiC and HCl, the silicon carbide progressively infilling the spaces between the fibres. As stated above CMCs of this type are still under development but offer the potential of being suitable materials for use in aero engines in the future.

CMCs with particulate reinforcement are under development also. Both titanium carbonitride and zirconia have been used as reinforcing particles within alumina for the manufacture of cutting tool tips with higher hardness and better tool wear characteristics.

19.9 Cermets

The dispersion-hardened materials normally contain less than 15 per cent of dispersed ceramic. The term *cermet*, however, refers to ceramic–metal composites containing between 80 and 90 per cent of ceramic.

The earliest cermet, also known as *hard metal*, was cemented tungsten carbide. This material was first developed in the early part of the twentieth century for use in wire drawing dies, and for metal-cutting tool tips. Originally, tungsten carbide powder was hot pressed with cobalt and iron, but subsequent developments have resulted in hard metal composed of a mixture of several carbides, such as tungsten carbide, titanium carbide and tantalum carbide, in a matrix of cobalt. Cobalt is an exceptionally good binder for the carbides as it forms a thin continuous film around all the carbide particles. With cobalt contents of up to 15 per cent, the cermet is a wholly elastic and brittle material. If the cobalt content is increased beyond 15 per cent, the brittleness of the cermet is reduced, but there is an accompanying loss of hardness.

The possibility of other cermets being useful as engineering materials, particularly as potential gas-turbine engine materials, led to considerable investigation into oxide-based and silicon carbide-based composites with nickel, cobalt and stainless steels. The major difficulties that were encountered in the development of cermets as possible turbine-blade materials were their very low impact strengths and low resistance to thermal shock. At present, research into cermets appears to have lost momentum.

19.10 Sandwich structures

Sandwich construction composites are being used increasingly in situations where high stiffness coupled with low structural weight is a major requirement. One of the first successful applications of sandwich construction for highly stressed structural components was the use of plywood skins with light balsawood as the sandwich filling to manufacture the main airframe structure of the Mosquito aircraft of World War II.

Sandwich structures generally are composed of two skins of high strength with a light weight core. They can be considered as miniature I-beams and possess a very high stiffness/weight ratio.

Multiple layer combinations are also used widely in engineering. This arrangement provides a material with low density and high specific stiffness, where the maximum tensile and compressive stresses are carried by the skins. Examples of skin materials are aluminium and titanium alloys and fibre-reinforced polymers and examples of light weight core materials are expanded polymers, such as polyurethanes and honeycombs (Figure 19.7) fabricated from metals or reinforced polymers, which resist shear stresses in the sandwich. The latter are more expensive, but provide superior properties. In addition to providing light weight structural materials for transport vehicles, particularly aircraft, sandwich materials offer useful thermal and sound insulation.

GLARE is a structural laminate of metal and glass fibre composite material. It is constructed from layers of a glass fibre prepreg tape, typically 0.2–0.3 mm in thickness, sandwiched between aluminium alloy sheet, the thickness of the sheet material being in the range 0.3–0.5 mm. This sandwich has been developed as high strength, low density structural material for airframes and will most likely be used for the upper fuselage skin panels for the Airbus A 380.

Figure 19.7 *Honeycomb panel construction.*

A structural foam material which has been developed recently in Germany is the aluminium foam sandwich which consists of two conventional sheets of aluminium enclosing a solid aluminium foam. The precursor for the foam is a powder mix of aluminium, or aluminium alloy, powdered together with a foaming agent. The foaming agent is titanium hydride, TiH_2, and it is also possible to blend ceramic powders or fibres into the mix to modify certain properties, for example, to increase resistance to abrasive wear. The powder mix is compacted to about 75 per cent of theoretical density by powder rolling, isostatic pressing, or other method and it can then be roll bonded between sheets of conventional aluminium, or aluminium alloy, and this sandwich can be formed by conventional deformation methods into the approximate shape required in the final product. The foam precursor material is expanded by placing the preform into an expansion mould, similar to dies for die casting, and by heating the mould to a temperature close to the melting temperature of aluminium, the TiH_2 decomposes creating hydrogen gas which blows the aluminium powder compact into a foam. After expansion to completely fill the mould, the mould is cooled so that the foamed structure solidifies giving a rigid solid foam contained within continuous solid surfaces. This new foam material has the potential to become a useful lightweight structural material for many applications.

Honeycomb and structural foam-filled sandwich structures are being used increasingly as major structural elements in high-performance military and civil aircraft. This type of construction is expensive to fabricate but offers an excellent solution to design requirements in the aerospace industry.

There are very many other sandwich structures produced and used in a wide variety of applications. One of the simplest and one that has been used for very many years is the sandwich made with paper sheets bonded over a core of corrugated paper and used for the manufacture of cardboard cartons.

Many sandwich materials are used in building construction because in addition to a high stiffness/weight ratio, they may also possess good thermal and sound-insulation characteristics. A variety of combinations are used. Skin materials include sheet metal, plywood, plastics (including glass-reinforced plastics), concrete and plasterboard, with cores that may be metal or paper honeycomb structures, rigid plastic foams, chip board or low-density porous masses of glass fibre or rock wool bonded with a plastic resin.

SUMMARY

This chapter gives a summary of the main groups of composite materials. The general principle is that fibres or particles of a strong or hard material are used to reinforce a softer, more yielding matrix and the two main classes are fibre reinforced and particulate reinforced materials. Timber can be regarded as a composite in that the basic structure is fibrous cellulose molecules in a matrix of lignin.

In a fibre reinforced material the greatest reinforcing effect is when the fibres are continuous and parallel, but this gives a material which is anisotropic and which is weak when stressed in a direction normal to the fibres. This can be countered by laying the fibres in the form of a two-dimensional woven mat but the maximum strength in any direction is only about one-third of the maximum strength when all fibres are parallel.

Many different types of fibre can be used as reinforcement, including glass and ceramics, carbon, organic materials such as polyamide, polyester and aramid, and metal filaments or wires. Generally, matrix materials are polymers, particularly polyester and epoxide resins, but many thermoplastic materials have also been tried as matrices. Both metals and ceramics can be used as matrix materials also, giving rise to a range of metal matrix composites (MMCs) and ceramic matrix composites (CMCs). Cermets form another group of composite materials, the most common members being composed of hard carbides bonded by cobalt for the manufacture of cutting tool tips.

Composite sandwich structures are used increasingly in those applications where a low structural weight and high stiffness are required. In this type of construction a honeycomb structure or solid foam is sandwiched between two skin panels.

Questions

19.1 (a) For a uniaxial parallel fibre composite calculate the maximum theoretical fibre volume fraction.

 (b) Plot the modulus of a continuously aligned boron fibre/epoxy composite as a function of the fibre volume fraction. Assume E for boron fibre is 350 GPa and E for the matrix is 4 GPa.

19.2 Calculate the strengths of composites containing glass fibre volume fractions of (a) 40 per cent and (b) 60 per cent, given that the tensile strengths of glass fibre and polyester resin are 3000 MPa and 30 MPa respectively. State any assumptions made in answering this question.

19.3 Axially aligned parallel continuous glass fibres provide the reinforcement in a nylon subject to a tensile load. The fibre diameter is 20 μm and the volume fraction is 0.45.

 (a) What fraction of the load applied is carried by the glass?

 (b) What is the stress in the glass when the average stress in the composite is 14 MPa?

 Assume E for the glass fibres is 70 GPa and E for nylon is 2.8 GPa.

19.4 Discuss the role of composite materials in engineering, illustrating your answer by reference to either reinforced concrete or high stiffness reinforced polymers.

19.5 The use of fibre reinforced plastics (FRP materials) for engineering applications is increasing at a rate of about 20 per cent per annum. Suggest reasons for this and discuss, briefly, the main advantages and limitations of FRP materials.

19.6 A component is to be manufactured from a fibre reinforced epoxy resin with the reinforcing fibres in the form of a two-dimensional woven mat. Three possible reinforcing materials have been suggested: e-glass, Type II carbon fibre and an aramid fibre Kevlar 49, with a 0.4 volume fraction of fibre in each case. Which type, or types, of reinforcing fibre would be suitable if the minimum tensile strength requirement for the composite is 350 MPa and it is desired to have a specific modulus, E/ρ, of at least 12 MPa kg^{-1}m^3. Data for the materials are given in the table below:

Material	Tensile strength (MPa)	E (GPa)	ρ (kg/m³)
Epoxy resin	70	3	1300
e-glass fibre	2000	70	2500
Type II carbon fibre	3300	220	3200
Kevlar 49 fibre	3600	130	1400

PART IV

FORMING AND FABRICATION OF MATERIALS

CHAPTER 20

Forming processes for metals

Why learn about forming processes?

Metals, and other materials, have to be processed into a great variety of shapes in order to make component parts of every type. The shapes that are required vary enormously, both in size and complexity, ranging from very small components with masses measured in milligrammes up to very large castings and forgings of, perhaps, several hundred tonnes mass. An engineer must have some knowledge of the wide range of manufacturing processes available and their applicability, and be aware of the advantages and limitations of the various processes. Also the type of processing used, and the conditions existing during processing, can exert a major influence on the properties of the material, and engineers need to be aware of this.

LEARNING OBJECTIVES

1. Know how components are made by melting and casting.
2. Know how metals are shaped by plastic deformation processes and understand the differences between hot and cold working.
3. Know about commercial annealing processes.
4. Know how metal components can be shaped from powders.
5. Have an appreciation of the influence that the forming process has on the final structure and properties of the component.

20.1 | Introduction

Metals and alloys need to be processed into a wide variety of shapes, both simple and complex and ranging from very small to extremely large, for the manufacture of a major range of components and products. Very many different forming processes are available for this and these can be classified broadly into four categories:

(a) casting, namely the pouring of liquid into moulds,

(b) manipulative processes, involving plastic deformation of the metal,

(c) powder techniques in which a shape is made by compacting and sintering a powder, and

(d) material removal processes, that is cutting and grinding operations. (This last category is covered in Chapter 23.)

Most metallic materials are produced initially in the liquid phase and the liquid is cast into shapes. The liquid metal may be cast into a regular shape with a circular, square or rectangular cross-section, termed an *ingot*, or into prepared moulds, shaped so that the finished casting will have the shape of the required product. Some casting processes will give a product which may require no further processing while others will give castings which will require some machining to obtain the final saleable product. Regular shaped ingots are intermediate products which, subsequently, are shaped by plastic deformation processes such as forging, rolling or extrusion.

All metal-forming processes incur costs and involve the consumption of energy, but some to a much greater extent than others. A metal or alloy must be heated to above its melting point if a casting is to be made, and a cast ingot must be reheated to a high temperature prior to processing by hot forging or rolling. In many processes a considerable quantity of scrap is generated. In the manufacture of sand castings, for example, around 50 per cent of the metal which is melted originally is returned as scrap for reprocessing. The bulk of this arises in the foundry as the metal which formed the feeder channels, runners and risers in the mould. The remainder of the scrap material is machine swarf generated during machining operations. On the other hand, in die casting and investment casting up to 90 per cent of the molten metal yields finished castings as the amount of metal in feeder sprues can be kept at a low level. Also, these castings can be made to closer dimensional tolerances and with better surface finish than sand castings, thus greatly reducing the need for machining. Similarly, there may be considerable scrap metal formed during hot and cold working operations, for example, in hot extrusion only about 85 per cent of the ingot can be extruded, the remainder having to be reprocessed. It is advantageous if components can be produced to *net shape* or *near-net shape*, thus reducing the number of costly finishing operations required, the amount of scrap generated and, hence, the amount of energy used during processing. Many processing methods have been developed which will achieve this end. In one of these, the production of finished components from metal powders, generally 95 per cent or more of the raw material yields a satisfactory product.

20.2 | Melting and alloying

The industrial melting of metals and the manufacture of alloys is not simply a case of raising the temperature of the metal to its melting point. During the process care must be taken to minimise, or avoid, contamination in order to produce material of the desired quality at an acceptable cost. Furnaces for the melting of metals may be heated by hydrocarbon fuels or by electrical power and may vary in size from units with capacities of a few kilogrammes to large furnaces holding many tonnes.

Most metal melting is carried out in furnaces open to the atmosphere. Many metals react with oxygen to a greater or lesser extent. Also, water vapour in the atmosphere will dissociate at a hot metal surface and the hydrogen formed may enter into the metal. In order to reduce oxidation and contamination generally, many metals are melted under the protective covering of a molten flux or slag. The type of melting furnace capable of giving the cleanest materials free from contamination is that in which the metal is melted in a crucible contained within a high frequency induction coil. This type of crucible furnace can be operated *in vacuo*. This type of melting is used often to produce the very high purity oxygen-free copper that is needed for many high conductivity electrical applications. on the other hand, open-hearth furnace melting is perfectly satisfactory for the production of lower purity copper for general engineering purposes. Titanium and its alloys are highly susceptible to contamination by oxygen and these materials are melted and cast *in vacuo*. Vacuum melting and casting is used also for some other materials, including the high strength nickel-based alloys used for the manufacture of turbine blades for aero engines.

For the manufacture of alloys the basis metal is first melted and the required alloying additions are made to the bath of molten metal. Generally, the additions are made in the form of *hardener alloys*, that is pre-prepared alloys rich in the alloying element. For example, in the manufacture of an aluminium alloy containing 5 per cent of copper, the copper is added in the form of an aluminium–copper hardener alloy containing 50 per cent of copper. Similarly, hardener alloys such as ferro-chrome and ferro-manganese are used to make alloying additions to steels. The use of hardener alloys greatly aids the dispersion and solution of the alloying element in the basis metal. In some instances it is possible to make the alloying addition simply by adding amounts of the pure alloying element to the bath of molten basis metal. Magnesium can be added to aluminium in this way.

Once melting and alloying have been completed, often it is necessary to treat the liquid metal prior to casting. It is quite common for the liquid metal to contain dissolved gases. The maximum gaseous solubility limit in a liquid metal is generally much higher than the maximum gas solubility limit in the solid state. The solidification of a liquid metal with a high gas content would result in much of the gas being rejected from solution. Much of this rejected gas would be unable to escape to atmosphere, as a solid skin quickly forms on the casting. A porous casting would result. Degassing treatments are necessary to cleanse the liquid metal before casting, thus allowing for the production of sound castings. The exact type of treatment given will depend upon the metal and the dissolved gas but the following examples will give an indication.

(a) Much steel is made by refining blast furnace iron. The refining is an oxidation process and the refined steel is in a highly oxidised state. Deoxidisers, such as aluminium, ferro-manganese or ferro-silicon, are added to the melt. Oxides of aluminium, manganese, or silicon are formed which separate to the surface of the liquid metal.

(b) Molten aluminium absorbs hydrogen readily. The removal of hydrogen from aluminium can be effected by bubbling chlorine gas through the melt. Chlorine is insoluble in aluminium and will combine with hydrogen to form hydrogen chloride gas, which is also insoluble in aluminium, thus removing hydrogen from the metal. Also, the gas bubbles will attach themselves to non-metallic inclusions, such as particles of aluminium oxide, and float them to the surface of the melt where they may be removed by skimming.

(c) Copper, and many copper alloys, absorb oxygen during the melting operation. One method for the removal of oxygen is to add phosphorus, in the form of a copper–phosphorus alloy, to the melt. Oxides of phosphorus are formed which separate to the surface of the melt, from which they may be removed by skimming. This technique cannot be used for deoxidising high conductivity copper, as any residual phosphorus in the copper would reduce the electrical conductivity greatly. High conductivity copper may be deoxidised with calcium boride and lithium.

20.3 | Casting

The casting of liquid metal into a shaped mould is an effective and convenient way of making many types of component as it offers a direct route from liquid to the final shape of the required component. The decision on whether or not to produce a component by casting and, if so, by which of the many casting processes, is made after a consideration of both practical and economic factors. The shape and/or size of the component required may be such that a casting process is the only suitable method available even though the crystal structure and mechanical properties may be inferior to those of a wrought product.

The main casting methods available are *sand casting* in which liquid metal is poured into a shaped cavity moulded in a sand, *die casting* in which the mould cavity is machined within metal die blocks, *investment casting* and *centrifugal casting*. Products made to close dimensional tolerances and with an

Table 20.1 *A comparison of casting methods.*

	Sand casting	Gravity die casting	Pressure die casting	Centrifugal casting	Investment casting
Alloys that can be cast	Unlimited	Copper-base, aluminium-base and zinc-base alloys	Copper-base, aluminium-base and zinc-base alloys	Unlimited	Unlimited
Approximate maximum possible size of casting	Unlimited	50 kg	15 kg	Several tonnes	25 kg
Thinnest section normally possible (mm)	3	3	1	10	1
Relative mechanical properties	Fair	Good	Very good	Best	Good
Surface finish	Fair	Good	Very good	Fair	Very good
Possibility of casting a complex design	Good	Good	Very good	Poor	Very good
Relative cost for production of a small number	Lowest	High	Highest	Medium	Low
Relative cost for large-scale production	Medium	Low	Lowest	High	High
Relative ease of changing the design during production	Best	Poor	Poorest	Good	Good

excellent surface finish can be achieved using both die casting and investment casting. The products of these casting processes will require little, if any, subsequent machining and finishing operations. On the other hand, sand castings will have a poorer surface finish, less close dimensional tolerances and may require extensive machining and finishing to give the final product. Moulding sand has a fairly low thermal conductivity so that the rate of solidification of liquid metal within a sand mould is fairly slow, giving rise to a coarse crystal grain size. In contrast, rapid solidification, giving a fine grain size and better mechanical properties, will occur with die casting. A broad comparison of the major casting processes is given in Table 20.1.

20.4 | Sand casting

The moulds for sand castings are made from particular types of sands. These sands, which may be either naturally occurring or synthetic, must possess a sufficient strength to retain the mould shape both before and during the entry of liquid metal to the mould. They must also be refractory and be permeable enough to allow for the escape of steam and gases.

The naturally occurring material used for mould making is *greensand*. This is a siliceous material containing a proportion of clay. The exact composition of greensands varies with the location of the deposit and the clay content may range from 5 per cent to 20 per cent. The clay, when moist, acts as a binding agent around the silica grains to give the sand a certain strength. The moisture content must be closely controlled to give a mix that is readily mouldable but not so moist that an excessive amount of steam would be generated when liquid metal is poured into the mould, as this would lead to defective castings. Synthetic sands are made by blending a dry silica sand with a clay. A sand mould is only used once, as it has to be broken to retrieve the finished casting, but the sand is reusable. However, the sand that was at the surface of the mould cavity will have undergone a change. The clay binder in this region will have been heated to a temperature sufficiently high to convert it into a hard non-plastic state, the same process that occurs when clay is fired in a kiln to make bricks or pottery. This altered material should be removed from the sand and not reused.

Moulds for the production of small castings are normally made from a fine-grained sand as this will ensure a good surface finish. Moulds for large castings are generally made from a coarse-grained sand to give good permeability for steam and gases to escape readily. A thin layer of a fine-grained facing sand would be used at the mould cavity surface to ensure that the casting has a good surface finish.

Cores have to be inserted within the mould cavity if a hollow casting is to be made. A core has to be strong enough to stand largely unsupported within the mould cavity but not so strong that it cannot be removed easily from the finished casting. Core sand may be a greensand with an additional binder. It was common practice to use linseed oil as a binder. Linseed oil-bonded sand cores need to be baked in an oven for several hours to render them sufficiently strong for use. The use of sand-bonded cores has been largely superseded by other types of core that do not require to be baked before use. Cores are now generally made by shell moulding techniques, or by the carbon dioxide process. Shell moulding sand is a mixture of silica sand and a thermosetting resin. The core moulds are made of metal and are heated. When the sand–resin mix is put into the core mould the heat causes the resin to 'cure' and bind the sand grains together. Because of the low thermal conductivity of the sand–resin mixture the curing is effective only for a distance of about 3 mm from the heated metal surface. Surplus sand is then removed leaving a hollow, but strong, shell core (see Figures 20.1 and 20.3). In the carbon dioxide process a greensand or a pure silica sand is mixed with a small amount of sodium silicate solution. After making the core shapes in a core-moulding box, carbon dioxide is passed through the cores. The effect of the carbon dioxide is to convert the sodium silicate into silica gel which firmly binds the sand grains together.

Figure 20.1 *Shell mould and cores.*

The first stage in the production of sand castings must be the design and manufacture of a suitable pattern. Casting patterns are generally made from hard wood and the pattern has to be made larger than the finished casting size to allow for the shrinkage that takes place during solidification and cooling. The extent of this shrinkage varies with the type of metal or alloy to be cast. For all but the simplest shapes the pattern will be made in two or more pieces to facilitate moulding. If a hollow casting is to be made the pattern design will include extension pieces so that spaces to accept the sand core are moulded into the sand (see Figure 20.2). These additional spaces in the mould are termed *core prints*.

Sand moulds for the production of small and medium-sized castings are made in a moulding box. The mould is made in two or more parts in order that the pattern may be removed. With a two-part mould the upper half of the moulding box is known as the *cope* and the lower half is termed the *drag*. The various stages in the production of a complete sand mould are described below and illustrated in Figure 20.2.

The drag half of the mould box is placed on a flat firm board and the drag half of the pattern placed in position. Facing sand is sprinkled over the pattern and then the mould box is filled with moulding sand. The sand is rammed firmly around the pattern. This process of filling and ramming may be done by hand but mould production is automated in a large foundry with the mould boxes moving along a conveyor, firstly to be filled with sand from hoppers and then to pass under mechanical hammers for ramming. When ramming of the sand is complete excess sand is removed to leave a smooth surface flush with the edges of the moulding box.

The completed drag is now turned over and the upper, or cope, portion of the moulding box positioned over it. The cope half of the pattern is placed in position, correct alignment being ensured by means of small dowel pins. Patterns for the necessary feeder, runner and risers are also placed in

1 Bottom half of pattern is placed on mould board in bottom half of flask.

2 The bottom half of mould is now prepared as shown above.

3 The bottom half of mould is now rolled over, the top half of pattern and flask is placed in position.

4 The top half of mould is prepared in the same manner as before.

5 The flask is separated, the patterns are removed, the core is set in place and the flask closed.

6 When the flask is closed and clamped together it is ready for pouring of metal

Figure 20.2 *Stages in making a small sand mould.*

position. Feeder and runner channels should be placed so as to give an even distribution of metal into the mould cavity. The risers should coincide with the highest points of the mould cavity so that the displaced air can readily escape from the mould. The sizes of risers should be such that the metal in them does not freeze too rapidly. An important function of a riser is to act as a reservoir of liquid metal to feed solidification within the mould. A thin coating of dry parting sand is sprinkled into the mould at this stage. This is to prevent the cope and drag sticking together when the cope half is moulded. The cope is now filled with moulding sand and this is rammed firmly into shape in the same manner as in the making of the drag.

After the ramming of sand in the cope is completed and excess sand has been removed from the top surface the two halves of the moulding box are carefully separated. At this stage *venting* of the mould can be done, if necessary. The venting operation is the making of holes in both cope and drag with the aid of a thin wire. This is to increase the permeability of the mould allowing for easy egress of steam and other gases from the mould when the liquid metal is poured in. After venting the patterns are carefully removed from both cope and drag. Minor repairs are now made to the surface of the mould cavity, if necessary, and a gate or gates are carefully cut to connect the runner channel with the main cavity. Gates should be sited to allow for entry into the mould with a minimum of turbulence. Any loose sand is gently blown away and if a core is to be used it is placed in position. The mould is reassembled by placing the cope upon the drag and it is then ready for use. Liquid metal is poured smoothly into the mould via the feeder. Pouring ceases when liquid metal appears at the top of the risers and the feeder channel is also full.

The moulds for large castings are made in a different way. The mould is generally made in a pit on the foundry floor and the basic outline of the mould is built in brick as a firm basis for the moulding sand.

When the metal that has been poured into a sand mould has fully solidified the mould is broken and the casting is removed. The casting still has the runner and risers attached to it and there will be sand adhering to portions of the surface. Runners and risers are cut off and returned to the melting furnace. Sand cores are broken and removed and adherent sand is cleaned from the surface by vibration or by sand blasting with dry sand. Washing with a high-pressure water jet may also be used with non-ferrous castings. Any fins or metal flash formed at mould parting lines are removed by grinding and the castings are then ready for inspection.

Lack of attention to design or bad practice in either moulding, melting of the metal or in pouring can lead to defects within a casting. Most metals undergo a considerable volume shrinkage during solidification, and it is the function of the risers to provide reservoirs of liquid metal to feed this shrinkage. Adequate provision of risers should largely eliminate the possibility of major solidification shrinkage zones within the casting but finely divided interdendritic porosity is inevitable. Another type of porosity that may occur is gas porosity. If the liquid metal is incompletely degassed before pouring, gases will be rejected from solution as the metal solidifies and will remain as gas-filled blow-holes within the casting. Reaction gas porosity will tend to give blow-holes just under the skin of a casting. This type of defect can occur if the moulding sand is too moist or if the mould has been so hard rammed that it has a very low permeability so that steam generated as hot metal comes into contact with the mould blows back into the metal. Inclusions within a casting may arise from one of two sources. Bad melting practice may give a liquid metal containing slag inclusions, but it is also possible that loose sand may be washed away from the surface of a poorly prepared mould to form sand inclusions within the casting. The removal of sand in this way will also give a casting with unsightly lumps of excess metal on the surface. Another type of defect is the *cold shut*. This is a discontinuity within the casting. The basic cause is that the temperature of the liquid metal is too low so that two streams of metal may meet within the mould but not fuse together completely. Cold shuts may also occur if the pour is irregular. Bad design of a casting may be responsible for the creation of large thermal stresses as the casting cools and contracts under constraint. Such stresses may cause the casting to warp or may cause *hot tears* to occur. Hot tears are fractures within the casting due to the generation of a thermal stress that exceeds the strength of the hot casting just after solidification.

Figure 20.3 *Shell mould casting: (a) sand/resin mix dumped over heated metal pattern plate, (b) shells assembled to receive liquid metal.*

Despite some disadavantages sand casting is suitable for the production of castings in almost any metal and of almost any size from a few grammes up to several hundred tonnes.

Shell moulding has already been mentioned as a method of producing cores for sand moulds. The process can also be used for making moulds for the production of small castings. Shell moulds have the advantage that they are relatively strong and can be transported and stored without damage. This type of mould can be made to close dimensional tolerances and with a good surface finish, ensuring that castings of high quality are made. Prior to casting the two halves of a shell mould are fixed together, placed in a box and surrounded by sand for support (see Figure 20.3).

20.5 | Die casting

Gravity die casting is very similar in principle to sand casting, in that the die design must include riser heads to feed solidification shrinkage. The main advantages of the process are that a more rapid solidification rate produces a finer crystal grain structure than in a sand casting, and in consequence the die casting will possess a higher strength than a corresponding sand casting. Also, better dimensional control and surface finish will be obtained, as compared with a sand-cast product. The cost of dies is high and the use of the process is restricted to metals and alloys with melting temperatures not exceeding values of about 1000–1100°C. Gravity casting into metal moulds is often referred to as *permanent mould casting,* and the term *die casting* tends to be reserved for the *pressure die process.*

The dies or moulds are made of steel. There is a variety of die steels available, most being either nickel/chromium/molybdenum steels or chromium/tungsten/cobalt steels. The choice of die steel will be determined by the melting temperature of the metal to be cast and the length of the production run required. Steel cores may be used with gravity dies if it is required to cast hollow shapes of simple design. Hollow shapes of more complex design can be made by gravity die casting with sand-core inserts. An example of a product made in this manner is the body of a domestic water tap.

The injection of liquid metal under pressure into a mould offers several advantages. It means that very thin sections can be cast, that metal can be forced into the recesses of a mould of very complex shape, and that rapid solidificaton under pressure will considerably reduce porosity. Also, the need to provide risers in the mould is eliminated. In place of risers all that is necessary is a series of vents, of about 0.1 mm in thickness, at the parting line of both halves of the die. These will be sufficient to allow for the escape of air from the mould cavity as metal enters, but will be too small to permit the passage of metal. In die casting there is little metal wastage and a casting is obtained that requires little, or no, machining. The process is, however, restricted to the lower melting point metals and alloys, the principal ones being zinc, magnesium, aluminium, copper, and their alloys (see Figures 20.4 and 20.5).

Figure 20.4 *Principle of pressure die casting: (a) hot chamber machine; (b) cold chamber machine.*

Figure 20.5 *A saw-grinding attachment making use of nine zinc-alloy die castings. The plate shows the individual castings and the complete assembly. The knobs are die cast onto the adjusting screws. (Courtesy of The Zinc Development Association and Hollands and Blair Ltd)*

There are two main types of process, namely *hot chamber die casting* and *cold chamber die casting*. The former process uses a 'goose-neck' type machine in which the goose-neck is partially immersed in the liquid metal bath in a holding furnace. In each cycle, liquid metal enters the goose-neck and is forced into the die attached to the furnace by either a mechanical piston or pneumatic pressure (Figure 20.4(a)) giving injection pressures of up to 40 MPa. This type of machine is used principally for zinc and magnesium based alloys. In the cold chamber process, the charge of liquid metal is poured into a transfer chamber from a holding furnace immediately prior to injection, under a pressure of up to 150 MPa, into the die (Figure 20.4(b)). The cold chamber process is suitable for the die casting of aluminium and brass as well as for magnesium and zinc.

An interesting feature of die casting is that 'insert' castings can be made, that is, an insert can be placed in the die and the liquid metal cast around it. This is particularly useful for including a threaded bolt or stud in a cast component, as in the examples shown in Figure 20.5. It is, of course, necessary that the part of the thread within the casting be mutilated before insertion into the die to prevent it unscrewing from the casting in use. Pressure die casting is used for the manufacture of such components as automotive parts, including carburettor and petrol pump bodies, door handles and fitments, power tool casings and accurately scaled toys.

20.6 | Investment casting

Another important casting process is the *investment*, or *lost wax*, process. This process has its origins in the early Chinese and Egyptian civilisations, but it has become a valuable part of twentieth century technology. Some metal components have to withstand very high temperatures and stresses in service and, in the case of such components as aero engine turbine blades, the parts have to be made to very high standards of dimensional accuracy. Many materials sufficiently strong at high temperatures are virtually impossible to machine or to shape at ordinary temperatures. A feasible solution would be to use a precision casting process. Pressure die casting is not possible in these cases, as the melting point of the alloy would be too high, and investment casting can provide the answer.

In the investment process, a master mould is produced in a readily machineable alloy, such as brass, and this mould is used for the production of accurate patterns in wax or a low melting point alloy. The pattern, together with a feeder sprue, is dipped into a ceramic slurry and the coating formed allowed to dry. In the case of small components several patterns may be assembled around a central sprue as a 'tree' before coating with slurry. The dipping process may be repeated several times until the required coating thickness, usually 8–10 mm, is achieved. (For the production of investment castings in metals of relatively low melting point, plaster of Paris may be used in place of a highly refractory material to make the mould.) After the completed coating has fully dried out the assembly is heated and the molten wax or low melting point alloy allowed to run out. This is followed by heating to a higher temperature to burn out all traces of wax and to sinter the refractory, forming a perfect ceramic mould. The thin-walled investment mould is placed in a mould box, backed by sand for support, and is then ready to accept molten metal. In order to produce precision castings it is necessary to force the metal into the mould under pressure and this can be accomplished by mounting the investment moulds in a centrifuge (Figure 20.8(a)). Examples of components made by investment casting are shown in Figure 20.6.

High-performance components, for example aero gas turbine blades, are generally cast into investment moulds *in vacuo* and using slow directional solidification. This is the technique used for the manufacture of blades in DS eutectic alloys and, with minor modifications, for single-crystal blades. In the latter case single crystal solidification is achieved either by using a prepared 'seed' crystal as a nucleus or by using a mould with a spiral extension (see Figure 20.7). Modern gas turbine blades are cast with a network of fine cooling channels within them. These are obtained by inserting a series of soluble ceramic cores in the master mould used to produce the wax patterns. After the castings have been made, the cores are removed by dissolving them out using a strong solution of sodium hydroxide, molten sodium carbonate or other solvents.

Figure 20.6 *Typical examples of parts made by investment casting. The casting shown at the top left is in steel and of 220 mm diameter and 5 kg mass. (Courtesy of British Investment Casters' Technical Association)*

Figure 20.7 *Arrangement for vacuum casting to give a single-crystal product. Solidification commences below spiral. One growing crystal only emerges from spiral section.*

20.7 Centrifugal casting

A method of casting that can be used for parts with an axis of rotational symmetry is *centrifugal casting*. This process is particularly suitable for the manufacture of large-diameter pipes for water and gas

Figure 20.8 *(a) Investment moulds in centrifuge. (b) Centrifugal casting machine.*

mains. A cylindrical mould shell, lined with moulding sand, is rapidly rotated about its longitudinal axis and liquid metal poured in (Figure 20.8(b)). Under the action of centrifugal force, the liquid metal will spread evenly along the length of the mould. Rotational speeds are such as to give accelerations of the order of $60\,g$ (g here refers to acceleration due to gravity). The properties of the finished casting are very good as the centrifuging action ensures that any slag particles, which are almost invariably less dense than the metal, segregate to the bore of the pipe from where they may be easily removed by machining. Similarly, any gases will also escape to the bore leaving a sound casting. Centrifugal casting can also be used for producing cylindrical shell bearings in white metal, and in bronze and other copper alloys.

A comparison of the major casting processes is summarised in Table 20.1.

EXAMPLE 20.1

A component is to be made in aluminium alloy LM6 by either sand or die casting. From the following data determine the minimum number of components that would be required before die casting becomes economically viable:

	£
Cost of making a wooden pattern	150
Cost of making a sand mould	4.40
Cost of producing a set of casting dies	48,000

Assume that the costs of melting, casting and finishing operations are approximately the same for both processes.

SOLUTION

If the number of castings produced is x then the unit cost of a sand casting is

$£(150/x + 4.4)$

Similarly, the unit cost of a die casting is

$£(48,000/x)$

At the breakeven point: $150/x + 4.4 = 48,000/x$

so
$$4.4x = 48,000 - 150$$

therefore
$$x = 47,850/4.4 = 10,875$$

The number of identical components required would have to exceed 10,875 to make die casting economically viable.

EXAMPLE 20.2

What are the main reasons for the choice of investment casting for the manufacture of components such as blades for aero gas turbines?

SOLUTION

Gas turbine blades are precision parts made in high strength, high temperature resistant alloys. Because of the nature of the alloys they cannot be readily machined by conventional metal-cutting processes. The machining processes which can be used are grinding or a non-traditional process such as electro-chemical machining, both methods being expensive processes. Consequently, these components need to be formed to net shape by a precision process. The precision casting processes are die casting and investment casting, but the former method is only suitable for metals of relatively low melting points. The only suitable precision forming process for components such as gas turbine blades is, therefore, investment casting.

20.8 Ingot casting

In very many instances it is necessary to shape metals by deforming them plastically. The starting point for these working processes is a large regular-shaped casting, termed an *ingot*. Ingots for further working are generally square, rectangular, or circular in cross-section, and may vary in size from about 10 kilogrammes for some non-ferrous alloys, up to 100 tonnes for some large steel forging ingots.

Ingots may be cast by pouring the liquid metal into large permanent moulds made from cast iron. Cast iron moulds for the production of steel ingots are tapered to allow for easy removal of the solidified ingot from the mould, and are termed either *big end up* or *big end down* depending on the direction of the taper. At the end of the steel-refining stage the liquid steel is in a highly oxidised state. Additions are made to the bath to deoxidise the steel before pouring, or *teeming*, the steel. Fully deoxidised steel is termed *killed steel*, and most killed steels are poured into moulds of the big end up variety. In order to minimise the formation of pipe defect as a result of solidification shrinkage, the moulds are generally provided with a refractory collar, or *hot top*, to act as a reservoir for liquid metal to feed the shrinkage (Figure 20.9). Liquid metal poured into a large mould from the top can splash considerably, and the splashes would freeze instantly on the upper mould walls. The splash surface would then rapidly oxidise and not weld satisfactorily to the main body of the ingot when the mould is filled with liquid metal. This would give rise to surface defects in the subsequent rolled or forged product. In order to avoid this some steel ingots are bottom poured. This process, which is considerably more expensive than straight teeming into moulds, is normally used for only the higher-grade steels.

Some grades of low carbon, or mild, steels are not deoxidised before teeming. These are termed rimming steels and they are normally cast into moulds of the big end down type. There is considerable gas evolution during the solidification of rimming steels. When solidification first begins at the mould

Figure 20.9 *Ingot casting: (a) big end up type; (b) big end down type, showing secondary pipe; (c) big end up with hot collar; (d) section of rimmed steel ingot; (e) transverse section of ingot.*

walls, carbon monoxide is evolved according to the reaction

$$FeO + C = CO + Fe$$

The stream of gas bubbles tends to carry impurities away with it and the outer layers of the ingot solidify as almost pure, clean iron, with the impurities concentrated in the central portion of the ingot. Some of the gas bubbles are trapped between growing crystals of iron giving a widely distributed porosity, and this porosity compensates for the solidification shrinkage of the metal. If the internal surfaces of the gas pores are clean they will readily weld together under rolling pressures, giving a sound product. Rimmed steel is very suitable for the production of mild steel sheet because the pure iron rim gives a ductile skin of good surface quality, and this makes the steel ideal for deep drawing and pressing operations.

Large ingots cast in permanent moulds possess coarse grain structures, and there will also be variation in crystal structure across the ingot. A typical transverse section of an ingot is illustrated in Figure 20.9(e). There are three main crystal zones shown. The initial chill zone, which is of small thickness, possesses small crystals. The second zone is the columnar zone, and this contains large crystals, which have grown preferentially inwards from the mould wall along the direction of the major temperature gradients. Finally, when the temperature of the central liquid zone has fallen to around the freezing point of the metal and temperature gradients are small, randomly orientated, or equiaxial, crystal grains form.

Before 1940 ingots of all metals were cast, almost exclusively, into permanent cast iron moulds, but considerable attention was given to the development of casting processes that would give ingots of a better and more consistent quality and, in particular, a casting process that would be continuous in operation. The introduction of continuous casting, on a production scale, took place in the aluminium and copper alloy sectors of industry. The continuous casting of steel was a much later development, owing to the greater difficulties that had to be overcome, with molten steel temperatures of about 1550°C, compared with temperatures of about 700°C and 1000°C respectively for molten aluminium and molten brass. Today, almost all large ingots produced in the aluminium industry, and much of the ingot production in the copper and steel industries, are continuously cast.

Figure 20.10 shows the lay-out of a semi-continuous casting unit for the production of aluminium ingots. The mould is shallow, and the distance from the metal surface to the bottom of the mould skirt is about 75 mm. The ingot withdrawal rate must be carefully controlled, and this is done by means of a hydraulic ram. Withdrawal speeds vary considerably with alloy composition and the sectional size of the ingot, and may be between 20 and 200 mm per minute. The liquid metal solidifies very quickly and this means that not only does the ingot possess a fairly fine grain size, but also that segregation of impurities is largely eliminated and there is no possibility of piping defect occurring. In semi-continuous casting the maximum length of ingot possible is determined by the length of stroke of the hydraulic ram. This is usually of the order of 5 m. The fully continuous casting process is similar, but in this case steady withdrawal of the ingot from the mould is accomplished by having a pair of pinch rolls situated about 1 m below the mould. The continuous ingot is cut into suitable lengths by either a high-speed saw, or by flame cutting.

Much material is continuously cast in the horizontal, or near horizontal, mode on belt casting machines (Figure 20.11). This method is suitable for the production of flat strip and plate. Dams are fitted between the two moving belts to control the width of the cast product. Rod and bar sections can be cast in a similar manner but with shaped channels, rather than flat belts, incorporated in the casting machine. In some cases, particularly in the non-ferrous production industries, continuously cast rod or strip may feed directly into a multi-stand rolling mill.

Figure 20.10 *Semi-continuous casting of aluminium.*

Figure 20.11 *Moving belt machine for continuous casting.*

20.9 | Hot working

The term *hot working* refers to plastic deformation carried out at temperatures in excess of the recrystallisation temperature of the metal (refer to section 8.7). Almost without exception, cast metal ingots are hot worked, in the first instance, and the three major hot working processes are rolling, forging and extrusion. The ingots are pre-heated in soaking pits or furnaces to bring them up to the desired temperature, usually about 100–200°C below the melting range. At such a high temperature the metal is much more plastic and requires smaller deformation forces than would be necessary at low temperatures. During the first stages of hot deformation the coarse, 'as cast', structure of the ingot will be broken up and distorted, but, almost immediately, recrystallisation will occcur and further deformation can be carried out. A sequence of rolling passes, or forging blows, will generally be planned so that the final deformation takes place at a temperature close to the recrystallisation temperature (the critical temperature range, in the case of steels) so that recrystallisation without excessive grain growth occurs.

In *rolling* operations the ingot is passed between two large cylindrical rollers, and the roll surfaces have to be provided with flood cooling to prevent overheating of the rolls and possible welding between ingot and rolls. Water, or a water/soluble oil emulsion, is used, depending on the nature of the material being rolled. If the rolls are plain cylinders the ingot can be processed into flat plate material. Rolls with grooves machined into the surface are used for the production of round bars and sectional shapes, such as rails and structural beam sections.

The product of the hot rolling mill may be the finished product, requiring only trimming to size, or it may be an intermediate product that will be deformed further by cold working operations. Structural plate and sections are examples of products used in the hot worked condition.

Generally, hot rolling mills are of the 2-Hi reversing type with a motorised roller table at either side (see Figure 20.12). The rolling mill operator has control over the direction of both rolls and roller tables and, of course, alters the distance between rolls after each rolling pass.

In a slab rolling mill the diameter of the cylindrical rolls will be of the order of 1 m and the barrel length may be up to 3 m. The hot ingot, in a highly plastic state, will be passed backwards and forwards between the rolls several times with a major reduction in thickness being brought about with each successive pass through the mill. For example, an aluminium ingot with the dimension length 2 m, width 1.5 m and thickness 0.3 m may enter a hot rolling mill at a temperature of 550°C and eventually emerge, after nine successive reduction passes as plate material with a thickness of 10 mm and approximately 60 m in length. There is very little lateral spread during rolling and virtually all the plastic deformation is in the direction of rolling. At the end of this series of hot rolling operations the temperature of the plate will probably be down to about 250°C.

The rolls, which are made of forged or cast steel, are not perfectly cylindrical but are ground to possess a calculated profile or camber. Massive though they are, the rolls will deflect when operating under load and, if perfectly cylindrical, this deflection will cause the cross-section of the rolled

Figure 20.12 *2-Hi hot rolling mill.*

material to be barrel-shaped (see Figure 20.13(a),(b)). If the rolls are given a convex profile or positive camber of the correct amount this will compensate for roll deflection and a parallel-sided rolled product should be obtained (Figure 20.13(c),(d)). This is the case for cold rolling operations but matters are further complicated in hot rolling. The ingots that are hot rolled are always somewhat smaller than the full width of the mill, so that during rolling the central portion of the rolls will become hotter than the edges, giving an uneven expansion. Rolls for hot rolling are surface ground to give a concave profile or negative camber when cold. During rolling operations the greater expansion at the centre will cause the rolls to have a positive camber (Figure 20.13(e)).

The process of hot rolling may be used to produce square, round or shaped sections such as rails and I-sections. For this type of rolling the rolls are not plain cylinders but have matched grooves and shaped profiles to enable the desired section to be produced. Each set of grooves is carefully designed so that the required shape is fully formed but with no excessive fin being produced (see Figure 20.14).

As mentioned earlier there is very little lateral spread during rolling. The metal is accelerated as it passes through the roll gap and leaves the mill at a higher velocity than it enters. An ingot of width w and thickness t_1, entering a rolling mill at velocity v_1, will emerge from the roll gap with a reduced

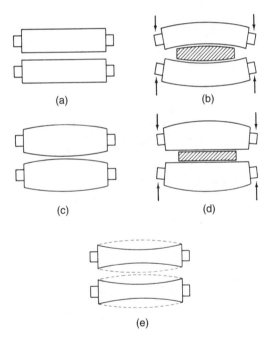

(a) (b)

(c) (d)

(e)

Figure 20.13 *Roll surface contours. Plain cylindrical rolls (a) would deflect under load (b). (c) Rolls with positive camber deflect to give a parallel-sided rolled slab (d). (e) Hot mill rolls ground with negative camber so that a convex profile is obtained when rolls are heated to operating temperatures.*

Figure 20.14 *Roughing and finishing pass for rails from square section.*

thickness t_2 but at a higher velocity v_2, (see Figure 20.15). Since the volume flow rate of metal must be the same at any point within the mill then this means that at only one point, N, within the roll/metal contact area will the velocity of the metal be equal to the peripheral velocity of the rolls. N is termed the *neutral point*. The forces that act on the metal within the roll gap are a radial force P_r exerted by the rolls and a friction force F which acts tangentially to the roll surface. The friction force acts to draw the metal into the roll gap between the entry point and the neutral point but its direction is reversed between the neutral point and the point of exit. The vertical component of P_r is the rolling load but as this is equal to the force exerted by the metal on the rolls it is often referred to as the roll separating force.

If the coefficient of friction between rolls and work-piece, μ, were zero no rolling would be possible. As μ increases so it becomes possible to take a greater 'bite' on an ingot and hence to give a substantial reduction in thickness in one rolling pass. Referring to Figure 20.16 the force tending to pull the work-piece into the roll gap is F and its horizontal component is $F \cos \alpha$. For the work-piece to enter the roll gap $F \cos \alpha$ must be equal to or greater than the horizontal component of the normal force, P_r, which acts in the opposite direction.

Therefore, at the limit $F \cos \alpha = P_r \sin \alpha$

or $F = P_r \tan \alpha$

But $F = \mu P_r$

therefore $\mu = \tan \alpha$

Figure 20.15   *Rolling mill forces.*

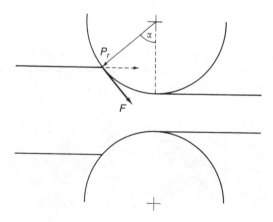

Figure 20.16   *Effect of friction on bite angle.*

In hot rolling operations the roll surfaces are ground so that μ is of the order of 0.2–0.3 and this, in conjunction with the use of large-diameter rolls, means that a large reduction in thickness per rolling pass can be achieved.

Forging means the shaping of metal by a series of hammer blows, or by slow application of pressure. The simplest example is a blacksmith's forging of a hot piece of metal by hammering the work-piece on an anvil. *Heavy smith's* forging is fundamentally similar, differing only in the scale of the operation. The work-piece may be an ingot of 100 tonnes and the deforming force provided by a massive forging hammer, but the whole process is controlled by the master smith, who decides each time where, and with what force, the blow should take place. Another form of forging is *closed die* forging, in which the hammer and anvil possess shaped recesses which effectively form a complete mould. The hot metal work-piece is caused to flow, under pressure, into the die cavity to produce the desired shape.

Forging operations may be carried out using either forging hammers or forging presses. With a press, as opposed to a hammer, pressure is slowly applied and plastic deformation tends to occur fairly uniformly throughout the material.

Forging hammers are of two basic types with the large hammer either dropping onto the work-piece accelerated by gravity alone or, as in the steam hammer, accelerated by both gravity and steam pressure (see Figure 20.17). A gravity drop hammer can deliver a rapid succession of forging blows but as the drop distance is constant all blows are delivered with the same force. For any given hammer size the power-assisted steam hammer can give a greater forging capacity. It is also possible to vary the

Figure 20.17 *Forging machinery: (a) gravity or board drop hammer; (b) steam hammer; (c) crank press; (d) hydraulic press.*

forging force by varying the steam pressure. Forging hammers are rated by the hammer weight and these tend to vary between 5 MN and 250 MN.

The use of a forging press tends to give forgings of greater accuracy than is normally possible with hammer forging. Forging presses may be of the mechanical or hydraulic type. Most mechanical presses are of the crank-operated type (see Figure 20.17(c)) and have capacities ranging from 3 MN to 120 MN. Hydraulic presses generally have capacities in the range from 5 MN to 200 MN although some very large machines with a 500 MN rating exist. A hydraulic press is more expensive than a mechanical press of the same capacity, but offers greater flexibility in that the force applied can be varied.

Smith's or open-die forging is used for making very large forgings or when the number of forgings of a particular design is small. Open-die forging is also used to pre-form metal prior to final shaping by means of closed-die forging. For Smith's or open-die forging the hammer and anvil are either plain or possess simple shapes.

Closed-die forging involves the use of accurately machined die blocks. The capital cost of a pair of die blocks is high, but this is allowable when it is required to make forgings to close dimensional tolerances and in large quantity, as would be the case for, say, connecting rods for the engine of a popular car. For such a component the die blocks would contain two pairs of cavities. The first pair, roughing or blocking impressions, would produce a forging approximating to the final shape. This would then be placed in the finishing cavity in which it would be forged to the final exact shape. The pre-heated blank or billet for closed-die forging is generally pre-formed to some extent before being placed in the blocking impression.

It is important in closed-die forging that a full and complete component is made and to ensure this the billet used will have a slightly greater mass than the mass of the finished forging. The die blocks are designed to allow for any excess metal to be extruded as a fin or flash (Figure 20.18) and this flash can be easily removed from the finished forging by a blanking operation.

Upset forging, or *upsetting*, is a forging operation in which the direction of plastic flow in the workpiece is changed from that of earlier operations and is carried out so as to obtain the best flow pattern or fibre structure for the component in question (see section 20.10). Generally, upsetting shortens the length and increases the diameter of the workpiece.

A variant on the forging process is *roll forging*. This is a deformation process for producing straight or tapered sections with a varying shape and section along the length of the workpiece. The forging rolls are grooved, conforming to the shape of the product, and in each cycle rotate through less than a complete revolution, and then reverse. Roll forging is used for forming such components as automotive drive shafts, axles and aircraft propeller blades.

The *Pilger process* is a hot roll forging process for the manufacture of seamless pipe and tube. A Pilger mill is a 2-Hi mill and each roll contains a groove of semicircular section. The size of the groove is not constant and decreases around the roll circumference (see Figure 20.19(b)). A previously pierced ingot is mounted on a mandrel, to control the bore diameter. The rolls rotate through almost 360°, rolling

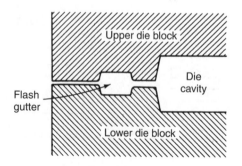

Figure 20.18 *Flash gutter.*

and forging a section of the ingot and reducing the external diameter. The rolls are then reversed, the ingot fed further into the roll gap, and the process repeated roll forging a further section of the ingot. Ingots for tube manufacture are pierced initially by rotary piercing, as for example in the Mannesman process. In this process, a heated steel billet of cylindrical section is rotated between two slightly tapered rolls whose axes of rotation are inclined at a small angle on either side of the billet rotation axis (see Figure 20.19(a)). The effect of the rolling is to deform the billet into an elliptical shape and rotation of this cross-section causes shear and the formation of a crack in the centre of the section. As the rolling proceeds, this centre crack is forced over a pointed mandrel, which both enlarges and shapes the central opening.

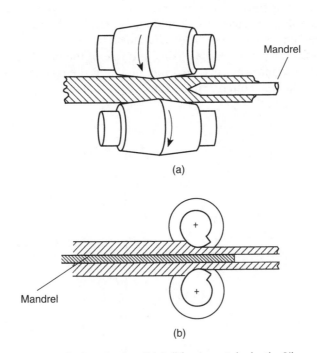

Figure 20.19 *(a) Mannesman process for hot piercing; (b) Roll forging a tube by the Pilger process.*

In the process known as *hot extrusion* the hot ingot is forced to flow, under pressure, through a shaped orifice, or die, in a similar manner to the flow of tooth paste from a collapsible tube (Figure 20.20(a)). It is possible to extrude an almost infinite variety of sectional shapes, including hollow sections, from solid ingots (see Figure 20.24). For many years hot extrusion was restricted to the lower melting point materials, principally aluminium and copper alloys, but in recent years considerable development has taken place in the extrusion of steels.

Extrusion of hot ingots is carried out in large hydraulic extrusion presses and these are generally in the horizontal mode. Extrusion press capacities vary considerably and the ram force may range from

Figure 20.20 *Principle of extrusion: (a) direct extrusion; (b) indirect extrusion.*

2 MN in a small press up to 150 MN in a very large press. Most hot extrusion is performed by the direct method in which a pre-heated ingot is placed in the chamber of the press and a force applied by the piston, through a pressure pad, causing the ingot to flow plastically and extrude through the die aperture (see Figure 20.20(a))

In indirect extrusion (Figure 20.20(b)), there is no relative movement between the billet and container. The extrusion die is attached to a hollow ram and is forced into the billet. The extrusion flow is through the die and hollow ram in a reverse direction to the ram movement. This system, also referred to as backward or inverted extrusion, is not used to a great extent.

The dies for extrusion must be extremely hard and high temperature resistant and are made from chromium/vanadium/tungsten steels. A die lubricant must be used to minimise the rate of die wear. For aluminium alloy extrusion molybdenum disulphide is used as a lubricant while for steel extrusion the lubricant is glass. The hot steel ingot or billet is wrapped in a glass fibre mat before insertion in the press chamber and a glass fibre pad is placed against the face of the die. At the temperatures involved the glass softens and acts as a lubricant between the ingot and the chamber walls and also at the die. The glass mat at the die surface softens and a film of glass about 25 μm thick separates the steel extrusion from the die.

Figure 20.21 *Force variation during (a) direct extrusion; (b) indirect extrusion.*

Figure 20.22 *Formation of extrusion defect.*

A high pressure is needed for extrusion, but in the direct extrusion process the pressure is not uniform throughout the ram stroke. When the hydraulic ram starts moving the pressure on the ingot increases rapidly to a maximum value at which time metal will start to flow through the die aperture. The pressure needed to continue extrusion reduces during the process (see Figure 20.21(a)). This is because friction between the billet and the chamber wall is reduced as the length of the billet within the chamber decreases.

Towards the end of the stroke the pressure for extrusion begins to increase again. Extrusion is normally ceased when about 90 per cent of the billet has been processed. The remaining 10 per cent is not extruded but discarded. This is not only because of the high pressure that would be needed to extrude the last portion but also because of the strong possibility of defect formation within the extrusion. The type of defect that could occur is known as extrusion defect or 'back-end' defect. The outer surface of the billet will be chilled by the chamber wall and so the surface layers of the billet will be less plastic than the core. During extrusion these surface layers may wrinkle as shown in Figure 20.22 and if extrusion is continued these oxidised surface zones could be pushed through the die giving a defective product.

In indirect extrusion, the pressure needed is fairly uniform throughout the whole stroke and the incidence of extrusion defect is almost nil as there is no relative movement between the billet and the container wall.

The range of sectional shapes that can be produced by extrusion is very wide, and includes both solid and hollow sections. Typical examples of shapes produced in aluminium are shown in Figure 20.23.

Figure 20.23 *Typical examples of extruded sections.*

Figure 20.24 *Bridge die for making hollow extrusions.*

Hollow shapes may be made from solid billets by extruding through a bridge die or porthole die. In a bridge die (Figure 20.24) the plastic metal is split into two streams which flow around the mandrel that gives the bore of the tube. After being split by the bridge the two streams of metal converge and pressure-weld together. As there is no air present to oxidise the metal streams flowing past the bridge, perfect welding occurs within the die to produce a seamless hollow section. A porthole die is similar in principle but the plastic metal is split into three or four streams at the entry side of the die.

A major advantage of any hot working process is that it breaks up the coarse grain structure of a cast ingot. Also, under the pressures involved, clean cavities such as minor shrinkage porosity, will be closed up. Consequently, the strength of a hot worked metal product will be considerably better than that of a casting. Some disadvantages of hot working are that very close control of dimensions is not always possible, and the product will have a poor surface finish owing to the effects of high-temperature oxidation. Subsequent cold working operations will give close dimensional tolerances and a good, clean surface.

20.10 | Fibre structure

The final properties of a hot worked product depend to a certain extent on the type of deformation process that was used. Commercial materials are not completely pure and will contain in addition to the alloying and trace elements, small non-metallic particles derived from the fluxes and slags involved in the melting operations. Within the cast ingot these particles will have a fairly random distribution. When the ingot is deformed these foreign particles will flow with the metal and will form fibre lines within the structure. After each hot deformation the flowed metal crystals will recrystallise, but the fibre particles will remain unchanged. In rolled and extruded products all fibre lines will lie parallel and will be in the direction of working, but in a forging the metal flow pattern will be more complex. Figure 20.25(a) shows the fibre structure of two bolts, one of which has been formed by machining from a solid hot

Figure 20.25 *Fibre lines. (a) Bolt shaped by (i) machining from rolled stock, (ii) forged bolt head; (b) Gear blank formed by (i) blanking from rolled plate, (ii) parting off from rolled bar, (iii) forging from small-diameter bar.*

rolled bar, and the other of which has been formed by forging the bolt head from small-diameter hot rolled bar stock. The difference in fibre structure will give the components differing properties even though all other factors may be equal. Another example of different production routes giving different fibre structures is the case of the small gear blank (Figure 20.25(b)) formed by (i) blanking from hot rolled plate, (ii) parting off a thin slice from a large-diameter hot rolled bar, and (iii) parting off a length from a small-diameter hot rolled bar and forging this into a thin disc. The fibre lines may constitute lines of weakness within the material, and lines along which fracture may occur. In the case of the gear blank formed from rolled plate the fibre lines are tangential at two points on the disc and teeth in these positions could be subject to failure. In case (iii) all fibre lines within the disc are radial and the risk of failure of the gear is lessened. The term *upsetting* is given to the type of forging that would be used in case (iii), as the former flow pattern is completely reversed and the arrangement of the fibre lines is totally altered.

20.11 | Cold working

Cold working is plastic deformation performed at temperatures below the recrystallisation temperature of the metal. During cold working the crystal structure becomes broken up and distorted and the material becomes strain, or work, hardened. The mechanical strength is increased by cold working, and the material becomes harder, but more brittle. The electrical resistivity is also increased. Eventually, the metal becomes so hard that further attempts at cold working would cause fracture. After cold working the material may be softened by heating it to a temperature above the recrystallisation temperature and allowing the distorted crystal grains to recrystallise. This type of process is termed *annealing*. Very good dimensional control can be exercised in cold working processes and good surface finishes can be achieved. The surface of the hot worked product may be covered by an oxide scale that was formed at high temperature. This layer of oxide scale must be removed before cold working operations can be commenced. Oxide scale may be removed from hot worked steel by pickling the steel in hot sulphuric acid.

Cold rolling is used for the production of sheet and strip material (the term 'strip' is applied to any material rolled from coils, irrespective of the width). The starting point for cold rolling is soft annealed material. During cold rolling the material becomes harder and in order to continue rolling it would be necessary to increase the roll separating force. The equipment used for cold rolling is different from that needed for hot rolling (see section 20.9). It is customary for cold rolling mills to have rolls of comparatively small diameter. In this way the contact area between strip and roll surface is reduced, with a consequent reduction in the total roll separating force. One drawback of small diameter rolls is that they could have a large deflection under load. To obviate this the work rolls are backed by larger-diameter rolls. One of the most popular arrangements is the 4-Hi mill (Figure 20.26(a)) in which the small-diameter work rolls are powered and the large back-up rolls are not. Cold rolling mills of the 4-Hi type are used for strip rolling with the material being unwound from a drum on the entry side of the mill and recoiled after passing through the mill. It is then possible to put tension into the strip at both sides of the roll. Both the back and front tensions will be acting in opposition to the friction forces between the strip and the rolls at entry and exit. This will have the effect of reducing the roll-separating force and so reducing the power required for rolling.

Frequently, a number of 4-Hi mills are arranged in tandem for the rolling of continuous strip material. Up to six or seven mill stands may be arranged in a strip mill train.

Another type of rolling mill suitable for the cold rolling of thin hard strip material is the cluster mill in which small diameter work rolls are backed by a series of rolls arranged in arcs (Figure 20.26(b)). The planetary mill configuration (Figure 20.26(c)), in which a set of small work rolls is mounted around a large diameter roll in a manner similar to a roller bearing, will permit very large reductions in thickness to be given in a single pass.

Another difference between cold rolling and hot rolling operations is that the roll surfaces are ground to a finer finish to give a reduced coefficient of friction. A lubricating oil is also sprayed on to the rolls

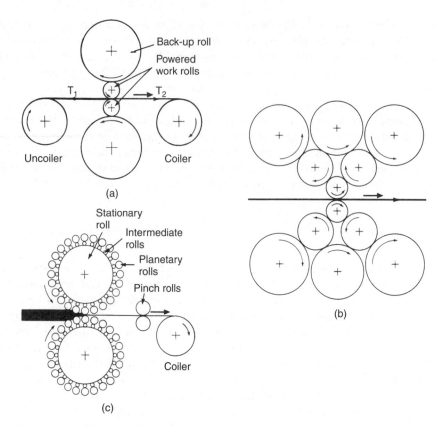

Figure 20.26 *(a) 4-Hi mill; (b) Cluster mill; (c) Planetary mill.*

to further reduce friction. If the final cold rolled product is required to have a smooth polished surface the last cold reduction passes are through highly polished rolls. When rolling a very thin product such as thin foil the system of pack-rolling, in which two layers of foil are rolled together, is used for the last few rolling passes. Aluminium foils with thicknesses as small as 0.005 mm can be made in this way. Pack-rolled foil is characterised by having one highly polished surface with a matt finish on the other surface, this latter being the interface between the two layers.

As has been mentioned any metal will work harden during cold rolling and one or more intermediate annealing operations may be necessary to soften the material before any further reductions can be obtained. If the finished product is required in a soft and ductile condition a final annealing treatment will be necessary. The annealing of cold rolled material is usually done in controlled-atmosphere furnaces to avoid surface oxidation of the strip. After cold rolling the coiled material may either be uncoiled, levelled and cut into sheets or circular blanks of the desired size, or supplied as coil for some other manufacturing process.

The final properties of cold rolled material may vary between those of the fully work hardened condition and those of the annealed, or soft condition. Properties intermediate between these two extremes may be obtained by carefully controlling the amount of cold reduction given to the material after an annealing operation or by partially annealing fully work hardened material (refer to Table 8.2).

Much rod and bar material is also produced by cold rolling operations. In this case, the rolls possess semicircular grooves in their surfaces and each pass through the rolls is a 'closed' pass. Generally, rod mills contain several stands, up to about 15, arranged in tandem. The product from a rod mill may be marketed as rod but much is used as feedstock for wire drawing.

Another useful cold working process is *drawing*. This is used for the production of wire, rod and tubing. The annealed material is pulled through a die with a reducing diameter. The die materials must be very hard to resist abrasive wear. Drawing dies are made from high carbon or alloy steels. Tungsten carbide is frequently used as a die insert.

Not all metals can be drawn successfully. Some metals may be highly *malleable*, that is they deform readily when subjected to forces that are largely compressive such as hammering, but they are not necessarily highly *ductile*. For good ductility, that is, the ability to be drawn well, it is necessary that the metal work hardens sufficiently rapidly so that the reduced diameter material at the exit side of the die is able to withstand the tensile force that is required to cause plastic deformation of the soft undrawn metal. Low carbon steels are highly ductile materials and a reduction in cross-sectional area of about 40 per cent can be achieved in a single pass through a cold drawing die.

A die for the drawing of round rod or wire will contain a conical hole tapering down to a cylindrical section (Figure 20.27). The diameter of the hole on the inlet side is greater than the diameter of the stock to be drawn and the space between die mouth and stock acts as a reservoir for the drawing lubricant.

Rod material is drawn in straight lengths on a draw bench, but the stock for wire drawing is usually a large coil and the driven coiling drum on the exit side of the die provides the necessary tensile force to pull the wire through the reducing die.

In tube drawing, the bore dimensions are controlled by means of a mandrel. Several techniques are used. When relatively large-diameter tubing is to be drawn in a series of straight lengths on a draw bench a fixed mandrel may be used (Figure 20.28(a)). Drawing using a floating mandrel (Figure 20.28(b)) is much more suitable for small-diameter tubing and unlike the fixed-mandrel method there is no limit to the length of tubing that can be drawn. Small-diameter drawn tubing may

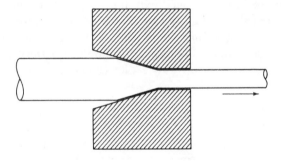

Figure 20.27   *Principle of wire drawing.*

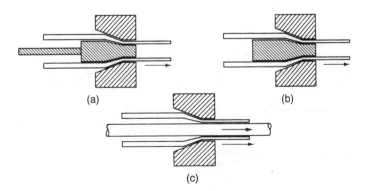

Figure 20.28   *Tube drawing: (a) with fixed mandrel; (b) with floating mandrel; (c) with moving mandrel.*

be coiled on to a drum. The third technique is the use of a moving mandrel (Figure 20.28(c)). In this case a hard straight rod with a diameter equal to the required bore of the tubing and of the same length as the finished tube length is used as the mandrel. Frictional resistance during drawing is less in this case because the mandrel moves with the tube. This technique may be used for both large-diameter and very small-diameter tubing. For example, the very fine steel tubing used for hypodermic syringe needles is drawn using hard steel wire as a moving mandrel.

Sheet and strip metal may be formed into an infinite variety of shapes. *Roll forming, deep drawing, pressing, rubber forming, stretch forming* and *spinning* are some of the processes that can be used for the production of complex and hollow shapes from flat sheet and strip material (Figure 20.29).

Roll forming is the shaping of flat strip metal into sections by passing it through a series of contoured forming rolls. The process can be used for forming angle and channel sections and a wide range of complex shaped sections.

In a true deep drawing operation for the production of a hollow cylindrical shape, such as a saucepan body, the sheet material is drawn through the aperture between punch and die, and there is, generally, a reduction in wall thickness occurring in the process. *Pressing* is the term given to the production of complex shapes, such as car-body panels. In this type of process the nature of, and the amount of, deformation may vary considerably from one part of the shape to another. It may be simple bending in some parts, and bending with some drawing taking place in other areas of the component. Metal spinning is the most suitable means for making conical shapes or re-entrant shapes, provided there is an axis of rotational symmetry for the latter. A circular sheet metal blank is rotated in a lathe-type machine, and the sheet metal is then forced into shape around a former, in much the same manner as a potter moulds clay shapes on a potter's wheel. Leather-faced wooden tools are normally the means of applying pressure to the metal sheet, and these may be hand held. In the manufacture of re-entrant shapes, such as the bodies of teapots, the former is segmented in order to facilitate its withdrawal from the moulded shape.

When a large amount of energy is supplied extremely rapidly to a metal, the metal may deform to an extent, and in a manner that would not be possible if force were applied slowly. One process involving high-energy rate forming is *impact extrusion*, and this process is used for the production of aluminium collapsible tubes and cigar tubes, and for zinc dry battery cases. Another high-energy process of note is *explosive forming*, in which an explosive charge is detonated under water and the

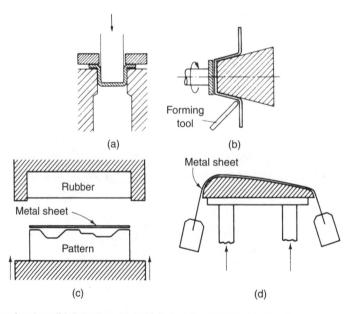

Figure 20.29 *(a) Deep drawing; (b) Spinning; (c) Rubber forming; (d) Stretch forming.*

pressure wave shapes metal sheet or plate by forcing it to flow and follow the contours of a female die. This process may be used for the manufacture of small or large components. It has been used for forming denture plates out of stainless steel sheet, and also for making the large dish ends for the underwater laboratory, Sealab II.

Electrohydraulic forming is a high-energy rate forming (HERF) process in which electrical energy stored in a bank of capacitors is suddenly discharged within a liquid. The shockwaves generated within the fluid cause a metal blank to be deformed into a shaped die, as in explosive forming. This method is suitable for small scale production and is used for the forming of fairly small items.

20.12 | Annealing

In many instances, cold rolled, or cold drawn, material has to be annealed, either as an intermediate stage in the processing, or as a final process. Normally the annealed material is required to have a fine crystal grain size, so that recrystallisation without excessive grain growth is necessary. Many annealing furnaces operate on the batch principle, in which a large batch of material is placed in the furnace and annealed in one lot. For this type of annealing the maximum temperature in the cycle is only marginally above the recrystallisation temperature. The surface layers of the batch furnace load may reach annealing temperature several hours before the inner layers. With only a small temperature excess above the recrystallisation temperature, there will be little grain growth occurring in the outer layers of material. An alternative annealing technique is flash annealing. A flash annealing furnace has a moving conveyor travelling through the hot zone. Single metal components are placed on the conveyor, and these travel fairly rapidly through the hot zone of the furnace and then through a cooling zone. The temperature of the hot zone may be several hundred degrees above the recrystallisaton temperature of the metal, but the components being annealed may be in the hot zone for only two or three minutes, so that full recrystallisation without excessive grain growth occurs.

With many metals oxidation occurs rapidly at high temperatures, and it is, therefore, necessary to provide the annealing furnace with an oxygen-free atmosphere. Many types of inert, or controlled, atmospheres may be used in conjunction with heat-treatment furnaces. One commonly used furnace atmosphere is produced by burning a hydrocarbon fuel, such as propane or butane, with slightly less than the theoretical quantity of air required for complete combustion. The burnt gases are then passed into the furnace.

EXAMPLE 20.3

Both hot and cold working processes are used for the manufacture of sheet metal, for example, aluminium sheet of 1.5 mm thickness. Why is it necessary to use both types of processing?

SOLUTION

Sheet metal is a high-volume production item and the starting point for this type of product will be a large cast ingot with a thickness generally in the range of 0.3–0.5 m. The only practicable way to reduce the thickness of such a large ingot is by hot rolling, with the ingot pre-heated to a temperature of about 100–150°C below its melting temperature. A series of successive reductions will be made, down to a thickness of 7–10 mm, by which time the metal will have cooled to about the recrystallisation temperature. Further thickness reductions down to the required final thickness of 1.5 mm will be made by a series of cold rolling operations, with or without intermediate annealing, depending on the final properties required. Cold rolling is used for the final stages as it gives much better control over dimensions and a better surface finish than is possible with hot rolling.

20.13 Powder metallurgy

The manufacture of finished components from a particulate material is a well-established method for both metals and ceramic materials and it involves preparation and grading of powders, compaction of the powder in a die, followed by sintering the compact at an elevated temperature to give a strong coherent product. (See also Chapter 22.) *Powder metallurgy* is a useful process for the manufacture of parts in metals that possess very high melting points. It is both difficult and expensive to melt these metals. Tungsten, for example, melts at 3410°C, but a powdered tungsten compact may be sintered into a fully coherent part at temperatures of the order of 1600°C. Another useful application of powder metallurgy is in the production of 'hard metals', namely sintered mixtures of hard metallic carbides and powdered cobalt, nickel, or other strong metals. These 'hard metals' are used for the manufacture of cutting tool tips, percussion tools and die inserts. For many years porous bearings have been made from metal powders. If a fairly low compaction pressure is used, a powder compact with a controlled degree of porosity can be obtained. Porous bronze bearings, with some 40 or 30 per cent porosity are made in this way. The bearings are then soaked in lubricating oil before they are put into service, and they can give long periods of service without requiring attention.

Some metal matrix composites are also made by compaction and sintering.

In recent years powder processes have been used increasingly for the production of small machine components of complex shape. The advantages of fabricating components from powders are considerable. One powder compaction operation produces an accurately dimensioned component that does not require machining and there is no scrap material produced. The only other operation necessary to produce the completed component is the sintering. In order to produce a similar component by conventional methods it may be necessary to commence with stock material produced by hot working and cold working processes and to give this stock several expensive machining operations, with, in consequence, some scrap arising at each production stage. Although the cost of metal powders is high, in comparison with the cost of ingot metal, it may well be cheaper to produce components by powder metallurgy techniques than by conventional machining.

There are several methods available for the production of metal powders, but not all methods apply to all metals. The more brittle metals may be powdered by pulverising in a hammer mill. Softer metals may be converted to powder by atomising a stream of the liquid metal in a jet of air, or an inert gas. Many metals, including aluminium, iron, tin, and zinc, are converted to powder in this way. Magnesium and zinc, metals with comparatively low boiling points are sometimes obtained in powder form by condensing the metal from the vapour phase. Some pure metal powders are obtained by electrolytic deposition of metal from a solution of a salt of the metal. The electrolytic cell operating conditions are arranged so that the cathode deposit is soft and spongy, rather than a hard coherent plating. This method is particularly suitable for the production of copper powder. Chemical methods for the production of metal powders include the reduction of heated powdered metallic oxide to metal in a stream of a reducing gas, such as hydrogen, and the dissociation of volatile compounds, such as carbonyls. Some metals combine with carbon monoxide to form gaseous compounds, termed *carbonyls*. Heating the carbonyl causes it to dissociate into metal and carbon monoxide. The carbon monoxide may be recirculated. Pure nickel and iron powders can be made in this way.

Powders require to be size graded before compaction. The powder mix for pressing is made up of different size fractions so that small particles fill the void spaces between larger particles and minimum porosity is achieved. The graded powder may be die pressed or isostatically pressed. In the former process, powders are compacted in alloy steel or cemented carbide dies at pressures up to 750 MPa. In isostatic pressing, the dry powder mix is placed in a flexible container, usually made of rubber, and subjected to a hydrostatic pressure of up to 500 MPa. The pressures used in each case will depend upon the nature of the powder, the average particle size, and the degree of porosity required in the compact. Higher and more uniform densities can be achieved by isostatic pressing than by die pressing.

During the compaction of metal powders, there will be some plastic deformation of individual powder particles, allowing for some mechanical interlocking of particles. There will also be some cold pressure welding of particles occurring and the compact will possess sufficient green strength to be handled. The powder compacts are then sintered in an inert atmosphere at some temperature below the melting point of the metal or alloy. The sintering temperature for iron is about 1100°C, and for bronze it is about 800°C. During the sintering operation, there is a diffusion of atoms across interfaces between particles bonding the particles together as one coherent mass. The driving force for the process is the reduction in surface energy achieved by the coalescence of powder particles. (Typically, the surface energy of a grain boundary is about one-half that of a free metal surface.) As the total surface area of a fine metal powder can be of the order of 1000 m², the reduction of energy which accompanies sintering is significant.

Pressing and sintering need not be carried out as two separate operations. Hot die pressing and hot isostatic pressing involve the simultaneous application of heat and pressure. These processes give sintered products of higher density and finer grain size than can be achieved by separate pressing and sintering and products with virtually zero porosity can be made. There is an increased driving force for sintering caused by the stresses established at the particle contact points. Some typical examples of metal parts made by powder metallurgy are shown in Figure 20.30.

Newer developments in powder technology include the production of sheet and strip material by compacting metal powders between rollers. This has been successfully accomplished on a pilot-plant scale, for pure aluminium, some aluminium alloys, and for mild steel. This technique probably has great potential as it offers a short route from raw material to finished product with a considerably lower level of capital investment than in conventional metal processing plant.

Another direct method for the production of sheet metal that is being developed is the direct spraying of atomised liquid metal on to a substrate. The deposited metal is then peeled from the substrate and is further processed by conventional hot and cold rolling.

Figure 20.30 *Typical examples of parts made by powder metallurgy. (Courtesy of Sintered Products Ltd)*

EXAMPLE 20.4

A sample of nickel powder has an average particle diameter of 0.8 μm. From the data given, estimate the available energy, per kilogramme of powder, to drive the sintering process.

Surface energy of nickel = 0.25 J m^{-2}

Grain boundary energy of nickel = 0.13 J m^{-2}

Density of nickel = 8900 kg m^{-3}

Assume that the powder surface area and grain boundary area are the same.

SOLUTION

Assume that all powder particles are spherical and of 0.8 μm diameter.

Volume of one particle = $4/3\pi r^3 = 4/3\pi (0.4 \times 10^{-6})^3 = 0.268 \times 10^{-18}$ m^3

No. of particles per m^3 = 3.73×10^{18}

No. of particles per kg = $(3.73 \times 10^{18})/8900 = 4.19 \times 10^{14}$

Surface area of a particle = $4\pi r^2$

Surface area per kg = $4\pi \times (0.4 \times 10^{-6})^2 \times 4.19 \times 10^{14} = 843$ m^2

Surface energy per kg powder = $0.25 \times 843 = 211$ J

Grain boundary energy per kg = $0.13 \times 843 = 110$ J

Therefore the energy available to drive the sintering process = $211 - 110 = 101$ J/kg

20.14 | Injection moulding

A fairly recent development for the forming of shapes from metal powders is injection moulding (Figure 20.31). This moulding process is similar to the injection moulding of polymer materials (refer to section 21.3). Parts of greater complexity can be made in this way than can be produced by conventional pressing and sintering. In order to render the metal powders suitable for processing by this method they are first mixed with a thermoplastic polymer powder as a binder. Typically, the quantity of thermoplastic is in the range from 35 to 50 per cent by volume. The metal/polymer powder mixture is fed to the barrel of an injection moulding machine, in which it is heated to plasticise the polymer binder, and then injected into the mould. The binder material is removed from the mouldings either by solvents or by preliminary heating prior to sintering. During the sintering process the component shrinks and densifies but retains the shape of the original moulding. The sintering shrinkage for a given powder mix is predictable. The moulds, of course, are made larger than the required final dimensions of the required component to allow for the shrinkage and dimensional tolerances of 0.2–0.3 per cent of the nominal final size can be achieved.

20.15 | Surface engineering

While the properties of bulk materials are important, there are many instances in engineering in which the effectiveness of a component is controlled by the quality and properties of the surface. For example, oxidation and corrosion are surface phenomena and it may be a major requirement that the component has a high resistance to oxidation and/or corrosion. Or it may be a requirement that the component be highly resistant to abrasive wear and so the surface needs to have a very high hardness. Often it is not necessary that the component material possesses the same properties throughout, in fact it could be a disadvantage for this to be so. In the case of gears, while the meshing surfaces need

Figure 20.31 *Injection moulded metal parts.*

to have high hardness and a low coefficient of friction, the gear material as a whole should be tough and ductile.

The term *surface engineering* relates to a range of processes used to create the required properties in a thin layer at the surface of materials and components. The creation of surface layers with specific properties applies to all classes of materials, metals, polymeric and ceramic materials, but I have chosen to group them all in one section in this chapter rather than fragment the subject by dealing with the different classes of material separately.

The methods which are used to either modify the condition of a surface or to apply a thin surface layer include thermal diffusion, painting, electroplating, electrophoresis, thermal spraying, chemical vapour deposition (CVD), physical vapour deposition (PVD) and ion implantation. Brief descriptions of these methods and some of their applications are given below.

The thermal diffusion of foreign atoms into the surface layers of a material is a method used for many applications including the introduction of dopant elements into silicon semiconductor material and for the surface hardening of many steels. Surface hardening methods for steels based on thermal diffusion involve the diffusion of carbon, nitrogen or silicon into the surface layers at elevated temperatures. The main processes are carburising, cyaniding, carbonitriding, nitriding and siliconising, and these are discussed in sections 12.5 and 16.10. Other methods, which do not involve the diffusion of foreign atoms, but the modification of surface structures by heat treatment can be used for the surface hardening of some steels. The surface layers are converted into a hard martensite structure by a process of rapid heating and cooling. These methods, flame hardening and induction hardening are described in section 16.10. The same principle for hardening applies also to a modern process in which the source of heat is provided by a laser beam.

There are many different methods used to produce thin layers on the surfaces of metals for the purpose of improving resistance to corrosion and the coating materials include paints, polymers and other metals. It is possible also to generate passive films on some metal surfaces by chemical means.

Paint films, many of which are polymer based, can be applied to metals but often some preparation of the metal surface is necessary to ensure that the paint layer keys well into the metal surface. Frequently the preparation takes the form of a chemical reaction to form a film of a metal chromate, phosphate, or other salt on the surface prior to the application of paint. Polymeric films or layers can also be applied to metals by dipping the heated metal component in a bath of molten polymer or in a fluidised bed of polymer powder.

Metal films or layers can be applied to metals by a variety of methods including hot dipping (for tin and zinc on steels), electroplating (for copper, nickel chromium, cadmium, silver and other metals onto many metal substrates) and thermal spraying (for many metals and alloys onto various metal substrates).

The principle of electrophoresis is the transport of electrostatically charged particles in an electric field and deposition on a substrate with an opposing electrostatic charge. The coating material may be in the form of a colloidal solution or a suspension of small particles in a carrying fluid. This principle is used for the deposition of metal, ceramic or polymer layers on various substrates. Examples are the deposition of aluminium on steel, the deposition of polymer coatings on metals and electrophoretic spray painting of components of complex shape. In many cases this type of deposition must be followed by a heat treatment to sinter the coating to the substrate.

Electroless or chemical plating is a process in which a metal coating is deposited on a substrate from a solution of a metallic salt. Coatings of several metals, including copper, chromium and nickel, can be deposited by this method. One advantage of the process is that a uniform thickness coating can be produced in the bore of tubes and in holes. Examples of the use of this process are the deposition of a metal layer onto a polymeric substrate prior to electroplating, and the deposition of a conductive layer through holes in printed circuit boards.

Many coatings can be applied to materials using a thermal spraying technique. *Thermal spraying* is a generic term for a number of processes in which a coating material is heated in a gaseous medium and projected onto a surface at a high velocity. In flame spraying the coating material in the form of wire or powder is fed into an oxyacetylene or oxyhydrogen flame, where it is melted and then sprayed onto the substrate surface by a gas jet. Typical coating materials applied by this method include aluminium, copper, zinc, nickel, bronze, alumina, chromium oxide and zirconia. In electric arc spraying an arc is struck between the ends of two wires of the coating material, causing the ends of the wires to melt. The molten material is atomised and directed at the substrate surface by a high pressure gas jet. Unlike flame spraying, only metal coating materials can be applied by this method. Another technique is plasma spraying. In this process a d.c. arc is struck between a tungsten electrode and a water-cooled copper electrode and a stream of inert gas is passed through the arc to create a high temperature plasma stream. The coating material is fed into the plasma, where it is melted and projected out of the plasma torch at high velocity. The very high temperature of the plasma flame makes this method suitable for spraying the refractory metals and high melting point ceramic oxides. A development of the last decade has been the use of high velocity oxy-fuel (HVOF) spraying. HVOF can be used to deposit a similar range of materials to plasma spraying and also carbide containing coatings such as tungsten carbide/cobalt mixes. Thermal spray processes are used for a wide variety of coating applications in general engineering, aerospace, automotive and bio-medical engineering, mainly to provide wear resistant and high temperature resistant surfaces.

Surface coatings may be produced using physical vapour deposition (PVD). The principle of this process is that a substance such as a metal or an oxide is vaporised by an intense source of heat. The atoms or molecules of the vaporised substance move away from the source and condense on the surface of the material to be coated. The process is carried out in a vacuum chamber, at a pressure of about

10^{-4} Pa, and the rate of deposition is improved by applying a d.c. field, with the material to be coated made the anode. There are several variants of PVD, one of which is termed 'r.f. sputtering' in which the use of a radio frequency generates a plasma around the target material. The systems can be used to apply thin layers of either metals or ceramics onto both metal and ceramic substrates. Examples of the applications of PVD coatings are titanium nitride and other ceramics onto high speed steel and cemented carbide cutting tools to reduce friction and increase tool life (see section 21.6), and yttria/zirconia layers onto nickel-base superalloys to provide thermal barrier coatings which enable gas turbine blades to operate at higher temperatures

Chemical vapour deposition (CVD) involves the deposition on a substrate of an element or compound as a result of a vapour phase reaction. Examples of its use are the formation of titanium nitride coatings on high-speed tool steels to increase cutting tool life, and the deposition of boron onto a thin wire substrate to give boron fibres

Another surface engineering technique is ion implantation. In this atoms of the element to be introduced into the surface of the workpiece are ionised and the ions accelerated towards the surface to be treated through a potential difference of the order of 50–100 kV. When these high-energy ions strike the surface they penetrate into the material to some extent. Ion implantation is used for the introduction of dopant elements into silicon wafers. The process is selective in that areas of the surface which are not to receive dopants can be masked. The method has a major advantage in that the ions are implanted at room temperature as compared with doping by thermal diffusion, in which the silicon has to be heated to a temperature of about 1100°C.

SUMMARY

Metals can be formed into a wide variety of shapes, both simple and complex, by a large number of processes. These processes can be placed into four main categories:

casting a liquid metal into a shaped mould,
plastic deformation by either hot or cold working,
forming a shape from metal powder, followed by sintering, and
removal of metal by machining.
(The last category listed is covered in Chapter 23.)

The required final shape is produced in a single operation in a casting process, although the casting may require finishing operations, such as machining. Some casting methods, die casting and investment casting, are precision processes which produce components to close dimensional tolerances.

Hot working is plastic deformation at temperatures above the recrystallisation temperature and the product emerges in the annealed or normalised state. Control of final grain size is obtained by controlling the finishing temperature of working. Hot working can produce finished products such as hot rolled sections, die forgings or extruded sections, or may give an intermediate product to be finished by cold working processes.

Cold working is plastic deformation at temperatures below the recrystallisation temperature and the product is in the work hardened condition. It may be used in this condition or softened by an annealing heat treatment.

Finished component shapes can be produced directly from metal powders by compaction in dies followed by sintering or by injection moulding a metal/thermoplastic powder mix, followed by sintering.

All processing methods incur cost. Often it is advantageous in both cost and energy consumption terms if the process will give a product of net or near net shape with little in the way of finishing operations, such as machining required.

Questions

20.1 Compare and contrast sand and die casting as methods for producing metal components.

20.2 State, giving reasons, the preferred casting method for producing the following:
  (a) cast iron cylinder blocks for internal combustion engines,
  (b) aluminium alloy cylinder heads for internal combustion engines,
  (c) 10 m lengths of cylindrical steel water pipe, with a wall thickness of 20 mm.

20.3 List the main stages in the preparation of a mould for investment casting.
  Explain why this casting method, of ancient origin, became a major production process during the 20th century.

20.4 From the data given below, calculate the unit cost of producing finished castings of 5 kg mass in aluminium alloy LM6 for
  (a) sand casting, if the total number of castings required is 300, and
  (b) die casting, if the total number of castings required is 15,000.
  What is the minimum number of castings required for die casting to become economically viable?

	£
Cost of LM6 ingot (£/tonne)	1,130
Value of LM6 scrap (£/tonne)	680
Cost of melting and melt preparation (£/tonne)	95
Cost of making a casting pattern	200
Unit cost of making a sand mould and casting	3.4
Cost of making a set of casting dies	45,000
Casting machine operating cost/casting	1.1
Mass of finished casting as % of liquid metal poured	sand casting – 55　　die casting – 85
Fettling, machining and finishing costs per casting	sand casting – £3.10p　　die casting – £0.95p

20.5 A sand casting, a die casting, a hot-rolled product and a cold-rolled product are all produced from the same composition alloy. Give an outline of the differences between the four products in terms of
  (a) their grain structures, and
  (b) the properties of tensile strength and ductility.

20.6 Distinguish clearly between the terms *hot working* and *cold working* as applied to metals and give the main advantages and disadvantages of both types of process.

20.7 Commencing with a cast ingot, list the main processing stages for the production of each of the following products:
  (a) lengths of steel rail,
  (b) aluminium foil of 0.010 mm thickness,
  (c) lengths of copper tubing of 13 mm OD and 11 mm ID,
  (d) saucepan bodies in aluminium,
  (e) low alloy steel connecting rods for an internal combustion engine, and
  (f) aluminium alloy window glazing bars.

20.8 What is meant by the term *fibre structure*? In what way may the fibre structure affect the properties of a finished component?

What steps should be taken in the production of steel blanks for the manufacture of spur gears to ensure uniformity of properties?

20.9 Describe the main stages in the production of finished components from metal powder. What are the main advantages and limitations of powder processing as compared with other methods for the production of metal components?

CHAPTER **21**

Forming processes for polymer materials

Why learn about forming processes?

In order to meet design requirements there is a need to be able to process and shape materials into a great variety of shapes. The shapes that are required vary enormously, both in size and complexity, ranging from very small components with masses measured in milligrammes up to quite large polymer mouldings with masses of many kilogrammes. An engineer must have some knowledge of the range of manufacturing processes available and their applicability, and be aware of the advantages and limitations of the various processes.

LEARNING OBJECTIVES

1. Have some knowledge of the principal forming processes for thermoplastics and thermosets.
2. Understand how thermoplastic sheet materials can be formed into shapes.
3. Gain some knowledge of how fibre reinforced plastic materials can be produced and shaped.

21.1 | Introduction

The nature of polymerisation and a brief description of polymerisation methods are given in Chapter 4. The pure polymers which are made are not suitable for shaping and forming into finished articles but have to be further processed to produce moulding compounds.

The polymer formed is ground into powder or broken up into granules, but, in order to produce a suitable moulding material, it is then compounded with a variety of additives. The additives necessary may include lubricants and plasticisers, fillers and reinforcements, pigments and dyes, stabilisers, and flame retardants. It is necessary to add a plasticising agent to many polymers to improve their moulding performance and the toughness of the material. Plasticisers are generally non-volatile solvents that partially dissolve the polymer and increase the flexibility of the material. Another form of plasticising agent is a non-solvent oil, highly dispersed throughout the material. In addition to improving moulding characteristics a plasticiser will also modify the mechanical properties of the plastic material. An extreme example of the effect of plasticiser content on properties is seen with PVC (polyvinyl chloride). Rigid unplasticised PVC is a tough horny solid and is widely used for such products as rainwater pipes and guttering but plasticised PVC, containing up to 50 per cent plasticiser, is a tough, flexible, rubbery material used for the manufacture of rainwear, as simulated leather in upholstery and for adhesive-backed electrical insulating tape. Fillers are used to add bulk to the moulding compound and to cheapen the product but they will also modify properties. Wood flour is widely used as a filler and it will give the material an improved impact strength. Chopped glass fibre is used as a filler and reinforcement (see Chapter 17 for the properties of glass-filled polymers). Pigments and dyes are used to colour the moulding compound and one of the attractions of polymer materials is that they are available in a wide range of colours. A stabiliser is sometimes needed to resist degradation of the material during its service life and stabilisers are generally antioxidants. Polymers, being carbon-based compounds, are often flammable materials and a flame-retarding agent should be incorporated during the compounding stage.

21.2 Forming of thermoplastics

The majority of thermoplastic materials possess comparatively low softening and melting temperatures and are termed *melt processable*. Some, such as PEEK, the polyimides and PTFE are difficult to process and require to be moulded at temperatures of the order of 350°C. Components in PTFE are generally formed by hot pressing a sintering powder, in a manner similar to that described for powder metal components (section 20.13).

Casting. Some thermoplastic materials may be cast into shape in moulds and allowed to solidify. Casting is not a widely used process for thermoplastics but is used for some polyamide and acetal products. It is also suitable as a production process when only a limited number of items of any particular design is required and the small number does not justify the high cost of making a set of dies for injection moulding. Sheet material may be made by a process termed *casting*, but this is not casting in the true sense. It involves pouring a solution of the polymer material on to a moving conveyor. Subsequently, the solvent evaporates, leaving behind a solid sheet of material.

21.3 Injection moulding

Injection moulding is the most important fabrication method for thermoplastics and is analogous to pressure die casting for metals. The compounded polymer material, in the form of granules or pellets, is fed into a cylinder, or barrel, where it is plasticised and the hot melt then injected into a relatively cool mould or die where it conforms to the shape of the cavity and is allowed to solidify. It is then ejected as a finished moulding when it is rigid. The polymer material is moved through the cylinder and injected into the mould by either the action of a simple plunger or by a screw mechanism. The principles of the two types of machine are shown in Figure 21.1. Generally, the cylinder of a simple plunger machine contains a spreader or torpedo to permit a more uniform heating and mixing of the polymer. Most industrial machines are of the screw type. The shearing action of the screw gives a better mixing and plasticising than is possible in the plunger-type machine. The sequence of events in a moulding cycle

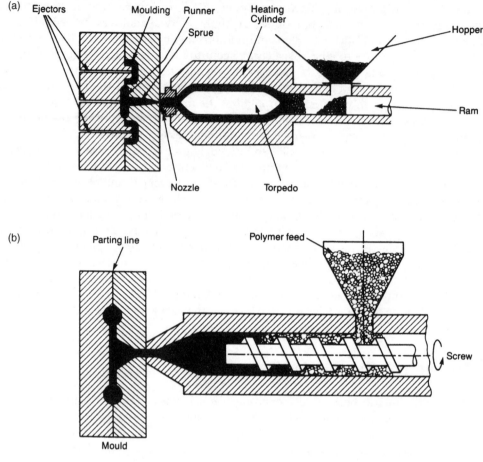

Figure 21.1 *Principle of injection moulding: (a) plunger-type machine; (b) screw-type machine.*

is closing of mould, after which the screw rotates. During this period, plasticised material builds up in front of the screw causing the screw to move axially backwards. Rotation is stopped and the screw moves forward as a ram to inject material into the mould cavity. The mould is then opened and the component ejected.

The screw in an injection moulding machine serves several functions and, generally, comprises three sections, feed, compression and metering. The feed section transports polymer granules from the feed hopper to the heating section of the barrel. The screw flights are of constant dimension in the feed section. In the compression section, the material changes from a granular form into a melt. The volumes of the flights decrease in this section to compensate for density change during the process. The final mixing and heating creating a homogeneous melt takes place in the metering section where again the flight dimensions are constant. There are several designs of screw to cater for various types of polymer material and these are illustrated in Figure 21.2. A highly crystalline polymer, such as a polyamide, has a sharp melting point and the screw best suited for this material has a very short compression section (Figure 21.2(a)). Semi-crystalline polymers, such as polyethylene, melt over a small range of temperature and require a screw with a longer compression section than highly crystalline materials (Figure 21.2(b)). Amorphous polymers such as PVC have no true melting point and the best screw for these materials shows a gradually increasing compression over its whole length (Figure 21.2(c)).

Figure 21.2 *Types of injection moulder or extruder screw: (a) for highly crystalline polymers; (b) for semi-crystalline polymers; (c) for amorphous polymers.*

21.4 Extrusion

Rod, tubing, sheet, film and solid or hollow sections in a wide variety of shapes can be produced by extrusion. The most common type of plastics extruder is the single screw machine in which a rotating screw conveys granular polymer material along a barrel in which it is heated, plasticised and compressed into a homogeneous melt and then extruded through a shaped die orifice (Figure 21.3). As for screw injection moulding machines, there are several types of screw designed to cater for various polymer types (Figure 21.2), but unlike injection moulding, the screw is in continuous rotation at constant speed during extrusion. Twin screw extruders are also available but are much more complex and expensive and their use is restricted to a few applications, one being the extrusion of large diameter piping in UPVC.

Figure 21.3 *Principle of screw extruder.*

The hot extruded plastic must be cooled immediately after emerging from the die to increase its rigidity and prevent loss of shape. This is accomplished by means of cold air jets or water. A necessary part of any extrusion equipment is the 'haul-off' table. This is a moving conveyor, with its speed geared to the rate of extrusion, to give support to the extrudate and to move it uniformly through the cooling zone. For some products, for example, sheet and film, the 'haul-off' is accomplished by a series of rolls.

Sheet and film are produced by extrusion through a flat die with a wide, straight, slit orifice. (Generally, plastic thicknesses of <0.25 mm are referred to as film while thicknesses of 0.25 mm or greater are referred to as sheet.) On emerging from the extrusion die, film or sheet extrudates are usually passed through a series of water cooled rolls with polished surfaces. This not only cools the material but also imparts a good surface finish. Some film extrusion is *co-extrusion* in which two or more layers of film are extruded simultaneously from separate extruders to produce a laminate. This is done for the purpose of improving properties, particularly the degree of impermeability and the barrier resistance of the film.

Filaments are made by extrusion, in this case through a multi-hole die with a number of very fine holes. The extruded filaments are usually quenched in water and then stretched between two sets of rolls to give full molecular orientation within the fibres. This method is used for polyamides, polypropylene and other thermoplastics.

There are several interesting variations on extrusion for the manufacture of a number of products. These are *extrusion coating* and *wire covering, film blowing* and *blow moulding*.

Extrusion coating is a means of applying a molten plastic film, usually polyethylene to a continuous sheet of paper, card or other flexible backing. This produces a waterproof layer and is used in the manufacture of paper sacks and cartons to hold liquids. In some cases a laminate is formed with a thin polyethylene film between two paper layers.

Plastic-coated wire and cables are produced by an extrusion process. The extruder nozzle is designed to turn the flowing polymer melt through 90° and to admit the wire to be coated (Figure 21.4(a)).

Thin film may be made by the film blowing technique, the principle of which is illustrated in Figure 21.4(b). During the major deformation which occurs during film blowing there is significant molecular orientation in both the longitudinal and transverse directions giving the film high strength in all directions.

Blow moulding is used for the production of bottles and other hollow shapes. A hollow tube, or *parison*, of plastic material is hot extruded. Immediately after emerging from the extrusion die, the hot parison is clamped in a mould and air is blown into it enlarging the parison so that it conforms to the contours of the mould. After a suitable cooling period, the mould is opened and the moulded article ejected. Materials which have been moulded successfully by this process include polyethylene, polypropylene, polycarbonate, polystyrene, ABS, acetal, polysulphone and PEEK.

Figure 21.4 *(a) Wire coating by extrusion; (b) Principle of film blowing.*

21.5 Calendering

Calendering is a process for forming thermoplastic sheet material and it is used mainly for PVC and copolymers containing vinyl chloride. Polymer powder and fillers and other additives are first blended together cold and then thoroughly mixed and agitated within an enclosed heated mixer. In one common type of mixer, the Banbury mixer, the two mixing rotors are hollow and contain channels for steam heating and the mixer jacket is also hollow with channels for both steam, for heating the mix, and water, for cooling. The hot and partialy gelled material from the mixer is then fed between a pair of heated blending rolls, in which the peripheral speed of one roller is about 10 per cent faster than the other, emerging as a continuous sheet at a uniform temperature in the range 150–170°C. At this stage, the sheet may still have some inhomogeneities with small areas of imperfectly mixed polymer and filler. The hot sheet from the blending rolls is cut up and extruded through a fine mesh straining screen to remove coarse particles and fed into the calender. A calender consists of either three or four steel or cast iron rolls, often referred to as bowls. Figure 21.5 shows the principle of calendering. The calendering rolls have highly polished surfaces and the barrels contain drilled channels for a heating fluid, oil or water. Sheet with thicknesses ranging from 0.1 mm to several millimetres and about 1 m in width can be produced by this process.

Figure 21.5 *Principle of calendering.*

21.6 Thermoplastic sheet forming

Flat thermoplastic sheet may be formed into a variety of shapes. The sheet is heated to soften it and it may then be moulded into the required shape. A shaping process which is used very extensively is *vacuum forming* in which the flat sheet is clamped over a mould box, heated, and the intervening space then evacuated causing the hot sheet to conform to the shape of the female mould (see Figure 21.6).

The production of shapes which involve deep formings within female moulds can be aided by giving the heated sheet some preforming before vacuum is applied. This is termed *plug-assist*. A male forming

Figure 21.6 *Principle of vacuum forming.*

Figure 21.7 *Preforming sheet by plug-assist.*

Figure 21.8 *Principle of drape forming.*

tool, or plug (Figure 21.7), is used to produce an approximate shape immediately before evacuation of the chamber.

Drape forming is a modified type of vacuum forming in which heated plastic sheet is draped over a male mould and mechanically prestretched, then fully formed under vacuum (Figure 21.8).

Bubble-assist is a prestretch technique for sheet in drape forming comparable to plug-assist for vacuum forming. In this case, after the sheet has been heated to the forming temperature, the air pressure in the drape moulding box is increased causing the sheet to stretch upwards in a bubble. The drape table is then raised and vacuum applied. The prestretching reduces the chance of the sheet thinning too much at high or sharp points of the male mould.

Vacuum and drape forming techniques are used for forming a wide range of thermoplastic materials including LDPE, HDPE, polypropylene, polystyrene, ABS, PMMA, CAB, PVC and polycarbonate.

These sheet forming techniques are used to produce a wide range of products, from small to very large, in both thin and thick sheet material. Products made include food packaging trays, refrigerator linings, containers, light fittings, advertising signs, baths and shower trays, boat hulls, lorry and tractor cabs, and so on. Sheet sizes as large as 3 m × 9 m have been formed by these methods.

21.7 | Rotational moulding

Rotational moulding is a process for forming hollow shapes in thermoplastics. A metered quantity of polymer moulding compound is placed in a split hollow mould. The mould, which is heated, is rotated

Figure 21.9 *Principle of rotational moulding.*

about two axes at right angles to each other (see Figure 21.9). The polymer in contact with the hot mould surface melts and flows to take up the shape of the hollow female mould. When sufficient time has elapsed for all the material to have melted, a cooling sequence is commenced and biaxial rotation continued until the moulding has cooled sufficiently to be removed from the mould without distortion. The process cycle time may be high, up to 30 minutes for large mouldings. The method is suitable for both large and small mouldings and there is very little molecular orientation within the product, thus minimising the risk of stress cracking. Very large mouldings with sizes up to 4 m × 4 m × 4 m are possible. The female moulds are relatively cheap to make and may be made from mild steel sheet or cast aluminium as they do not have to withstand high pressures. Technically, the method is suitable for use with any powdered thermoplastic but it is used mainly for PVC, polyethylene, polystyrene and CAB.

21.8 Expanded plastics

Expanded, or foamed, plastics and rubbers are used to a considerable extent for packaging, heat and sound insulation, and as cushioning materials. Foams may be formed by a number of methods, including fully aerating an uncured resin in liquid form, or mixing the plastic resin with a volatile material that vaporises at the temperature of the curing process. Another method of producing a foam is to add a chemical blowing agent. This is a chemical that decomposes with the evolution of a gas, such as carbon dioxide, when it is heated. This technique is used for the production of expanded polystyrene.

Much polystyrene is used in the form of foam. Expandable polystyrene beads are composed of polystyrene compounded with a gas-releasing ingredient, *n*-pentane being a common ingredient. Mouldings for such shapes as expanded foam cups, food trays, and packaging shapes are made by *in-situ* expansion of polystyrene beads within a mould. The mould is heated to about 120°C, at which temperature the polystyrene softens and the expansion agent decomposes.

Foamed polymers may be used as the basis of structural composite materials, as, for example, sandwich structures with a foamed centre. A manufacturing method for this type of composite is *foam reservoir moulding*. In this process, an open-cell polyurethane foam is impregnated with a controlled amount of epoxy resin and the foam faced on both sides with sheets of glass or carbon fibre fabric. The assembly is then placed in a mould which is closed under pressure. The resin saturates the facing fabric and the foam and is cured to form a rigid, lightweight composite.

21.9 | Forming of thermosets

The thermosetting plastics are produced by condensation polymerisation and the reaction must be stopped before completion, as the raw plastic for moulding is required in a part-polymerised condition. The polymerisation is completed during the moulding process to give a full network rigid structure. The raw plastic material is normally compounded with fillers in order to produce a moulding compound. There are many filler materials that may be used, and the purpose for adding them is to improve the properties of the material, and also to reduce cost, as the fillers are often cheaper than the polymer. Wood flour is a commonly used filler in Bakelite type resins, but cotton fibres, asbestos fibres, and mica, may also be used (refer to Chapter 17).

The principal forming processes for thermosetting plastics are *compression moulding* and *transfer moulding*. In compression moulding, the moulding powder is compressed between the two parts of a heated metal mould, or die. The powder becomes plastic and flows into the recesses of the mould. The pressure is maintained until the polymer has cured, that is, until full polymerisation has occurred and the material has set rigid. Compression moulding is not suitable for the moulding of shapes of thick section, or with large changes of section. The resins possess low thermal conductivities, so that the centre portion of a thick section may not become fully heated and only partially cure. This problem is overcome in transfer moulding. The resin powder is heated in an ante-chamber, and when plastic it is injected into the main mould, where curing occurs (Figure 21.10). It is, however, a more expensive process than simple compression moulding as there is some scrap material formed each time.

The cold setting rigid plastics may be readily formed into shape by casting. Also, these castable materials may be used for encapsulating small electrical components, and for mounting metallic and biological specimens.

Figure 21.10 *Transfer moulding: (a) moulding powder in ante-chamber; (b) resin forced into mould for curing.*

21.10 | Reaction injection moulding (RIM)

This is a plastics forming process, developed initially for moulding polyurethanes, in which two rapidly reacting constituents are mixed together just prior to injection into a mould, or co-injected into a mould. The time for curing may be some 30–60 seconds after the initial mixing. Reaction injection moulding is used for some thermoplastic and thermoset materials. Fibre reinforcement may also be added, in which case the process is termed *reinforced reaction injection moulding* (RRIM). Some reacting mixes may include an agent in their formulation which releases a gas to produce a foamed product. In these cases, the foam bubbles collapse at the mould surface to form a solid skin to the moulding.

21.11 Moulding of fibre composites

The majority of the fibre reinforced composites used are based on thermoset or thermoplastic matrix materials with glass, carbon or aramid fibre reinforcement. Continuous fibres are supplied in extremely long lengths wound on a reel, referred to as *rovings* for glass or *tow* for carbon fibre. Glass rovings comprise a number of *strands*, each strand composed of about 200 fine filaments. Carbon fibre tows may contain from several hundreds to several thousands of individual fibres. Strands of glass or carbon fibre may be woven into mats or fabrics. Fibre rovings or tows may also be chopped into short lengths for use as reinforcing fillers, the individual fibre length in chopped fibres generally ranging from about 1 mm up to about 25 mm.

The simplest method for production of FRP shapes is *hand lay-up*. A mould, which may be made of wood, with a varnished surface, plaster or fibreglass, is first given a *gel coat*. This is an initial layer of resin to give a smooth surface free from protruding fibres. The reinforcing material, in the form of cloth, mat or fibre strands, is placed in position manually while the gel coat is in a tacky condition and further resin applied by pouring on, by brush or by spraying. Full impregnation of the fibres by resin is accomplished by using a roller or squeegee. Successive layers of reinforcement and resin are added until the required thickness is achieved. In the *spray lay-up* technique, a special gun is used to spray both resin and reinforcement onto a mould which has been given a gel coat. Resin and rovings are fed into the gun, the rovings being chopped into short lengths within the gun. The component fabricated by either hand or spray lay-up is then allowed to cure and harden. The curing time may be accelerated by heating. Greater consolidation and densification of a moulding made by hand or spray lay-up can be achieved by means of pressure bag moulding, or vacuum bag moulding. In the former, pressure is applied to the free surface of the moulding via a shaped rubber bag, while in the latter method the mould and moulding are enclosed in an impermeable bag and the air evacuated. The external air pressure then compacts the moulding. *Autoclave moulding* is used for resins that require an elevated temperature for cure. This is similar to pressure bag moulding except that the cure is carried out in a steam autoclave. Autoclave moulding is widely used for the manufacture of structural shells and sections in the aerospace industries.

The methods described above use an open mould but it is possible to produce reinforced components within a closed mould. The two most widely used processes are *resin transfer moulding* and *reinforced reaction injection moulding* (RRIM). In resin transfer moulding, the reinforcing fibres, frequently in the form of *preforms* (see section 21.12), are placed in the mould before closure and resin is injected to fully impregnate the fibres and fill the mould. Curing occurs within the heated mould. RRIM is similar to RIM (see section 21.10) but with chopped fibre being mixed in with one of the reactants. RRIM is used mainly in connection with polyurethanes.

Filament winding is used for producing hollow cylindrical or spherical components. Continuous fibre rovings or tow are passed though a bath of resin and then wound around a shaped former, or mandrel. Fibres are wound onto the mandrel in a helical pattern and the helix angle and winding pattern can be designed to give the best strength characteristics for the intended application. Alternatively, *prepreg* tapes (see section 21.12) may be wound around the mandrel.

Pultrusion and *pulforming* are forming methods for producing rod, tubing and sections in fibre reinforced plastics. In pultrusion, a continuous length of fibre rovings or tow is passed through a resin bath and then through a series of carding plates to give full penetration of resin into the fibre bundles, and to remove excess resin. The impregnated fibres are pulled through a heated die where full consolidation occurs and the resin gels and then through a tunnel oven to complete the cure (see Figure 21.11).

In pultrusion, sections of uniform and constant cross-section are formed. Pulforming is similar to pultrusion but it gives a product with a varying cross-sectional geometry but of constant cross-sectional area. These processes give composites with a high axial strength and stiffness.

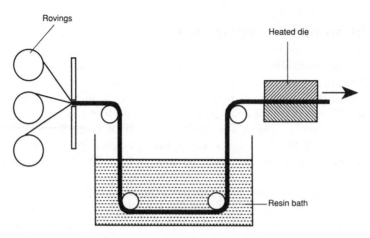

Figure 21.11   *Principle of pultrusion process.*

21.12 | Preforms, prepregs and moulding compounds

Various measures can be taken to aid manufacture of fibre reinforced composites, including the use of preforms and prepregs. A preform is used in connection with resin transfer moulding (section 21.11). Fibres laid within a mould could be moved out of position during the injection of resin. This problem can be obviated by use of a preform, which is a mass of fibres, or fibre fabric, shaped to fill the mould cavity and held in shape by a small amount of a dilute resin as a binder.

Prepreg is the shorthand term in general use describing pre-impregnated fibre/resin intermediate products which can be formed, at a later stage, into the required shapes. Prepregs are available in several forms and can use glass or carbon fibre and either thermoset or thermoplastic matrices. When thermosets are used, the fibres, either as a woven fabric or as continuous unidirectional fibres, are fully impregnated with resin and partially cured. In this form, the resin is slightly tacky and the sheets or tapes are protected with paper coated with a silicone barrier resin for storage. (The term *warp sheet* is used for unidirectional prepreg sheet.) Prepregs may be shaped by conventional processes such as vacuum or drape forming, pressure bag moulding and autoclave moulding, and thermoset prepregs are fully cured during or after the moulding operation.

Thermosetting polyesters with glass reinforcement are available as moulding compounds. These contain uncured resin, chopped glass fibres and an inert filler in approximately equal proportions. They are available in three forms, as sheet moulding compound (SMC), dough moulding compound (DMC) and bulk moulding compound (BMC). SMC, produced in various sheet thicknesses from 3 to 12 mm, can be formed by conventional vacuum and drape forming methods, the resin being cured during the process. DMC and BMC can be injection moulded, the resin curing in the mould.

21.13 | Laminating

Laminating is a process used for the production of stock material such as sheet, rod and tubing but it is also used for the manufacture of more complex shaped components such as safety helmets. Laminates may be made using a range of thermoset materials as the base material but the majority are based on phenolic resins with paper or cloth. The general manufacturing process is to pass paper or woven cotton or glass fibre fabrics through a bath of resin to give full impregnation of the material and partially cure to gel the resin. (Figure 21.12) For sheet manufacture, the resin impregnated materials are layered until the required thickness is obtained and the layered sheet is then compressed in a heated press to cure the resin thoroughly. For the manufacture of rod or tubing, impregnated tape is wound to a cylindrical shape before curing, while for moulded shapes the required number of layers are assembled and press formed into shape and cured. A proprietary name for paper and cloth laminates is *Tufnol*.

Laminated sheets of glass or carbon fibre reinforced resin can be built up by bonding several thin layers or plies together, in a similar manner to the manufacture of plywood. The properties of a sheet with continuous, parallel fibre reinforcement are highly anisotropic. Properties which approach isotropic within the plane of the sheet can be achieved by bonding several layers together, with a differing orientation of reinforcing fibres in successive layers. When constructing laminated sheet in this manner, there should be an odd or an even number of plies in the assembly although, unlike the manufacture of plywood laminates, it is more usual to have an even number of plies. The arrangement of sheets in the laminate should be symmetrical about the centre. A possible arrangement of layers in a laminate, showing the orientation of the parallel fibres, is shown in Figure 21.13.

Figure 21.12 *Principle of production process for laminated plastics.*

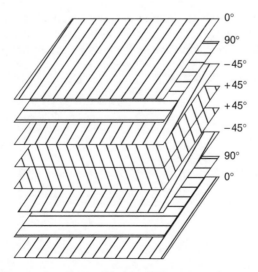

Figure 21.13 *Possible build-up of successive plies in a laminated fibre composite.*

SUMMARY

Many thermoplastic materials are *melt processable* and can be formed into a wide variety of shapes by the processes of injection moulding and extrusion. Modifications of the extrusion process can be used for making thin films and for putting plastic coatings on wire and cable.

Hot thermoplastic sheet material can be formed into three-dimensional shapes by processes such as drape forming and vacuum forming.

The range of processing methods available for thermoset materials is mainly restricted to casting, compression moulding and reaction injection moulding, and the material is cured in the mould.

A range of processes is available for forming fibre reinforced composites but most involve impregnating the fibres with the resin, forming the required shape before finally curing the material.

Questions

21.1 What are the main differences between compression moulding and transfer moulding? For what types of materials may these forming processes be used? What are the main considerations when choosing which of the two processes is to be used?

21.2 Suggest, giving reasons, a suitable forming process for the manufacture of the following products:
(a) bottles in PET to hold carbonated drinks,
(b) cylindrical pipe of 100 mm OD in UPVC,
(c) a water container of 0.8 m diameter and 1.2 m height in PP,
(d) PVC floor tiling, and
(e) automobile rear light lenses in PMMA.

21.3 Crystalline polymers, such as polyamides, and amorphous polymers, such as UPVC, can be formed by injection moulding or extrusion. What are the differences, if any, required for the processing of these two types of material?

21.4 Describe briefly the processes of vacuum forming and drape forming. State the main materials for which these methods are suitable and the type of product that can be made.

21.5 Suggest two alternative methods that could be used for making lengths of cylindrical tubing in a carbon fibre/epoxy composite. What are the advantages and disadvantages of each method proposed, and what differences, if any, would there be in the final properties of the product?

CHAPTER **22**

Forming processes for ceramics and glasses

Why learn about forming processes?

Like other materials, ceramics and glasses are processed into a variety of shapes in order to make component parts. Clay materials, when moist, are plastic and the plastic mass can be manipulated and formed into the required shape. Most of the newer industrial ceramics are shaped from powder by pressing and sintering operations but highly complex shapes can be made in some ceramics by an injection moulding process. Glasses, on the other hand, are shaped when they are hot and can flow as viscous fluids. An engineer must have some knowledge of the range of manufacturing process available and their applicability, and be aware of the advantages and limitations of the various processes.

LEARNING OBJECTIVES

1. Know how shapes can be made in plastic clays.
2. Know the principles of pressing and sintering, including reaction sintering and how shapes in industrial ceramics can be produced.
3. Have some knowledge of the ways in which glass products are formed.

22.1 | Forming of clay ceramics

Ceramics have been used since earliest times, and many ceramic articles are produced today by methods that are basically the same as those used several thousand years ago. Clay materials are normally

413

shaped from a moist soft plastic mass. For articles with rotational symmetry, the clay may be *thrown*, or moulded, on a potter's wheel. *Hand throwing* in the age old manner is still used on a small scale, but for mass production the process is automated and the potter's hands have been replaced by moulds and templates. Plastic clay may also be readily formed by extrusion, and this process is used for the forming of bricks, tiles and pipes.

After shaping, clay articles are dried and then fired in a kiln at temperatures in the range 800–1500°C. The firing causes expulsion of water of crystallisation, and a recrystallisation, creating a rigid, though brittle, material. There is a partial melting of some of the constituents of the clay during firing and a glass phase is formed between the grains, cementing the grains together. The extent of this vitrification during firing depends on both the composition of the clay and also on the firing temperature, higher temperatures increasing the glass content of the fired product. The higher the glass content, the stronger will the body be. The changes which occur during firing are accompanied by a shrinkage. Fired clay products are porous, and this porosity may be as high as 15 or 20 per cent in some earthenware products. In many cases a glaze is applied to the surface of a fired article and this is converted to a hard, vitreous and non-porous layer during a second, or *glost*, firing operation.

Another forming method that may be used for forming clay shapes is *slip casting*. The slip, which is a thick slurry or suspension of clay in water, is poured into a porous mould. Water is rapidly absorbed into the mould from the slip, and the surface layers of the slip become solid. When the required wall thickness has solidified, excess slip is poured from the mould, leaving a hollow casting. The casting may then be removed from the mould and fired. Slip casting is suitable for the production of complex shapes, such as wash basins and sanitary ware, and it is also used for the production of shapes in some of the newer industrial ceramics.

Another method used for forming ceramics from a slip is *tape casting*. This method is used to produce thin ceramic wafers for use in the electronics industry. The slip is filtered and fed onto a continuous moving belt, the thickness of the deposited film being controlled by an adjustable gate. The continuous tape produced is dried and pieces of the required shape and size can be blanked from the cast tape before firing.

22.2 | Pressing

Generally, oxide, carbide and nitride industrial ceramics are formed from powder by processes of compaction and sintering in the same general way as that used for powder metal parts (see section 20.13). One major difference from powder metallurgy is that the ceramic powders are non-ductile and show no plastic deformation when compressed. To overcome this, ceramic powders are often mixed with a small amount of a binder and a lubricant so that a coherent compact with sufficient green strength to permit handling can be obtained.

Green ceramic pressings may be produced by either *die pressing* or *isostatic pressing*. In die pressing, graded ceramic powder is fed into a metal mould or die and compacted by either one male tool, or plunger, forced into the die under pressure or two plungers applied from opposite directions. After compaction, the green compact is ejected from the die and sintered. The process is used for powder mixes of low plasticity and, although the powder may contain up to about 8 per cent water, it is also referred to as *dry pressing*. Isostatic pressing is capable of producing compacts which possess higher and more uniform densities than is possible with die pressing. The dry ceramic powder, or a powder preform made by die pressing, is contained in a flexible mould, usually made of rubber. There are two major variants, these being 'wet-bag' or hydrostatic pressing or 'dry-bag' pressing. In the former process, the rubber mould is placed in an enclosed pressure vessel and hydrostatic pressure is applied using a water/soluble oil mix as the transmitting fluid. In the latter process, the pressure fluid is a gas. Pressures in the range of 50–500 MPa are applied in isostatic pressing.

Hot pressing and hot isostatic pressing (HIP) are processes in which heat and pressure are applied simultaneously and sintering, under pressure, occurs (see section 22.3).

22.3 | Sintering

Sintering is the term for the process which converts a powder compact into a solid polycrystalline material. It is a thermally activated process and the driving force is the reduction in surface energy which occurs as the powder particles coalesce. In solid phase sintering, which is carried out at temperatures below the melting points of the powder constituents, the first stage in the process is the transport of some material by diffusion and its deposition in the inter-particle spaces immediately adjacent to particle contact points. This deposit is termed a *neck*, and the formation of necks increases the particle contact area. During sintering, the necks grow and merge with one another until, eventually, the original particle structure is replaced by a polycrystalline grain structure with a network of porosity between grains. There is an increase in density, and a volume shrinkage, during sintering and this shrinkage may be as high as 20 per cent.

EXAMPLE 22.1

Estimate the energy, per kilogramme of powder, available to drive the sintering process of alumina powder.
Assume: average particle diameter of alumina powder sample is 10^{-6} m
       density of alumina powder is 3900 kg/m^3
       surface energy of alumina powder is 1 J/m^2
       grain boundary energy of alumina is 0.5 J/m^2
       grain boundary area after sintering is same as original surface area of powder.

SOLUTION

Volume of one particle of diameter 1×10^{-6} m $= \pi/6\, d^3 = 5.24 \times 10^{-19}$ m^3
Number of particles/m$^3 = 1/(5.24 \times 10^{-19}) = 1.91 \times 10^{18}$
Number of particles/kg $= (1.91 \times 10^{18})/3900 = 4.90 \times 10^{14}$
Surface area of one particle $= \pi d^2 = 3.142 \times 10^{-12}$ m^2
Surface area of 1 kg of powder $= (3.142 \times 10^{-12}) \times 4.90 \times 10^{14} = 1539$ m^2

Surface energy available to drive sintering $= (1 - 0.5) \times 1539 = 769.5$ J/kg.

EXAMPLE 22.2

A ceramic powder is fabricated by pressing and sintering. Assuming that shrinkage during sintering is uniform in all directions, what linear shrinkage should be allowed for if the green powder compact has 23 per cent porosity and the total porosity of the sintered product is 2 per cent?

SOLUTION

The green compact has 23 per cent porosity, so 77 per cent of the total volume is material. Ceramic material makes up 98 per cent of the finished product.

So, the volume of the product, as a percentage of the green volume, is 77/0.98 = 78.6 per cent.
Therefore, the volume shrinkage during sintering = 100 − 78.6 = 21.4 per cent.
Assuming uniform shrinkage, the linear shrinkage is the cube root of 21.4 = 2.78 per cent.
The linear shrinkage is 2.78 per cent or 27.8 mm/m.

Sintering under pressure will produce higher density materials with finer grain sizes than is possible with sintering at atmospheric pressure. The stresses at particle contact points caused by the pressure will give an increased driving force for sintering. The two forms of pressure sintering are *hot pressing*, where the compact is sintered under pressure in a heated graphite or ceramic die, and *hot isostatic pressing* (HIP), in which a preformed compact is sintered at high temperature in an autoclave with argon or nitrogen as the pressure transmission medium. HIP is capable of giving a sintered product with uniformity of properties and virtually zero porosity.

Another form of sintering which may be used to produce dense ceramics with little or no porosity is *liquid phase sintering*. A small amount of an additive in the ceramic powder mix may react with the ceramic forming a compound which is liquid at the sintering temperature. An example is the addition of a small amount of magnesia, MgO, or yttria, Y_2O_3, to silicon nitride powder. During sintering in a nitrogen atmosphere, a liquid oxynitride phase is formed. This phase aids densification by means of a solution–diffusion–precipitation mechanism and the final product is a dense silicon nitride with an oxynitride glass as a grain boundary phase. The presence of the glass phase limits the maximum temperature at which liquid phase sintered components can be used.

22.4 | Reaction bonding and reaction sintering

Reaction bonding is the conversion of a powder compact into a polycrystalline ceramic by means of a chemical reaction between the powder particles and a gas or a liquid. This occurs in the production of some grades of silicon carbide and silicon nitride ceramics.

Reaction bonded silicon carbide (RBSC) is made by heating a porous powder compact of silicon carbide and graphite in contact with molten silicon. Silicon enters the open pores by capillary attraction and reacts with the graphite to form silicon carbide, which deposits on the surface of the original silicon carbide particles, bonding them together. All the carbon reacts and excess silicon remains, the final structure being of silicon carbide grains with interconnecting silicon-filled pores. This material is used for seals, bearings and engine components.

Reaction bonded silicon nitride (RBSN) is made by heating a compact of silicon powder in an atmosphere of nitrogen at a temperature in the range 1100–1450°C. Silicon is converted into silicon nitride according to the reaction

$$3Si + 2N_2 \rightarrow Si_3N_4$$

This reaction bonding process is also referred to as *nitriding*. There is virtually no change in the dimensions of the compact because, although silicon nitride occupies a greater volume than the original silicon, the nitride formed fills some of the pore space of the original silicon compact. However, even after nitriding, the product still has about 20 per cent porosity. A major advantage of this process is that the silicon compact may be partially nitrided into a coherent compact that can be machined to any required shape and to accurate final dimensions using conventional material cutting methods and tooling. Finally, the machined part is reaction bonded to fully nitride it into the very hard silicon nitride. RBSN is used as pouring tubes and pumping equipment for liquid metals and for high temperature engineering components.

Reaction sintering is the term describing the sintering of a powder compact into a polycrystalline ceramic involving a chemical reaction between two solid components of the powder. The product of reaction sintering is a dense single phase or multi-phase ceramic. An example is the reaction sintering of a mixture of alumina, Al_2O_3, and zircon, $ZrSiO_4$, to produce a dense zirconia–mullite composite ceramic.

22.5 | Injection moulding

In recent years there has been considerable research and development into the forming of ceramic components by injection moulding. The ceramic powders, which in themselves are non-plastic, are mixed with various plasticisers and binders. Very close control of both the chemical composition and powder particle size is necessary if a powder mix is to possess the rheological properties necessary to give a good injection moulding. Use is being made of injection moulding techniques to produce ceramic components of complex shape. A far greater degree of complexity can be achieved in injection moulding than is possible by die pressing. Injection moulding has been used successfully for the production of ceramic engineering components, including the soluble cores used in the investment casting of superalloy turbine blades with cooling channels.

22.6 Manufacture of glass

The raw materials for the manufacture of common soda–lime–silica glass are soda ash, Na_2CO_3, limestone, $CaCO_3$, and silica sand, SiO_2. These ingredients are crushed and ground to a fine particle size, typically 0.1–0.6 mm, and blended together before charging to a melting furnace, also known as a glass tank. During melting, the carbonates decompose, liberating carbon dioxide, and react with the silica. The gas evolution agitates the melt and aids the mixing to achieve a uniform composition. When fully molten, the glass contains many gas bubbles and the final stage, termed *refining*, involves raising the temperature of the melt to about 1550°C to reduce viscosity and allow the gas bubbles to come to the surface. This stage is helped by making additions of agents such as sodium sulphate, arsenous oxide or antimony oxides to the melt.

Glass tanks are divided into two zones, the melting and refining zone, and the working area, a bridge wall separating the two areas. The glass in the working area is maintained at the appropriate temperature for the type of processing which is to be carried out (see section 22.7).

22.7 Forming processes for glass

The main forming processes for glass are sheet rolling, tube making, hot pressing, float moulding and blow moulding. Different viscosities are required for these various processes and the working section, or forehearth, of the glass tank is maintained at the appropriate temperature. The range of viscosities for glass working is from 10^3 to 10^6 Pa s. Low viscosities, of the order of 10^3–10^4, are required for float and blow moulding, while for sheet rolling and tube making viscosities in the range 10^4–10^5 Pa s are needed. A high viscosity of 10^5–10^6 Pa s is suitable for forming by pressing.

Much glass sheet, in thicknesses from about 1–10 mm is made by a rolling process in which glass is drawn from the forehearth of the glass tank and flattened by passing through either a pair of rolls or between a pair of moving continuous belts (Figure 22.1(a)). The glass sheet formed by this method contains some imperfections and the surfaces are not perfectly flat and parallel. It is, however, suitable for many applications including use as window glass.

Distortion-free glass sheet and plate is made by the float-glass process, developed by Pilkingtons in the 1950s. In this process, molten glass flows from the forehearth of the glass tank in a continuous strip to float on the surface of a bath of molten tin (Figure 22.1(b)). It is heated sufficiently for the glass to flow so that both surfaces are flat and parallel.

Glass tubing is made by causing glass to flow from the glass tank onto the surface of a rotating mandrel, through which air is blown (Figure 22.1(c)). As the glass cools and its viscosity increases, it is drawn out to smaller dimensions.

Pressing is a process analogous to the closed-die forging of metals. A measured quantity of hot glass is put into a steel or cast iron die and the upper half of the die set pressed in position causing the glass to flow into the contours of the dies.

Blow moulding is used for the manufacture of bottles and other hollow shapes and, as in the blow moulding of thermoplastics, air pressure is used to blow a 'gob' or parison of glass so that it conforms to the shape of a mould. For the production of some shapes a pre-shaped parison is formed by an operation

(a)

(b)

(c)

Figure 22.1 *(a) Glass rolling; (b) Float-glass process; (c) Tube making.*

in which a quantity of hot glass is pressed in a die to give a hollow preform. This is reheated to give the required viscosity for blow moulding and then blown in a mould.

Hand-blowing of glass is still carried out but this is mainly for art glasswork and not for mass-produced glass articles.

Glass products need to be annealed after forming and this is carried out in an oven or *lehr*. An annealing lehr is an oven containing a conveyor belt which carries the glass products through the heated zone. The annealing temperature is that at which the viscosity of the glass is about 10^{12} Pa s (see section 18.13) and the glass is slowly cooled from the annealing temperature to the *strain point* (viscosity 10^{13} Pa s).

SUMMARY

Ceramic materials can be classified into two main groups: clay ceramics and industrial ceramics. With clay ceramics the basis material is a moist, soft plastic mass which can be readily moulded into the required shape. Parts with rotational symmetry are moulded by spinning on a wheel. More complex shapes can be made by the process of slip casting, in which a fluid clay slurry is poured into a porous mould. The clay form is then *fired* in a kiln. A second firing is needed if the product is to have a non-porous, glazed surface.

Generally, industrial ceramics are formed by the pressing of powders, followed by sintering, but parts of greater complexity can be produced by injection moulding followed by sintering.

Shrinkage occurs during both the firing of clay ceramics and the sintering of powder compacts. Generally, this shrinkage is predictable and is allowed for when making the pre-fired shape.

The forming of glass into shapes is conducted with the glass in a fluid state. The working range of viscosities is 10^3–10^6 Pa s, the actual viscosity being dependent on the forming process.

Questions

22.1 The processing of a clay ceramic body takes place in three stages: forming, drying and firing.
(a) Describe three methods used for the forming of clay shapes.
(b) Why is it necessary to interpose a drying stage between forming and firing?
(c) Describe the changes which occur during firing.
(d) Frequently a clay body is subjected to a second firing. What are the reasons for this?

22.2 Briefly describe the structural changes which take place when:
(a) wet clay is processed to make a solid ceramic article,
(b) an oxide powder is pressed and sintered to form a solid ceramic component,
(c) a molten silicate is cooled to form a glass.

22.3 What advantages result from the use of hot pressing as opposed to separate pressing and sintering for the forming of ceramic articles?

22.4 Why is it necessary to mix additives with a ceramic oxide powder prior to pressing or injection moulding?

22.5 Alumina powder is pressed to give a 'green' compact with a total porosity of 25 per cent. After sintering the total porosity was found to be 3 per cent. Assuming that the shrinkage is uniform, determine the linear shrinkage that occurs during sintering. Assume that shrinkage is uniform and the density of alumina powder is 3900 kg m^{-2}.

22.6 Alumina discs with final dimensions of 50 mm diameter and 4 mm thickness are to be made by the pressing and sintering of alumina powder. What should be the dimensions of the die if the powder compacts are made with a porosity of 24 per cent and the total porosity of the sintered article is 1 per cent? Assume that shrinkage is uniform and the density of alumina powder is 3900 kg m^{-2}.

22.7 Give an account of the way in which components can be made in the ceramic silicon nitride and explain why it is possible to achieve extremely close dimensional tolerances.

22.8 Explain the differences between *reaction bonding* and *reaction sintering* and give examples of each type of process.

CHAPTER 23

Material-removal processes

Why learn about material removal?

Although much progress has been made in developing manufacturing processes to produce net or near-net shapes in materials, there is still a need for machining operations. It is possible, through the use of machining, to create shapes, surface finishes and dimensional accuracies that may not be achievable by other means. Also, shape creation by machining may be the only viable method for making a prototype or where the number of identical items required is small.

The range of machining processes includes both traditional cutting techniques such as drilling, milling, turning and grinding in which chips are formed, and a number of non-traditional chipless machining processes, including electrochemical and electro-discharge machining.

The majority of machining is carried out on metals and alloys but plastics materials and composites are machineable, and some material-removal processes are applicable to ceramics.

LEARNING OBJECTIVES

1. Understand the cutting process and appreciate the significance of *rake angle* and clearance angle.
2. Understand the process of *chip formation* and the conditions for continuous and discontinuous chip formation.
3. Be able to use Taylor's law relating cutting tool life to cutting speed and other parameters.
4. Distinguish between the various cutting tool materials and know the basic characteristics of each type.
5. Understand the factors that determine the machineability of a material.
6. Distinguish between *forming* and *generating*, and understand the action of single-point and multi-point tools.
7. Gain a broad appreciation of the range of both cutting processes and non-traditional chipless machining processes.

23.1 Introduction

The majority of the forming processes described in the three preceding chapters give products in their final, or near-final shape. However, in many instances, there will be a need to perform one or more machining operations in order to produce an article to the correct dimensions and with the required surface finish.

23.2 The cutting process

Metal cutting is a cold working process in which a cutting tool forms chips. The cutting tool presents a wedge-shaped point to the work. In ideal orthogonal cutting, the cutting edge of the tool is normal to the direction of motion. There is severe deformation by shear in a plane, the shear plane, resulting in the formation of a chip (Figure 23.1). Considerable compression occurs in the deformed material and the thickness of the deformed chip produced will be greater than the depth of cut. There are two aspects of the tool geometry which are of importance, these are the *rake angle* and the *clearance angle*. The rake angle of the cutting face, which is measured from the normal to the work-piece surface, influences the shear angle and the amount of chip compression. The clearance angle is necessary to reduce friction between the tool and the machined surface.

The geometry of a single-point cutting tool has a major influence in determining the efficiency of a cutting operation. Increasing the rake angle has the effect of reducing both tool forces and friction and giving a less deformed and cooler chip. However, the provision of a large rake angle gives a smaller tool section, thus reducing the strength of the tool. The most effective rake angle for any cutting operation will be determined by the nature of both the work-piece and cutting tool material. The rake angles used in tools for cutting the softer materials, such as aluminium and its alloys, can be larger than those employed in the machining of steels. Carbide and ceramic cutting tools, although extremely hard, are also brittle. Consequently, the rake angles used in connection with this type of tool are usually either zero or negative to provide added tool strength. The same principle of rake and clearance angles applies to all multipoint tools, such as broaches, milling cutters and saws, in fact, any tool in which the cutting action is one of presenting a wedge-shaped point to the work.

Figure 23.1 *Orthogonal cutting with a single-point tool.*

In many practical instances, oblique cutting, rather than orthogonal cutting, is adopted. In this, the cutting edge of the tool is inclined at some angle other than 90° to the cutting direction (Figure 23.2). This offers significant advantages. This chip leaves the work-piece at an angle equal to the cutting edge angle and, in the case of a continuous chip, curls into a helical form rather than the closed spiral shape of a chip from orthogonal cutting. The chip then tends to move away from the work.

The other advantage of oblique machining is that the effective rake angle of the tool is somewhat greater than the rake angle measured in the cutting direction. This higher effective rake angle gives a reduced cutting force without weakening the tool.

Generally, single-point tools are provided with both end and side cutting edges and this is shown in Figure 23.3.

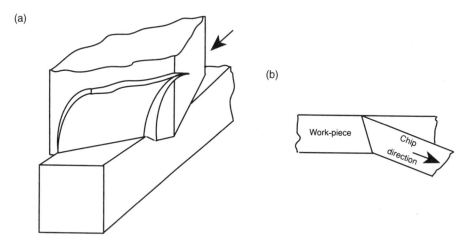

Figure 23.2 *(a) Oblique cutting; (b) Direction of chip travel.*

Figure 23.3 *Single-point cutting tool showing rake and clearance angles.*

23.3 Chip formation

The type of chip formed in cutting processes may be continuous or discontinuous. Soft, ductile materials will deform plastically to a very large extent and a continuous chip will be formed. With a less ductile metal, severe work hardening of the cut material will lead to a series of fractures and give rise to a discontinuous chip (Figure 23.4). Under some conditions, a *built-up edge* will form. This is the adherence of material to the cutting tool face and it alters the geometry of the tool, effectively increasing the rake angle (Figure 23.4(c)) and reducing the cutting force.

How does a built-up edge occur and how does its presence affect the efficiency of the machining process? When cutting at relatively low speeds using a cutting fluid, a smooth machined surface is formed and the chip surface is also smooth and slides across the rake face of the tool. As the cutting speed is increased, there is an increase in the temperature of both chip and tool face and, above a certain critical value of cutting speed, there will be some welding of chip material to the tool, the *built-up edge*, altering the tool cutting profile. Under certain circumstances, this may be advantageous in that the larger effective rake angle reduces cutting power requirements and the built-up edge, if stable, affords some protection to the tool face. Frequently, with a small, stable built-up edge, a machined surface finish of acceptable quality can be produced. However, in many cases, the built-up edge is not stable, may break off from time to time and a machined surface of poor quality results. As the cutting speed is increased further, and even higher chip temperatures occur, a second critical speed is reached at which a built-up edge does not form and a machined surface with a good finish is again achieved. The critical speeds are a function of several factors, the nature of the work-piece, the cutting tool material and the type of cutting fluid. For the machining of mild steel with a high-speed steel cutting tool, the two critical speeds are of the order of 320 mm/s and 650 mm/s respectively.

In high speed machining, particularly with automatic machine tools, it is desirable to have discontinuous chip formation as these can be removed easily from the cutting zone. Long continuous chips pose considerably greater problems. They are more difficult to remove from the cutting zone and they

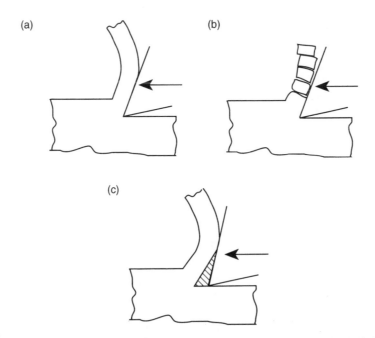

Figure 23.4 *Chip formation: (a) continuous chip; (b) discontinuous chip; (c) continuous chip with built-up edge.*

could wrap around the work-piece or tool, creating a potentially hazardous situation. The situation can be improved by incorporating a *chip breaker* into the tool. This is a device for breaking up a continuous chip into shorter lengths. One form of chip-breaker is a groove cut into the rake face of the tool. This forces the chip into an increased curvature causing fracture of the chip. An alternative is to attach an 'obstruction-type' chip-breaker to the rake face. Again, this causes the chip to assume an increased curvature. The so-called *free-cutting* materials contain internal chip-breakers within their microstructures, namely second phase particles which are structural discontinuities and aid chip break-up, for example the discrete small globules of lead in the microstructure of a leaded free-machining brass.

23.4 Cutting tool life

Because of the frictional forces involved in cutting, wear and loss of material from the tool face will occur with time. The general types of tool wear that occur are flank wear and crater wear (Figure 23.5). As tool wear occurs there will be a decrease in the cutting efficiency and in the quality of the machined surface. When the amount of wear has reached a pre-determined value, based on inability to produce the required dimensional accuracy or machined surface quality, the tool will have reached the end of its 'life' and must be withdrawn. Either the cutting face may be reground to permit reuse, or the tool disposed of.

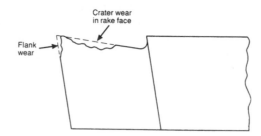

Figure 23.5 *Main forms of tool wear.*

Tool life is an important parameter. The highest practicable cutting speed, consistent with the attainment of good surface quality and dimensional accuracy, is sought. This is necessary to obtain a high output of finished parts from very expensive machine tools. Cutting speeds of up to 600 m/min are possible with some high-speed machine tools. But cutting tools are exposed to high pressures and temperatures in high-speed machining, and possibly shock loading conditions as well, and these factors tend to reduce tool life. Tool lives, generally, are of the order of 50–100 minutes but in some high-speed cutting operations made with complex high-cost machine tools, a tool life of the order of 5–10 minutes is common. Tool life, therefore, has a significant effect on the economics of metal-cutting operations as every tool change will necessitate a certain amount of machine down-time.

F.W. Taylor developed an empirical law relating cutting tool life, T, to cutting speed, V, and other parameters. The general form of Taylor's tool life equation is

$$VT^n = C$$

where C is a constant based on feed rate and other parameters and the exponent n is characteristic of the cutting tool material. The relationship can also be written in the form $\log V + n \log T = \log C$, and a graphical plot of $\log V$ against $\log T$ gives a straight line with a slope equal to $-n$. Typical values of the exponent n are 0.14 for high-speed steel, 0.25 for carbide cermets, 0.3 for coated carbides and 0.4 for sintered alumina and some other ceramics.

The extent of the reductions in cutting tool life brought about by an increase in cutting speed is better indicated by expressing Taylor's equation in terms of tool life, T. This gives $T = C^{1/n} \div V^{1/n}$. It can be seen clearly now that, for a high-speed steel cutting tool ($n = 0.14$), the tool life reduces according to the seventh power of cutting speed whereas, for cemented carbide tools, the life reduces according to the fourth power of speed and, for ceramic tools, it reduces according to $V^{2.5}$.

EXAMPLE 23.1

The life of a carbide cutting tool, when used for cutting a low-carbon steel at a speed of 100 m/min, is 100 minutes. Estimate the maximum cutting speed that can be used for a tool life of

(a) 30 minutes,

(b) 10 minutes.

Assume the Taylor exponent, n, for a carbide tool is 0.25.

SOLUTION

Using the Taylor equation $V\,T^n = C$
At $V = 100$ m/min and $T = 100$ min $C = 100 \times 100^{0.25} = 316.2$

(a) For a tool life of 30 min $V_{max} = C/T^n = 316.2/30^{0.25} = 135.1$ m/min.

(b) Similarly, for a tool life of 10 min $V_{max} = C/T^n = 316.2/10^{0.25} = 177.8$ m/min.

EXAMPLE 23.2

Experimental data obtained from a series of cutting tests using coated carbide tools for cutting 0.4 per cent carbon steel was as follows:

Cutting speed (m/min)	Tool life (min)
180	50
225	24.3
250	17.3
300	9.6

Using a graphical method, estimate the values of the Taylor constant, C, and the exponent, n.

SOLUTION

A graph of $\ln V$ against $\ln T$ should give a straight line with a slope equal to $-n$ with $\ln C$ being the intercept on the $\ln V$ axis.
From the above data:

$\ln V$	$\ln T$
5.193	3.912
5.416	3.190
5.521	2.851
5.704	2.262

These data are plotted in Figure 23.6. From the graph the slope, $-a/b = -0.4/1.28 = -0.31$ and $\ln C = 6.40$. So, the Taylor constant $C = 602$ and the exponent, n, is 0.31.

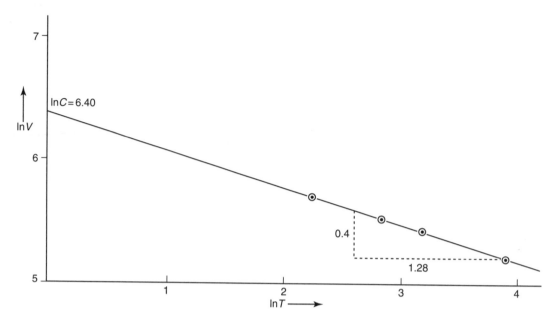

Figure 23.6

23.5 | Cutting fluids

Although, in a cutting operation, the majority of the heat energy liberated is carried away in the chip, there is a significant rise in the temperature of the cutting tool, particularly at high cutting speeds. The majority of machining operations are carried out with use of a cutting fluid and such cutting fluids fulfil a number of functions. These can be listed as:

(a) to act as a coolant and conduct heat away from tool and work,

(b) to act as a lubricant and reduce friction between the chip and tool,

(c) to carry chips away from the cutting point,

(d) to aid production of a good surface finish,

(e) to reduce energy consumption, and

(f) to extend the life of the tool.

Two main types of cutting fluid are used, coolants and lubricants. In high-speed machining, it is the cooling and heat-removal aspects which are the most important and water/soluble oil emulsions are widely used. They are good coolants and give some lubrication. Some complex machining operations, such as gear cutting, require a greater lubrication than can be provided by soluble oil emulsions and a suitable oil will give a good surface finish and a prolonged tool life. Most of the lubricant type cutting oils are based on mineral oils, often with chlorinated or sulphurised additives to increase lubricity.

23.6 | Cutting tool materials

The properties required in a cutting tool material are high hardness, the ability to retain hardness at elevated temperatures and to possess a degree of toughness sufficient to minimise the incidence of chipping and cracking. The main categories of cutting tool materials in use are: carbon tool steels, high speed tool steels, cemented carbides, coated carbides, ceramics and diamond.

Carbon tool steels

Carbon tool steels contain between 0.6 and 1.5 per cent carbon and are used in the hardened and tempered condition (refer to sections 16.8 and 16.9). They lose much of their hardness, through tempering when heated above 300°C. Nowadays, carbon tool steels are used mainly for the cutting of plastics and timber and have been almost entirely superseded for metal cutting operations by other materials.

High-speed steels

High-speed steels are highly alloyed steels containing tungsten, chromium, vanadium and molybdenum, the most widely used composition being 18 per cent tungsten, 4 per cent chromium, 1 per cent vanadium and 1 per cent carbon. Steels of this type contain hard tungsten and chromium carbides in their microstructures and require a three-stage heat treatment comprising oil quenching from a high temperature and two tempering treatments (see section 16.13). These steels retain their hardness at reasonably high temperatures and do not soften appreciably until they are heated above 600°C. High-speed steels are used for the manufacture of many metal-cutting tools, including drills, reamers, milling cutters, hacksaw blades and single-point tools for general purposes. They are not used to a great extent in connection with high rate production operations using automatic machine tools, having been largely replaced by carbide and ceramic materials for this type of application.

Cemented carbides

Cemented carbides, also known as cermets and hardmetal, are composed of carbides of tungsten, titanium, niobium and other metals, in a matrix of cobalt (see also section 19.9), and are made by powder metallurgy processing involving compaction and sintering. Cemented carbides are produced as both tool tips, which are then brazed to a steel tool shank, or as 'throw-away' inserts which can be secured to a tool shank by a set screw.

Coated carbides

These are cemented carbides provided with a thin coating, typically about 5 μm thick, of a harder ceramic material. The coating may be applied by chemical vapour deposition (CVD) or by sputtering. Titanium carbide, titanium nitride and alumina are among the hard coatings used and in some cases a multi-layer coating may be applied, for example, titanium carbide, followed by alumina and titanium nitride. These coatings give the tool tip a wear resistance approaching that of a ceramic tool tip while retaining the greater toughness of a cemented carbide.

Ceramic tools

A number of ceramic materials are used as disposable tool tips. These include alumina (section 18.7), cubic boron nitride (section 18.11), sialons (section 18.9) and polycrystalline diamond (PCD). These ceramics are harder than cemented carbides but are of lesser toughness. They are capable of cutting very hard materials such as *superalloys*, and of operating at much higher cutting speeds than is possible with cemented carbides (section 23.4). PCD, although extremely hard, interacts with iron at high temperatures and is not suitable for cutting steels. The ceramic materials are formed by powder compaction and sintering into tool tips and 'throw-away' inserts.

23.7 Machineability

Machineability is the relative ease with which a material may be machined by a cutting process. A very soft and ductile metal may spread under the tool cutting pressure, with the result that the tool tends to become buried in the work-piece and a tearing, rather than a clean cutting action, may occur. A similar effect can occur with soft thermoplastics. In the case of a less ductile metal, the severe work hardening effect will lead to the creation of a series of fractures giving discontinuous chip formation. It is this type of chip formation which is desirable for high-speed machining with automatic machine tools.

For the best machining characteristics, the requirements are for low hardness, coupled with low ductility. The machineability of a metal can be improved by any means that will decrease the ductility and increase the susceptibility to fracture. This includes work hardening and alloying to give solution or dispersion strengthened materials. Of course, the material must not be made too hard, as then it will increase the difficulties of cutting and lead to very rapid tool wear. Multi-phase alloys often machine well because the additional phases provide discontinuities within the microstructure and assist in the formation of discontinuous chips. The machining characteristics of some common materials is given below.

Plain carbon steels

The hardness and ductility of plain carbon steels is affected by both carbon content and the type of heat treatment (see Chapter 16). Low-carbon steels in the annealed or normalised condition are of high ductility and have poor machineability. In the work hardened condition, the machineability of these materials is improved considerably. A very large number of engineering components are manufactured from low-carbon steels by machining rod and bar stock and there is a major requirement for low carbon steels possessing good machineability. The so-called *free-cutting* steels are low-carbon steels with either an increased sulphur content or added lead. Their microstructures contain either manganese sulphide (MnS) or lead as globular second phase particles which act as internal chip-breakers. Medium carbon steels, with their higher carbon content, are harder and of lower ductility and are best machined in the annealed or normalised condition. High-carbon steels are best machined after being spheroidise annealed, as otherwise their hardness would be too great and result in rapid tool wear.

Alloy steels

The low alloy pearlitic steels are similar to plain carbon steels of corresponding carbon content in their machining characteristics. Many high-alloy steels, including stainless steels are of low machineability.

Cast irons

Grey cast irons are, in general, free-machining materials as the graphite flakes act as internal chip-breakers. The graphite also acts as a dry lubricant. White and mottled irons, however, are difficult to machine owing to the very high hardness of cementite.

Aluminium and its alloys

Pure aluminum is extremely soft and ductile and it is difficult to obtain a good machined surface unless the material is in a work hardened condition. A tearing of the metal surface rather than a clean cut may occur. This also applies to the softer alloys. The strong heat treatable alloys have good machining characteristics.

Copper and its alloys

Pure copper and some of the α-phase alloys are soft and ductile and are best machined in the work hardened condition. Free-machining alloys are available, for example, tellurium copper (section 15.5) and leaded brasses.

Magnesium and its alloys

These materials possess excellent machineability as they possess both a low hardness and low ductility. The main problem is the hazard posed by fine magnesium chips, as these could ignite. It is essential that there is no interruption to the flow of cutting fluid during the machining of these materials.

Plastics materials

Most plastics materials can be machined using conventional wood-working or metal-cutting equipment. The main problems with thermoplastics are that their properties are highly heat sensitive and temperature build-up must be avoided. This problem is somewhat aggravated by their low thermal conductivities. The soft thermoplastics, with T_g values below room temperature, give continuous chip formation but discontinuous chips are formed in materials, such as polystyrene and PMMA, with high T_g values. Polymers possess low elastic moduli and can be deformed relatively easily by the cutting tool, making the machining less precise than metal cutting. A small depth of cut is necessary to keep the tool force to a minimum. Glass-filled thermoplastics and most thermosets, including the phenolic laminates, although much harder than unfilled thermoplastics, can be readily machined with discontinuous chip formation.

Ceramics

Ceramics are of very high hardness and extremely difficult to shape by machining. Generally, of the traditional machining operations, only abrasion and grinding processes are suitable. The non-traditional methods of electron beam and laser beam machining can be used in conjunction with ceramics. An exception is silicon nitride. Partially nitrided silicon compacts (see section 22.4) can be machined using conventional metal cutting equipment.

23.8 Forming and generating

Forming is the production of a shape by a cutting process when the cutting tool has the reverse shape or contour to that required in the machined product. A form tool is a lathe or milling tool with a profiled cutting edge. In *generating* a surface, the shape and accuracy of the machined surface is determined by the movements of the machine tool slides rather than by the shape of the cutting tool. For example, the obtaining of a cylindrical or conical shape by lathe turning using a single point tool is generating. In practice, many machined surfaces are produced by a combination of generating and forming. A good example is the cutting of a screw thread on a lathe. The thread profile is *formed*, as it conforms to the contour of the single-point tool, but the helical thread is *generated* along the length of the work-piece by the movement of the tool.

23.9 Single-point machining

The main machining processes involving use of a single point tool are *shaping* and *planing*, which generate flat surfaces, and *turning* on a lathe to generate curved surfaces (Figure 23.7), although other operations such as *boring* and *facing* can also be carried out on a lathe.

Shaping and planing

Both shaping and planing are machining processes for producing flat surfaces by the straight line cutting action of a single-point tool. Most shaping is in the horizontal plane with a horizontal push-cut, the

Figure 23.7 *(a) Shaping or planning; (b) Turning; (c) Facing.*

cutting tool being pushed across the work by a ram. During the return stroke of the ram, the tool slides over the work surface. Work-piece feed between cutting strokes is at right angles to the cutting direction. In planing, it is the work-piece which moves in a horizontal plane relative to a stationary cutting tool. Planing is used for work-pieces which are too large to fit onto shapers. Both planing and shaping are slow processes and productivity is low in comparison with other types of machine tools. Their use has been largely replaced by other more efficient cutting processes such as milling.

Turning, boring and facing

Turning is a rotational cutting process in which the work rotates and tool feed is parallel to the axis of rotation. Turning is only one of several cutting operations which may be made on a lathe. Boring is the generation of the internal surface of a hollow cylindrical part using a single-point tool. The tool may be stationary and the work rotated, or vice versa, with the feed parallel to the axis of rotation. Facing is the generation of a flat surface on a component which is rotated in a lathe. The cutting tool is fed towards the centreline and the cut follows a spiral path on the work, giving a flat surface normal to the lathe centreline (Figure 23.7(c)).

Lathes

Centre lathes are probably the most widely used types of machine tools. There are many different types and sizes of lathe but basically a lathe provides a rotary motion to the work. The cutting tool, in its holder, can move along sideways in two directions, parallel to and normal to the axis of rotation (Figure 23.8). In addition to turning, facing and boring, other operations such as drilling, reaming and screw-cutting can be conducted on a lathe. A centre lathe requires an operator who is a skilled craftsman and this machine is used for prototype manufacture and small production runs. The *capstan lathe* is a

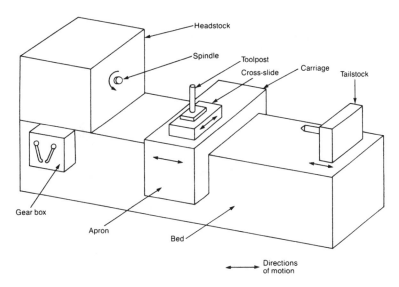

Figure 23.8 *Schematic representation of a centre lathe.*

production lathe for small to medium sized components and can be operated by semi-skilled labour. The machine carries a hexagonal capstan or turret in place of the tailstock and this carries up to six tools or special attachments. The operator can move the turret head to and from the work very quickly and the provision of stops enables each tool to cut to the required depth and length. The turret indexes automatically on withdrawal from the workpiece to bring the next tool into position.

In addition to manually operated lathes, there are many types of automatic or autolathes, some with turret heads and with single spindles or multi-spindles. Fitted with cams and a variety of rapid-acting electromechanical devices to control spindle speed and tool movement, they are used to machine accurately large batches of components. Special types include screw-cutting lathes and copying lathes in which tool movement is controlled by the movement of a stylus following a template or copy of the required product.

23.10 Multi-point machining

In multi-point machining, two or more cutting edges of the same tool are in operation at the same time. The main multi-point machining processes are *drilling, milling, broaching* and *sawing*.

Drilling

The provision of cylindrical holes in components is a major requirement and the majority of machined holes are produced by drilling. The spade drill, with a relatively simple shape, was the traditional type of drill in use for thousands of years for drilling holes in wood. The most common type of drill in use today is the twist drill. A twist drill has two cutting edges with two helical grooves, termed *flutes* (Figure 23.9(a)). The flutes allow for the access of cutting fluids and also a path for the removal of chips. The helix angle for most twist drills made from high-speed steel is 24° but smaller angles may be used for drilling plastics and soft metals, while large angles can be used for materials which can be drilled very rapidly. Drills with straight flutes are generally used for drilling holes in thin sheet. The spade drill design has been revived and this type, with cemented carbide cutter inserts (Figure 23.9(c)), is being used increasingly in automatic mass production work.

Figure 23.9 *(a)Twist drill; (b) Simple spade drill; (c) Spade drill with insert cutter.*

Figure 23.10 *(a) Conventional milling; (b) Climb milling; (c) End milling on a vertical mill.*

Milling

Milling is a metal cutting process in which a multi-point tool rotates and the work feeds past the tool. In horizontal milling the cutter rotates about an horizontal axis, parallel to the work-piece surface. The cutter is mounted on a shaft which is supported at both ends. In vertical milling, the cutter rotational axis is perpendicular to the work-piece surface and the cutter is supported at one end only. Horizontal milling may be of two types. In *conventional* or *up-milling*, the work-piece feed is in the opposite direction to the motion of the cutting tool (Figure 23.10(a)) while in *climb milling*, also known as *down-milling*, the feed is in the same direction as the direction of cutter rotation (Figure 23.10(b)). In the former method, the cutter teeth engage in the work at minimum depth and starting cutting forces are lower than in climb milling. However, the machined surface may have a greater degree of waviness than that generated by climb milling. In climb milling, the chip thickness is greatest at the start of the cut and decreases as the cut proceeds. The initial cutting forces are high but a good machined surface quality can be achieved. Formerly, much milling was of the conventional type until the introduction of more advanced and rigid machine tools allowed climb milling to be carried out with ease. In vertical milling operations, there are two main types of cutter used, the *face mill*, with cutting edges on the surface normal to the axis of rotation, and the *end mill*, with cutting edges on the cylindrical surface as well as on the surface perpendicular to the axis (Figure 23.10(c)).

CNC machining

For modern-day mass production, machine tools are computer controlled and NC is the universally used abbreviation for numerical control. In CNC (computer numerical control), the milling machine is controlled by a dedicated mini or microcomputer attached or assigned to it. Instead of using relatively simple milling machines, much production work today is carried out on *machining centres*. These are CNC milling machines with an extended capability and possessing the ability to perform operations such as boring, drilling and tapping, in addition to milling. They may possess more than one machining head and be capable of performing machining operations on five sides of the work-piece.

Many CNC machine tools are designed and built to perform certain specific functions and some are extremely large. Figure 23.11 shows a computer-controlled mill specially designed for the purpose of

Figure 23.11 *Large computer-controlled mill, with machined wing panel. (Courtesy of Airbus UK Ltd)*

milling upper and lower wing panels from aluminium plate for the whole series of Airbus aircraft. This, the largest mill currently in existence, has bed dimensions of 3.5 m width by 87 m in length.

Broaching

This is a metal cutting operation in which a tapered multi-toothed tool, a *broach*, is pushed or pulled through (internal broaching) or over (external broaching) the work. A broach has a series of cutting edges projecting from a rigid bar with successive teeth protruding further as one proceeds along the length of the tool. This rise, or step, between each successive cutting edge determines the chip size. Figure 23.12 is a representation of an internal broach. An internal broach may have any cross-section and can be used to cut holes of any profile, for example, a circular hole with an internal keyway, to a high degree of accuracy.

Figure 23.12 *Internal broach.*

23.11 | Abrasive machining and finishing

Abrasives are very hard materials and they are used for the removal of material by scratching, grinding or polishing. The abrasive machining processes enable components to be produced to close dimensional tolerances and with good surface finishes. Until the development of some of the non-traditional machining techniques, the only methods available for the machining of extremely hard metals and ceramics were abrasive processes.

Abrasives

Many different materials are used as abrasives, the common feature being that they are all of high hardness. The materials used include silica, alumina, silicon carbide (carborundum), cubic boron nitride and diamond. Corundum is a naturally occurring form of alumina and an impure variety of this is known as *emery*. Other abrasives used include magnesia, chromic oxide and ferric oxide (jeweller's rouge). Abrasive powders may be consolidated and bonded into a solid shape, such as a grinding wheel, made into *coated abrasives*, that is, bonded to a flexible paper or cloth backing, or used as a paste or powder within a carrying fluid. The abrasive powders are sized and graded to produce a wide range of wheels, papers and pastes capable of giving high or low rates of material removal and enabling a variety of surface finishes, from coarse to fine polished, to be achieved.

Grinding

This is material removal by abrasion and may be effected by using a grinding wheel or a coated abrasive. Each abrasive particle acts as a mini-single-point cutting tool, producing very small chips. There are several types of grinding operation. Surface grinding is used for generating a flat, precise surface using either the cylindrical surface of a grinding wheel or by bringing the surface of the work into contact

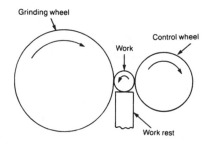

Figure 23.13 *Principle of external centreless grinding.*

with a rapidly moving flat belt of coated abrasive. Cylindrical grinding is used for the generation of cylindrical surfaces and may be carried out either by rotating the work between centres or by centreless grinding. In both cases, the work rotates at a slower rate than the grinding wheel. The principle of centreless grinding is shown in Figure 23.13. The small control wheel is made of abrasive particles bonded with rubber and the rotational speed of the work is essentially that of the slower moving control wheel. This process is often automated and used for mass production. The principle can also be applied to the internal grinding of cylindrical bores.

Form grinding is grinding using a profiled grinding wheel and is used for producing formed shapes, for example, screw threads.

Grinding wheels

There are many different types of grinding wheel available as there are many variable parameters, including type of abrasive, grain size and packing density of abrasive, type of bonding and amount of bonding material. The abrasives used may be silica, alumina (or the natural form corundum), silicon carbide, cubic boron nitride or diamond, while the bonding may be a vitreous bonding, shellac, rubber, or a polymer resin. The majority of grinding wheels are vitreous bonded and made by mixing the abrasive with a clay, pressing, drying and then firing in a kiln to give a glass bond around all the particles. Bonding with rubber gives a wheel with some flexibility while the use of a shellac bonding tends to give a ground surface with a polished finish. Both rubber and shellac have now largely been superseded by a range of polymer resins as bonding agents. The efficiency of a grinding wheel will reduce during use due to a rounding of the abrasive grains and the filling of spaces between grains with metal chips and other wear debris. Also, general wear may alter the grinding wheel profile. These problems can be rectified by dressing and trueing. Dressing is a cleaning operation, using a hardened steel tool, which removes or fractures rounded grains and removes foreign particles, thus exposing sharp cutting edges. A diamond tool may be used to restore the wheel profile, an operation termed trueing.

Abrasive jet machining

This is a process which is used for cutting holes or slots in very hard materials by means of a jet of high pressure air or carbon dioxide carrying entrained dry abrasive particles.

Ultrasonic machining

This is an abrasive machining process using an abrasive slurry in connection with a form tool. The form tool holder is vibrated by a piezoelectric transducer at a frequency of about 20 kHz and the tool is fed slowly into the work. The abrasive particles at the tool/work-piece interface are accelerated by the vibrating tool and perform the cutting. This process is suitable for machining holes, slots and shaped cavities in hard metals, ceramics and composite materials.

Barrelling

This is a general term covering several abrasive cleaning and finishing processes. In *barrel cleaning*, or *tumbling*, components are loaded into a cylindrical drum together with an abrasive such as sand, alumina, or granite chips and the drum rotated. Flashings, fins and oxide scale are removed by this process. *Barrel finishing* is similar but employs finer abrasive particles and is used to remove surface scratches and produce an even finish. In some cases, components may be passed successively through several finishing barrels with a finer grade of abrasive in each drum. In *barrel burnishing* or *barrel finishing*, the components are tumbled in a drum together with steel balls or shot. This is usually done wet, with water containing soap or other cleaning agent. A smooth even surface is produced.

Honing, lapping and polishing

These are abrasive finishing processes and only a small amount of metal removal occurs. *Honing* is a technique for sizing and finishing bored holes. The honing head carries four or six honing stones. These are made of abrasive particles bonded by a resin or wax. The honing head moves in the hole with both a rotating and a reciprocating motion, the stones being held against the work by a light pressure. Surface finishes as good as 0.075 μm CLA can be achieved. *Abrasive flow machining* is a finishing process used for internal surfaces but, in this case, unlike honing, it can be used for small holes and for non-cylindrical holes. A semi-solid medium, charged with fine abrasive particles, is forced through a hole. One of the mediums used is methyl polyborosiloxane, or 'potty putty'. *Superfinishing* is a process similar to honing but used for external surfaces. *Lapping* is a finishing process in which the surface is in relative sliding motion with a lap, the abrasive being a slurry containing very fine particles. Laps may be made from a variety of materials, including pitch, plastics, lead, copper and cast iron, but it is always softer than the surface to be finished. The lap surface may be flat or curved. Lenses and curved mirrors are finished by lapping on a lap with a curved surface. *Polishing* is an abrasive finishing process in which very fine abrasive particles are embedded in a soft backing, or polishing cloth, and there is relative motion between the surface and cloth with a fluid lubrication. Very little material is removed and the main effect is to produce a smooth surface of high reflectivity. In the polishing of a fine ground metal surface, the effect of frictional heating during polishing is to cause a softening of 'peaks' causing them to flow into 'troughs'.

23.12 | Gear and thread manufacture

Gears and screw threads are complex shapes which are required to be manufactured in great quantity to an accurate size with close dimensional tolerances. Both types of component may be made by a variety of methods, many of which involve material removal.

Gear making

Manufacturing methods for gears include metal cutting, plastic deformation, casting, polymer moulding techniques and processing from powder. The major metal cutting processes used are:

(a) *Gear hobbing*, which is a continuous cutting process using a multi-point tool, the hob, and is the fastest of the gear generating processes. The hob and blank are connected by means of gearing so that they rotate in mesh. The hob is fed into the blank in full depth and then feeds across the face of the blank, generating involute teeth. This is a very versatile process suitable for straight and helical gears, splines and sprockets.

(b) *Form cutting* using a profiled form tool in conjunction with a milling machine or a machine tool with a reciprocating cutting action.

(c) *Broaching*, in which a blank is pulled or pushed through a pot broach, a female tool as opposed to the usual male broach.

(d) *Cutting* on special-purpose gear cutting machines whose action involves both forming and generating. This is the only cutting method which can be used for bevel gears, whether straight, helical or hypoid.

Gear forms in metals may be made by sand, die or investment casting. Sand casting is used for the large gears in some low speed machinery and finish machining operations are usually required. Gears in some plastic materials are made by casting but, generally, injection, compression or transfer moulding techniques are used. The plastic deformation processes of forging and roll forming of metals can be used for the production of most gear forms and spur gears can be made by parting off lengths from an extrusion having a spur gear cross-section. Spur gears for low-load applications, as in analogue instruments and clocks are made by blanking from cold rolled metal strip. Also, many small gears are made from powder by compaction and sintering.

Gear finishing

For many high-performance applications, it is necessary to use one or more finishing processes, either to give improved meshing surfaces or to manufacture to very close dimensional tolerances. Many gears are made from hardenable steels and gear cutting is carried out prior to hardening and tempering. It may be necessary, then, to use more than one finishing process, one before and one after hardening and tempering. The main finishing processes used are shaving, grinding, burnishing and lapping:

(a) *Gear shaving* is the most widely used finishing process and is applied prior to hardening. The shaving tool, which is a meshing gear, the teeth of which have a number of grooves or slots, is run in mesh with the gear. The shaving tool acts as a broach and a small amount of metal, of the order of 0.025 to 1 mm, is removed from the gear faces.

(b) *Gear grinding* is a finishing process for hardened gears and may take one of two forms. One form involves a straight-sided grinding wheel and an accurate ground surface is generated. The other process uses a profiled grinding wheel and is a forming operation. Gear grinding is a slow and expensive process.

(c) *Gear burnishing* is a rolling process and an unhardened gear is rolled in mesh and under pressure against three hardened gears. There is some plastic deformation of the gear surface.

(d) *Gear lapping* is a finishing process for hardened gears in which the gear is run in mesh with a cast iron lapping gear in conjunction with a slurry of fine abrasive powder in oil.

Screw thread manufacture

The main principles involved in the manufacture of thread forms, both external and internal are cutting, plastic deformation and direct casting or moulding to shape. The main thread cutting techniques used are:

(a) *Use of taps and dies.* A solid tap is a multi-point cutting tool, similar in appearance to a threaded bolt but with a number of flutes, usually four, to give the cutting edges, and is used for cutting internal threads in pre-drilled holes. Taps may be used manually or on lathes, drilling machines and other machine tools. Also, collapsible taps are available for use with fully automatic machine tools and the cutting heads collapse inwards after the full thread depth has been cut to facilitate rapid withdrawal of the tool. A threading die is similar to a threaded nut with several, usually four, longitudinal grooves which expose multiple cutting edges. For manual thread cutting, dies are held in a die-stock. After cutting an external thread, the die is removed by unscrewing. Dies for automatic production work are of the self-opening type with a series of multi-point cutters arranged either radially or tangentially within a die-head, and designed to open automatically when the required length of thread has been cut. This permits the die-head to be removed rapidly from the work.

(b) *Screw cutting on a lathe* is suitable for both internal and external threads. The cutting tool is a single-point tool which conforms to the thread profile. Longitudinal movement of the tool is obtained through use of a lead screw which permits the correct movement of the lathe carriage in relation to work-piece rotation.

(c) *Cutting on a special-purpose machine.* A number of types of automatic machines dedicated to the cutting of either external or internal screw threads have been designed. These may be single-spindle or multi-spindle machines.

(d) *Milling.* External screw threads may be cut to a very high degree of accuracy by a milling operation using specially designed form cutters. Milling is often used for the production of threads on large diameter stock.

(e) *Grinding* is used to produce accurate external thread forms, particularly in very hard materials. The grinding wheel is shaped and is, essentially, a form cutting tool. The grinding wheel may be shaped with one rib, in which case the process is similar to thread cutting on a lathe with a single-point cutting tool, or it may have several ribs and the action is analogous to thread milling with a multi-point form tool. It is also possible to use a centreless grinding technique for forming threads and this last method is particularly useful for the manufacture of headless set screws.

A very high proportion of external threads required in large quantity are made by *thread rolling*, a plastic deformation process. The rolling between grooved dies or rolls, with consequent work hardening of the material, gives a smooth thread profile with a hard, wear-resistant surface. Internal threads in ductile metals may also be made by a plastic deformation process using a tool which is basically a tap without flutes. There is no cutting action and the thread is formed by plastic deformation in a manner analogous to the rolling of external threads.

In some instances, threads may be formed directly in die-cast or investment-cast products and many threaded components in thermoplastic materials are produced by injection moulding.

23.13 | Non-traditional machining processes

There is a range of material-removal processes which do not involve the formation of chips. These non-traditional, or chipless, machining processes are suitable for use with a variety of materials, including very hard metals which, otherwise, would be difficult to machine, and some can be used in connection with ceramics.

Chemical machining (CM)

CM is the removal of material by controlled chemical attack. One form of chemical machining, *engraving*, has been in use for hundreds of years. The basic principle of chemical machining is to cover those parts which are not to be attacked by a mask or resist, so that only portions of the surface are etched away. The term *chemical milling* is used when there is overall removal of material and the formation of etched pockets, contours or thinned down sections, as in the manufacture of integrally stiffened structural aircraft panels. *Chemical blanking* is the term used when the process involves etching completely through thin sheets, as in the manufacture of printed circuits. This latter process is also known as *photochemical machining* when used in connection with a photo-sensitive resist. The thin sheet or foil surface is coated with a light-sensitive emulsion, placed in contact with a film negative of the master pattern, and exposed to light. After development and washing away of unexposed emulsion, the photo-sensitive resist or mask remains. Lines as thin as 2 or 3 μm can be produced in the chemically blanked product using this type of resist.

Electrochemical machining (ECM)

ECM is a metal-removal process based on the solution of anode material in an electrolytic cell. The work-piece is made the anode and a tool, shaped so that it is a negative of the required component shape, is made the cathode of the cell. The tool is fed into the work at a constant rate, and equal to the rate of metal removal, thus maintaining a gap of about 0.01 mm between the tool face and the work. The electrolyte is not static, but is pumped around the system at high velocity, 30 m/s or greater. The high electrolyte flow rate is necessary as a major function of the electrolyte is to conduct heat away from the work. ECM is an expensive process but is of particular use in the machining of very hard complex alloys, such as the superalloys used for high temperature components in gas turbines. Surface finish values as good as 0.1 µm CLA can be achieved. There are two other variants of ECM, these being electrochemical hole drilling, suitable for the production of round or shaped holes in alloys which are difficult to machine, and electrochemical grinding (ECG). ECG, which is also known as electrolytic grinding, is used widely for grinding carbide tool tips. It offers several advantages over conventional grinding, giving higher material-removal rates and lower wear rates of the abrasive material. The process is similar to conventional grinding, but the grinding wheel is electrically conductive and is made the cathode, with the work-piece made the anode of an electrolytic cell. A continuous stream of electrolyte is directed into the interface between wheel and work. During the process, about 90 per cent of material removal is by electrolytic action, the remainder by abrasive cutting. Surface finishes as good as 0.1 µm CLA are possible.

Electrodischarge machining (EDM)

EDM, also referred to as *spark erosion*, is a machining process based on the erosive effect of an electric spark. Of the two electrodes used to produce a spark, one is the tool and one is the work-piece. The work is made positive because erosion of the positive electrode is greater than that of a negative electrode, other things being equal. The tool is produced to a shape which is a negative of the required machined component profile. The work and electrode tool are contained in a tank filled with a dielectric fluid, such as kerosene, white spirit, transformer oil or a glycerol/water mixture, and sparks at a high frequency are generated between tool and work. The gap between work and tool is kept at about 0.05 mm and the spark frequency may be between 400 Hz and 200 kHz, depending on the nature of the tool and work-piece, rate of metal removal and surface finish required. As erosion occurs, the tool is moved into the work by a closed-loop servocontrol mechanism so as to maintain a constant gap size. EDM is widely used for the manufacture of dies for casting, forging, stamping and extrusion, and surface finishes as good as 0.2 µm CLA can be achieved.

Electron beam machining (EBM)

In this process, a high-energy electron beam is focused onto the work-piece, which is contained in a vacuum chamber. The heat energy generated when the beam strikes the work is sufficient to melt and vaporise the material. EBM is used with extreme precision for making small-diameter holes and slots in relatively thin materials. Holes with diameters as small as 2 µm can be made in very thin materials. The process may be used to drill holes in sapphire and other ceramics for use as bearings in delicate instruments.

Laser beam machining (LBM)

LBM is similar to EBM in that the intense heat generated when a finely focused laser beam strikes a material is sufficient to melt and vaporise the material. A laser beam can be focused down to a very small spot size, about 0.02 mm, and can be used for the drilling of very small holes in almost any metal in thicknesses of up to 2.5 mm. Unlike EBM, a laser beam does not have to be operated *in vacuo*. The energy of a laser beam can be used for the cutting of materials and almost any material can be cut. For the cutting of many metals, a jet of an inert gas or oxygen is directed at the area of cut to blow molten metal away. In the case of an oxygen jet, oxidation effects tend to accelerate the cutting process.

Plasma jet machining

A plasma is a highly ionised gas produced by heating a suitable gas, such as argon, to a very high temperature in a confined space. Plasmas can be created by passing the gas through a constricted electric arc. The plasma jet produced in a plasma arc torch can be used for metal cutting. The extremely high temperatures and jet action of the plasma will give relatively smooth cut surfaces. Very high cutting speeds are possible and cutting rates of 2.5 m/min have been achieved in steel plate of 12.5 mm thickness.

Jet machining

This is a slitting process suitable for cutting soft materials, including paper and paper products, some plastics, leather, rubber and GRP material, using a very high velocity fluid jet as the cutting medium.

SUMMARY

Material-removal processes can be classified into the traditional cutting techniques, in which chips are formed, and chipless machining processes. In the cutting process a wedge-shaped tool causes severe plastic deformation by shear in a plane within the work-piece and the formation of a chip. The efficiency and effectiveness of the operation are affected by many factors including the nature of the work-piece, the nature of the cutting tool, cutting tool geometry, depth of cut and cutting speed.

The chip formed during cutting may be continuous or discontinuous. Discontinuous chip formation is an advantage in high-speed machine tool operation and can be achieved by modifications to the cutting tool design and by use of work-piece materials whose microstructures aid the formation of discontinuous chips.

Wear of the cutting tool will occur, reducing the cutting efficiency. The useful life of a cutting tool is related to cutting speed, cutting tool material and other factors, and an empirical law relating tool life to these factors was developed by F.W. Taylor.

Cutting tool materials require to be of high hardness, coupled with sufficient toughness to resist cracking. The main materials used are carbon tool steels, high-speed tool steels, cemented carbides, ceramics and diamond.

The term *machineability* relates to the relative ease with which a material can be machined in a cutting process. In general terms, work-piece characteristics of low-hardness coupled with low ductility will give good machineability. The machineability of a soft, highly ductile metal can be improved by incorporating an insoluble second phase in the microstructure, as in the addition of lead to brass to give free-machining characteristics. Ceramics, being of high hardness and with no ductility, cannot normally be machined, other than by grinding and some of the chipless machining processes.

There are many different material cutting techniques using both single-point and multi-point cutting tools, including turning, planing, drilling, milling, broaching and grinding. Some products required in very large numbers are still best made by machining processes. Various gears and screw threads fall into this category and although some of these, in some materials, may be made by other processes, very many are made using cutting processes.

There is a range of non-traditional or chipless machining processes which are suitable for the machining of very hard materials which, otherwise, would be extremely difficult to shape by cutting processes.

Questions

23.1 (a) What is meant by the terms *rake angle* and *clearance angle*, as applied to cutting tools?

(b) Discuss the influence of both work-piece material and cutting tool material on the selection of the most effective rake angle for a cutting operation.

23.2 What are the main advantages of oblique cutting as opposed to orthogonal cutting in metal-cutting operations?

23.3 The generation of a continuous chip in high-speed metal cutting operations can create a hazard. Outline the steps which can be taken to minimise continuous chip formation.

23.4 A carbide cutting tool is used for the machining of a low-carbon steel. Assuming that the exponent, n, is 0.25 and the Taylor constant, C, is 300 m/min, estimate the maximum cutting speed that will give a tool life of 1 hour.

23.5 Estimate the reduction in cutting tool life when the cutting speed is increased by 50 per cent for the following tool types: high-speed steel, cemented carbide and coated carbide. Assume the Taylor exponents are: high-speed steel 0.14, cemented carbide 0.25 and coated carbide 0.3.

23.6 Explain the difference between *forming* and *generating*.

23.7 (a) What alloying additions may be made to:

(i) low-carbon steel,

(ii) copper, and

(iii) brass,

to improve the machineability of these alloys?

(b) (i) Why is it unnecessary to make similar additions to magnesium?

(ii) What precautions should be taken during the machining of magnesium?

Joining processes

Why learn about the joining of materials?

There are many instances in the fabrication of products where individual manufactured components have to be joined to make a permanent assembly or structure. Many assemblies are made using fastening devices such as bolts, rivets and screws but these mechanical forms of assembly will not be discussed in this book. Much assembly work involves joining by welding and the allied processes of brazing, soldering and diffusion bonding. Most of these processes involve heat and it is important that engineers have not only an appreciation of the wide range of joining processes available and their applicability, but also how the heating may affect the properties of the basis work-piece and the resulting joint. The manufacture of many assemblies may also involve the use of adhesives and it is important to know the main features of these.

LEARNING OBJECTIVES

1. Distinguish between soldering, brazing and welding.
2. Distinguish between fusion and pressure welding processes.
3. Possess some knowledge of the main welding processes for metals and plastics.
4. Have an appreciation of the metallurgical factors governing the structures of weldments and heat affected zones.

24.1 | Introduction

The assembly of individual manufactured components into a complete fabrication is an important aspect of engineering production. There are many methods available for this, including the use of fastenings such as bolts, rivets and screws, soldering, brazing, welding, diffusion bonding and adhesive bonding. The use of fastenings will not be discussed in this book.

Soldering and brazing are joining techniques involving the use of a molten filler alloy of a different composition and lower melting point from the metals being joined. The parent metal is not melted in these processes. In fusion welding, on the other hand, some part of the materials being joined is melted during the process and, generally, a molten filler of similar composition to the parent material is needed to complete the joint. In pressure welding no filler material is used. Most metals and thermoplastics can be joined by some form of welding.

Diffusion joining processes are solid-state joining techniques. The parts to be joined are heated *in vacuo* or in an inert atmosphere and there is diffusion of material across the interface, thus effecting a good joint. These methods can be used for the joining of dissimilar metals, some ceramics, and some metal/ceramic components.

24.2 | Soldering and brazing of metals

Soldering and brazing processes are similar in principle in that the filler material melts at a comparatively low temperature and this liquid filler is drawn by capillary action into the small gap between the parts being joined. One major advantage of soldering and brazing operations is that the joint is made at fairly low temperatures so that there is little heat distortion of parts and little change to the microstructure of the parent metal. Soft solders are comparatively weak but hard solders (brazes) may have tensile strengths in the range 400–500 MPa. Figure 24.1 shows the soldering process diagrammatically.

Figure 24.1 *The soldering process.*

Soft soldering

The traditional soft solders are alloys of tin and lead or tin/lead with antimony, cadmium or bismuth. These soft solders have melting temperatures in the range 70–300°C (see Table 24.1). However, there has been concern in recent years over the use of these materials because of the toxicity of lead. Many lead-free solders have been developed and the use of solders containing lead is being phased out gradually. Most of the major Japanese electronics companies are already only using lead-free solders in their products and the remainder aim to be lead-free by 2005. European legislation requires that lead and other heavy metals, including cadmium, be phased out of new equipment by 2008. Table 24.2 gives a list of lead-free solders currently available.

Table 24.1 *Some soft solders.*

Composition (per cent)						Melting range or point (°C)	Comments
Pb	Sn	Sb	Bi	Cd	Ag		
4.5	95	0.5				183–223	Electrical instruments
34	65	1				183–186	High strength – quick setting
38	62					183	Tinman's solder
49.5	50	0.5				183–214	General-purpose solder
60	40					183–236	Soldering of tin cans
35	40		25			96–160	Soldering of glass to metal
66	32	2				185–245	Cable wiping solder
25	12.5		50	12.5		70	Very low melting point solder
97.5					2.5	304	High-strength joints in copper

Table 24.2 *Lead-free solders.*

Constituents	Melting range (°C)	Industrial use
Sn–Cu	227	Telecommunications, consumer products
Sn–Ag–Cu–Sb	216–222	
Sn–Ag	221–226	Automotive
Sn–Ag–Bi	206–213	Aerospace, military, consumer
Sn–Ag–Bi–Cu	213–218	Aerospace, military
Sn–Ag–Cu	217	Automotive, telecommunications
Sn–Zn	198.5	Telecommunications, consumer products
Sn–Bi	138	Consumer products

For soldering to be effective the liquid solder must 'wet' the surface of the metal to be soldered. This means that some alloying must take place between a constituent of the solder and the metal to be joined. The type of alloying that occurs may cause formation of a solid solution or the formation of an intermetallic compound. In soft solders tin will form intermetallic compounds with both copper and iron. A liquid solder will only wet a metal surface that is clean and grease free so all surfaces to be soldered must be perfectly clean. A thin oxide film will quickly form on a fresh metal surface and so a soldering flux is necessary to dissolve this oxide layer. The flux used is either zinc chloride solution or of the resin type. When the former flux is used the part must be washed after soldering to remove any remaining flux, otherwise corrosion could occur. When a resin flux is used any flux residue left after soldering is not corrosive. Soft solder wire with a resin core is invariably used for electrical jointing work.

The most commonly used heat source for soft soldering is a soldering iron with an electrically heated copper bit. The size of the heated bit may vary from large to very small depending on the type of work to be undertaken. A gas torch may be used as a source of heat for some soldering operations, for example, when making joints in copper pipework using 'Yorkshire' fittings. 'Yorkshire' fittings are capillary fittings with a solder insert.

Pre-tinning of a component to be soldered is frequently carried out. This will ensure that when the soldered joint is made a rapid penetration of the joint with solder will occur.

If too high a temperature is used in soldering, or if a soldered joint is maintained at elevated temperature during service, an excessive amount of intermetallic compound may be formed and the joint could become brittle. To avoid this type of brittleness in soft-soldered joints in copper, a lead/silver solder is used instead of a lead/tin alloy.

Hard soldering

Hard soldering, or *brazing*, makes use of copper-base alloys with melting temperatures ranging from 620°C to 900°C for making the joint (see Table 24.3). Again, the materials to be joined must be clean and a flux must be used. The most commonly used flux is borax. Borax melts at 750°C and is a good solvent for many metal oxides. For brazing at lower temperatures than this alkali metal fluorides are used as fluxes.

As with soft soldering the parts to be joined are prepared and fitted together. During brazing the molten filler metal is drawn by capillary action into the joint. The most commonly used heat source for brazing is the oxy-gas torch. The gas used is propane or acetylene. Generally, a neutral or slightly reducing flame is used, but when brazing copper a slightly oxidising flame should be used to reduce the possibility of hydrogen embrittlement.

Table 24.3 *Some brazing alloys and silver solders.*

Composition (per cent)						Melting range (°C)	Comments
Cu	Zn	Ag	Si	Ni	Cd		
50	50					860–870	General-purpose alloy
60	39.5		0.5			880–890	High-fluidity filler
59	40	1				880–890	High-fluidity filler
50	38		0.5	10		860–900	High-strength brazing
16	4	80				740–795	High electrical conductivity
29	10	61				690–735	Braze with good conductivity
15	16	50			19	620–640	Low melting point hard solder
37	20	43				700–775	General-purpose hard solder

Figure 24.2 *Assembly for furnace brazing.*

Furnace brazing

Furnace brazing is a technique used for the mass production of small assemblies (see Figure 24.2). The assemblies are made up and the filler placed in the joint, either as a powder or a preformed ring or disc. The parts are then placed in a controlled atmosphere furnace and heated to a temperature just above the melting temperature of the brazing filler.

Hard solders

Brasses, alloys of copper and zinc, are widely used as hard solders, hence the term brazing. The brasses used contain more then 40 per cent zinc but some zinc is lost during the operation and the final jointing metal possesses an α or $(\alpha + \beta)$ structure. The brazing alloys may contain small amounts of silicon or silver to increase the fluidity of the filler, or nickel and manganese to increase the strength of the joint. Silver solders, while much more expensive, are very strong jointing materials and have the additional advantage that they possess much lower melting temperatures than brasses (see Table 24.3).

Braze welding

This is a metal joining process in which the filler metal has a lower melting point than the metals to be joined but, unlike brazing, the molten filler is not drawn into the joint by capillary action. There is a wider gap between parts to be joined than in a brazed joint and the filler metal, typically either a brass or a bronze, is melted by means of an oxy-gas torch or an electric arc. When the filler material is a bronze the process is termed *bronze welding*. Cast irons may be successfully joined by braze or bronze welding. Because of the relatively low melting point of the filler, the iron is not melted and the tendency for thermal stress crack formation in the cast iron is minimised.

Aluminium soldering

Lead–tin alloys will not 'wet' an aluminium surface and it is not possible to solder aluminium using soft-solders. There are two main groups of aluminium solders, these being:

(a) aluminium–zinc alloys, with varying quantities of zinc and small additions of tin or cadmium, and possessing melting ranges close to 400°C,

(b) aluminium–silicon alloys; the 7.5 per cent silicon alloy has a melting range 580–620°C and the eutectic alloy melts at about 580°C.

Aluminium can be soldered successfully with these alloys using a chloride flux. For the manufacture of aluminium cans the canning material is commercially pure aluminium sheet with a thin layer of aluminium silicon alloy roll-bonded to the surface. This layer is put on during an early stage in manufacture by placing a cladding plate of aluminium–silicon alloy on the surface of an aluminium ingot and hot rolling them together.

24.3 | Fusion welding of metals

As stated earlier, in fusion welding a filler material of similar composition to the metals being joined is generally used and during the process a portion of the metals being joined is also melted. There are very many fusion welding processes and the heat necessary for welding may be obtained from a chemical reaction or by using electrical power (see Figure 24.3). One of the first successful fusion welding processes to be developed was gas welding using heat energy from the combustion of acetylene in oxygen in a special torch.

Gas welding

The main type of gas welding is oxyacetylene, in which acetylene (ethine) and oxygen are mixed and burnt in a torch. The full and complete combustion of acetylene is

$$2C_2H_2 + 5O_2 = 4CO_2 + 2H_2O \ (53.4\,MJ/m^3 \text{ acetylene})$$

In this reaction the ratio oxygen volume/acetylene volume is 5/2. In practice the volume ratios of oxygen/acetylene entering the torch are much less than this and normally range from 0.85/1 to 1.7/1. For a ratio of 1/1 the reaction equation would be

$$C_2H_2 + O_2 = 2CO + H_2 \ (18.3\,MJ/m^3 \text{ acetylene})$$

The carbon monoxide and hydrogen reaction products of this equation then burn with atmospheric oxygen at the outer envelope of the flame. By controlling the relative amounts of oxygen and

Figure 24.3 *Fusion welding processes.*

acetylene, flames with differing characteristics can be obtained. A chemically neutral flame is obtained when the oxygen/acetylene ratio is about 1.1/1, with a flame temperature of about 3250°C. If the oxygen/gas ratio is increased a shorter, fiercer, oxidising flame with a temperature of about 3500°C is produced, while an oxygen/gas ratio of about 0.9/1 will give a chemically reducing flame with a flame temperature of about 3150°C.

Gas welding is less widely used than the electric arc processes but it is still an important process. The gas torch can be used to preheat the joint area and to reduce the rate of cooling after welding. This is a great advantage when welding hardenable steels, as it reduces the possibility of brittle martensite formation in the weld zone.

The chemically neutral flame is used for the welding of plain carbon steels and most ferrous and non-ferrous alloys, but an oxidising flame is used for welding copper, bronze, brass and nickel-silvers. This is to avoid any possibility of hydrogen absorption which would lead to brittleness. A reducing flame is used for welding aluminium alloys and some stainless steels.

Flame cutting

For flame cutting, the torch-nozzle design is different from that used for welding. The cutting nozzle consists of a series of small holes arranged around a large central jet. The oxyacetylene mixture emerges through the small holes while a high-velocity jet of pure oxygen passes through the central hole. When cutting steel the oxyacetylene flame heats up the surface of the steel to a sufficiently high temperature (900–1000°C) so that the iron will ignite and burn in the strong oxygen jet according to the reaction

$$3Fe + 2O_2 = Fe_3O_4 \text{ (4.9 MJ/kg iron)}$$

The heat generated by this strongly exothermic reaction melts some of the surrounding metal and this is blown away by the strong oxygen jet. In this way steel thickness of up to 1 m can be cut. In addition to cutting, this principle can be used for planing, turning and boring large steel sections.

Arc welding

Arc welding in one form or another is the most widely used form of welding. The electrical supply is low voltage but high amperage and may be either alternating or direct. In many cases the supply for welding operations is obtained from the secondary of a step-down transformer connected to the mains supply, with rectification of the secondary output if direct current is required. Also, many portable welding units are available in which current is generated by a direct-current compound-wound generator driven by a petrol engine. The latter are particularly useful for working on construction sites.

The earliest forms of arc welding used carbon electrodes but nowadays the arc is struck between a metal electrode and the work-piece. The electrode may either be of tungsten or be a consumable metal electrode that melts, acting as a source of filler metal. This latter form is the most widely used type of welding (see Figure 24.4).

An alternating-current arc is broken and re-established at each half cycle and this leads to arc instability although the use of arc-stabilising agents in the flux coating of electrode wires can overcome this problem. When direct current is used there is a choice of polarity, with the electrode being either positive or negative with respect to the work-piece. The choice of polarity is important and both types of polarity are used in practice. The most widely used system is with the electrode negative (DCEN). In this case the electron flow is from electrode to work-piece. A concentrated arc issues from the electrode tip and heating of the work-piece is largely confined to the very small area beneath the electrode. The tip of the electrode and the metal directly under the arc melt and mix thoroughly with considerable turbulence. The pressure produced by the arc stream causes a crater to form in the liquid metal pool. As the arc progresses along the joint, liquid metal flows back into the crater. The depth to which melting of the work-piece occurs is called penetration of the arc, and the depth of penetration is greatest with DCEN welding. If the electrode is positive with respect to the work-piece (DCEP) the electron flow is from work-piece to

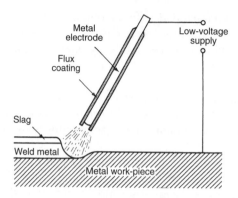

Figure 24.4 *Metallic arc welding.*

electrode. The emission of electrons from the work-piece has the effect of breaking up and scattering any tenacious oxide film that may be present on the work-piece metal. Also, greater heat is generated at the electrode tip giving more rapid electrode melting than with DCEN. The extent of heat penetration into the work-piece is less with DCEP than with DCEN, but there is a smoother transfer of liquid metal from the electrode. DCEP polarity is the preferred system for the arc welding of those metals that possess tenacious oxide films, namely, aluminium alloys, stainless steels and some copper alloys.

Transference of molten metal from the electrode tip to the weld pool may occur in one of several ways, these being *dip transfer, free-flight arc* and *spray transfer*. In dip transfer, when an arc is established the metal droplet at the end of the electrode wire begins to grow and the arc gap closes. When the drop touches the weld pool it creates a short circuit and the arc is extinguished. The high short circuit current causes resistance heating in the electrode wire until a molten drop leaves the electrode tip, re-establishing the arc. This sequence repeats with a relatively constant frequency. The conditions for effective dip transfer are dependent on arc voltage, wire feed rate and wire size. This form of transfer is used mainly for the welding of steels with carbon dioxide shielding, but it can be used also with other metals.

The arc condition in which the droplets form and detach themselves from the electrode tip to transfer to the weld pool without bridging and short circuiting is termed *free-flight*. Changes in the voltage and arc current will affect the droplet transfer frequency and as the transfer frequency increases, so the droplet size reduces. At a high transfer frequency, the metal transfer takes the form of a stream of small droplets projected axially from the electrode wire. This last condition is termed spray transfer.

Electrode coatings

Uncoated welding rods can be used for arc welding, but it is more usual to use flux-coated electrodes. When steels are welded using uncoated electrodes, oxides and nitrides can form and remain in the weld with a consequent loss of toughness.

Flux-coated electrodes are widely used. The composition of the coatings is complex and a variety of different coatings are used to cater for different types of welding application. However, in all cases the coating is formulated to satisfy three objectives. These are: (1) to form fusible slags, (2) to stabilise the arc, and (3) to produce an inert gas shield during welding.

Manual metal-arc welding uses flux coated electrodes in short lengths, ranging from 250 to 500 mm. For automatic flux-shielded arc welding, continuous electrode wire is required and this must be constructed in such a way that electrical contact can be made between the flux coated wire and the electrode holder. One method of doing this is to wrap the core electrode wire with helical wire spirals, with the spaces between the spirals being filled with flux. The electrical contacts in the welding head touch the outer wire spiral, where it protrudes through the flux coating, and current is conducted to the core wire. Another solution to the problem is to manufacture the electrode in tubular form and have the welding flux filling the tube. Flux-coated welding wires are much more expensive than bare electrode wire and there are two types of automatic flux shielded arc welding in which a bare metal electrode wire is used. These are submerged arc welding and magnetic flux/gas-arc welding.

Submerged arc welding

Submerged arc welding (see Figure 24.5) is an automatic form of metallic arc welding and is suitable for making long straight-line welds in the horizontal mode in plate thicknesses ranging from 6 to 30 mm. The surfaces to be joined are prepared with chamfered edges for a vee-butt joint. The welding head consists of a hopper and delivery tube for powdered flux together with the wire feed mechanism for the welding electrode wire. Powdered flux is delivered into the prepared joint ahead of the welding wire. The melting tip of the electrode wire is below the flux surface. The heat generated by the arc melts not only the electrode wire and some of the work-piece but also the flux, which forms a molten protective slag over the weld. When cool, the solidified slag can easily be chipped away from the metal surface. Submerged arc welding is used mainly in connection with low carbon and low alloy steels but the process can also be used for welding copper, aluminium and titanium alloys.

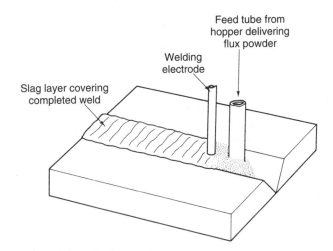

Figure 24.5 *Submerged arc welding.*

Magnetic flux/gas-arc welding

This is an automated continuous welding technique in which a magnetisable powdered flux is carried in a stream of carbon dioxide through a nozzle surrounding a bare electrode wire. When current is flowing, the electrode wire has a magnetic field surrounding it. This attracts the flux, which adheres to the wire. This method which can be used for making both vertical and horizontal weld runs in mild steel, has advantages over submerged arc welding. These are that the weld is visible and that welding is not confined to the horizontal mode.

Inert gas shielded arc welding

It is possible to produce a complete blanket of an inert gas around a weld area and by this means arc welding without fluxes, and using uncoated electrodes, can be readily accomplished. The most widely used shielding gas is argon, but nitrogen can be used as a shield for the welding of copper and carbon dioxide can be used with steels. The welding electrode may be of tungsten and non-consumable, in

Figure 24.6 *Principle of TIG welding.*

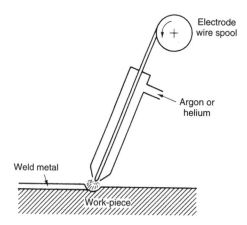

Figure 24.7 *Principle of MIG welding.*

which case a separate filler rod has to be fed into the arc to provide a source of weld metal, or a consumable metal electrode can be used. The former is termed the TIG (tungsten–inert gas) process (Figure 24.6) and the latter is known variously as MIG (metal electrode–inert gas), MAGS (metallic-arc gas shielded) or GMA (gas metallic arc) welding (Figure 24.7).

In both types of electrode holder (welding torch) the electrode is surrounded by a tube and nozzle, through which inert gas is fed and blown over the weld area. Inert gas welding techniques lend themselves to automation and automatic TIG and MIG welding processes are used to produce welds of consistent high quality.

When carbon dioxide, CO_2, is used as a shielding gas, dissociation of the CO_2 can occur at high temperatures according to the reaction

$$2CO_2 = 2CO + O_2$$

When using CO_2 as a shielding gas for the welding of steels the oxygen produced by the above reaction could oxidise some of the iron and the iron oxide thus formed could then react with some carbon in the molten steel weld pool: $FeO + C = Fe + CO$. This generation of carbon monoxide could lead to porosity in the final weld.

The CO_2 welding process is widely used for the welding of mild and low-alloy steels, but in order to produce consistent high-quality welds the electrode wire is of a modified composition. The welding wire must contain relatively high proportions of manganese and silicon as de-oxidising agents to remove FeO from the weld pool. In this way the CO_2 process can be both effective and economic.

There are some variants on TIG and MIG welding. The use of a pulsed d.c. in conjunction with an inert gas shielded tungsten arc, known as *pulsed arc welding*, can offer advantages for the welding of thin sheet material. A low current is used to maintain the arc and a square wave pulse is superimposed on this. As the welding torch is moved along the joint, a series of weld nuggets is made and these may be made overlapping or separate, depending on the speed of travel of the torch.

Plasma arc welding

This can be regarded as a specialised form of TIG welding. The plasma is a mass of extremely hot ionised gas formed by passing argon, or some other suitable gas, through a constricted electric arc. A constricting orifice in the plasma torch gives a stable, high-energy, concentrated arc. When the arc is squeezed, the voltage gradient, current density and heat transfer intensity all increase. An annular layer of gas flows around the arc and this relatively cool, non-conductive layer gives electrical insulation, stabilises the arc, and protects the water-cooled constricting nozzle. The plasma arc process gives a high energy concentration and, consequently, higher temperatures than in conventional arc processes. This gives deep penetration with a narrow heat affected zone, less distortion and permits fast welding speeds. Most metals and alloys can be welded by means of this process. The variant *plasma GMA welding* uses a consumable metal wire electrode fed through a nozzle which also guides a plasma stream and can be regarded as a specialised form of MIG welding. As with plasma arc welding, this process permits fast welding speeds.

24.4   Other fusion welding processes for metals

There are numerous other fusion welding processes, and some of these are briefly described below.

Electroslag welding

Electroslag welding was first developed in the former USSR as a means of joining large castings or forgings to produce large complex assemblies, for example, the housing for a large rolling mill or forging press, and the method has been used to successfully join pieces with sectional thicknesses of up to 1 m. The process, which can be used only in the vertical mode, is now used for welding material with thicknesses of 50 mm and greater (see Figure 24.8).

The two pieces to be joined are placed close together and water-cooled copper shoes are placed across the joint to make the third and fourth sides of a mould. A special flux is placed in the bottom of the joint and two welding wires dip below the surface of the flux. This is not an arc process. A circuit is completed through the flux when the electrode wires are energised. The heat generated in overcoming the resistance of the flux is sufficient to melt the flux into a protective slag and to start melting the ends of the electrode wires. As the process proceeds a pool of weld metal is produced beneath a protective slag layer. There is sufficient heat generated to begin melting the edges of the work-pieces so that total fusion occurs. As the weld metal solidifies in the lower portion of the joint so the water-cooled copper shoes are moved slowly up the sides of the joint. It is customary when welding very thick sections to oscillate the wires across the weld zone to achieve uniformity of temperature and weld deposition.

Electroslag welding can be used for welding plate thicknesses from about 10 mm upwards, but is used chiefly for sectional thickness greater than 50 mm. For the lower end of the thickness range *electrogas welding* is preferable and gives faster welding speeds. The main applications for electroslag welding are for the fabrication of thick-walled pressure vessels and the joining of castings or forgings to produce large structural assemblies.

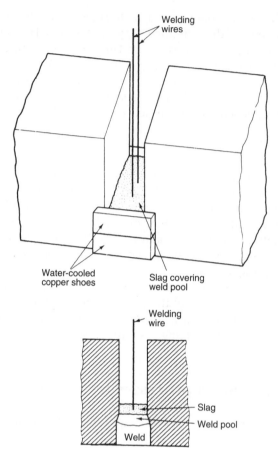

Figure 24.8 *The electroslag welding process.*

Electrogas welding

This is an automated shielded arc welding process for making vertical, or near-vertical, weld runs in steel plate of thicknesses ranging from 12 to 75 mm. It can also be used for joining plates with slightly curved surfaces. There is a superficial similarity to electroslag welding in that water cooled copper shoes are used to contain the weld metal within the joint. A flux-cored electrode is used to maintain the arc and this provides a thin protective slag cover. Additional shielding is given by carbon dioxide or argon. It is a rapid automatic method capable of consistently producing good welds. Typical applications for electrogas welding are for the construction of large gas and petroleum storage tanks and for shipbuilding.

Electron beam welding

The heat source here is a finely focused beam of high-energy electrons. The electron beam can be finely focused and the extent of heat penetration is confined to a narrow zone. The energy of an electron beam is dissipated rapidly when not in a vacuum. For much electron beam welding, the assembly to be welded is placed in a chamber which is then evacuated before the beam is energised and focused on the joint area (Figure 24.9). In high-vacuum electron beam welding, the chamber pressure is reduced to between 10^{-4} and 10^{-5} torr (0.0133–0.00133 Pa). The advantages of electron beam welding are that a deep penetration fusion weld with low distortion and a very thin heat affected zone is made. Almost

Figure 24.9 *Electron beam welding.*

all metals can be welded. The absence of contaminating gases means that reactive metals can be welded. Dissimilar metals may be joined to one another and it is possible to weld some ceramic materials – alumina, magnesia and quartz can be joined, both to themselves and to metals.

Electron beam welding in a partial vacuum and in a non-vacuum (in air or inert gas) has also been developed. Cycle times can be reduced using a partial vacuum chamber working under pressures in the range of 0.05 to 0.3 torr (6–40 Pa). A high voltage set has to be used if out-of-vacuum electron beam welding is to be carried out and the work-piece must be sited very close to the source of the beam.

There are many industrial applications for electron beam welding and these include nuclear fuel elements, gas turbine rotors, finned superheater tubing, gears, bellows, diaphragms and honeycomb structures.

Laser beam welding

A disadvantage of electron beam welding, namely the rapid dissipation of the beam in air, is not present when a high intensity laser beam is used as the energy source. The power outputs of lasers used for welding are in the range from 1 to 20 kW. The laser beam can be very finely focused and, like electron beam welding, this is a precision process with a deep penetration and little lateral spread of energy. In consequence, there is a very small heat affected zone. Welding may be carried out in air or under an inert gas atmosphere and high welding speeds are possible. It is possible to weld virtually all metals and joints between dissimilar metals can be made, for example, plain carbon steels can be welded to stainless steels, nickel or titanium. It can also be used for making welds in very small components and with very fine wires. Applications of the process include high precision watch and instrument components, bellows, diaphragms and dental bridgework.

Flash butt welding

Flash butt welding is a fusion process which was developed from resistance butt welding (see section 24.5). The parts to be joined are made part of an electrical circuit, brought lightly together, and then separated slightly to strike an arc. Arcing is continued until the surfaces are uniformly heated and beginning to melt. The current is switched off and the parts are forged together under pressure. Molten metal is expelled to form a ragged fin around the joint (Figure 24.10). The process is used for joining bars, tubing and sections. Dissimilar metals can also be joined by this process, as for example, in the joining of a mild steel stem to an alloy steel head in the manufacture of internal combustion engine valves.

A further development of this process is MIAB (magnetically impelled arc butt) welding and this is particularly suitable for the butt welding of hollow sections. A small concentrated arc is created between the two parts and, under the influence of a rotating magnetic field, this arc moves uniformly

(a) (b)

Figure 24.10 *Flash butt welding: (a) arc struck; (b) parts brought together under pressure.*

around the complete section at a frequency of 50 Hz. The arc is ceased when the surfaces begin to melt and the two parts are brought together under pressure to forge the weld. The decrease in length due to upsetting when the two parts are brought together is less, and more predictable, than in flash butt welding. Also the size of fin formed at the joint is less. This process is used as a fabrication technique for the manufacture of such assemblies as rear axle housings and propellor shafts for road vehicles.

Arc stud welding

It is frequently necessary to attach studs and fixing attachments to plate or sheet material. Arc stud welding is similar to flash butt welding in that an arc is struck between a stud and a plate for a pre-set time, usually between 150 and 500 ms. This is sufficient to melt the end of the stud and to form a small crater in the plate. The two are then brought into contact under pressure and a weld is made. The process is used for the placing of studs and fixing attachments with diameters ranging from 3 to 25 mm in steels, including stainless steels, brass and aluminium alloys for a wide range of applications in civil engineering, shipbuilding and general engineering. No elaborate surface preparation is necessary other than shot blasting or wire brushing.

Percussion welding

This process, which is also known as *spark discharge welding* or *capacitor discharge welding*, is used for joining wires to larger parts and for welding studs of up to 6 mm in diameter to sheet metal. The energy source is the sudden discharge from a bank of storage capacitors. For stud welding, the stud is shaped with a small cylindrical tip projecting from the end. Within the first millisecond of current discharge, the very high current (about 1000 A) causes the stud tip to vaporise and an arc is initiated. The arc causes some melting of the stud and sheet surfaces. The stud moves forward under a percussive force to make the weld and the whole process is complete in about 10 ms.

Thermit welding

In this form of welding the source of heat is the chemical reaction between iron oxide and aluminium:

$$8Al + 3Fe_3O_4 = 9Fe + 4Al_2O_3 \ (15.4 \ MJ/kg \ Al)$$

The thermit reaction can be used for the site welding of steel sections such as rails, in which a few kilogrammes of weld metal are required, and for the joining of very heavy sections where several tonnes of weld metal might be needed. It can also be used for the repair of large castings and forgings (see Figure 24.11).

The thermit powder, when ignited, produces pure liquid iron at a very high temperature (3500°C). In order to keep temperatures down and produce a liquid steel of similar composition to the steel being joined, steel scrap and alloying additions are made to the thermit mixture. When the reaction is complete the liquid steel produced is allowed to run into the prepared mould space.

Figure 24.11 *Thermit welding a rail.*

24.5 | Pressure welding of metals

The oldest form of pressure welding is forge welding in which two pieces of wrought iron are heated to red heat and hammered together to make a weld. The slag within the wrought iron acts as a flux. Similarly, mild steel can be forge welded with a little silica sand used as a flux. Most of the pressure welding processes used industrially today are of the electric resistance type. The range of pressure welding processes is shown in Figure 24.12.

Figure 24.12 *Pressure welding processes.*

Spot welding

In this process the parts to be joined are tightly clamped between a pair of electrodes and a large current passed through the metal for a short, but accurately controlled, interval of time (see Figure 24.13). The electrodes, which are made from a strong chromium–copper or beryllium–copper alloy, are hollow and water-cooled to prevent the possibility of welding between the electrode and the work-piece. The pressure exerted by the electrodes is generally in the range 70–100 N/mm^2, and the welding current density is up to 75 A/mm^2 at a voltage of 10–20 V.

The maximum resistance in the welding circuit occurs at the interface between the two parts to be joined. The total heat generated is given by I^2Rt, where I is the current, R the total resistance and t the welding time. R is made up of the resistance of the work-piece and interface resistances. Some of the heat generated is dissipated from the joint area by conduction through the work-piece. Spot welding is far more suited to steels and other materials with low electrical and thermal conductivities than to high-conductivity metals such as aluminium and copper.

A type of resistance spot welding which is suitable for welding metals of high conductivity and those which are difficult to weld by normal techniques is *magnetic force spot welding*. In this method, the force is applied to the joint electromagnetically, using the welding current.

Cooling
water

Weld
nugget

Cooling
water

Figure 24.13 *Principle of spot welding.*

A spot-welding cycle comprises three parts, firstly the bringing together of the electrodes on the work-piece, creating a pressure load within the material. Secondly, the welding current flows for a pre-set time. This is frequently of the order of 0.5 s. The welding time is accurately controlled by means of an electronic timer. The timers are generally capable of giving pre-set times of 0.1–2 s with an accuracy of ±0.01 s. During this part of the cycle the weld is made and some plastic deformation of the hot metal occurs as a result of electrode pressure. This shows as a surface depression on both sides of the work-piece. Finally, electrode pressure is maintained for a short time after the welding current has ceased. This aids cooling of the weld area and ensures formation of a fine crystal grain size within the weld nugget.

Spot welding is mainly used for the joining of thin-gauge sheet metal, generally mild steel, although other materials can also be joined by this method. It has become a major fabrication technique for the assembly of such items as pressed-steel radiator panels, car-body panels, washing machines, refrigerator cabinets and many others. Much factory-assembly work employs either specially designed multi-spot welding machines capable of making up to 100 simultaneous spots or computer-controlled robotic spot welders.

While spot welding is generally used in connection with light-gauge metal sheet, it is also possible, using heavy-duty equipment, to produce satisfactory spot welds in plate thicknesses of up to 25 mm and this has been used as an alternative to arc welding or rivetting for the assembly of girder frameworks for civil-engineering constructions.

Projection welding

This is a modified form of spot welding in which one of the parts to be joined is embossed before welding. Projection welding offers advantages over spot welding when the two parts to be joined differ in composition or thickness (see Figure 24.14).

Seam welding

This is a process for making a continuous seam between two overlapping sheets of metal. In this case the two electrodes are discs, between which the sheets to be joined are passed (see Figure 24.15). As the overlapping sheets move between the discs an intermittent current flows between the electrode discs through the work, creating a series of overlapping spot welds.

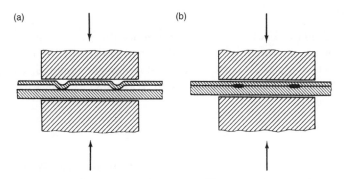

Figure 24.14 *Projection welding: (a) electrical circuit completed through embossed projections; (b) after welding current ceases pressure causes plastic deformation, flattening the projections.*

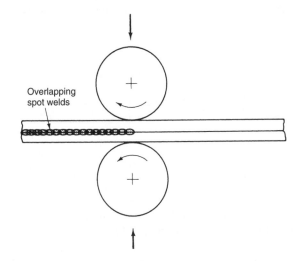

Figure 24.15 *Seam welding.*

Butt welding

In butt welding the two parts to be joined are held together under pressure and a current passed through the joint area until the temperature is sufficiently high to create a good pressure weld (see Figure 24.16). This method is used for joining tubes and rods. The surfaces to be joined must be clean, plane and parallel otherwise imperfect welding will occur.

Continuous seam-welded steel tubing is frequently made by the continuous butt weld process. In this technique flat steel strip is cold formed into cylindrical shape and passed through the welding head where a continuous welding current is supplied via conducting rolls. A high frequency source is required for this type of resistance butt welding. The electrical supply is at a voltage of about 100 V and at a frequency of 400 to 450 kHz.

Figure 24.16 *Butt welding: (a) parts ready for welding; (b) parts brought together and current passed.*

Induction welding

This is a solid-state welding process which has been used for the creation of a continuous butt weld in a seam welded steel tubing manufactured from cold formed steel strip. The necessary heat for welding is generated by the effects of electrical currents induced within the material. This process, while effective technically, is probably less efficient than the high frequency resistance welding process described above.

Friction welding

In this process one of the parts to be joined is rotated at a high angular velocity in a chuck, while the other part is held in a stationary chuck. The two parts are then brought together under pressure and seizure and welding occurs between the two surfaces owing to the frictional heat generated (see Figure 24.17). Friction welding has been successfully used for joining shafts of up to 0.3 m diameter. The method is also very suitable for joining dissimilar metals, for example, steel to titanium. This process is also known as *inertia welding* and *spin welding*.

Figure 24.17   *Friction welding.*

Explosive welding

Explosive welding was first developed in the late 1950s. It has been mainly used for cladding the surface of one metal with another, but it can also be used for making butt and lap welds. In the process one portion of metal hits the other at a very high velocity at an oblique angle causing welding between the two surfaces (see Figure 24.18). For explosive cladding the moving or 'flyer' plate is inclined at a small angle to a target plate resting on a firm base. A sheet of explosive is separated from the flyer plate

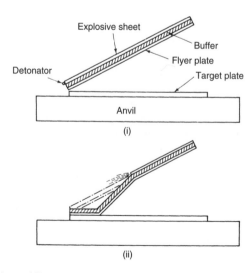

Figure 24.18   *Principle of explosive welding.*

Figure 24.19 *Sketch showing the microstructure of the ripple interface in an explosive weld.*

by a buffer sheet. The buffer sheet, made of a material such as PVC, protects the surface of the flyer plate from damage. When the explosive is detonated the flyer sheet is deformed and strikes the target at velocities of up to 300 m/s, generating pressures of up to 5000 N/mm^2. At these very high pressures the flyer plate and target plate surfaces behave as fluids and a weld interface with a ripple-type surface is produced (see Figure 24.19).

Examples of the explosive-cladding method are the bonding of a thin layer of aluminium, titanium or stainless steel to a mild steel base.

Cold pressure welding

Cold pressure welding is a solid-phase welding process carried out at ambient temperature by the application of pressure across the interface. The process is applicable for soft, ductile metals, particularly aluminium and copper. Both lap joints and butt joints can be made. This process is used for sealing aluminium cans and for joining wires and bar stock.

Ultrasonic welding

This is a method for producing spot welds in which the two pieces to be welded are clamped between an anvil and a probe, or *sonotrode*, which is vibrated at frequencies within the range 10 to 50 kHz. This causes a small lateral movement between the mating, or faying, surfaces which disrupts surface oxide films, giving an increase in temperature and creating a weld. Most metals can be joined and joints between dissimilar metals are possible. This method is used mainly for delicate work, including the making of electrical connections to electronic devices using fine aluminium or noble metal wires. The welding time may range from 5 to 10 ms for very fine wires up to about 1 s for thicker gauges.

24.6 Diffusion joining

Diffusion joining is a generic title which covers the three solid-state joining processes of *diffusion bonding*, *diffusion brazing* and *diffusion welding*. A common feature of these closely related processes is that the joint is made by diffusion across the interface at elevated temperatures *in vacuo* or in an inert atmosphere. The mating surfaces have to be prepared carefully by grinding or abrasive polishing. These methods were developed primarily for the fabrication of some of the newer high technology materials such as beryllium, nickel and cobalt based superalloys, and titanium alloys, particularly for aerospace and nuclear engineering applications.

Diffusion bonding is the process in which a very thin layer of another metal separates the two parts to be joined. This layer may be electroplated or sputtered onto the surfaces to be welded or may take the form of a thin foil. When the assembly is held at high temperature, this intermediate layer diffuses into both parts of the assembly and the final joint shows little or no evidence of the layer.

Diffusion brazing is similar to diffusion bonding in that an intermediate layer of another metal is placed as a filler in between the two parts to be joined. During the heating stage, there is an interdiffusion of filler and parent metal to the extent that the properties of the material in the joint approach those of the parent material.

Diffusion welding is the joining of two mating parts by high temperature diffusion but without the presence of a film or shim of another metal in the joint.

Dissimilar metals, some ceramics and some ceramic–metal joints can be made by the diffusion joining processes described above.

24.7 | The welding of plastics

Welded joints can be made in most thermoplastics and many can be joined using adhesives. Fully cured thermosets cannot be welded and any joints required are made using either adhesives or mechanical fastenings. The main welding methods for thermoplastics are given below.

Hot gas welding

Hot gas welding is a fusion welding process for thermoplastics which resembles the oxy-gas welding of metals but the heat source is a jet of heated gas rather than a flame. Air or an inert gas is passed through a torch, where it is heated by either gas or electrical heating, and used to melt the joint surfaces and a thermoplastic 'welding filler rod'. At the same time, pressure is applied at the joint helping to 'forge' the joint. The filler rod is of the same material as the parts to be welded. The bond strengths which are achieved are, typically, about 70 per cent of that of the parent material.

Hot plate welding

This method, also known as *heated tool welding*, relies on the application of heat and pressure. The parts to be joined are heated by an electrically heated resistance strip or a heated metal plate, the softened areas brought together and cooled under pressure. In some cases, a strip or rod of the thermoplastic can be used as a filler material in order to improve the joint quality. Materials most commonly joined by hot plate welding are polyethylene, polypropylene, PVC, ABS and ABS/polycarbonate alloys. The *heat sealing* of thin polymer film and film laminates, as for example, in the welding of polyethylene film in the manufacture of bags, is a type of hot plate welding.

Spin welding

This is another name for *friction welding* (see section 24.5). Generally, the term friction welding is reserved for the joining of metal parts and the welding of thermoplastics is referred to as spin welding. It is a welding technique whereby the frictional heat generated by the rapid rotation of one thermoplastic component against a second, stationary thermoplastic part causes localised surface melting of both parts. Usually, pressure is applied to the joint as soon as the spinning stops, thus producing a good weld. A typical example of the use of the process is the welding of rims to can bases in the manufacture of polypropylene industrial paint pots.

Vibration welding

The frictional heat generated at the mating surfaces of thermoplastic parts when one of the parts is subject to a low-frequency oscillation is sufficient to cause localised melting and the two parts fuse together. Vibrational frequencies of the order of 90 to 120 Hz are used and the method is used for the manufacture of butt joints. When very much higher frequencies are used, the process is termed *ultrasonic welding*.

Ultrasonic welding

This is similar in principle to vibration welding, described above, but vibrational frequencies in the range 10 to 50 kHz are employed. Ultrasound energy is transmitted into the thermoplastic component by means of a transducer assembly or horn and the small relative lateral movement at the mating surfaces causes local heating and melting of the thermoplastic. Two other processes involving the use of ultrasonic vibrations are *ultrasonic spot welding* and *ultrasonic staking*. The first process is used for making spot welds in thermoplastic sheet material. The most common type of equipment consists of a high-energy pistol with a vibrating pointed probe. In the welding of sheet the probe is forced through the top sheet and into the lower sheet. A flow of fused plastic from the top sheet into the cavity formed in the base gives a spot weld. The best results are obtained with polyethylene, polypropylene and high impact polystyrene. *Ultrasonic staking* is a method which can be used for the assembly of metal parts to polymers. A thermoplastic component with moulded studs is assembled to a metal part so that the plastic studs protrude through holes in the metal. The studs are then headed using ultrasonic energy, resulting in a rivet type joint.

High-frequency welding

Thermoplastics are dielectrics and dielectric heating occurs when they are placed in a high-frequency electromagnetic field. Although, theoretically, frequencies in the range 1 to 200 MHz are feasible, permissible wave bands have been specified by international agreement and, of these, the only practicable frequency available is 27.12 MHz. This form of welding is suitable for polyamides, PPVC, PVA and ABS.

Induction welding

The principle of electromagnetic induction heating can be used for the welding of thermoplastics. Although thermoplastics are insulators, weldable materials can be made by incorporating fine magnetic particles within the polymer. Generally, these are of the order of 1 μm in size. When subject to a high-frequency field the particles are heated causing a heating and melting of the surrounding plastic. The heated surfaces are then brought together under pressure to create the weld.

Solvent welding

Some thermoplastics can be softened and joined with the aid of specific solvents. The solvent partially dissolves the surfaces to be joined allowing them to fuse into one another when pressed together. Evaporation of the solvent leaves a sound bond. Acrylics, polystyrene and PVA are the main polymers which are joined in this manner.

24.8 Adhesive bonding

Adhesives, or glues, are polymeric materials which can be used to make joints between similar or dissimilar materials. The use of adhesive bonding offers a number of advantages, the main ones being that it can offer a low-cost assembly method, the joint is made at ambient or only slightly elevated temperatures, and it may be the only method for joining dissimilar materials. The nature of adhesion between an adhesive and a metal surface is not fully understood, but there is no alloying and it must rely on secondary bonding forces. The adhesive must fully wet the surface of the adherend, and surface preparation of a metal adherend may be necessary. Surface films and loosely bonded surface oxide need to be removed but a strongly bonded oxide coating, particularly if of a porous nature, can improve adhesion. A roughened surface may also help adhesion as it provides a greater surface contact area and can permit some mechanical interlocking. Surface treatment with some chemical compounds, termed adhesion promoters, can aid adhesion. In the case of porous materials, such as many ceramics, and fibrous materials, such as timber, the adhesive can penetrate into the pores or between fibres giving a good mechanical interlocking. Many adhesives tend to be relatively strong in both tension and shear but weak in cleavage. The design for adhesive joints needs to take account of this.

Animal and vegetable glues, made from hoof, bone and vegetable starches, have been used from earliest times. Today there is a very wide range of adhesives, many of them specific in application, based on both thermoset and thermoplastic polymer materials and they fall into five general categories. These are: solvent adhesives, dispersion adhesives, hot melt adhesives, thermosets, and monomers which polymerise *in situ*.

Solvent adhesives have been covered already under the heading *solvent welding* in section 24.7.

Dispersion adhesives are dispersions of a finely divided thermoplastic material in a fluid and the fine particles bond as the carrying fluid evaporates. This group includes water borne emulsions of PVA and plastisols, emulsions of PVC in an organic fluid. The former are only suitable for use with porous adherends, for example, paper, while the latter are used mainly for bonding PVC.

Hot melt adhesives are thermoplastics of low softening temperature. The polymer is heated to some temperature above its T_g and applied to the surfaces to be joined. Rapid cooling under pressure establishes a bond.

Several *thermoset* resins including phenolics, amino resins, polyesters and epoxies are used as adhesives. A resin and hardener mix will react together within the assembled joint. Epoxy materials possess excellent adhesion to most materials and this, coupled with their high strength, make them excellent structural adhesives.

There are several liquid *monomers* which can be used as adhesives, these polymerising *in situ*. Typical of this type are the cyanoacrylates which are low viscosity liquids which polymerise rapidly in the presence of moisture, which is usually present on the surface of an adherend. The so-called 'super-glues' are based on the cyanoacrylate family.

24.9 | Metallurgical considerations for welding

In general, the conventional oxy-gas and arc welding processes are used for joining parts of the same or very similar compositions. In order to obtain a good joint the filler metal should normally be of the same composition as the parts being joined so that, for example, a plain carbon steel filler or electrode is used for welding a plain carbon steel while for welding a 1 per cent chromium low-alloy steel a 1 per cent chromium steel filler would be used. However, for oxy-gas and unshielded arc welding the carbon content of the filler or electrode should be slightly higher than the carbon content of the steel being welded to compensate for the small carbon loss that occurs through oxidation during welding.

Weldability of metals

Not all materials can be easily welded by all processes. Mild steels are readily weldable by almost all processes but high carbon steels are difficult to weld with oxy-gas or unshielded arc techniques. Some of the carbon content of the steel may oxidise producing carbon monoxide which may be trapped within the weld pool giving porosity. A shielded arc process is much better for these steels. It is possible to weld stainless steels with an oxyacetylene torch, providing that a reducing flame and a suitable flux are used, but these days it is very unusual to use anything other than MIG or TIG welding for these materials. Similarly, aluminium and its alloys are generally welded by means of the TIG process even though they are readily weldable with oxy-gas using a reducing flame and a fluoride flux. If aluminium is gas welded it is most important that all traces of residual flux are removed from the material after welding as the fluxes are highly corrosive.

Structure of weldments

A fusion weld is a small casting and the actual weld area will tend to have the structure of a casting. During the welding process the temperature of the surrounding material will be raised and this may cause microstructural alterations in the area adjacent to the weld, the heat affected zone. Some examples are given below.

(a) Weld between two pieces of normalised mild steel (see Figure 24.20). The weld itself will solidify to give an initial coarse cast structure but rapid cooling through the critical temperature range will produce very fine ferrite and pearlite with a high strength. The heat affected zone will be that area adjacent to the weld in which the temperature rose to or above the critical temperatures. Here, with subsequent rapid cooling, a fine pearlite and ferrite structure will be formed. This is generally finer than the original normalised structure. The result is that the heat affected zone and the weld will usually have better mechanical properties than the parent metal.

(b) Weld in a low-alloy steel. Low-alloy steels possess a much greater hardenability than plain carbon steels and the rate of cooling experienced after welding may exceed the critical cooling velocities for these materials giving formation of brittle martensite in the weld and heat affected zones. A good flux cover will reduce cooling rates to minimise martensite formation but a post-weld heat treatment may be necessary.

(c) Weld in austenitic stainless steel. A different type of problem may occur with this type of steel. The normal heat treatment condition for an austenitic steel is water quenching from 1000°C to ensure that all carbon is in full solution in austenite. During welding grain-boundary precipitation of carbides may occur in that portion of the heat affected zone where the temperature was raised to 500–700°C, and this carbide precipitation can lead to subsequent grain-boundary corrosion of the type known as *weld decay*. This type of corrosion can be avoided by fully heat treating the component after welding. Alternatively, a more expensive welding quality stabilised stainless steel should be used (see section 16.13).

(d) Weld in a non-ferrous material such as aluminium or copper (see Figure 24.21). The weld zone will normally possess the coarse-grain structure of a small casting. Within the heat affected zone grain growth may have occurred. This means that both the heat affected zone and the weld zone will be softer and weaker than the basis material. The relative effects will be even greater if the original material was in the work hardened condition.

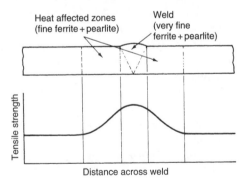

Figure 24.20 *Variation in properties across butt weld in mild steel.*

Figure 24.21 *Variation in properties across weld in non-ferrous metal.*

24.10 | Defects in welds

With the correct choice of welding method and good welding practice, defects within welds should not occur. However, nothing in this world is perfect and defective welds are sometimes made. The most common types of welding defect are:

(a) Lack of penetration. Cause: bad welding practice.

(b) Oxide and slag inclusions. Cause: incorrect choice of welding method or poor execution of weld.

(c) Gas porosity. Cause: incorrect choice of method or poor execution of weld.

(d) Distortion and residual stresses. Distortion of parts during welding may be caused by poor design of parts for welding or it may be caused by bad welding practice. Residual stresses may frequently occur in weldments after cooling. Occasionally, owing to poor design for welding, the stresses may be sufficiently large to cause immediate cracking. For many welds, particularly in components of complex shape, it is necessary to give the welded article a post-weld stress-relief heat treatment.

SUMMARY

A wide range of processes is available for the joining of materials, including the use of adhesives, soldering, brazing and a variety of welding techniques embodying fusion and solid-state methods.

Both soldering and brazing involve the use of a fusible filler material but the parts to be joined are not melted during the process. There is little, if any, change to the structure of the parent material and distortion and thermally induced stresses are at a minimum.

In fusion welding a molten filler material is used and some of the parent material is melted during the process. The weldment is, in effect, a small casting. Heat flow from the joint area into the parent material will cause microstructural changes in part of the parent material, the heat affected zone. Heat distortion and thermally induced stresses are likely and post-weld heat treatments may be necessary.

In pressure welding there is no filler material but there will be some melting at the weld interface. In most pressure processes the extent of the heat affected zone is small.

Many of the welding processes, both fusion and pressure, can be fully automated and joints of consistently high quality can be made.

Most metals and thermoplastic materials can be welded but diffusion joining may offer the only practicable method for some metals and for metal/ceramic joints.

Adhesive bonding can be used with most materials, including thermosets, ceramics and glasses.

Questions

24.1 (a) Distinguish clearly between oxy-gas welding and gas brazing as joining methods for low carbon steels.

(b) Discuss the main advantages of brazing for effecting a joint between two steel parts.

24.2 (a) Discuss the use of an oxyacetylene torch as a heat source for fusion welding. Include in your answer the flame characteristics and how these may be varied.

(b) Explain why it is necessary to use a different type of flame when gas welding aluminium as compared with mild steel.

24.3 (a) Explain the meaning of the terms DCEP and DCEN in connection with arc welding.

(b) In what welding situations would it be preferable to use DCEP rather than DCEN?

24.4 Discuss the principles involved in making spot welds to join mild steel sheet material. Why is it necessary to maintain electrode pressure for a period after the cessation of the welding current? Sketch the microstructure across the completed weld nugget.

24.5 (a) It is required to make a manual arc Vee-butt weld for a length of 0.3 m between two 20 mm thick plates of 0.25 per cent carbon non-alloy steel.
    (i) What electrode polarity should be used?
    (ii) What type of welding electrode should be used?
    (iii) Should the weld be made in a single pass or not?
    (iv) What post-weld treatment may be necessary?

    (b) The same quality steel as in (a) is to be welded in a fabrication shop where many butt welds need to be made, each weld being in the horizontal mode and 10 m in length. Suggest, giving reasons, how this may be done.

24.6 Suggest, giving reasons for your choice, suitable joining processes that could be used for each of the following:
    (a) butt joints between 6 mm thick plates of an Al–Mg alloy,
    (b) repair of a fractured grey iron casting,
    (c) joints between lengths of 13 mm diameter copper pipe,
    (d) joining two lengths of polypropylene pipe of 200 mm OD and 190 mm ID, and
    (e) joining lengths of railway track.

24.7 Describe the variation in microstructure and properties that exist across completed butt welds in
(a) low-carbon steel and (b) aluminium alloy 2024. What post-weld treatments, if any, may be needed in both of these cases?

24.8 A large pressure vessel, 5 m × 2 m diameter, is to be fabricated from 20 mm thick plate of 0.25 per cent carbon steel, roll formed into a cylindrical shape, and two forged dish-shaped ends in the same quality steel. A longitudinal seam weld and two circumferential welds are needed to complete the vessel (see Figure 24.22). Suggest, giving reasons, and describe, with the aid of sketches, a suitable welding technique for making these welds.

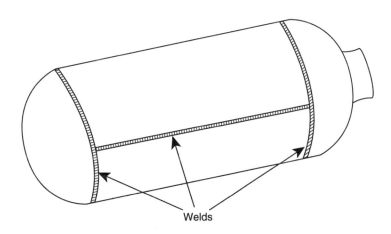

Welds

Figure 24.22

PART V

BEHAVIOUR IN SERVICE

CHAPTER 25

Failure, fatigue and creep

Why learn about failure?

It is the aim of engineers and designers to produce mechanisms and structures that will not fail catastrophically in service. The collapse of a bridge or tower block, or the break up of an aircraft in flight could result in many fatalities. While it may never be possible to prevent all catastrophic failures it must be the aim to reduce the chance of such failures occurring to an absolute minimum. This can only be achieved if engineers and designers possess a good understanding of the various mechanisms of failure. The general principles of ductile and brittle failure under static loading were discussed in Chapter 10. In many real-life situations structures and mechanisms are subject to dynamic loads, with the level of applied stress varying in some cyclical pattern. Such conditions can lead to failure by fatigue. At elevated temperatures materials subject to a constant level of stress may continue to strain slowly, or creep, and eventually fail. The mechanisms of fatigue and creep, and the various factors that influence these phenomena, are discussed in this chapter.

LEARNING OBJECTIVES

1. Define failure by fatigue.
2. Define alternating, fluctuating and repeating stress cycles and be able to use the empirical relationships between stress range, mean stress and fatigue life.
3. Be able to use Miner's law to estimate fatigue life when more than one loading cycle is involved.
4. Understand the mechanisms of crack nucleation and growth and be able to calculate the time taken for a crack to grow to critical size.
5. Define creep and know the relationships between applied stress, temperature and strain rate.
6. Use the relationships to determine limiting strain rates, stress and temperature values.

25.1 | Failure

A material, component or structure is deemed to have failed when its ability to fully satisfy the original design function ceases. This may be due to a variety of causes. It may be due to fracture, either partial or complete, plastic buckling, dimensional change with time, loss of material by corrosion, erosion or abrasive wear, or an alteration of properties and characteristics with time due to environmental or other effects. These various modes of failure may be dependent on stress, time, temperature, environment or a combination of any of these.

The conditions for failure by brittle fast fracture were considered in Chapter 10. Failure by fatigue is stress/time dependent and the effect of a cyclical stress system is to cause nucleation and propagation of cracks within the material. Final failure is by fast fracture when one or more cracks have grown to critical size.

Creep is a continued straining of a material under constant stress conditions and is stress/time/temperature dependent. The creep process will lead to eventual fracture but a component subject to creep may be deemed to have failed when its dimensions have changed by some pre-determined value.

25.2 | Fatigue

If a material is subjected to repeated, or cyclic, stressing, it may eventually fail even though the maximum stress in any one stress cycle is considerably less than the fracture stress of the material, as determined by a short-term static test. This type of failure is termed *fatigue failure*. (The term fatigue is sometimes used to describe the failure of a plastic material in a long-term static load test although this would be more correctly described as creep.)

Very many components are subjected to alternating or fluctuating loading cycles during service, and failure by fatigue is a fairly common occurrence. The mechanism of fatigue in metals has been thoroughly investigated. When a metal is tested to determine its fatigue characteristics, the test conditions usually involve the application of an alternating stress cycle with a mean stress value of zero. The results are plotted in the form of an *S–N* curve (Figure 25.1), where *S* is the maximum stress in the cycle and *N* is the number of stress cycles to failure. Most steels show an *S–N* curve of type (i), with a very definite *fatigue limit*, or *endurance strength*. This fatigue limit is usually about one-half of the value of the tensile strength, as measured in a static test. It means that if the maximum stress in the stress cycles is less than this fatigue limit, fatigue failure should never occur. Many non-ferrous materials show *S–N* curves of type (ii) with no definite fatigue limit. With these materials it is only possible to design for a limited life, and a lifing of 10^6 or 10^7 cycles is often used.

Figure 25.1 *S–N curves for (i) metal showing fatigue limit, (ii) metal showing no fatigue limit.*

There are, essentially, two forms of fatigue failure, *low-cycle fatigue* and *high-cycle fatigue*. In the former case, the maximum stress in any cycle, while below the tensile strength of the material, is greater than the yield stress and the number of stress cycles to failure is low, generally less than 1000. High cycle fatigue, which requires some 10^5–10^6 cycles, is the result of a stress regime in which the maximum stress in a cycle does not exceed the yield stress of the material.

Failure by fatigue is the result of processes of crack nucleation and growth or, in the case of components which may contain a crack introduced during manufacture, the result of crack growth only brought about by the application of cyclical stresses. This is evident from the appearance of a typical fatigue fracture surface. The appearance of a fatigue fracture surface is distinctive and consists of two portions, a smooth portion, often possessing conchoidal, or 'mussel shell', markings showing the progress of the fatigue crack up to the moment of final rupture, and the final fast fracture zone (Figure 25.2). The striations visible in the smooth portion, sometimes referred to as 'beach marks', indicate stages in the growth of the crack. The final fracture is usually of the cleavage type.

There are several types of stress cycle and cycles are described as *alternating*, when the mean stress, σ_m, is zero, *repeating*, when the minimum stress, σ_{min}, is zero, and *fluctuating*, when the mean stress, σ_m, has some value other than zero. These cycle types are shown in Figure 25.3. The stress range of a cycle, $\Delta\sigma$, is $(\sigma_{max} - \sigma_{min})$, the cyclic stress amplitude, $\sigma_a = \frac{1}{2}(\sigma_{max} - \sigma_{min})$, and the mean cycle stress, $\sigma_m = \frac{1}{2}(\sigma_{max} + \sigma_{min})$.

Much fatigue testing is carried out using alternating stress cycles in which σ_m is zero but, in many practical situations where components or structural assemblies are subject to cyclical stresses the mean stress, σ_m, is not zero. Several *laws* of an empirical nature have been proposed to give the relationships

Figure 25.2 *Fatigue failure of large shaft showing the two zones.*

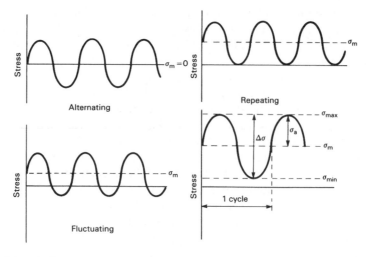

Figure 25.3 *Types of stress cycle.*

between stress ranges, mean stresses and fatigue life. These are:

The modified Goodman equation: $\dfrac{\sigma_a}{\sigma_{FL}} + \dfrac{\sigma_m}{\sigma_{TS}} = 1$

The Gerber parabolic equation: $\dfrac{\sigma_a}{\sigma_{FL}} + \left(\dfrac{\sigma_m}{\sigma_{TS}}\right)^2 = 1$

The Soderberg equation: $\dfrac{\sigma_a}{\sigma_{FL}} + \dfrac{\sigma_m}{\sigma_y} = 1$

In the above equations, σ_{FL} is the fatigue strength as determined in tests with a mean stress of zero, σ_{TS} is the tensile strength, and σ_y is the yield strength of the material. These empirical laws are not very precise but they are used for some preliminary design calculations.

EXAMPLE 25.1

A steel having a tensile strength of 500 MPa and a fatigue limit of 250 MPa is to be subjected to a fluctuating stress regime in which the mean stress in tension is 80 MPa. Using the Goodman relationship, determine the maximum and minimum stresses in the stress cycles if failure by fatigue is to be avoided.

SOLUTION

The Goodman relationship is $\sigma_a/\sigma_{FL} + \sigma_m/\sigma_{TS} = 1$
so $\qquad\qquad\qquad \sigma_a = (1 - \sigma_m/\sigma_{TS}) \times \sigma_{FL}$
Substituting $\qquad\qquad \sigma_a = (1 - 80/500) \times 250 = 210\,\text{MPa}$
and so the maximum stress in a cycle should be $\sigma_m + \sigma_a = 80 + 210 = 290\,\text{MPa}$
and the minimum stress should be $\sigma_m - \sigma_a = 80 - 210 = -130\,\text{MPa}$ (compressive).

There are also many practical situations in which a component or structure may be subjected to more than one type of loading system during its lifetime with the various stress cycles having different stress ranges, $\Delta\sigma$. Miner's law of cumulative fatigue is an empirical rule which can be used to estimate the fatigue life of a component when subjected to a series of different loading cycles. For example, if stressed for n_1 cycles in a regime which would cause failure in a total of N_1 cycles and for n_2 cycles in a regime where failure would occur after N_2 cycles, then n_1/N_1 of the fatigue life of the component would be used up in the first case and the fraction n_2/N_2 of the fatigue life used in the second instance. Miner's rule can be expressed by stating that failure will occur when $\Sigma(n/N) = 1$.

EXAMPLE 25.2

Tests on a steel show that its fatigue life is 10^6 cycles at a maximum stress of 200 MPa, 10^5 cycles at a maximum stress of 230 MPa and 5×10^4 cycles at a maximum stress of 240 MPa.

A shaft made of this steel is subjected to 10,000 cycles at a maximum stress of 240 MPa followed by a further 30,000 cycles at a maximum stress of 230 MPa. What is the remaining life of the shaft if it is then to be operated at a maximum stress of 200 MPa?

SOLUTION

Using Miner's law $\Sigma(n/N) = 1$
we have $(10^4/5 \times 10^4) + (3 \times 10^4/10^5) \times (x/10^6) = 1$
so $x = (1 - (0.2 + 0.3)) \times 10^6 = 5 \times 10^5$
At a maximum stress of 200 MPa the shaft should withstand a further 5×10^5 cycles.

25.3 | Crack nucleation and growth

In low-cycle fatigue, the maximum stress in a cycle is greater than the yield stress and some plastic deformation occurs. The process of slip creates slip bands which roughen the surface. This effect creates points of stress concentration and a crack is initiated which, in the first instance, grows along a slip plane.

Although the maximum stresses in high-cycle fatigue conditions are nominally below the elastic limit of the material, it has been established that some plastic slip takes place. During continued cyclic stressing, slip bands appear on the material and some extrusive and intrusive effects are associated with the slip bands (Figure 25.4). The extrusions and intrusions so formed are extremely small, being of the order of 1 μm in size, but once an intrusion has formed on a highly polished surface it can act as point of initiation for a fatigue crack. The intrusion, with a very small root radius, acts as a point of stress concentration and a crack can propagate through the material until, eventually, the remaining sound portion of the section is too small to sustain the maximum load in a stress cycle and sudden fast fracture of the component occurs.

Figure 25.4 *Intrusions and extrusions formed in early stages of fatigue.*

The process of fatigue crack nucleation by the generation of intrusions and extrusions is not always necessary. The starting points for fatigue crack development may be present already. For example, the presence of a surface flaw or an accidental scratch on a polished surface, or the troughs of a machined surface will all be points for the initiation of fatigue cracks.

Cracks, once initiated, propagate in materials by one of two methods, either by ductile tearing or by cleavage. In the former process, which applies to ductile metals and many thermoplastic materials, a lot of energy is absorbed in plastically deforming the material just ahead of the advancing crack. The plastic deformation tends to blunt the crack tip, so reducing its stress concentrating effect. However, during the severe plastic deformation, small cavities tend to form in the deformed area ahead of the crack and these eventually link up, so extending the crack length. In non-ductile materials such as ceramics, glasses, brittle metals, thermoplastics at temperatures below their T_g values, and thermosets, little, if any, plastic deformation can occur ahead of a crack tip to blunt the crack. When the local stress at the crack tip exceeds the fracture stress of the material, the interatomic bonds will be broken and the crack develops catastrophically as a fast brittle fracture.

Once fatigue crack initiation has occurred the crack will continue to grow in size during cyclic stressing. The rate of crack growth, da/dN, is related to a cyclic stress intensity factor, K, by an empirical relationship

$$da/dN = A\ \Delta K^m \tag{1}$$

where A and m are constants for the material. ΔK is the range of stress intensity factor in a stress cycle $(K_{max} - K_{min})$. The exponent m usually has some value between 1 and 6 (the value of m for most steels is 3) and, for most materials, the value of A is of the order of 10^{-11}–10^{-12}.

The relationship given above can be rewritten in logarithmic form:

$$\ln\ (da/dN) = \ln A + m \ln \Delta K$$

A plot of $\ln\ (da/dN)$ against $\ln \Delta K$ gives a curve as shown in Figure 25.5. It is in three sections. In section I, with low values of stress intensity factor ΔK, cracks will not grow during cyclic stressing. Section II shows a linear relationship between $\ln\ (da/dN)$ and $\ln \Delta K$, the slope of this section being m, while section III shows that crack growth rate is accelerating, this being the situation immediately prior to final rapid fracture.

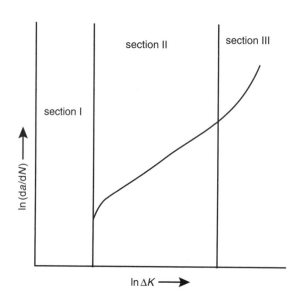

Figure 25.5

The depth of surface crack, a, stress intensity factor, ΔK, and stress range, $\Delta\sigma$, are all related by the expression

$$\Delta K = M\Delta\sigma(\pi a)^{1/2} \tag{2}$$

This is the same as the general fracture mechanics equation (see section 10.4). The term M in the equation is a geometry factor to take into account finite boundaries and varying flaw shape. It has a value of 1 for a centre crack in an infinite plate. If $\Delta\sigma$, the cyclic stress range, remains constant, then as the stress intensity factor, ΔK, increases, so also will the crack depth, a, increase until they both reach a critical value, at which time catastrophic fast fracture will occur. (Note that a is the length of a surface crack, but half the length of an internal crack.) We can use these relationships both to determine the critical crack size for fast fracture and also to estimate the number of stress cycles needed for a crack to grow to the critical size. Combining equations (1) and (2) above we have

$$da/dN = A(M\,\Delta\sigma(\pi a)^{1/2}))^m = AM^m\,\Delta\sigma^m\,\pi^{m/2}\,a^{m/2}$$

For a constant cyclic stress range, $\Delta\sigma$, the only variables are a and N. Separation of these gives

$$a^{-m/2}\,da = AM^m\,\Delta\sigma^m\,\pi^{m/2}\,dN \tag{3}$$

If this equation is integrated between the limits of a_1, the initial crack size, and a_c, the critical crack size for fast fracture, and, for the right-hand side, from zero to N_f, the number of cycles for failure, the result is

$$\frac{2m}{2-m}\,\{a_c^{1-m/2} - a_1^{1-m/2}\} = AM^m\,\Delta\sigma^m\,\pi^{m/2}\,N_f \tag{4}$$

Note that this integration result does not apply when $m = 2$. In this case the result is

$$\ln(a_c - a_1) = AM^2\,\Delta\sigma^2\,\pi N_f \tag{5}$$

If the critical defect size is known the number of cycles to failure can be determined using equations (4) or (5).

EXAMPLE 25.3

An aluminium plate is subject to repeating stress cycles ranging from 50 MPa in compression to 120 MPa in tension. It is found to contain an edge crack of size 3 mm. Calculate the remaining fatigue life of this plate if the fracture toughness (stress intensity factor) of the material is 30 MPa m$^{1/2}$. Consider the plate as semi-infinite and take the value of M as 1.12. Also take the exponent m as 3 and the constant A as 10^{-12}.

SOLUTION

It is necessary first of all to determine the critical crack size, a_c, for fast fracture.
Use equation (2) $\Delta K = M\,\Delta\sigma\,(\pi a)^{1/2}$
This can now be reordered as $a_c = \Delta K^2/M^2\,\Delta\sigma^2\,\pi$
The stress intensity factor is 30 MPa m$^{1/2}$. For $\Delta\sigma$ we can take the value of 120 MPa, the maximum tensile stress in the cycle, as no crack extension will occur when there is a compressive stress operating.

So $a_c = 30^2/(1.12^2 \times 120^2 \times \pi) = 0.0159\,m = 15.9\,mm$

Now using this value of a_c in equation (4) and substituting the values of $m = 3$, $A = 10^{-12}$ and $\Delta\sigma = 120$ MPa, we have

$$\frac{6}{(2-3)}\{(15.9 \times 10^{-3})^{-1/2} - (3 \times 10^{-3})^{-1/2}\} = 10^{-12} \times 1.12^3 \times 120^3 \times \pi^{1.5} \times N_f$$

$$6\{18.257 - 7.930\} = 61.96 = 10^{-12} \times 1.405 \times 1,728,000 \times 5.568 \times N_f$$

from which $N_f = 4.583 \times 10^6$ cycles

The remaining fatigue life of the plate, that is the time for the crack to grow from 3 mm to the critical size of 15.9 mm, will be 4.583×10^6 cycles.

25.4 Factors affecting fatigue

There are many factors that affect the fatigue strength of a material, and these include surface condition, component design and the nature of the environment. Specimens for fatigue testing are usually prepared with a highly polished surface, and this condition will give the best fatigue performance. The fatigue limit for highly polished steels is approximately one-half of the tensile strength. If the surface of the specimen contains a scratch or notch, or is a ground rather than a polished surface, the fatigue limit of the material will be reduced. The presence of scratches or notches will provide small defects from which fatigue cracks can be initiated. Similarly, a sharp change in section with a small fillet radius can act as a stress raiser, and fatigue cracks can commence from such points. Keyways and oil holes in shafts are often points at which fatigue commences. The effect of a notch or scratch is not the same for all materials. As in the case of brittle failure, a ductile metal is much less sensitive to the presence of surface flaws than a brittle material. If conditions are such that corrosion can occur, not only is the fatigue limit very greatly reduced but also the rate of corrosion is increased. For some materials, including some steels, there is no fatigue limit in a corrosive environment, and failure will eventually occur, even when the stress levels are very low.

The resistance of a metal to failure by fatigue will be highest when the surface of the metal is highly polished and free from scratches and other defects. The fatigue life may be improved by treating the metal surface by *peening*. Peening involves lightly hammering the surface of the material with a round-nosed hammer, or by bombarding the surface with small steel pellets. The result of the peening treatment is to induce residual compressive stresses within the surface layers of the material.

The rate of stress cycling has no effect on the fatigue strength for frequencies up to about 150 Hz, but at higher frequencies there appears to be a small increase in the fatigue strength. This increase is of the order of 10 per cent at frequencies of up to 15,000 Hz.

Materials other than metals are also subject to failure by fatigue, but comparatively little work has been done in this area. For concrete and polymers, as with metals, the number of stress cycles necessary for failure is increased as the maximum stress in the loading cycles is decreased, but there does not appear to be a definite fatigue limit with these materials. There are difficulties in the fatigue testing of polymers because, owing to the low thermal conductivities and high damping capacities of these materials, there is an increase in the temperature of a polymer test-piece during a test.

Composite materials are also subject to fatigue. The growth of a crack in a composite material in a direction normal to the line of reinforcing fibres may be stopped at a fibre. The high stress concentrated at the crack tip may cause a rupture of the bond between matrix and fibre, thus blunting the crack tip so that fracture cannot proceed into the fibre. This does not happen if crack development is parallel to the fibre direction. Aligned (uni-axial) fibre composites may give S–N curves with a fatigue limit and often show, under cyclic loading conditions, a behaviour which is superior to metals such as aluminium. The greatest resistance to fatigue damage is provided by carbon and boron fibre reinforcement, rather than glass fibre reinforcement, since the higher elastic moduli of the former limits the amount of matrix strain. Random fibre composites usually exhibit larger matrix strains in the direction of applied stress and, hence, do not perform so well. True fatigue effects are only observed in polymer matrix composites at cyclic frequencies less than about 30 Hz, since cyclic frequencies of greater than 30 Hz lead to heating and thermal degradation of the polymer.

25.5 Creep

Creep is the continued slow straining of a material under the influence of a constant load. The amount of strain developed in a material becomes a function of time and temperature, as well as of stress. The phenomenon of creep can occur in all types of materials but there is a threshhold temperature, below which creep does not become a factor in stress–strain relationships. In the case of thermoplastic materials, creep is of significance at all temperatures above the T_g value. In metals and ceramics, creep is considered as a high temperature phenomenon, although lead creeps at room temperature, and occurs at temperatures above about 0.3–0.4 T_m in metals and above 0.4–0.5 T_m in ceramics, where T_m is the melting temperature in Kelvin.

A typical creep curve of strain against time at constant stress is shown in Figure 25.6. This shows three stages of creep.

In the first stage, *primary* or *transient creep*, the rate of creep strain is initially high but steadily reduces to the constant rate of the second stage, *secondary* or *steady-state creep*. Eventually this moves into the final stage, *tertiary creep*, where the creep rate increases and leads rapidly to failure. A number of mechanisms have been postulated for creep in polycrystalline materials but they fall into two main categories – those involving the lattice and those involving grain boundaries. Lattice mechanisms involve vacancy diffusion, dislocation climb and mobility (see section 8.11) while grain boundary sliding may also be important, particularly where second phases are present at grain boundaries.

At a constant stress level, the rate of secondary creep increases with increased temperature according to an Arrhenius type relationship:

$$d\varepsilon/dt = K\exp(-B/T)$$

where $d\varepsilon/dt$ is strain rate, T is temperature (Kelvin), and K and B are constants for the material.

The effect of increasing temperature on the creep rate is shown in Figure 25.7. It will be seen that there is primary creep only, with a secondary creep rate of zero, at a relatively low temperature, but at a very high temperature the primary stage merges in with tertiary creep.

At a constant temperature the creep rate is stress-dependent and follows a power law, the relationship being

$$d\varepsilon/dt = C\sigma^n$$

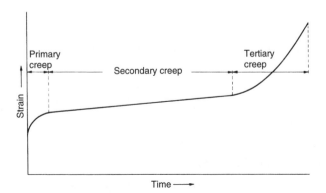

Figure 25.6 *Typical creep curve.*

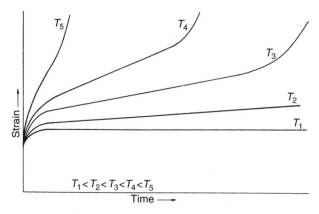

Figure 25.7 *Variations of creep rate with temperature.*

where C is a constant for the material and n is another constant, the time exponent. For many materials, the value of the exponent, n, lies between 3 and 8, while for glasses it is of the order of 10.

The above two relationships can be combined, to give a general creep equation:

$$d\varepsilon/dt = A\sigma^n \exp(-B/T)$$

EXAMPLE 25.4

When tested at a stress of 100 MPa, the steady-state creep rate of a material is $1 \times 10^{-4}\,s^{-1}$ at 900 K and is $7.5 \times 10^{-9}\,s^{-1}$ at 750 K. Calculate the maximum working temperature if the limiting creep rate is $1.0 \times 10^{-7}\,s^{-1}$.

SOLUTION

The Arrhenius expression $d\varepsilon/dt = A\exp(-B/T)$ applies.

$$1.0 \times 10^{-4} = A\exp(-B/900) \tag{1}$$
$$7.5 \times 10^{-9} = A\exp(-B/1750) \tag{2}$$

Dividing (1) by (2) we have

$$(1.0 \times 10^{-4})/(7.5 \times 10^{-9}) = 13.33 \times 10^3 = \exp\{B(1/750 - 1/900)\}$$

from which $B = 42.74 \times 10^3$

Substituting this value in equation (1) gives $A = 4.21 \times 10^{16}$

Let the temperature at which the creep strain rate is $1 \times 10^{-7}\,s^{-1}$ be T.

Then $1.0 \times 10^{-7} = 4.21 \times 10^{16} \exp(-42.74 \times 10^3/T)$

or $\ln(1.0 \times 10^{-7}) = \ln(4.21 \times 10^{16}) - 42.74 \times 10^3/T$

so $T = (42.74 \times 10^3)/(38.284 + 16.118) = 786\,K$

The maximum working temperature at a stress of 100 MPa, if the creep strain is not to exceed $1.0 \times 10^{-7}\,s^{-1}$, is 786 K.

EXAMPLE 25.5

Creep tests on a stainless steel at a temperature of 550°C produced a strain of 0.12 after 300 hours when subjected to a stress of 350 MPa and a strain of 0.08 after 1200 hours when stressed at 245 MPa. Assuming steady-state creep, calculate the time required to produce 0.1 per cent strain in a bar of the same material when stressed at 75 MPa at 550°C.

SOLUTION

At 550°C and a stress of 350 MPa the creep rate is $0.12/300 = 4 \times 10^{-4}$ hr^{-1}, and at 550°C and a stress of 245 MPa the creep rate is $0.08/1200 = 6.67 \times 10^{-5}$ hr^{-1}.
Using the expression $d\varepsilon/dt = C\sigma^n$ we have

$$4.00 \times 10^{-4} = C \times 350^n \text{ or } \log(4.00 \times 10^{-4}) = \log C + n \log 350 \tag{1}$$

$$\text{and } 6.67 \times 10^{-5} = C \times 245^n \text{ or } \log(6.67 \times 10^{-5}) = \log C + n \log 245 \tag{2}$$

Subtracting (2) from (1), we have $0.778 = 0.155n$
from which $n = 5.02$ and $\log C = -16.17$.

For a stress of 75 MPa let the strain rate be x hr^{-1}

So we have $\log x = -16.17 + (5.02 \times \log 75)$

giving $\log x = -6.757$ or $x = 1.75 \times 10^{-7}$ hr^{-1}

0.1 per cent strain is a strain of 0.001. At a creep rate of 1.75×10^{-7} hr^{-1} it will require a time of $0.001/1.75 \times 10^{-7} = 5714$ hours to produce a total strain of 0.001.

The change from secondary to tertiary creep is caused by the intervention of some other effect such as the formation of small internal cavities or localised necking of the material. When such effects occur, the effective stress in the fissured or necked zone will increase and cause an increase in the rate of strain. This is the onset of tertiary creep and, usually, there is a rapid acceleration in the creep strain rate leading to failure.

A creep limit, as such, does not exist. However, for design purposes, it is useful to know the stress/temperature combinations which will give some limiting value of total creep strain, say 10^{-3} in a time of 1000 hours or 10,000 hours, or which will give some limiting rate of creep strain, say 10^{-5} or 10^{-6} per hour. Summary curves, showing data derived from very many creep tests, are used to present information in these forms (Figure 25.8).

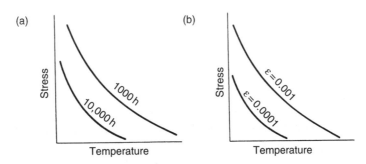

Figure 25.8 *Summary creep curves: (a) stress-rupture; (b) limiting strain.*

When creep occurs in ceramics under a tensile load the overall strain is usually less than 1 per cent and failure often occurs during steady state creep, before the tertiary creep stage is reached. However, under compressive stress, tertiary creep, with an increasing creep rate, is observed before final failure. In glasses, creep is due to the viscous flow of a Newtonian fluid of high viscosity. Creep of thermoplastic polymers also occurs by a process of viscous flow. The intermolecular bonding forces between the molecular chains are relatively weak van der Waal's forces and, even at low levels of stress, creep can occur in many thermoplastics. The creep resistance of polymers is increased substantially by incorporating glass or carbon fibre reinforcement

Materials with a high resistance to creep are needed for the 'hot' end of gas turbine engines. The requirement for the development of engines of greater thermal efficiency and higher operating temperatures provided the stimulus for research into creep and into the development of better creep-resistant alloys. Alongside this there have been also major developments in processing, for example, vacuum melting and casting, including the production of single-crystal castings. Refer to sections 15.13 and 20.6.

Stress-rupture curves for a number of metals and alloys are shown in Figure 25.9.

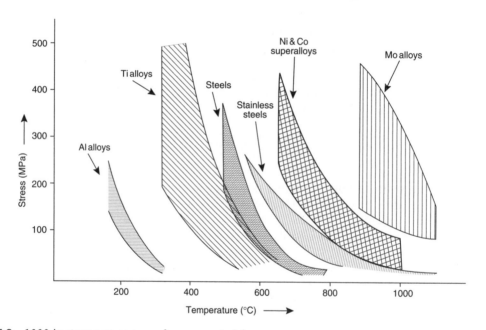

Figure 25.9 *1000-hr stress-rupture curves for some materials.*

25.6 | Relaxation

Relaxation is a phenomenon related to creep and is the slow reduction of stress with time in a material which is subject to a constant value of strain. This will occur in metals and ceramics at elevated temperatures but in thermoplastics it will often take place at room temperature. The rate of relaxation is exponential and the *relaxation time* is the time required for the level of stress to fall to $1/e$ of the original value. The general relationship for relaxation at a constant temperature is

$$\sigma_t = \sigma_0 \exp(-At)$$

where σ_0 is the initial stress, σ_t is the stress after time t, and A is a constant for the material.

An example of relaxation occurring within a material during service is the level of stress in a bolt. The bolt is strained to a particular level when the nut is initially tightened. During service, the total strain in the bolt remains constant but, if the temperature is such as to permit creep, then stress relaxation will occur within the bolt and retightening of the nut will be required at intervals.

SUMMARY

A common mode of failure is the type termed *fatigue*, and this may occur when a component or structure is subjected to a series of alternating, fluctuating or repeating cycles of stress, even though the maximum level of stress in any one cycle is below the yield strength of the material. The effect of the cyclic stressing is to cause cracks to nucleate and grow until, eventually, the crack reaches a critical size and final failure by fast fracture occurs. If a crack or surface defect is present already then the number of stress cycles required for failure will be reduced. It is the tensile component of a stress cycle which is responsible for crack growth and the resistance to fatigue will be increased if the mean stress in a cycle is compressive.

There are several expressions which relate to stress range, the cumulative effect of several different types of stress cycle, and the influence of the stress intensity factor on the rate of crack growth. Use of these expressions permits fatigue life to be estimated.

The are several factors which affect the fatigue life of a component or structure. A rough surface, small surface scratches or notches and internal flaws will all act as points for fatigue crack nucleation and reduce the fatigue life. Component geometry is also important as sharp sectional changes or keyways can act as points of stress concentration from which fatigue cracks may start.

The fatigue life of metal components can be improved by surface treatments such as *peening* which induce residual compressive stresses within the surface layers.

Creep is the continued slow plastic straining of a material under a constant load or stress. In metals this can occur at temperatures in excess of 0.3 or 0.4 of the melting temperature, measured in Kelvin. The rate of steady-state creep is increased by an increase in either the level of stress applied or an increase in temperature. Any increase in the strain rate will reduce the time for rupture or failure of the material.

For high temperature service in high stress conditions a group of metallic materials, known collectively as superalloys, are used. These materials, and specialist processing methods, have been developed to give high resistance to creep.

Relaxation is a phenomenon related to creep and is the reduction of stress with time for a material which is subject to a constant value of strain.

Questions

25.1 (a) What is meant by the term *fatigue*?

   (b) With the aid of sketches, explain the meanings of the following terms:
   (i) fatigue limit, (ii) alternating stress, (iii) mean stress, (iv) repeating stress and (v) fluctuating stress.

25.2 Using the Goodman law, calculate the maximum mean stress that a steel shaft is able to withstand when it is subjected to a fluctuating stress cycle with a stress range of 200 MPa. The tensile strength of the steel is 500 MPa and the fatigue strength, as determined in tests using alternating stress cycles, is 250 MPa.

25.3 A steel has the following properties: tensile strength 640 MPa, tensile yield stress 510 MPa and a fatigue limit of 320 MPa. The steel is to be used for a component which will be subject to a static tensile stress of 225 MPa with a superimposed cyclic stress. Estimate the maximum range of cyclic stress on the basis of (a) Gerber's law, (b) the modified Goodman law and (c) Soderberg's law.

25.4 The cyclic stress intensity, ΔK, for a steel is given by the relationship: $\Delta K = 1.12\, \Delta\sigma(\pi a)^{1/2}$, where $\Delta\sigma$ is the cyclic stress range and a is the depth of a surface defect. For a steel for which $\Delta K = 5$ MPa m$^{1/2}$, calculate the maximum safe cyclic stress range for the following surface conditions:
  (a) polished with surface roughness of 1 μm,
  (b) turned surface with roughness of 10 μm,
  (c) rough turned surface with roughness of 100 μm,
  (d) polished with a surface defect 1 mm deep,
  (e) a weld induced surface crack of 10 mm depth and
  (f) a large surface flaw of 100 mm depth.

25.5 (a) Define Miner's law of cumulative fatigue, clearly identifying each of the parameters.
  (b) A steel with a tensile strength of 900 MPa and a fatigue strength (for 10^6 cycles) of 450 MPa is subjected to a series of alternating stress cycles as follows: 10^4 cycles at $\sigma_{max} = 575$ MPa and 10^4 cycles at $\sigma_{max} = 550$ MPa. Estimate the number of additional stress cycles the steel can endure at a value of $\sigma_{max} = 500$ MPa.

25.6 Given that the growth rate of a fatigue crack, da/dN, occurs according to the relationship da/d$N = A\, \Delta K^m$, where $\Delta K = 1.12\, \Delta\sigma(\pi a)^{1/2}$, estimate the number of cycles required for a crack in a steel component to grow in depth from 2 mm to 10 mm when subjected to an alternating cyclic stress with a stress range of 225 MPa. The constants for the steel are: $A = 4 \times 10^{-30}$ and the exponent $m = 3$.

25.7 Estimate the maximum safe cyclic stress range that can be applied to a steel shaft of 20 mm diameter if complete failure is not to occur before a fatigue crack has developed to at least one-quarter of the diameter. Assume that the relationship $\Delta K = 1.12\, \Delta\sigma(\pi a)^{1/2}$ holds and that ΔK for the steel is 5 MPa m$^{1/2}$.

25.8 Explain what is meant by the term *creep*.

Sketch a typical strain against time creep curve and explain the significance of the various sections. How would (a) an increase in applied load and (b) a decrease in temperature affect the creep response of the material?

25.9 The steady-state creep strain rates for an alloy at several temperatures and at a constant stress of 105 MPa were determined and are given in the table below:

Temperature (°C)	180	220	260
Strain rate (s^{-1})	4.72×10^{-8}	2.08×10^{-7}	7.34×10^{-7}

Estimate the life of the material when stressed at 105 MPa at a temperature of 160°C if the total creep strain is not to exceed 0.01.

25.10 The data in the table below are those for the creep of an aluminium alloy at 200°C.

Time (h)	Strain at 60 MPa	Strain at 35 MPa
0	320×10^{-6}	108×10^{-6}
20	532×10^{-6}	215×10^{-6}
60	815×10^{-6}	332×10^{-6}
90	1000×10^{-6}	400×10^{-6}
120	1200×10^{-6}	477×10^{-6}
140	1320×10^{-6}	532×10^{-6}
170	1477×10^{-6}	610×10^{-6}
220	1800×10^{-6}	723×10^{-6}

From the data, plot creep curves and calculate the value of stress that would give a creep rate of $4.8 \times 10^{-6}\,h^{-1}$.

25.11 The steady-state creep strain rates for a steel, when tested at 510°C at several stresses, are: $1.69 \times 10^{-5}\,h^{-1}$ at 180 MPa; $3.22 \times 10^{-5}\,h^{-1}$ at 200 MPa; $7.55 \times 10^{-5}\,h^{-1}$ at 240 MPa; $1.41 \times 10^{-4}\,h^{-1}$ at 265 MPa.

(a) Does steady-state creep at 510°C obey a simple power law? If so, evaluate the exponent.

(b) Estimate the maximum stress to which the steel can be subjected at 510°C if the limiting strain rate is $1.65 \times 10^{-4}\,h^{-1}$.

CHAPTER 26

Oxidation, corrosion and other effects

Why learn about corrosion and other degrading processes?

Materials may react with the environment and the extent of the reaction may cause a deterioration in the material sufficient to necessitate replacement. The rusting of iron and steels is a good example of a reaction involving the corrosion and deterioration of a material in the atmosphere. If the mechanisms of the reactions are understood then it becomes possible to devise and use methods and systems which will prevent such reactions from occurring or, at least, reduce the rate of reaction. The costs of protecting materials and structures from corrosion are high, but the cost could be considerably greater if the component or structure had to be replaced because the extent of corrosion was unacceptable.

LEARNING OBJECTIVES

1. Understand that all metals will oxidise to a greater or lesser extent and that the rate of oxide formation is controlled by both the nature of the oxide formed and the temperature.

2. Describe the formation of galvanic cells and explain their role in corrosion.

3. List the main measures that can be taken to prevent or reduce corrosion.

4. Understand the main mechanisms in the degradation of polymeric materials.

5. List the main effects of exposing materials to radiation.

26.1 | Introduction

All metals have a tendency to oxidise although the affinity of a metal for oxygen varies considerably from those with a high affinity, such as aluminium and magnesium, to the noble metals such as gold and platinum which do not react with oxygen at ordinary temperatures. The ionisation of a metal, that is the loss of one or more electrons leaving a positively charged metal ion, is also an oxidation process and it is the relative tendency of metals to oxidise that permits the creation of electrochemical or galvanic cells when in contact with an electrolyte.

Many ceramic materials are pure or mixed metal oxides. As such they are already in a fully oxidised state and will not oxidise further. Thus, ceramics and glasses are almost inert and do not corrode, although nitride and carbide ceramics may oxidise to some extent at very high temperatures.

Polymeric materials tend to depolymerise or degrade when heated, and these changes will begin at comparatively low temperatures with some thermoplastics. They may also react with oxygen or ozone causing a degradation and change in properties, and these changes will be accelerated if the polymer material is exposed to strong sunlight or ultra-violet radiation.

26.2 | Oxidation of metals

Metals possess an affinity for oxygen and tend to oxidise, but with most metals the amount of oxidation that takes place at ordinary temperatures is not serious. In fact, in many instances, the oxide layer that rapidly forms on a freshly exposed metal surface tends to protect the metal from further oxidation.

The direct reaction between a metal and oxygen can occur at any temperature, but the rate of the oxidation reaction is temperature dependent and increases rapidly with an increase in temperature. The reactivity of a metal with oxygen varies very considerably from one metal to another, with some metals such as copper and lead showing low reactivities, and other metals such as aluminium and magnesium possessing very high reactivities with oxygen. However, the rate at which oxidation occurs at the surface of a metal is largely controlled by the nature of the oxide that is formed. This may form as a coherent and almost non-porous layer on the surface of the metal, in which case it can act as a protection against further oxidation, or it may form as a broken or porous layer, allowing oxygen relatively free access to an exposed metal surface allowing oxidation to continue at a constant rate. In the case of the metal molybdenum there is a worse phenomenon. At temperatures above 900 °C there is a catastrophic oxidation of the metal. No surface oxide layer can be formed as molybdenum oxide is a vapour at such temperatures.

The oxide layer that forms on the surface of a metal is termed either a film or a scale. There is no clear distinction between these names, apart from the fact that scale refers to a thick layer and films are very thin. Generally, the term film is used for an oxide thickness of 10^{-3} mm or less, while the term scale is used for greater thicknesses than this.

The Pilling–Bedworth (P–B) ratio defines the characteristics of an oxide layer on a metal surface and is expressed as

$$\text{P–B ratio} = \frac{M_o \times \rho_m}{n M_m \times \rho_o} = \frac{M_o \times V_o}{n M_m \times V_m}$$

where ρ_m and ρ_o are the densities of metal and oxide respectively, V_m and V_o are the respective volumes, M_m and M_o are the atomic mass numbers of metal and oxide, and n is the number of metal atoms in the oxide molecule. If the P–B ratio is less than 1, the oxide occupies a smaller volume than the volume of metal from which it was formed and, in this case, the oxide film is porous. When the P–B ratio is ≈1, the volumes of oxide and metal are nearly equal and the tendency is for the formation of a tenacious, non-porous, protective oxide film. When the P–B ratio is greater than 1, the oxide film formed is initially protective but, as the thickness of the oxide layer increases, stresses are developed within the oxide layer and these may cause the oxide to flake off from the surface.

EXAMPLE 26.1

A fresh tin surface will oxidise in air forming an oxide with the formula SnO_2. Using the data below, would you expect the oxide layer formed on the surface to be protective? Atomic mass numbers for tin and oxygen are 118.7 and 16.0 respectively. The density of tin is 7290 kg m^{-3} and that of the oxide is 6950 kg m^{-3}.

SOLUTION

Using the data given the Pilling–Bedworth ratio is $(150.7 \times 7290)/(118.7 \times 6950)$
$$= 1.33$$
This ratio of 1.33 is not much greater than unity and the oxide layer formed is likely to be protective.

Porous oxide films or scales are non-protective as there is free access of oxygen to the metal surface. In such cases oxidation will continue at a constant rate at any given temperature and the thickness of the oxide layer will increase linearly with time according to the expression $y = kt$, where y is the thickness of the oxide coating after time t and k is a constant. The value of k, however, varies with temperature according to the Arrhenius relationship, $k = A \exp(-B/T)$, where A and B are constants and T is the temperature (K).

When a non-porous protective film forms on the exposed surface of a metal it prevents the access of gaseous oxygen to the surface. Consequently, the diffusion of ions and electrons must take place through the oxide film if there is to be any further growth in the thickness of the film. Such diffusion processes are aided by the presence of defects within the crystalline structure of the oxide. The reactions that occur are as follows: (i) at the oxide film/air interface formation of oxygen ions takes place, $O_2 + 4e^- = 2O^{2-}$; (ii) at the metal/oxide interface ionisation of the metal occurs, $M = M^{2+} + 2e^-$. The electrons liberated by the ionisation of metal atoms will be conducted fairly freely through the oxide layer to combine with oxygen molecules forming negatively charged oxygen ions at the oxide/air interface. Both metal and oxygen ions can diffuse through the oxide film, but usually the rate of diffusion of metal ions is greater than that for oxygen ions as many species of metal ions possess smaller diameters than the oxygen ions. The metal ions meet the oxygen ions close to the oxide/air interface and there combine forming oxide and the oxide layer grows outwards (see Figure 26.1). (In those cases

Figure 26.1 *Oxidation of a metal through a protective film.*

where the oxygen-ion diffusion rate is greater than that for the metal ions the film growth occurs from the metal/oxide interface.) As the thickness of the oxide layer increases and diffusion distance increases there will be a corresponding decrease in the rate of oxide formation. The variation of film thickness with time follows a parabolic law expressed like $y^2 = kt$, where y is the thickness of the oxide layer after time t and k is a constant. As for linear growth, the value of the constant, k, varies with temperature according to the Arrhenius expression. This type of oxide growth is observed in copper, iron, nickel and cobalt.

Some metals have highly protective oxide films with a growth rate that decreases much more rapidly than given by the parabolic law. In these cases the growth rate follows a logarithmic law expressed as: $y = k \log(at + 1)$. Again, y is the film thickness at time t, and k and a are constants. Chromium and zinc are of this type. In the case of aluminium the oxide film possesses an extremely high value of electrical resistance and when the film thickness is about 100 atoms thick electron flow through the film ceases.

EXAMPLE 26.2

A metal sample is heated to a temperature of 500 °C and the thickness of the oxide coating formed is determined at various time intervals, the results being:

Time	Oxide thickness (μm)
20 min	1.15
50 min	2.93
200 min	11.70

Determine whether the oxidation process follows a linear, parabolic or logarithmic law. Estimate the oxide thickness after a period of 24 hours.

SOLUTION

Inspection of the data indicates that a linear expression, $y = kt$ is likely.

If this is so then $1.15 = k \times 20$ giving $k = 0.0575$
$2.93 = k \times 50$ giving $k = 0.0586$
$11.7 = k \times 200$ giving $k = 0.0585$

The agreement between these values of k confirms that oxidation follows a linear law.
24 hours = 1440 minutes. Using a value of $k = 0.0585$

$$y = 0.0585 \times 1440 = 84.2 \; \mu m$$

Alloys to resist oxidation are often based on an alloy system in which the solute has a much higher affinity for oxygen than the parent metal. A good example of this are alloys of copper and aluminium (aluminium bronzes). When these alloys are oxidised at high temperatures cuprous oxide, Cu_2O, forms very rapidly and copper ions cross the metal/oxide interface into the oxide layer. The concentration of aluminium on the metal side of the interface therefore increases until it is sufficiently high to form a continuous layer of protective aluminium oxide. This layer is highly impermeable to copper ions, which can no longer enter the cuprous oxide layer to increase its thickness. The higher the aluminium

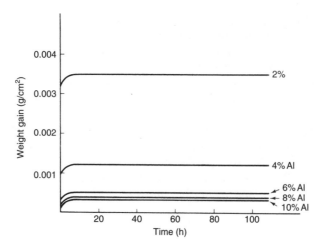

Figure 26.2 *Oxidation rate Cu–Al alloys at 800°C.*

content of the alloy, the quicker will the oxidation rate of copper be reduced (see Figure 26.2). This type of protection may not occur at low temperatures. Aluminium oxide is still formed but it tends to form as particles within the Cu_2O layer.

There is a similar behaviour in copper–beryllium alloys where a protective layer of beryllium oxide forms at elevated temperatures. The principle of protection against oxidation by formation of a continuous oxide layer also applies to stainless steels where protection is afforded by chromic oxide, Cr_2O_3, in all materials containing more than 12 per cent Cr. In ferrous materials containing several per cent of silicon the oxide that forms is a double oxide of iron and silicon, namely a silicate. The silicate layer has a glass-type structure, is relatively impervious to the passage of ions and is largely protective.

26.3 | Degradation of polymers

When polymeric materials are heated, there is a tendency for them to depolymerise, or *degrade*. These degradation reactions commence at comparatively low temperatures, and for this reason, accurate control of temperature is necessary in the plasticising cylinders of injection moulding and extrusion equipment. Polymers that are formed by condensation reactions involving the formation of water may depolymerise if heated for lengthy periods in an atmosphere containing steam. In a dry atmosphere they will tend to degrade when heated to a fairly high temperature, but in this case the reaction is essentially charring, with carbon formed as one of the degradation products.

The term *degradation* is used to refer to any change in a polymer which leads to some breakdown of its chemical structure or, in other words, causes it to depolymerise. This may be brought about by oxidation, hydrolysis, namely a reaction with water, heat, radiation, stress, or any combination of these. Most thermoplastics and elastomers can be affected by these factors and the stability of the materials is improved by incorporating antioxidants and stabilisers into the polymer moulding compounds.

Many polymers will age slowly in air with a consequent deterioration in some properties, particularly toughness and, in the case of transparent materials, optical clarity. This is the result of oxidation. The process is accelerated in the presence of sunlight, especially if there is radiation in the ultra-violet sector of the spectrum. This is termed *photo-oxidation*. The high energy incident photons stimulate the chemical reactions involved in the degradation. An increase in temperature increases the rate of degradation in the presence of oxygen, but heat alone can also bring about polymer degradation. Ozone is more reactive than oxygen and a number of elastomer materials, including natural rubber and SBR, are attacked by ozone in the atmosphere. The ozone attacks unsaturated C–C bonds in the molecular chains leading to a splitting of the chains.

The term *weathering* is used to describe the deterioration of polymer materials in an outdoor environment. Weathering includes the effects of water, together with oxygen, heat and ultra-violet radiation, and leads to general deterioration in appearance and properties with possible embrittlement and cracking together with degradation.

Environmental stress cracking (ESC) is a phenomenon associated with some polymer materials when stressed and in contact with certain active substances. For example, polyethylene is subject to this form of cracking if, when stressed, it is in contact with a surface-active substance such as an aqueous solution containing a detergent. Other materials which are prone to ESC are polystyrene, ABS and UPVC.

26.4 Oxidation of ceramics and composites

Very many ceramic materials are pure or mixed oxides and, as such, do not oxidise as they are already in a fully oxidative state. However, carbide and nitride ceramics can oxidise to some extent if heated in air at temperatures in excess of 100 °C. At high temperatures, in the presence of atmospheric oxygen, a layer of oxide can form on the surface of a ceramic. In the case of silicon carbide or silicon nitride, the oxide layer formed on the surface is silica, SiO_2. The exact nature of the oxide formed is determined by the composition of the carbide or nitride ceramic. Also, the composition and nature of the oxide layer is affected by presence of other constituents, for example, any sintering additives used to regulate the type of grain boundary phases.

The thermosetting resins used as matrix materials for many fibre reinforced composites are highly resistant to oxidation and chemical attack and, thus, protect the fibre reinforcement against environmental conditions. However, some composite materials designed for use at high temperatures may suffer from oxidising effects. Examples are nickel, reinforced with carbon fibre and ceramic matrix composites. In the former material, oxygen can diffuse through the nickel matrix to the fibres while, in the latter example, the presence of microcracks in the ceramic matrix will permit the ingress of oxygen to the fibres.

26.5 Corrosion

A metal in contact with an electrolyte ionises to a small extent, a typical reaction being

$$Zn \rightleftharpoons Zn^{2+} + 2e^-$$

The free electrons produced by the forward reaction remain in the metal and the ions pass into the electrolyte. At equilibrium, therefore, the metal will possess a certain electrical charge. Different metals ionise to differing extents and will possess different values of electrical charge when in equilibrium with their ions. The potential difference between a metal electrode and a standard hydrogen electrode is termed the *Standard Electrode Potential** for the metal. Table 26.1 gives the standard electrode potentials for some metals.

Much metallic corrosion is of the galvanic or electrochemical type. If two dissimilar metals, for example copper and zinc, are connected via an electrolyte and are also electrically connected external to the electrolyte (Figure 26.3) a galvanic cell will be set up. The potential difference of this cell will be 1.10 V, with the zinc being the anode and copper being the cathode of the cell. At the anode, the reaction $Zn \rightleftharpoons Zn^{2+} + 2e^-$ will be thrown out of equilibrium, as the electrons produced in the reaction flow through the external circuit to the copper, and ionisation of zinc will be accelerated in an attempt to restore equilibrium, that is, the zinc will be dissolved or corroded. Conversely, at the cathode, the ionisation of copper will be suppressed owing to the constant arrival of free electrons from the anode. Corrosion cells may be established in many ways in practice.

* Standard Electrode Potentials are determined with electrodes immersed in a Normal solution of hydrogen ions (1 gramme equivalent of hydrogen ions per litre).

Table 26.1 *Standard electrode potentials.*

Base metals	Metal	Ion	Electrode potential V	Anodic
	Sodium	Na$^+$	−2.71	
	Magnesium	Mg^{2+}	−2.38	
	Aluminium	Al^{3+}	−1.67	
	Zinc	Zn^{2+}	−0.76	
	Chromium	Cr^{2+}	−0.56	
	Iron	Fe^{2+}	−0.44	
	Cadmium	Cd^{2+}	−0.40	
	Cobalt	Co^{2+}	−0.28	
	Nickel	Ni^{2+}	−0.25	
	Tin	Sn^{2+}	−0.14	
	Lead	Pb^{2+}	−0.13	
	Hydrogen	H$^+$	0.000	
	Copper	Cu^{2+}	+0.34	
	Mercury	Hg^{2+}	+0.79	
	Silver	Ag$^+$	+0.80	
	Platinum	Pt^{2+}	+1.20	
Noble metals	Gold	Au$^+$	+1.80	Cathodic

Figure 26.3 *Galvanic cell between copper and zinc.*

EXAMPLE 26.3

(a) What would be the voltage developed if electrodes of iron and cadmium were connected both externally and through an electrolyte to make a galvanic cell?
(b) What inference can be drawn from the result for the possible usefulness of cadmium plating for the protection of steel from corrosion?

SOLUTION

(a) From the data given in Table 26.1 the difference in electrode potential values for iron and cadmium is 0.04 V, with iron being anodic with respect to cadmium.

(b) Cadmium, which is a metal which forms a protective oxide film on its surface, is a suitable metal for plating onto the surface of steel to give corrosion protection. If for some reason the protective coating is broken and a galvanic cell is created, technically the steel would corrode as it would be the anode of a cell. The rate of corrosion would be very small because the cell voltage is small. For this reason, cadmium plating is used for the protection of steel fastenings, bolts, nuts and washers. There is a likelihood of the plating layer being damaged by abrasion during the tightening of a nut on a bolt and if this happens and a galvanic cell is created the rate of corrosion will be very small.

The order in which the various metals are placed, in Table 26.1, is termed the *electromotive series*. It should be noted, however, that the potential difference between metals in equilibrium with their ions may differ from the standard values if the metals are not immersed in a standard 'normal' electrolyte but in some other electrolyte, for example, seawater. Also, the order of metals in the series may differ from the order of the standard electromotive series when metals are immersed in other electrolytes. The order of metals, when determined for some electrolyte other than a normal solution of hydrogen ions, is termed a *galvanic series*. The galvanic series for some metals and alloys in seawater is given in Table 26.2.

Usually, in a galvanic cell, it will be the anode material which is corroded and the cathode material is protected against corrosion. There are a few exceptions to this and cathodic corrosion, a type of corrosion associated with the formation of alkaline reaction products, may occur in some instances.

Galvanic, or corrosion, cells may be established in several ways, and the main types of galvanic cells are; (a) *composition* cells; (b) *stress* cells; (c) *concentration* cells. An obvious example of a composition cell is two dissimilar metals that are in direct electrical contact and also joined via a common electrolyte. For this reason, dissimilar metals should not be in contact in situations where corrosion conditions may

Table 26.2 *Galvanic series in seawater.*

Anodic end:	Magnesium
	Magnesium alloys
	Zinc
	Galvanised steel
	Aluminium
	Cadmium
	Mild steel
	Wrought iron
	Cast iron
	Austenitic stainless steel (active)
	Lead
	Tin
	Muntz metal
	Nickel
	Alpha brass
	Copper
	70/30 Cupronickel
	Austenitic stainless steel (passive)
	Titanium
Cathodic end:	Gold

prevail. It is also possible to have composition cells formed within one piece of material. In any two-phase alloy, or a metal containing impurities, one phase may be anodic with respect to the other and galvanic microcells may be established. Examples of this are pearlite in steels, and precipitate particles of chromium carbide in austenitic stainless steels. In pearlite, the ferrite is anodic with respect to cementite, and in austenitic stainless steels the grain boundary precipitate of chromium carbide, which is formed if the steel is incorrectly heat treated, causes the austenite solid solution adjacent to the pre-cipitate particles to be depleted in chromium. These low-chromium zones are anodic with respect to the normal composition austenite. This latter fact is responsible for the weld decay type of corrosion referred to in section 16.13. From the above, it follows that, normally, pure metals of single-phase alloys will possess better resistance to corrosion than impure metals or multi-phase alloys.

Atoms within a highly stressed metal will tend to ionise to a greater extent than atoms of the same metal in an annealed condition. Consequently, the stressed material will be anodic with respect to the unstressed metal. A stress type of galvanic cell may be established in a component, or in a structure, where the stress distribution is uneven. (This also applies to an uneven distribution of residual stresses with a cold worked metal.) Some alloys are particularly prone to stress corrosion failure, which is often a grain-boundary (intercrystalline) type of corrosion. This may occur in cold worked α brasses, if they are not properly stress relieved. Randomly arranged grain-boundary atoms tend to ionise more rapidly than atoms within a regular crystal lattice, and so a grain boundary tends to be anodic with respect to a crystal grain. Much corrosion is of an intercrystalline nature. It also means that a coarse-grained metal tends to possess a better resistance to corrosion than a fine-grained sample of the same metal.

When a metal is in contact with a concentrated electrolyte solution it will not ionise to as great an extent as when it is in contact with a dilute electrolyte. In other words, if one piece of metal is in con-tact with an electrolyte of varying concentration, those portions in contact with dilute electrolyte will be anodic with reference to portions in contact with more concentrated electrolyte. Galvanic cells of the concentration type may be encountered in situations involving flowing electrolytes in pipes and ducts. Another type of concentration cell can occur if there are variations in dissolved oxygen content throughout an electrolyte. But, before considering the oxygen concentration cell, let us consider the various reactions which can take place at anodes and cathodes.

At the anode:

(a) metal may go into solution

$$M \rightarrow M^+ + e^- \tag{1}$$

(b) anions, such as $(OH)^-$, will be attracted to the anode giving rise to the formation of an oxide, hydroxide, or an insoluble salt

$$2M + 2(OH)^- \rightarrow M_2O + H_2O + 2e^- \tag{2}$$

$$M + (OH)^- \rightarrow MOH + e^- \tag{3}$$

$$M + X^- \rightarrow MX + e^- \tag{4}$$

At the cathode:

(a) metal may be deposited

$$M^+ + e^- \rightarrow M \tag{5}$$

(b) hydrogen may be evolved

$$2H^+ + 2e^- \rightarrow H_2 \tag{6}$$

(c) oxygen may be reduced

$$O_2 + 2H_2O + 4e^- \rightarrow 4OH^- \tag{7}$$

A most important type of galvanic corrosion cell is the oxygen concentration cell in which the elec-trolyte has variations in its dissolved oxygen content. Metal in contact with electrolyte low in dissolved oxygen will be anodic with respect to metal in contact with electrolyte rich in dissolved oxygen. A principal reason for this is that in an area with water rich in oxygen the following reaction will take place: $2H_2O + O_2 + 4e^- \rightarrow 4(OH)^-$. This is a cathode-type reaction, in that it absorbs electrons. The electrons will be supplied from another part of the metal where there is less oxygen and so such areas act as anodes. This is one of the main features of the *rusting* of iron.

Rust is the hydrated oxide of iron, $Fe(OH)_3$. Both oxygen and water are required for the formation of rust and iron and steels will not rust in a dry atmosphere or when immersed in oxygen-free water.

Consider the situation with a drop of water on the surface of a piece of steel (Figure 26.4). There will be a greater amount of oxygen available near the edge of the water drop and this area will become cathodic with hydroxyl ions formed according to the reaction: $2H_2O + O_2 + 4e^- \rightarrow 4(OH)^-$. At the anode area under the centre of the drop electrons will be generated by the reaction: $Fe \rightarrow Fe^{2+} + 2e^-$. The Fe^{2+} cations will move through the electrolyte toward the cathode and the $(OH)^-$ anions will tend to move toward the anode. The very much smaller cations have much greater mobility than the large hydroxyl ions and the two ion types meet close to the cathode areas forming insoluble ferrous hydroxide, $Fe(OH)_2$, which settles as a deposit near the cathode area. This subsequently oxidises to rust, $Fe(OH)_3$.

This type of galvanic cell can form in many situations and can give rise to severe effects. Waterline corrosion is a good example of this. The oxygen concentration at a water/air interface will be higher than at lower levels in the water, making the steel at the waterline cathodic and with anode areas existing just below the water surface. The most rapid corrosion will take place just below the waterline with solution of the anode areas, while the greatest rust deposit will form at the waterline.

Once a rust deposit has formed on the surface of a steel, the area under the porous deposit will become oxygen deficient, and hence anodic, compared with bare metal. This means that dissolution of iron can continue, unobserved, below the covering of rust. Similarly, any covered or hidden area to which moisture as electrolyte can gain admittance, for example, a dirt-covered surface, a crack or crevice, or the gap between overlapping rivetted plates, can become an anode (Figure 26.5). A watertight caulking of seams, in the latter case, can minimise this type of corrosion.

Figure 26.4 *Oxygen concentration cell forming rust.*

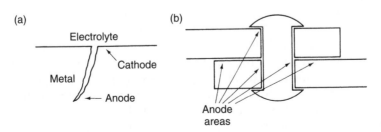

Figure 26.5 *Formation of anode and cathode areas in (a) crack and (b) rivetted joint.*

Corrosion due to this type of cell will be more severe in salt water than in fresh water as the presence of other dissolved ions greatly increases the conductivity of the electrolyte.

The extent to which a metal corrodes depends not only on its electrode potential relative to other materials present but also on the nature of the electrolyte and in particular on its pH value. Many different types of reaction may occur during corrosion. A reaction may be of the type that involves only electrons as, for example, $Fe^{2+} \rightarrow Fe^{3+} + e^-$. This type of reaction will be dependent on potential. A second type of reaction may involve hydrogen ions, but not electrons as, for example, $Fe^{2+} + H_2O \rightarrow Fe(OH)_2 + 2H^+$. Reactions of this type will be dependent on the pH of the electrolyte. Both electrons and hydrogen ions may be involved in a reaction, such as for example, $Fe^{2+} + 3H_2O \rightarrow Fe(OH)_3 + 3H^+ + e^-$. This type of reaction will be dependent on both potential and pH value. *Pourbaix* collected thermodynamic data on reactions between metals and water, together with data on the solubilities of oxides and hydroxides. All this information is combined to produce Pourbaix diagrams which indicate stable situations as a function of electrode potential and pH value. A simplified Pourbaix diagram for iron is shown in Figure 26.6.

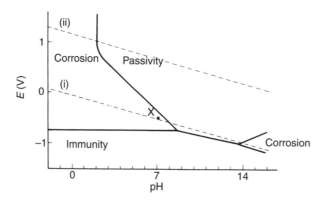

Figure 26.6 *Simplified Pourbaix diagram for iron.*

The two dotted lines in the diagram are (i) the hydrogen line, below which hydrogen will be evolved, and (ii) the oxygen line, above which oxygen will be evolved. The electrode potential of iron in water of pH 7 is given by point X in Figure 26.6, and it will be noted that this is within the corrosive zone. The diagram indicates ways in which the corrosion of iron may be reduced. By making the electrolyte alkaline, say to pH 9, the iron is moved into the passive region. One of the functions of inhibitors is to control pH values and maintain a value that will keep the metal in a non-corrosive regime. Another way would be to reduce the electrode potential value while keeping the electrolyte pH value constant. This would place the iron within the region of immunity on the diagram. This is the principle of cathodic protection which is achieved using an anodic metal such as aluminium in conjunction with iron. Thirdly, the iron could be placed in the region of passivity by increasing the potential and this could be done by applying a suitable external e.m.f.

Once a galvanic cell has been established there is a tendency for *polarisation* to occur, namely a reduction in the current flowing within the corrosion cell. When an anode starts dissolving, metal ions go into solution in the electrolyte. If the diffusion rate of these ions away from the anode area is relatively slow the concentration of such ions in the anode area increases and exerts a back-e.m.f. effect, suppressing the ionisation of the anode material. A different effect occurs at the cathode. Here cations are neutralised by electrons and the concentration of cations in the vicinity of the cathodes is greatly reduced if ion diffusion rates are low. Both anode and cathode effects combine to reduce the current flow in the cell and, hence, the corrosion rate. Since such polarisation is controlled by diffusion rates it follows that any change that will increase ion diffusion, such as an increase in temperature, an increase in concentrations or agitation of the electrolyte, will reduce the polarisation effect and maintain high rates of corrosion.

Passivation is the reduction in corrosion rate that occurs when products of corrosion reactions form a largely impervious layer over an anode surface. Passivity is due to the formation of a very thin impervious oxide layer on the surface and this often occurs in highly oxidising conditions. Materials such as aluminium, titanium and stainless steels readily form passive films but these may be destroyed in a reducing atmosphere and the metals will then become active for corrosion.

Corrosion of a metal can easily lead to failure. When the effective thickness of the metal is reduced the level of stress within the material is increased. This increase could accelerate corrosion in a localised area until the level of stress is sufficient to cause failure of the component or structure. The following paragraphs contain a brief description of some of the types of corrosion that could lead to early failure.

Pitting corrosion is a highly localised form of corrosion and is usually associated with a local breakdown of an otherwise protective film. Pits can start where there is a discontinuity in the passive oxide layer and they can also occur, owing to a differential aeration effect, where an external deposit has settled on a metal surface. Frequently the source for corrosion pits is a non-homogeneous metal and impurity inclusions within the structure may be the cause of a break in the oxide coating. A corrosion product often forms over the surface of a pit, but corrosion continues beneath this. Pitting can occur on any metal but it is much more common with metals and alloys that possess a high-resistance passive film. The presence of the film tends to prevent the spread of corrosion to a wider area and attack is concentrated in the series of small localities. Stainless steels are susceptible to this form of corrosion in the presence of chloride solutions.

It has already been stated that a metal is more susceptible to corrosion when it is highly stressed. *Stress corrosion* is a phenomenon in which there is the initiation of a crack and then propagation of the crack through the material as the result of the combined effects of a stress and a corrosive environment. Stress corrosion is generally found only in alloys, not in pure metals, and it occurs only in certain specific environments, for example, the cracking of brass in ammonia or the cracking of aluminium alloys in chloride solutions. The stress within the material must have a tensile component, and both stress and specific corrosive environment must be present for cracks to form and propagate. The stress corrosion cracks are generally of an intercrystalline type although transcrystalline cracking occurs in some cases.

Corrosion fatigue failure is the development and propagation of cracks in a material that is subjected to alternating or fluctuating cycles of load. The presence of a corrosive environment, even if it is only mildly corrosive, will accelerate the formation and growth of fatigue cracks, thus reducing the fatigue life of the material.

Impingement attack is a form of corrosion attack found in marine condensers and occurs when large bubbles in the turbulent cooling fluid burst on the tube surface. The effect of this is to produce large horseshoe-shaped pits. Where a large bubble impinges at high velocity it will remove any corrosion products that are causing a polarising effect, and hence re-activate the corrosion cell. The bursting of a bubble will also provide a good supply of oxygen. The problem may be minimised in two ways. Careful design to reduce turbulence to a minimum and ensure that any bubbles formed will be small is necessary. Also, the materials used in the construction of marine condensers must be highly corrosion resistant. Cupronickel alloys are about the best materials available for this application.

Cavitation attack occurs on ships' screws and steam-turbine blades and it can result in the formation of pits. As the blade moves through the fluid air bubbles will be formed at, or close to, those parts of the blade where there is a low fluid pressure. The later collapse of these bubbles at the surface of the blade will produce a compression wave. There is evidence to show that the pitting that occurs is due to a combination of the mechanical effect of collapsing bubbles and the chemical corrosion effect of aerated cell formation, as in impingement attack. Selection of suitable corrosion-resistant alloys and careful attention to blade design will minimise the problem.

Fretting occurs as a result of the minor amounts of relative motion between two parts in an assembly. Fretting is often associated with fatigue. When the surfaces abrade each other the very fine metal debris formed rapidly oxidises and this oxide debris can itself act as an abrasion agent. As the process continues the metal surface gradually wears away, forming more fine oxide. With steels the corrosion product is a fine brown powder known as 'cocoa'. The continued wastage of metal results in severe wear and may proceed to complete failure. Fretting is more pronounced in a dry atmosphere. The presence of moisture tends to reduce fretting, the moisture possibly acting as a lubricant.

26.6 | Corrosion protection

Many methods are used to minimise the possibility of corrosion and these may be grouped into the following categories: (1) the use of protective coatings, (2) cathodic protection, and (3) the use of inhibitors. There are many types of protective coating available ranging from paints and bituminous coatings, through plastic and vitreous enamel coatings to a variety of metallic coatings applied by hot dipping, electroplating, metal spraying or other methods. Also, by careful material selection and attention to design the formation of composition-type galvanic cells may be avoided.

The object of a protective coating is to isolate the metal from the corrosive environment. One of the more common forms of protection is a paint layer. Paints are easily applied, either by brush or spray, but are seldom fully effective. Most paints age and weather, particularly in strong sunlight, and the layer either becomes porous or flakes off the surface. Also paint surfaces are fairly soft and may be easily scratched and damaged. Once oxygen and moisture can reach the metal surface, corrosion can begin and it can even occur beneath portions of the paint-covered surface.

Metal parts are sometimes covered with a thermoplastic coating. This is frequently used for the protection of steel wire trays, as used in refrigerators and freezers. The heated-metal assembly is dipped into a container filled with powdered thermoplastic. The plastic adjacent to the dipped part melts and forms a continuous protective layer over the component.

Vitreous enamelling is used for the protection of metal parts that will be subjected to moderate heating during their service life, for example, gas and electric cookers. An enamel coating is hard but brittle. It is a wear-resistant coating but it is easily damaged by mechanical shock treatment.

Metallic coatings are very widely used for the protection of metals, particularly steels. The metal coating could be of a metal that is anodic with respect to the steel such as aluminium, zinc or chromium but is itself resistant to corrosion because of a passive oxide film. This type of coating may still afford some protection to the steel even if there are breaks in the coating, as the coatings will be anodic and will corrode in preference to the cathodic steel. Other coatings, such as tin, are cathodic with respect to steel and if the coating is broken the exposed steel will become the anode and corrosion will take place rapidly (Figure 26.7). Cadmium, which has an electrode potential value very close to that of iron, is very often used as a protective plated layer on parts such as nuts, screws and bolts. The plated layer could be broken by abrasion during the tightening of a nut on a bolt, but if this happens the e.m.f. of a galvanic cell will be small (0.04 V) and the rate of corrosion will be very low.

The maintenance costs of protecting large steel structures, whether by regular painting, or using cathodic protection tend to be high. A range of structural steels, termed *weathering* steels has been developed and these contain both chromium, about 0.6 per cent, and copper, up to 0.4 per cent, as alloying elements. With exposure to weather, these steels develop an adherent protective oxide film

Figure 26.7 *Effect of break in protective coating on steel. (a) Tin (cathodic) coating. Exposed steel becomes anodic and is corroded; (b) Zinc (anodic) coating. Coating is anodic and corrodes, thereby protecting the steel.*

instead of the usual flaky rust and, after a period of about two years, this has a pleasing purplish copper colour. The cost of weathering steels is about 20 per cent greater than general structural steels but this extra cost may be offset by savings in other protective treatments and maintenance.

Cathodic protection is widely used to protect structures in which the effects of severe corrosion could be serious or where the renewal of a corroded structure could prove very expensive. It is particularly useful for the protection of ships' hulls, steel piers and harbour installations, buried steelwork in building foundations and buried pipelines. A block of a highly anodic material, magnesium, aluminium or zinc, is placed adjacent to the structure to be protected and connected electrically. A galvanic cell is deliberately created and the sacrificial anode will corrode, protecting the structure. The sacrificial anodes are inspected periodically and renewed when necessary. For the protection of the hulls of ships the anode material is strapped to the hull near the stern and below the waterline (Figure 26.8). A variant on the use of a sacrificial anode of another metal is to use an impressed voltage. The steelwork will be protected as long as it receives a steady supply of electrons. When using a direct-current supply as a means of protection it is common to bury scrap iron or steel adjacent to the structure, with connections that render the scrap metal anodic and the steel requiring protection cathodic.

Underground pipework is usually given a thick bituminous coating before installation but this coating may be damaged during pipe laying and cathodic protection is used to minimise corrosion of any exposed surface. Uncoated pipes are not used as the very large surface area needing protection would result in too rapid a solution of sacrificial anode material. The risk of corrosion will vary considerably with soil type. In low-risk conditions anodes may be widely separated but in high-risk conditions anodes would be placed closer to one another. Corrosion would be minimal in dry sandy soils but there could be a high risk in wet clays. The presence of bacteria in soils or clays with a sulphate content could result in the breakdown of sulphates and the liberation of oxygen, and this could lead to the establishing of oxygen concentration cells. There is also a major corrosion risk at the junction of two different soil types.

Inhibitors and *passifying agents* are widely used for the reduction of corrosion in closed cooling systems where cooling water is recirculated. They may take the form of pH regulators or they may help to form a protective film on metal surfaces within the system. Chromate salts are frequently used as inhibitors within car radiator systems. Any iron ions generated at a steel-anode surface will combine with chromate ions to form an insoluble iron chromate which will create a protective coating over the steel surface. Certain phosphates and tungstates may also be used as inhibitors.

Figure 26.8 *Cathodic protection. (a) Protection of ship's hull below waterline; (b) Protection of buried pipeline with magnesium anode; (c) Protection of buried pipeline using impressed voltage.*

26.7 | Microbial attack

Micro-organisms do not attack metals or ceramics and glasses. Thermoplastic and thermosetting materials are highly resistant to attack by micro-organisms, but some thermoplastics, because of their very low hardness, may suffer attack from borers. Some materials of animal or vegetable origin, including casein and some cellulose derivatives, are subject to some forms of bacterial or fungal attack. One of the few synthetic materials which may suffer microbial attack is PPVC and in this it is the plasticiser which is affected rather than the PVC itself. This high resistance of plastics materials to organisms poses a major problem from the point of view of waste disposal and research is taking place into the possibility of microbial methods for the breakdown of plastic refuse. Some biodegradable varieties of plastics materials are currently available.

Timbers are subject to bacterial and fungal attack, and numerous organisms feed on moist wood fibre. Thoroughly seasoned timber is not such good food value as moist wood, and the rate of attack is very greatly reduced in dry wood. Preservatives may be used to retard the decay of timber. A common and cheap preservative is creosote, as this is poisonous to the wood-attacking organisms. There are also many other wood preservatives available that contain specific fungicides.

There are also larger organisms that feed on certain types of wood. A number of types of beetle cause damage to timber. They lay their eggs on the surface or in small crevices and the grubs that hatch tunnel their way into the interior. The most common types that affect well-seasoned timber are the furniture beetle, the death-watch beetle and the house longhorn beetle. The grubs of the furniture beetle, woodworm, feed on the sapwood of softwoods used in building structures, as well as in furniture. The death-watch beetle is common in old hardwood timber, particularly if moist or subject to some fungal decay. In some parts of the world, attack on timber by termites can be a problem. The teredo is a small marine animal that attacks timber immersed in sea water, and the lower hulls of old sailing ships were coated with copper sheeting to afford protection, hence the use of the phrase 'copper-bottomed' to refer to a sound proposition.

26.8 | Effects of radiation

Radiation may have an effect on materials. Electromagnetic radiation, of various frequency ranges, will have an effect on some materials. A good example of this is the effect of radiation in the infra-red, visible light, ultra-violet and X-ray ranges on photographic emulsions. The radiation causes dissociation of silver from the silver salts in the emulsion. Similarly, visible light, ultra-violet light, or X- and γ-radiation will stimulate numerous chemical processes. An example is the effect of X- or γ-radiation on a linear polymer, such as polyethylene. The radiation increases the energy of the polymer and promotes branching and cross-linking reactions, thus producing thermoplastics of greater rigidity. This process is utilised in the production of some plastic components. Bombardment of a thermoplastic with electrons (β-rays) will also produce similar effects. The effects of high-frequency electromagnetic radiation on some compounds is to excite electrons to higher states of energy. When the excited electrons subsequently fall back to lower energy states, energy is emitted, often in the visible-light range, producing fluorescence and phosphorescence. A similar effect is noticed when some compounds are subjected to β-radiation (electrons), and this is the principle used in cathode ray tubes. High-frequency radiation or electron bombardment will not affect metals, but the radiation will be partially absorbed by the material. A considerable amount of radiation will pass through the metal, and this principle is utilised in the radiography of metals and in electron transmisson microscopy. Because the wavelengths of X- and γ-radiation approach the sizes of individual atoms, there will also be diffraction effects and diffraction patterns are used in the identification and analysis of crystals.

The irradiation of a material with neutrons is invariably damaging. A neutron is a small particle possessing no electrical charge, and a fast neutron will not be attracted to or deflected by charged particles, such as electrons and protons. A fast neutron may travel through a large number of atoms before it collides with an atomic nucleus. When such a collision occurs the atom, or ion, that has been hit is displaced from its normal position, and the path of the neutron will be deflected. A neutron will lose some energy at each collision and, eventually, a lower-energy neutron will be 'captured' by an atomic nucleus. When a nucleus captures a neutron, the nucleus will become unstable and will emit α-, β- or γ-radiation, and in doing so may be changed into a radioactive isotope of another element (refer to section 2.10).

The irradiation of polymers with neutrons at low neutron flux densities will cause some cross-linking and branching to occur, in a similar manner to activation by X-rays, but at higher neutron flux densities the effect of neutron collisions is to cause degradation of polymers. The effect of a fast neutron colliding with the nucleus of an atom in a metallic crystal lattice is to cause the atom to be knocked out of its equilibrium position within the lattice. The knocked-out atom moves into an interstitial lattice position, leaving behind a vacancy. The lattice defects created by neutron collisions make the material harder and increase the yield strength. They also make the material more brittle. In the case of those metals that show ductile–brittle transition behaviour, the transition temperature is increased. (The neutron irradiation of steels will raise ductile–brittle transition temperatures by up to 100 °C.) This type of radiation damage may be removed by annealing the material. Irradiation with slow neutrons will give a different type of radiation damage. Slow neutrons may be captured by atomic nuclei and the irradiated metal will become radioactive or 'hot'. Also the transmutation product may alloy with, or react chemically with, the other atoms of the metal.

SUMMARY

Metals will react directly to form oxides. The oxide may form as a porous or non-coherent layer, in which case oxidation will continue unimpeded, or it may form as an impervious layer on the metal surface and prevent further oxidation taking place. Depending on the nature of the oxide layer which is formed, the rate of growth of the oxide may vary with time either linearly or according to a parabolic or logarithmic law.

Generally, the oxidation of metals is not a serious problem at ordinary temperatures. The rate of oxidation increases exponentially with temperature and oxidation may be a problem at elevated temperatures.

Metallic corrosion is electrochemical in nature. All metals tend to ionise, but some more readily than others, and a metal, when in equilibrium with its ions, will possess an electrical potential. When two dissimilar metals are in direct contact and also in electrical contact via an electrolyte, a galvanic cell will be established, the potential difference of the cell being the difference between the electrical potentials of the two metals. Galvanic corrosion cells can be established also within a single material if there are differences in the level of stress from one portion to another, or if the metal is in contact with an electrolyte in which there are differences in concentration from one area to another.

Some metals and alloys, under certain environmental circumstances, can become passive and cease to corrode. This passivity is associated with the formation of a protective film on the metal surface. A change in environmental conditions could cause the passive film to be dissolved and the metal again become active as regards corrosion.

Many methods are used to protect against corrosion and these can be classified into three main groups:

(a) the use of protective coatings to isolate the metal from the environment,

(b) cathodic protection, and

(c) the use of inhibitors and passifying agents for pH regulation and passive film formation.

Ceramics and glasses are, generally, inert and do not oxidise except, for some nitrides and carbides, at very high temperatures.

Polymeric materials tend to degrade, that is depolymerise, when heated and they may also suffer degradation when exposed to strong sunlight, ultra-violet or ionising radiation. Timbers and some polymeric materials are attacked by micro-organisms. Efforts are being made to develop a range of polymeric materials which are biodegradable, as a solution to the major problem of waste disposal.

Questions

26.1 An oxide layer forms on the surface of a metal which is held at an elevated temperature and the gains in weight with time are shown in the table.

Weight gain (g/m²)	Time (min)
2.92	15
5.69	60
7.90	120
10.30	200

Does the oxidation reaction for this metal follow a linear, parabolic or logarithmic law? What is the expected weight gain after 2 days at high temperature?

26.2 Estimate the thickness of the oxide film formed on the surface of a metallic alloy assuming growth according to a parabolic law, $x^2 = kt$, under the following conditions: (a) at 350°C after 1 hour, after 150 hours and after 1000 hours; (b) at 600°C after 1 hour, 150 hours and after 1000 hours. Assume that k varies temperature according to $k = \exp(-16 \times 10^3/T)\,\text{cm}^2\,\text{s}^{-1}$, where T is temperature (K).

26.3 Define the Pilling–Bedworth ratio and discuss its significance in relation to the degree of protection offered to a metal surface by an oxide layer

Cadmium oxidises to the oxide, CdO. Given the following data, deduce whether or not the oxide layer formed is likely to be protective: M for $Cd = 112.4$, M for $O = 16.00$, ρ for $Cd = 8640\ \text{kg/m}^3$, ρ for $CdO = 8100\ \text{kg/m}^3$.

26.4 (a) What is meant by the term *Standard Electrode Potential?*

(b) Give an account of the reactions which occur if aluminium is directly connected to iron and also connected by a continuous electrolyte.

(c) Explain why the corrosion resistance of wrought iron is superior to that of mild steel when both are immersed in seawater.

26.5 What is a galvanic series and how does it differ from the electromotive series?

26.6 The surface of mild steel may be coated with either a layer of tin or zinc as protection against corrosion. Give a comparison of these two methods and explain the effects of a small break in the coating on the subsequent corrosion of the steel.

26.7 What steps can be taken to minimise the corrosion of the following:
  (a) the steel hull of a ship,
  (b) the cooling system for a liquid cooled car engine,
  (c) the steel body shell of a car?

26.8 It is proposed that marine marker buoys be made from aluminium alloy 5154A (Al–3.5% Mg), and that the buoys be anchored to the sea-bed by means of wrought iron chains. Is this proposition feasible?

  What precautions should be taken if an aluminium alloy is to be used successfully for this application?

PART VI

EVALUATION OF MATERIALS

CHAPTER 27

Property testing

Why learn about property testing?

The engineer and designer need to know the hardness, strength and other characteristics of the materials that they will be using. They should know also about the way in which these properties are determined and that the tests need to be conducted according to standard procedures so that one can have confidence in published test results. Some of the basic property tests that are conducted on materials are described in this chapter.

LEARNING OBJECTIVES

1. From indentation measurements, compute the Brinell or Vickers hardness values for materials and use the results to estimate the tensile strength.

2. Know how the hardness of other materials, including ceramics, glasses, rubbers and polymeric materials, can be measured.

3. From tensile test data, calculate the yield strength, tensile strength and percentage elongation values for materials.

4. Distinguish between *nominal* stress and strain and true stress and strain, and calculate true stress and true strain values from test data.

5. Describe methods for determining the tensile strength of ceramics and glasses and understand what validity may be placed upon the results obtained.

6. Describe methods for notch-impact testing.

7. Describe the type of tests which can be used for assessing the fatigue and creep strengths of materials.

27.1 | Introduction

Testing is an essential part of any engineering activity. Inspection and testing must take place at many stages in the complex process of producing engineering materials, be they metals, polymers, ceramics or composites, and during the forming of these materials into components and assembling the components to create an engineering product to satisfy some specific requirement. The requirement for testing does not automatically cease when the product has been manufactured. It is frequently necessary to check and test the article during its service life in order to monitor changes, such as the possible development of fatigue or corrosion damage.

The types of test used can be broadly classified into two categories:

(a) tests to establish the properties of the material, and

(b) tests to determine the integrity of the material or component.

Those tests in the first category are generally of a destructive type. They are performed on samples of a material and the test-piece is damaged or broken in the process, as is the case when determining the tensile strength of a material in a tensile test to destruction. If the sample test-piece is correctly chosen and prepared, the results should be indicative of the properties of the bulk material represented by the sample.

The tests in the second category are of a non-destructive nature and are used to detect the presence of internal or surface flaws in a material, component or finished product. By their very nature, these tests do not damage the parts being tested and sampling is not required as, if necessary, every item can be checked. The tests of this type are described in Chapter 28.

Many test procedures have been devised for the determination of some of the characteristic properties of materials, for example, the tensile test. For the results of any test to have value, it is important that the tests be conducted according to certain set procedures. To this end, standardised test procedures have been evolved. In the United Kingdom, it is the British Standards Institution which publishes the standards and codes of practice which cover most aspects of the testing and utilisation of materials. All the major developed countries possess standards organisations and while standards for a particular type of test may differ slightly on points of detail from one country to another, the broad principles will be similar.

27.2 | Hardness tests

The hardness of a material may be determined by using either a scratch test, or by making a surface indentation. This latter type of test may be used only for materials that are capable of plastic deformation, namely metals and thermoplastic materials. The resistance of a material to indentation is not necessarily the same as its resistance to abrasion, but a hardness measurement obtained from an indentation test can be used as an empirical check for abrasion resistance. In general, metals possessing high hardness will have a high resistance to abrasive wear. Indentation-type hardness tests are widely used for the checking of metal samples as they are easy to make and yield information on heat treatment condition. There are also empirical relationships between the hardness of a metal and its tensile strength.

The hardness of a ceramic material may be determined using a scratch test. Mohs' scale of hardness, which was devised for assessing the relative hardness of minerals and rock, is based on ten naturally occurring minerals ranging from very soft talc to diamond (Table 27.1). In this test an attempt is made to scratch the surfaces of the standards with the material under test. The hardness of the unknown lies between the number of the mineral it just fails to scratch and that of the mineral it just scratches.

Of the various types of indentation test available for the hardness testing of metals the most commonly used are the Brinell test, the Vickers diamond test and the Rockwell test.

The Brinell test

The Brinell test was the first static indentation test to come into general use. In its original form, it utilised a hardened steel ball indentor of 10 mm diameter forced into the surface of the metal being tested under a static load of 3000 kg (29.43 kN) and the load maintained for 10 to 15 seconds. A load of 3000 kg, in conjunction with a ball indentor of 10 mm diameter, is suitable for use with steels and cast irons. For softer non-ferrous metals and alloys, lower values of static load are used (see Table 27.2).

Table 27.1 *Mohs' scale of hardness.*

Number	Mineral
1	talc
2	gypsum
3	calcite
4	fluorite
5	apatite
6	orthoclase felspar
7	quartz
8	topaz
9	corundum
10	diamond

Table 27.2 *Recommended ratios of F/D^2 for the Brinell hardness test on various materials.*

Material	F/D^2
Steels and cast irons	30
Copper alloys and aluminium alloys	10
Pure copper and aluminium	5
Lead, tin and tin alloys	1

The diameter of the resulting impression is then measured with the aid of a calibrated microscope. The Brinell hardness number, H_B, is given by

$$H_B = \frac{\text{applied load (kg)}}{\text{surface area of the impression (mm}^2)}$$

The hardness number obtained from a Brinell test is not independent of the load used and if two tests are made on the same material, one using a large static load and one using a small load, the hardness values obtained may differ. The recommended values of F/D^2, where F is the load in kg and D is the diameter of the ball indentor in mm, to be used in the testing of various materials are given in BS 240 (1986) and ASTM E 10-84 (see Table 27.2).

After making a Brinell hardness impression, the parameter that is measured is the diameter, d. Accurate Brinell hardness values will only be achieved if the impression diameter, d, is between the limits of $0.25D$ and $0.5D$, where D is the diameter of the ball. The surface area of an impression of diameter d, made by a spherical indentor of diameter D, is given by

$$\text{Area} = \pi D/2\{D - \sqrt{(D^2 - d^2)}\}$$

and so

$$H_B = \frac{2F}{\pi D\{D - \sqrt{(D^2 - d^2)}\}}$$

The Brinell test is not suitable for the testing of very hard materials. As the hardness of the metal approaches that of the ball indentor, there will be a tendency for the indentor to deform. The Brinell test will produce reliable results up to H_B values of around 400 and it is not recommended that the test be used on metals which would give values of H_B greater than 500. A comparison between H_B values and hardness values obtained from other tests is given in Table 27.4. H_B values and Vickers diamond hardness values, H_D, are comparable up to $H_B = 300$ but for higher hardness values there is a divergence between the two scales, with the H_B values being less than the H_D values for a given material.

The Vickers diamond test

In the Vickers diamond test, the indentor used is a pyramidal shaped diamond, and as in the Brinell test, the indentor is forced into the surface of the material under the action of a static load for 10 to 15 seconds. As in the Brinell test, the Vickers diamond hardness number, H_D, is given by

$$H_D = \frac{\text{applied load (kg)}}{\text{surface area of impression (mm}^2)}$$

The Vickers diamond hardness test is covered by BS 427 Part 1 (1981) and ASTM E 92-82. The standard indentor is a square pyramid shape with an angle of 136° between opposite faces. One advantage of the Vickers test over the Brinell test is that the square impressions made are always geometrically similar, irrespective of size. The plastic flow patterns, therefore, are very similar for both deep and shallow indentations and, in consequence, the hardness value obtained is independent of the magnitude of the indenting force used.

After an impression has been made, the size of the impression is measured accurately using either a calibrated microscope or by projecting a magnified image of the impression onto a screen and measuring this image. Both diagonals of the impression are measured and the mean value of D, the diagonal length, is used in the determination of the hardness number:

$$H_D = \frac{2F \sin \theta/2}{D^2} \quad \text{where } \theta = 136°$$

$$H_D = \frac{1.8544F}{D^2}$$

where F is the applied load in kg and D is the mean diagonal length in mm.
Convenient loads to use for some common materials are

steels and cast irons	30 kg
copper alloys	10 kg
pure copper, aluminium alloys	5 kg
pure aluminium	$2\frac{1}{2}$ kg
lead, tin, tin alloys	1 kg

EXAMPLE 27.1

Brinell and Vickers diamond hardness impressions are made on a piece of steel. Using a 1 mm diameter indentor and a load of 30 kg the average diameter of several Brinell impressions is 0.308 mm, and the mean diagonal length of several Vickers indentations, using a 30 kg load, is 0.425 mm. Determine both the Brinell and Vickers hardness values for the steel and account for any differences in the two values.

SOLUTION

For a 1 mm diameter ball and a 30 kg load

$$H_B = \frac{60}{\pi\left\{1 - (1 - d^2)^{\frac{1}{2}}\right\}}$$

$$= \frac{60}{\pi\left\{1 - (1 - 0.308^2)^{\frac{1}{2}}\right\}}$$

$$= \frac{60}{\pi(1 - 0.951)}$$

$$= 393$$

Using a 30 kg load

$$H_D = (1.8544 \times 30)/D^2$$

$$= 55.63/0.362^2$$

$$= 425$$

The hardness value, as determined in the Vickers diamond test, is greater than the Brinell value. Hardness values of the order of 400 signify a hard material. With materials of this hardness the steel ball indentor of the Brinell test may itself begin to deform. For materials with hardnesses above 300, Brinell and Vickers hardness values diverge, the Brinell values being lower than Vickers values for the same material.

The Rockwell test

The Rockwell test machine is a rapid action direct reading machine. This provides a very convenient method for speedy comparative testing. In this test, or rather series of tests, it is the depth of the impression which is measured and directly indicated by a pointer on a dial calibrated, inversely, into 100 divisions (1 scale division = 0.01 mm of impression depth). Consequently, a low scale number indicates a deep impression, hence a soft material, and vice versa. There are a series of Rockwell hardness scales because there are several indentors and several indenting loads available and covered by standards (BS 891 (1989) and ASTM E 18-89a refer to Rockwell hardness testing). The indentors used are hardened steel balls of various diameters or a diamond cone with an included angle of 120°. The standard ball indentors are of $\frac{1}{16}$, $\frac{1}{8}$, $\frac{1}{4}$ and $\frac{1}{2}$ inch diameter. The standard indenting loads are 60, 100 and 150 kg. Each separate scale of hardness is designated by a letter – A scale, B scale and so on. The various scales and their applicability are given in Table 27.3.

There are other Rockwell scales of hardness available. These scales, known as the N and T scales, are variants of the Rockwell A and B scales but involve smaller indenting forces and are suitable for hardness tests on thin samples. BS 4175 (1989) and ASTM E 18-89a are the relevant standards for these tests. The scales using the diamond cone indentor are known as the 15N, 30N and 45N scales and those using a $\frac{1}{16}$ inch diameter steel ball indentor are known as the 15T, 30T and 45T scales. In each case, the initial number refers to the total indenting force in kg.

Microhardness testing

There are systems available for microhardness testing and the two most widely used methods are the *Vickers diamond* test and the *Knoop diamond* test. The principle of the Vickers diamond microhardness test is basically the same as for the standard Vickers test but the indenting loads used are measured in grammes rather than kilogrammes. This type of hardness testing is performed on a metallurgical microscope adapted for the purpose. The small pyramidal diamond indentor is embedded in the surface of a special objective lens. The surface of the test sample is prepared to a high polish and etched for micro-examination. When viewed under the microscope with a high magnification, usually some value

Table 27.3 *Rockwell hardness scales.*

Scale	Symbol	Indentor	Total indenting load	Material for which the scale is used
A	H_{RA}	Diamond cone	60 kg	Thin hardened steel strip
B	H_{RB}	$\frac{1}{16}$-inch diameter steel ball	100 kg	Mild steel and non-heat treated medium carbon steels
C	H_{RC}	Diamond cone	150 kg	Hardened and tempered steels and alloy steels
D	H_{RD}	Diamond cone	100 kg	Case hardened steels
E	H_{RE}	$\frac{1}{8}$-inch diameter steel ball	100 kg	Cast iron, aluminium alloys and magnesium alloy
F	H_{RF}	$\frac{1}{16}$-inch diameter steel ball	60 kg	Copper and brass
G	H_{RG}	$\frac{1}{16}$-inch diameter steel ball	150 kg	Bronzes, gun metal and beryllium copper
H	H_{RH}	$\frac{1}{8}$-inch diameter steel ball	60 kg	Soft aluminium and thermoplastics
K	H_{RK}	$\frac{1}{8}$-inch diameter steel ball	150 kg	Aluminium and magnesium alloys
L	H_{RL}	$\frac{1}{4}$-inch diameter steel ball	60 kg	Soft thermoplastics
M	H_{RM}	$\frac{1}{4}$-inch diameter steel ball	100 kg	Thermoplastics
R	H_{RR}	$\frac{1}{2}$-inch diameter steel ball	60 kg	Very soft thermoplastics

between $\times 200$ and $\times 2000$, any particular microconstituent or feature can be centred in the field of view and a micro-sized diamond indentation made using a small indenting load.

The *Knoop* test was developed in the United States of America and uses a diamond pyramid indentor designed to give a long thin impression, the length being seven times greater than the width and about thirty times greater than its depth. This shape offers an advantage over the square pyramid of the Vickers test for microhardness work in that the length, l, of a Knoop impression is about three times greater than the diagonal, D, of a Vickers impression and can be measured with a greater degree of accuracy.

The Knoop hardness number, H_K, is given by

$$H_K = \frac{\text{load}}{\text{projected area of impression}}$$

From the geometry of the indentor

$$H_K = \frac{10F}{l^2 \times 7.028}$$

The range of loads used with the Knoop indentor is similar to that used for Vickers microhardness tests. Knoop hardness test results are very similar to those obtained from the Vickers test but are consistently 20 or 25 numbers above Vickers values for the same material. Standards BS 5441 Part 6 (1988) and ASTM E 384-89 cover microhardness testing.

The Shore scleroscope test

This is a dynamic test and involves allowing a small diamond tipped weight to fall freely through a known height onto the test-piece surface and measuring the height of rebound of the weight. In the test apparatus, the small weight of a half ounce (14.2 g) falls freely through a height of 10 inches (250 mm) in a graduated glass tube onto the surface of the material being tested. The tube is graduated into 140 equal divisions. The rebound height of the weight is estimated, by eye, against the graduations on the tube. The Shore test is particularly useful for measuring the hardness of very hard metals and, because the test equipment is small and very portable, it is also very useful for the *in-situ* testing of parts such as gears and the surfaces of the large rolls used in metal working operations. The relationship between Shore hardness values and the results of static indentation tests is given in Table 27.4.

A dynamic test, such as the Shore test, can be used to gauge the elastic recovery response of a rubber type material. While this is not the same as hardness, when the test is used for metals, the rebound values obtained with rubbers and thermoplastics is termed their Shore hardness value and this

Table 27.4 *Approximate hardness conversions.*

H_D	H_B	H_{RB}	H_{RC}	Shore	H_D	H_B	H_{RB}	H_{RC}	Shore
20	19				520	482		51.1	66.5
40	38				540	497		52.4	68.0
60	57				560	512		53.7	69.5
80	76	31.9			580	527		54.8	71.5
100	95	52.5			600	542		55.7	73.0
120	114	66.3			620	555		56.7	74.5
140	133	76.1			640	568		57.6	76.0
160	152	83.4			660	580		58.5	77.5
180	171	89.2			680	592		59.3	79.0
200	190	93.8	14.0	31.5	700	602		60.1	80.5
220	209	97.5	18.0	34.5	720			60.9	82.0
240	228		21.8	38.0	740			61.7	83.5
260	247		25.1	40.5	760			62.5	85.0
280	266		28.2	43.0	780			63.3	86.5
300	285		30.0	45.5	800			64.0	88.0
320	304		33.4	48.0	820			64.8	89.0
340	323		35.7	50.0	840			65.5	90.5
360	342		37.8	52.0	860			66.3	92.0
380	361		39.8	54.0	880			67.0	93.5
400	380		41.7	55.5	900			67.7	94.5
420	399		43.5	57.5	920				96.0
440	418		45.1	59.5	940				97.5
460	437		46.7	61.0	960				98.5
480	452		48.2	63.0	980				100.0
500	467		49.7	64.5	1000				101.0

parameter is a good indication of the quality of the material. The Shore value for rubber and plastic materials is generally determined using a small instrument known as a *Durometer*, rather than using the standard Shore falling weight apparatus. A typical durometer is a compact hand-held device in which a rounded indentor is pressed into the material surface under the action of a spring or weight and a pointer registers a hardness value on a graduated scale. Various designs are available to cover the range of elastomers and plastics from the very soft to the very hard over the range Shore A to Shore D. The hardness of rubbers and plastics is often quoted on the IHRD scale (International rubber hardness degrees). The IHRD scale closely approximates to the Shore scale.

27.3 | Relationships between hardness and other properties

As a general rule it appears that, for metals, as the hardness value increases, so also do properties such as tensile, compressive and shear strengths. There is no specific relationship between hardness and strength which holds for all metallic materials but some empirical equations have been used to estimate the tensile strengths of some metals from a hardness value. One such relationship, which is valid for annealed and normalised steels, is

tensile strength $(MPa) = H_D \times 3.4$

This equation does not hold for either heavily cold worked steels or for austenitic steels. However, hardness testing, which is a rapid and relatively simple process, is used frequently as an approximate means for assessing the tensile strength of materials.

Similarly, there is a general trend between the hardness of metals and other properties, such as ductility and toughness. Most materials of high ductility tend to be relatively soft and the ductility decreases as the hardness of the material increases. Toughness, or its antithesis, brittleness, also tends to vary with hardness with very hard materials having a tendency to be very brittle, but these are only tendencies and there are no empirical equations which enable a hardness number to be converted into, say, an impact strength value.

A work hardening index for a metal, the *Meyer index*, can be derived from a series of Brinell hardness tests. For this a series of hardness measurements is made using varying indenting loads but with the same sized ball indentor. The Meyer relationship is

$F = ad^n$

where F is the indenting load (kgf), d is the diameter of indentation (mm), and a and n are constants of the material and its condition. The resistance to indentor penetration is represented by a, and n is the work hardening index. The expression may be written as

$\ln F = \ln a + n \ln d$

A graphical plot of $\ln F$ against $\ln d$ will give a straight line from which both a and n can be evaluated.

27.4 | Tensile, compressive and shear testing

A knowledge of the mechanical properties of materials in tension, compression, shear and bending is needed by design engineers and a series of tests, many of which have been standardised, have been developed for the determination of these properties. In the case of metals and many plastics materials, tensile testing is used to a much greater extent than compression or shear testing, mainly because the problems associated with this form of testing tend to be less. Materials such as concrete, ceramics and glasses are generally much stronger in compression than in tension (this is due mainly to the presence of porosity and other internal flaws) and tend to be used in situations where the service stresses are largely compressive. Compressive testing is widely used for this type of material. When it is required to know the tensile strength of such brittle materials, it is often determined by testing the material as a beam in bending using a three-point loading arrangement or by testing a cylindrical test-piece in diametral compression (see section 27.10).

27.5 | Testing machines

There are many different types of testing machine available. Some are designed to perform one type of test, for example, tensile, while others are of the 'Universal' type and as such are suitable for uniaxial testing in both tension and compression and also for three-point bend tests on beam type test-pieces. Some designs are small 'table-top' machines with maximum force capacities ranging between 500 N and 20 kN, while at the other end of the scale large machines with load capacities of 1 MN or greater exist.

There are certain features which must exist in any testing machine be it for tensile, compressive or shear tests, and irrespective of size. These are:

(a) a system for locating and holding the test-piece in a satisfactory manner,

(b) a mechanism for applying a force to the test-piece and for varying the force at a controlled rate,

(c) a system for accurate measurement of the applied force.

In addition, some designs of testing machines incorporate systems for the accurate measurement and recording of changes in test-piece dimensions but, generally, such measurements are made using separate devices, such as extensometers and torsionmeters, which can be attached to the test-piece (see section 27.6).

The system used within a testing machine for the application of force may be either mechanical or hydraulic. The mechanical system normally comprises a screw, or screws, attached to the load-applying cross-head, and the screw is moved by means of a rotating nut. In a hydraulic system, the load is applied by a hydraulic ram moving in an oil-filled cylinder. In many test situations, it is important that the rate of strain is kept constant. Most types of machine have the facility for operating at several specific strain rates but some are capable of infinitely variable rates of strain between fixed upper and lower limits. The results obtained during the testing of metal samples is not affected by the rate of strain employed but polymeric materials are generally strain-rate sensitive. Specific rates of strain for these materials are recommended in the relevant standards.

The force applied to the test-piece may be measured in one of several ways. The oldest system, but highly accurate, is that involving a mechanical lever with a movable jockey weight. The major disadvantage of the lever system is that it is bulky and occupies a lot of space, particularly for a large capacity machine. Some machine types in which the load is applied hydraulically use a hydraulic system for measuring the load on the test-piece. In this arrangement, the increasing oil pressure, as the force is applied to the test-piece, moves a piston in a small calibrated pressure cylinder against a weighted pendulum lever, moving the pendulum out of the vertical position. As the force, and hence the oil pressure, increases so the pendulum is moved further from the vertical. Movement of the pendulum causes a pointer to move around the face of a calibrated dial scale. This is also a measurement system capable of high accuracy. A third system of load measurement utilises the deflection of a spring. This is the principle used in the small Monsanto Hounsfield Tensometer table-top machine in which the elastic deflection of a steel beam under load causes a piston to move in a small cylinder containing mercury. The mercury is forced into a glass tube mounted alongside a calibrated scale. A range of beams of varying degrees of stiffness is available to give a series of load ranges. A fourth system of load measurement involves the use of load cells. A load cell may be either a transducer, or a carefully prepared piece of material fitted with sensitive strain gauges. Application of a load creates an electrical output signal from the cell and this is amplified and presented as a digital meter readout and/or a graphical display on a pen-recorder. The standards of accuracy for tensile testing machines are given in BS 1610 (1985) and ASTM E 4-89.

27.6 | Measurement of strain

Many testing machines are fitted with autographic recorders which give a graphical display of the force–deflection behaviour of the test-piece. However, the graphs produced must be treated with caution. For example, in a tensile test while an autographic record shows the correct shape of the force–extension diagram for the material and also shows accurate values of load applied, the readings on the extension axis refer to the separation distance between the test-piece holding grips and not the extension relative to the gauge length of the test sample. Many modern testing machines are interfaced with a microcomputer and after completion of a test a full tabulated and graphical printout of the test results is obtained including, if required, a comparison with previous test results.

The accurate measurement of test-piece dimensional change and, hence, strain is generally achieved by attaching a sensitive measurement device to the test-piece. The devices used for the measurement of longitudinal strain are termed *extensometers* and those designed to measure strain during a torsion test are termed *torsionmeters*.

One of the most commonly used extensometers is the Lindley type (Figure 27.1). This is a robust, yet sensitive device. The Lindley extensometer is attached to a test-piece by tightening two screw grips, which are set 50 mm apart. When a force is applied to the test-piece and strain occurs, relative movement between the gripping points is transmitted through a lever to a dial gauge. The dial gauge is calibrated in steps equivalent to an extension of 0.001 mm and the maximum amount of extension which may be measured is 2.5 mm.

Figure 27.1 *Lindley-type extensometer positioned on a test-piece.*

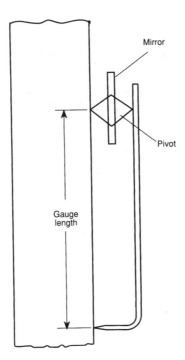

Figure 27.2 *Principle of Marten's mirror extensometer.*

Electronic extensometers are an integral feature of modern testing machines, the extensions being displayed as a digital LED or LCD display. Some extensometers operate on an optical principle. One such is Marten's extensometer (Figure 27.2). A change in the gauge length of the material will cause an angular movement of the mirror. A scale is viewed through the mirror by means of a telescope. Alternatively, a light source may project, by way of the mirror, a spot of light onto a graduated scale. Optical extensometers are highly sensitive and are often used for the measurement of small strains during long-term creep tests.

Another device which may be used for the measurement of strain is the electric resistance strain gauge. The resistivity of some conductors is very sensitive to variations in elastic strain. A resistance strain gauge consists of a zig-zag of very fine wire mounted between waterproof sheets. The gauge is securely bonded to the surface of the material under test and the ends of the wires connected to a bridge network so that changes in resistance can be measured.

In torsion testing, a cylindrical shaped test-piece is subjected to axial torsion and the strain measurement which needs to be taken is the angle of twist. A common type of torsionmeter is similar to the Lindley-type extensometer with angle of twist transmitted via a lever to a dial gauge. Such an instrument is capable of giving an angular reading to an accuracy of 0.001 radian over a gauge length of 50 mm.

27.7 | The tensile testing of metals

The metal test-pieces used for a standard tensile test are shaped in such a way that fractures will occur within the desired portion, that is, within the gauge length, and definite standards are laid down for their dimensions (Figure 27.3). In some instances, particularly in some fundamental experimental programmes, very small test-pieces may be used, but these small sizes are not quoted in the relevant testing standards as the results obtained from small test-pieces may not be truly representative of the properties of bulk material. The recommended dimensions for metal tensile test-pieces (Tables 27.5 and 27.6) are

(a)

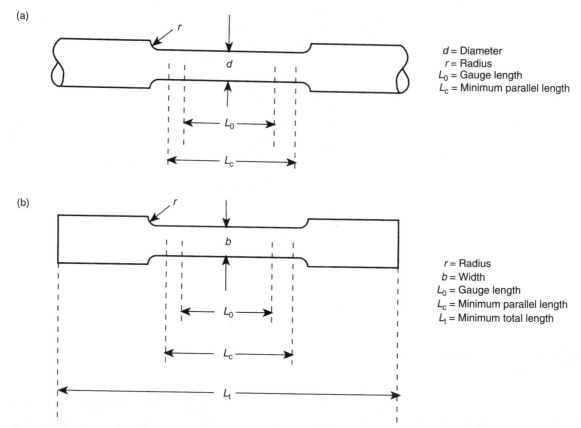

d = Diameter
r = Radius
L_0 = Gauge length
L_c = Minimum parallel length

(b)

r = Radius
b = Width
L_0 = Gauge length
L_c = Minimum parallel length
L_t = Minimum total length

Figure 27.3 *Types of tensile test-piece: (a) round test-pieces; (b) flat test-pieces.*

given in BS 18 (1987) and ASTM E 8-89b and E 8M-89b. (The testing of cast iron is covered in a separate publication, BS 1452 (1990).)

The shape of the load–extension curve for a non-ferrous metal is shown in Figure 27.4(a). The initial strain is elastic, but beyond point E, the elastic limit, strain is plastic. Point U is the maximum load, and this value of load is used for the determination of the tensile strength of the material. Point F marks the point of fracture. Although the applied load on the test-piece decreases beyond point U, the true stress acting on the test-piece, taking into account the reducing cross-sectional area, continues to increase until fracture occurs. (Refer to section 27.8.) In the commercial testing of metals, it is the load–extension curve, rather than a true stress–strain curve, that is plotted and strengths are calculated on the basis of the original cross-sectional area of a test-piece. The following information is determined in a routine tensile test.

Tensile strength (TS) (formerly known as ultimate tensile strength). This is based on the maximum load sustained by the test-piece, when the latter is tested to destruction, and corresponds to point U in Figure 27.4. The numerical value of tensile strength is calculated as a nominal stress and is given by

$$\text{TS} = \frac{\text{maximum load applied}}{\text{original cross-sectional area}}$$

The units in which tensile strength is normally quoted are megapascals (MPa), meganewtons per metre^2 (MN/m^2) or newtons per millimetre^2 (N/mm^2). Numerically these three values are equal. (In the construction industry, the units used generally are N/mm^2.)

Table 27.5 *Round bar tensile test-pieces. Table of standard dimensions.*

S_0 (mm²)	d (mm)	L_0 (mm)	L_c (mm)	r wrought materials (mm)	r cast materials (mm)
200	15.96	80	88	15	30
150	13.82	69	76	13	26
100	11.28	56	62	10	20
50	7.98	40	44	8	16
25	5.64	28	31	5	10
12.5	3.99	20	21	4	8

d = diameter of test-piece, L_0 = gauge length, L_c = parallel length, S_0 = original cross-sectional area, r = radius at shoulder (refer to Figure 27.3(a)). For proportional test-pieces $L_0 = 5.65\,S_0$, $L_0 \approx 5d$.

Source: Table 27.5 is an extract from BS 18 and is reproduced by permission of the British Standards Institution, 2 Park Street, London W1A 2BS, from whom complete copies of the standard can be obtained.

Table 27.6 *Flat tensile test-pieces. Table of standard dimensions.*

b (mm)	L_0 (mm)	L_c (mm)	L_t (mm)	r (mm)
25	100	125	300	25
12.5	50	63	200	25
6	24	30	100	12
3	12	15	50	6

b = width of test-piece, L_0 = gauge length, L_c = parallel length, L_t = total length, r = radius at shoulder (refer to Figure 27.3(b)).

Source: Table 27.6 is an extract from BS 18 and is reproduced by permission of the British Standards Institution, 2 Park Street, London W1A 2BS, from whom complete copies of the standard can be obtained.

Yield point or *yield stress*. There is a sharp discontinuity in the load–extension diagram for wrought iron and many steels (Figure 27.4(b)) and the material will suddenly yield with little or no increase in the applied load necessary (point Y in the figure). The extent of this sudden yielding is about 5–7 per cent of the original gauge length. The yield point Y is close to, or coincident with, the elastic limit. The yield stress of the material is given by

$$\text{Yield stress} = \frac{\text{applied load at yield point}}{\text{original cross-sectional area}}$$

Proof stress or *offset yield stress*. Most metals, unlike many steels, show a smooth transition from elastic to plastic deformation behaviour. It may be difficult to determine the value of elastic limit with exactitude and the parameter *proof stress* is determined. This is the level of stress required to produce some specified small amount of plastic deformation. Often the amount of plastic strain specified is 0.1 per cent (nominal strain = 0.001) but other values of plastic strain may also be used. Frequently, proof stress is referred to by the alternative term *offset yield stress*, or just *yield stress*. In Figure 27.4(c), the load–extension diagram for some test-piece is shown in curve OEPUF. If OA corresponds to some percentage x of the original gauge length, and AP is drawn parallel to OE, then the point P denotes the

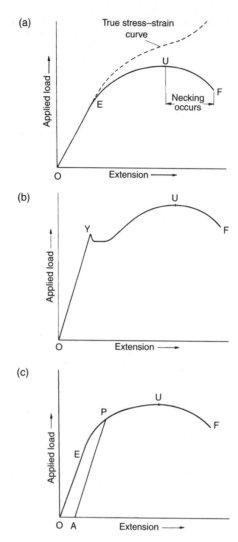

Figure 27.4 *(a) Load–extension curve and its relation to a true stress–strain curve. (b) Load–extension curve with sharp yield point (mild steel). (c) Determination of proof load from load–extension diagram.*

x per cent proof load. The value for the x per cent proof stress is given by

$$x \text{ per cent proof stress} = \frac{x \text{ per cent proof load}}{\text{original cross-sectional area}}$$

In the United Kingdom, values are normally quoted for either the 0.1 per cent proof stress, or the 0.2 per cent proof stress, but in the USA it is customary to quote 0.2 per cent proof stress values. There will be a fairly wide difference between the value of proof stress and tensile strength for an annealed metal, but in a work hardened metal, or, in the case of a fairly hard and brittle material, the proof stress and tensile strength values will be fairly close to one another. The extent of the separation between proof stress and tensile strength values gives a measure of the amount of cold work that may be performed on the material.

Limit of proportionality and *elastic limit*. The limit of proportionality is the value of the stress at which the stress–strain curve ceases to be a straight line, that is, the point at which Hooke's law ceases to apply. The elastic limit may be defined as the level of stress at which strain ceases to be wholly elastic. In most cases, these two values will be the same, but some polycrystalline metals do not completely obey Hooke's law. Such materials are termed anelastic, but the stress–strain curve does not vary greatly from linear. It is very difficult to determine a value for the limit of proportionality, as the values obtained depend on the sensitivities of the load-measuring system and the extensometer used. In general commercial testing, the limit of proportionality is not usually reported. In its place, for metals which do not show a marked yield point, a *proof stress* is quoted.

Modulus of elasticity, E (Young's modulus). The modulus of elasticity may be calculated from the slope of the straight-line portion of the load–extension curve. E is given by (gauge length/cross-sectional area) × slope. The units in which E is quoted may be gigapascal (GPa), giganewton/metre² (GN/m²) or kilonewton/millimetre² (kN/mm²).

Percentage elongation. A definite length, the gauge length, is marked off on the test-piece before testing. After fracture, the two portions of the test-piece are placed together and the distance between gauge marks is remeasured. The amount of extension, expressed as a percentage of the original gauge length, is then quoted as the elongation value:

$$\text{Percentage elongation on gauge length} = \left(\frac{L - L_0}{L_0}\right) \times 100$$

where L is the length between gauge marks after fracture and L_0 is the original gauge length. For an elongation figure to have any validity, the fracture must occur in the central section of the gauge length and the gauge length must be specified, for example, *the percentage elongation on 50 mm is 20 per cent*. As the amount of plastic deformation of the test-piece is greatest nearest to the point of fracture, the elongation value for any particular material will be much higher if measured over a short gauge length than if measured over a long gauge length. The percentage elongation value for a material will give a measure of its ductility.

Percentage reduction of area. The percentage reduction of area is often quoted for round-bar specimens instead of a percentage elongation value. There is a certain merit in this as the reduction of area value is largely independent of specimen dimensions and gauge length. It is the difference in area between the cross-sectional area of the test-piece at the point of fracture and the original cross-sectional area, expressed as a percentage of the original cross-sectional area:

$$\text{Percentage reduction of area} = \left(\frac{A_0 - A}{A_0}\right) \times 100$$

where A_0 is the original cross-sectional area and A is the cross-sectional area at the point of fracture.

EXAMPLE 27.2

The following data were obtained in a tensile test on a specimen of 15 mm diameter with a 50 mm gauge length:

Load (kN)	70	120	150	160	170	200	220	233	235	220
Extension (mm)	0.25	0.40	0.50	0.60	0.75	1.75	3.00	5.00	6.50	8.00

The specimen diameter after fracture was 12.45 mm.
Determine: (i) tensile strength, (ii) Young's modulus, (iii) 0.2 per cent proof stress, (iv) percentage elongation and (v) percentage reduction in area.

SOLUTION

From the data plot a force–extension diagram, as shown in Figure 27.5

Figure 27.5

Cross-sectional area of test-piece $= \pi/4 \times 15^2 = 176.7\,\text{mm}^2$

(i) From graph plot maximum force $= 235\,\text{kN}$
$$\text{Tensile strength} = 235 \times 10^3/176.7 = 1330\,\text{MPa}$$

(ii) Slope of elastic portion of curve $= 150 \times 10^3/0.5 = 300 \times 10^3\,\text{N/mm}$
$$\text{Young's modulus} = (300 \times 10^3 \times 50)/176.7 = 84.9\,\text{GPa}$$

(iii) 0.2 per cent of 50 mm $= 0.1$ mm. Draw a line parallel to the linear portion of curve from an extension of 0.1 mm. This cuts the curve at a load of 164 kN.
$$0.2 \text{ per cent proof stress} = 164 \times 10^3/176.7 = 928\,\text{MPa}$$

(iv) On the force–extension curve draw a line from the fracture point, parallel to the linear elastic portion. This cuts the abscissa at an extension of 7.3 mm.
The percentage elongation on a 50 mm gauge length $= (7.3/50) \times 100 = 14.6$ per cent.

(v) Diameter after fracture $= 12.45$ mm. Cross-sectional area $= \pi/4 \times 12.45^2 = 121.7\,\text{mm}^2$
Percentage reduction in area $= \{(176.7 - 121.7)/176.7\} \times 100 = 31.1$ per cent.

27.8 | True stress and true strain

Generally, the data obtained during a tensile test are plotted as a *force–extension* curve and sometimes this type of curve is referred to, erroneously, as a *stress–strain* curve. The force–extension data obtained from a tensile test may be plotted in the form of stress and strain to give a nominal stress–nominal strain curve in which nominal stress $\sigma_n = F/A_0$, where F is the force and A_0 is the original cross-sectional area (c.s.a.) of the test-piece, and nominal strain $\varepsilon_n = (L - L_0)/L_0$, where L is the extended length and L_0 is the original gauge length. Even so, such a nominal stress–strain curve is not a true stress–strain curve because it is based on the original test-piece dimensions and does not take into account the changes which occur during a test to destruction. A nominal stress–strain curve for a ductile metal, like a force–extension curve, shows a maximum before fracture. This is because, in a tensile test, there is a point of plastic instability, corresponding to the maximum on a force–extension or nominal stress–nominal strain curve, at which there is a sudden necking of the test-piece. If a true stress–true strain curve were drawn this would show that stress increases steadily to fracture.

The true stress $\sigma_t = F/A_i$, where A_i is the instantaneous cross-sectional area, and true strain

$$\varepsilon_t = \frac{dL}{L} = \ln\left(\frac{L}{L_0}\right)$$

But

$$\frac{L}{L_0} = (1 + \varepsilon_n)$$

so

$$\varepsilon_t = \ln(1 + \varepsilon_n)$$

where L_0 is the initial gauge length and L is the instantaneous gauge length. Assuming there is no volume change during plastic deformation, $A_i L = A_0 L_0$ or

$$A_i = \frac{A_0 L_0}{L}$$

so

$$\sigma_t = \frac{F L}{A_0 L_0}$$

But

$$\frac{L}{L_0} = (1 + \varepsilon_n)$$

so

$$\sigma_t = \frac{F}{A_0}(1 + \varepsilon_n)$$

For most metals the relationship between true stress and true strain during plastic deformation can be written as

$$\sigma_t = k\varepsilon_t^n \text{ or } \log \sigma_t = \log k + n \log \varepsilon_t$$

where k and n are constants for the material. The constant n is the strain hardening exponent and the true stress–true strain equation is an empirical expression for work hardening. Instability and necking of a ductile metal occur when $\varepsilon_t = n$ and at this point, the maximum on the nominal stress–strain curve, the maximum value of σ_n, the nominal tensile strength of the material, is given by

$$\sigma_{max} = k n^n (1 - n)^{(1 - n)}$$

EXAMPLE 27.3

From the data in Example 27.2 (Figure 27.5) determine the true stress at a nominal strain of 8 per cent.

SOLUTION

A nominal strain of 8 per cent (ε_n) on a 50 mm gauge length is equal to an extension of 4 mm. From the force–extension diagram the force needed to give an extension of 4 mm is 230 kN.

True stress $\sigma_t = (F/A_0) \times (1 + \varepsilon_n)$

$\qquad\qquad = \{(230 \times 10^3)/176.7\} \times 1.08$

$\qquad\qquad = 1406$ MPa

EXAMPLE 27.4

The following information was obtained when a metal test-piece of 7.5 mm diameter and with a gauge length of 50 mm was subjected to tensile loading.
   When the load was 10.5 kN the distance between gauge marks was 52 mm.
   When the load was 11.1 kN the distance between gauge marks was 59.5 mm.
Assuming that $\sigma_t = k\varepsilon_t^n$, determine:

(a) values for the coefficient k and exponent n,

(b) the amount of extension at maximum load,

(c) the maximum load applied, and

(d) the tensile strength of the metal.

SOLUTION

(a) Use the expressions $\sigma_t = (F/A_0) \times (1 + \varepsilon_n)$ and $\varepsilon_t = \ln(1 + \varepsilon_n)$
   The cross-sectional area $A_0 = 7.5^2 \times \pi/4 = 44.18 \, \text{mm}^2$

   When $F = 10.5 \, \text{kN}$ the extension is $(52 - 50)$ so $\varepsilon_n = 2/50 = 0.04$

   $\sigma_t = (10.5 \times 10^3/44.18) \times 1.04 = 247.2 \, \text{MPa}$ and $\varepsilon_t = \ln(1 + 0.04) = 0.039$

   When $F = 11.1 \, \text{kN}$ the extension is $(59.5 - 50)$ so $\varepsilon_n = 9.5/50 = 0.19$

   $\sigma_t = (11.1 \times 10^3/44.18) \times 1.19 = 299.0 \, \text{MPa}$ and $\varepsilon_t = \ln(1 + 0.19) = 0.174$

   So substituting these values in the expression $\log \sigma_t = \log k + n \log \varepsilon_t$
   we have

   $\log 247.2 = \log k + n \log 0.039$

   and

   $\log 299.0 = \log k + n \log 0.174$

   Solution of these two equations gives $k = 373.2 \, \text{MPa}$ and $n = 0.127$.

(b) At maximum load instability occurs and $\varepsilon_t = n$, so at maximum load

   $\varepsilon_t = 0.127 = \ln(1 + \varepsilon_n)$ from which $\varepsilon_n = 0.135$

   A nominal strain of 0.135 means an extension of $0.135 \times 50 = 6.75 \, \text{mm}$ on a gauge length of 50 mm.

(c) At maximum load $\sigma_t = k \varepsilon_t^n$, but $\varepsilon_t = n$ so $\sigma_t = 373.2 \times 0.127^{0.127} = 287.2 \, \text{MPa}$
   But $\sigma_t = (F/A_0) \times (1 + \varepsilon_n)$, so the maximum load $F = (287.2 \times 44.18)/1.135 = 11.179 \, \text{kN}$.

(d) The tensile strength $= F_{max}/A_0 = 11179/44.18 = 253 \, \text{MPa}$.

27.9 | The tensile testing of plastics

The tensile testing of plastics materials is conducted in a generally similar way to the tensile testing of metals. There are differences, however, and these are necessary because of the different nature of plastics materials and metals. Many thermoplastic materials do not show Hookean elasticity even at low stress levels but are viscoelastic. In other words, the strain developed is not dependent on the level of stress alone, as would be the case for an elastic solid, but also depends on the length of time for which the stress is applied. One consequence of this is that variations in the rate of strain can give differences in the force–extension values and, hence, different test results (see Figure 27.6). It is recommended in

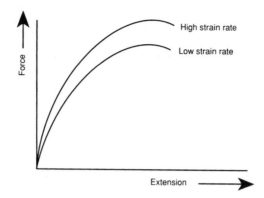

Figure 27.6 *Influence of rate of strain on the force–extension values for a thermoplastic material.*

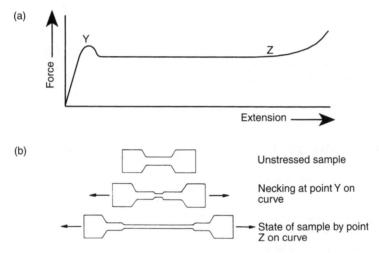

Figure 27.7 *(a) Typical load–extension curve for a thermoplastic which cold draws. (b) Stages in cold drawing of a test-piece.*

the relevant standards publications that testing be conducted at comparatively high rates of strain (refer to BS 2782, Part 3, Methods 320A to 320F (1986) and ASTM D 638-89 and D 638M-89).

Unlike metals, many thermoplastic materials cold draw during a tensile test, and in these cases values are quoted for the yield stress and the drawing stress, these being

$$\text{Yield stress} = \frac{\text{yield load}}{\text{original c.s.a.}} \qquad \text{Draw stress} = \frac{\text{drawing load}}{\text{original c.s.a.}}$$

The phenomenon of cold drawing can occur in some thermoplastic materials and this is illustrated in Figure 27.7. At a stress corresponding to point Y in Figure 27.7(a) the test-piece necks down considerably and thereafter further strain takes place at a constant stress, usually slightly lower than the yield stress, with undrawn material being drawn into the necked zone. This drawn material is much stronger than the original undrawn plastic material due to the alignment of polymer molecules which occurs during drawing. The drawn state, with a more crystalline structure, is also referred to as *stress induced crystallinity*.

The load–extension curves that are obtained for many plastic materials either show no initial straight line portion, or show a departure from Hooke's law at very low values of applied load. In these cases, it would be very difficult, or impossible, to obtain a value for Young's modulus. The value that is quoted

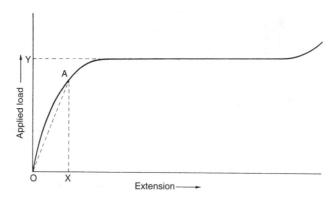

Figure 27.8 *Load–extension diagram for a thermoplastic. Y is the yield load. OX is an extension corresponding to a strain of 0.2 per cent. The slope of the straight line OA is used to determine the modulus of elasticity for the plastic, the secant modulus.*

as the modulus of elasticity, E, for plastics is in fact a *secant modulus*, and is obtained by determining the stress at a value of 0.2 per cent strain (see Figure 27.8).

Many thermoplastics extend to very considerable extents during a tensile test, but there is an almost instantaneous recovery of much of this strain as soon as the test-piece fractures. The elongation value quoted for thermoplastics is known as the *percentage elongation at break*, and is determined by noting the distance between the gauge marks at the moment of fracture, and not by placing the fractured portions together after the test, as is the case with tests on metals.

27.10 | Determining the tensile strength of brittle materials

The axial tensile testing of brittle materials, such as ceramics and glasses, would be extremely difficult, if not impossible, because of the problems of preparing suitable shaped test-pieces and those of effectively holding them within the testing machine. It is customary to determine the fracture strength from a three-point bend test. The flexural strength value determined in this type of test is also known as the *modulus of rupture* of the material. When a sample is subjected to bending, as shown in Figure 27.9, a compressive stress is generated in the upper surface and a tensile stress is generated in the lower surface. The tensile strength of these materials is less than the compressive strength. When a sample is tested in bending in this way fracture commences at the tensile surface so the breaking load, F, is related to the tensile strength of the material. The magnitude of the direct stress, σ, is related to the bending moment, M, by the general bending equation

$$\frac{\sigma}{y} = \frac{M}{I}$$

where y is the distance from the neutral surface (half the thickness for a sample of symmetrical section) and I is the second moment of area of the section.

$I = \dfrac{BD^3}{12}$ for a rectangular section of width B and thickness D

$I = \dfrac{\pi D^4}{64}$ for a circular section of diameter D

The maximum value of bending moment for a symmetrical three-point loading system as in Figure 27.9 is $FL/4$ and so

$$\sigma = \frac{My}{I} = \frac{FLD}{8I}$$

Figure 27.9   *Principle of modulus of rupture test by three-point loading.*

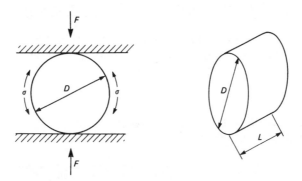

Figure 27.10   *Principle of split cylinder test for tensile strength.*

For a sample of rectangular section, therefore

$$\sigma = \frac{3FL}{2BD^2}$$

The modulus of rupture value for a material is approximately double the true tensile strength. This mode of testing is also used to determine the fracture strength of rigid thermoset materials including laminated plastics such as 'Tufnol', a phenol–formaldehyde resin laminated with either cloth or paper, and is covered by BS 2782, Part 3, Method 335A (1989) and ASTM D 790-86 and D 790M-86.

Another indirect method which may be used to determine a tensile strength is the split cylinder test. This type of test, shown diagrammatically in Figure 27.10, is widely used for determining the tensile strength of concrete and is detailed in BS 1881, Part 117 (1983) and ASTM C 496-86. As the compression force is imposed, circumferential tensile stresses are induced in the material, these having maximum value at the central horizontal plane. The cylinder will eventually fail and split at this plane. The value of the tensile stress, σ, at failure is given by

$$\sigma = \frac{2F}{\pi LD}$$

where F is the maximum force applied, L is the cylinder length and D is its diameter.

EXAMPLE 27.5

A plain concrete beam with the dimensions, length 2.2 m, width 150 mm, depth 250 mm, is simply supported, the supports being 1.8 m apart. A load is applied at mid-span and the maximum load at failure was 21.2 kN. Determine the tensile strength of the concrete.

SOLUTION

For a beam in three-point loading the tensile strength, or modulus of rupture, is given by

$3FL/2bd^3$

Substituting the test values we have

$(3 \times 21.2 \times 10^3 \times 1.8)/(2 \times 0.15 \times 0.25^2) = 6.1$ MPa

EXAMPLE 27.6

A cement cylinder of 100 mm diameter and 100 mm length is tested in diametrical compression and the maximum force recorded in the test is 55.8 kN. Determine the tensile strength of the cement.

SOLUTION

In a cylinder test the tensile strength is given by $2F/\pi LD$
So, tensile strength $= (2 \times 55.8 \times 10^3)/(\pi \times 0.1 \times 0.1) = 3.55$ MPa.

One test result alone is not sufficient for a brittle material, such as a glass or ceramic. There is a major variation in strength for such materials because of the presence and distribution of cracks, pores and other flaws of differing sizes. Many test results are necessary in order to obtain a statistical mean. Weibull devised a method for dealing with the statistical nature of the fracture strength of such brittle materials and he defined the probability of survival, $P_s(V_0)$, as the fraction of identical samples of volume V_0 which survive after the application of a tensile stress of σ. The Weibull expression is

$$P_s(V_0) = \exp(-\sigma/\sigma_0)^m$$

where σ_0 is the value of tensile stress at which $1/e$, that is 37 per cent, of the samples do not fail and the power m is the Weibull modulus.

The Weibull expression can be written in logarithmic form:

$$\ln P_s(V_0) = (-\sigma/\sigma_0)^m$$

A plot of σ/σ_0 against $\ln P_s(V_0)$ should give a straight line from which the modulus m can be evaluated. The higher the value of m, the less will be the variation in strength of a group of similar samples (see Figure 27.11). The value of the Weibull modulus for ceramic materials such as brick and pottery is about 5, while the value for some engineering ceramics, such as alumina and silicon nitride, is about 10. Most metallic materials possess Weibull moduli of around 100. With such a high value of m, the material can be considered as having a single failure strength.

27.11 | Compression testing

The compression testing of metals is little used. A tensile test sample is fairly long and waisted down to give a parallel-sided central portion, within which the tensile characteristics of the material can be accurately assessed. In compression testing, a long sample cannot be used because buckling failure would occur rather than direct axial compression. Test-pieces have to be short and the length not be greater than three times the diameter to avoid buckling. The other problem is that of friction between the ends of the test-piece and the platens of the testing machine. As compressive strain occurs, reducing

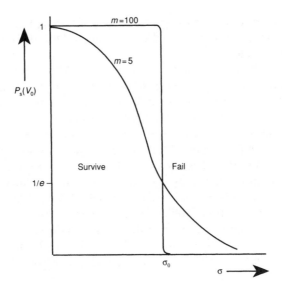

Figure 27.11 *Weibull curves showing probability of survival.*

Figure 27.12 *(a) Ductile compressive failure. (b) Brittle compressive failure.*

the test-piece length, so there is a corresponding increase in the diameter of the sample. Friction hinders lateral expansion of the diameter at the test-piece ends and once plastic deformation occurs the material tends to deform into a barrel shape (Figure 27.12(a)). This type of deformation generates tensile stresses at the surface of the material and failure is based more on the tensile strength of the material than anything else. Brittle materials, when tested in compression, normally fail by shear at 45° to the direct stress axis. The type of fracture may be either of the double cone type or straight shear (Figure 27.12(b)). These failure modes apply to all brittle materials, whether metallic or non-metallic.

Compressive test-pieces tend to be small and of cylindrical shape, as mentioned above, but compression test samples of cement or concrete are often cast into cube shapes with edge lengths of either 100 or 150 mm. (See BS 1881, Part 116 (1983) and ASTM C 39-86.)

27.12 | Testing in shear

There are two main types of test which may be used for the determination of the properties of a material in shear. One is based on the application of a torque to a cylindrical sample causing the test-piece to twist while the other involves direct shear. There are problems associated with both types of test. The former method causes the development of an almost pure shear stress within the material but the shear stress is not uniform throughout and varies from zero at the central axis to a maximum value at the surface of the cylinder. For pure torsion on a cylinder the general torsion equation applies within the elastic range:

$$\frac{\tau}{r} = \frac{T}{J} = \frac{G\theta}{L}$$

where τ is the shear stress at radius r from the central axis, T is the applied torque, J is the second polar moment of area, G is the modulus of rigidity of the material, and θ is the angle of twist in radians over the gauge length, L, of the test-piece.

The value of J for a solid cylindrical section is given by $\pi D^4/32$ and for a hollow cylindrical section by $\pi(D^4-d^4)/32$ where D and d are the outside and inside diameters, respectively.

As the applied torque, T, is increased so the angle of twist, θ, increases in direct proportion within the elastic range. The fact that the level of shear stress varies across a diameter is relatively unimportant when all stresses and strains are elastic. However, results become unreliable where the surface shear stress reaches the elastic limit and plastic strain commences. Plastic deformation tends to spread inwards in an irregular manner. Materials such as mild steel with a pronounced yield point tend to show greater irregularities than other metals. One method used to reduce the problem is to employ hollow cylindrical test-pieces. In this way, differences between the value of shear stress at outer and inner surfaces are reduced, but a further problem is encountered. The tendency for plastic buckling of the tubular specimen increases rapidly as the wall thickness is reduced.

The results of a torsion test may be plotted graphically in the form torque, T, against angle of twist, θ. The value of the modulus of rigidity, G, may be obtained from the slope of the linear elastic portion of the curve (T/θ).

Direct shear tests are sometimes used but it is very difficult to obtain accurate quantitative data from them. Several systems have been adopted, the main ones being: double shear for round bar test-pieces using a fork and eye device, double knife shear for samples with a rectangular section, and the shearing of a disc from sheet materials using a punch and die. The test equipment generally takes the form of attachments which can be used in conjunction with a universal testing machine. These three methods all suffer from the same disadvantage. Even when all parts of the attachments are made to close dimensional tolerances from very hard materials, there will be some bending of the test-piece and the stresses within the material will not be of the pure shear type but will include bending stresses. The results of such tests, while not providing fundamental values of shear strength, can be used to give a qualitative assessment and indicate the behaviour patterns of the material for fabrication processes involving blanking and shearing.

27.13 | Notch impact testing

In this type of test a bar specimen with a milled notch is struck by a fast-moving hammer, and the energy that is absorbed in fracturing the test-piece is measured. This type of test has the advantage of revealing a tendency to brittleness that is not revealed by the standard tensile test, or by a hardness test. As mentioned in Chapter 10, the ductile-to-brittle transition temperature is considerably higher for impact loading conditions than for slow straining conditions. It is also possible for two samples of a material to possess very similar properties, as determined in the standard tensile test, but to possess widely different properties when tested under impact loading conditions. This is illustrated in Table 27.7 for samples of a low-alloy steel subject to temper brittleness.

One major use of the notch impact test is to determine whether heat treatment of a material has been carried out successfully, as this type of test gives the most information about that. The impact strength should not, however, be judged alone but viewed in conjunction with the results of other tests, including the standard tensile and hardness values.

The types of notched bar impact test that are most widely used in the United Kingdom are the Charpy test and the Izod test. In both types of test a heavy pendulum is released and is allowed to strike a test-piece at the bottom of its swing. A proportion of the energy of the pendulum is absorbed in fracturing the test-piece. The height of the follow-through swing of the pendulum is measured, and the energy absorbed in fracture determined. In the Charpy test the test-piece is in the form of a simply supported beam and the sharp edge of the pendulum strikes at mid-span directly behind the milled notch.

Table 27.7 *Material with similar tensile properties but different impact loading properties, according to condition.*

Composition (per cent)	Condition	Tensile strength (MN/m²)	Elongation on 50 mm (per cent)	Charpy impact value (J/mm²)
C 0.3 Ni 3.2 Cr 1.0	Hardened, tempered, and cooled rapidly after tempering	855	28.6	1.31
	Hardened, tempered, and cooled slowly after tempering	836	26.5	0.15

Figure 27.13 *(a) Charpy test specimen and position of impact. (b) Izod test specimen and position of impact.*

An Izod test specimen is tested in a cantilever mode. The test-piece is firmly clamped in a vice with the prepared notch level with the edge of the vice. The impact blow is delivered on the same side as the notch (Figure 27.13). Impact strengths are generally quoted in J/mm² and based on the cross-sectional area of the test-piece below the notch.

The impact strengths determined in both types of test are roughly comparable, provided that the same type of notch is used (see BS 131, Parts 1, 2 and 3 and ASTM E 23-88).

A miniature Charpy-type testing machine is available for the impact testing of plastics.

27.14 ▌ Fatigue testing

A component or structure in service may be subjected to fluctuating or alternating cycles of stress, but rarely can it be found that one constant type of loading cycle applies during the whole of the life of a component. Laboratory tests to determine fatigue strength are usually based either on an alternating stress cycle, with a mean stress of zero, or on a fluctuating stress with some positive value of mean stress, rather than making any attempt to duplicate exactly the variable conditions that may occur during the life of the component.

The types of fatigue test that are often employed are those that involve axially loading the test-piece alternately in the tensile and compressive modes, and rotary bending tests. Of the latter, the two

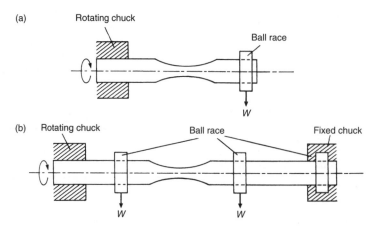

Figure 27.14 *Principle of rotary bending fatigue tests: (a) Wöhler cantilever system; (b) four-point loading.*

most common types are the Wöhler system, in which the test-piece is loaded as a cantilever (Figure 27.14(a)), and the four-point loading system (Figure 27.14(b)). With both of these loading systems, the test-piece is rotated so that, during one complete revolution, any element of the test-piece surface goes through a complete stress cycle of tension and compression. The four-point loading system offers the advantage of a constant value of bending moment over the central section of the test-piece. Because notches and sharp changes in cross-section act as points of high stress concentration, specimens for fatigue testing are prepared with polished surfaces, and the design is such that changes in section are gradual and well radiused. The various recommended types of fatigue test-piece, and their dimensions, are quoted in BS 3518, Parts 1 and 2 (1984) and ASTM E 466-82 and E 468-90.

The rate of cycling in fatigue testing is normally up to 150 Hz, but if a very rapid assessment is required for a material a high-frequency vibration technique may be used. With some high-frequency machines, tests may be carried out at up to 500 Hz. At this rate it takes only about five hours to complete 10^7 cycles. The results obtained, however, show fatigue limits as being slightly higher than those obtained from lower-frequency tests.

In certain cases there may be a need to conduct fatigue tests to failure on a complete engineering structure. Following the pressure-cabin failures in early Comet aircraft in 1951, a complete Comet structure was tested to destruction in a special test rig. The test simulated the cabin-pressure changes that occurred in each flight, and also wing flexing during flight. Since that date a similar test to destruction has been made on one air-frame of every civil-aircraft type destined for service. This has been a requirement of the airworthiness certification procedure.

27.15 | Creep testing

Almost all creep testing is conducted in the tensile mode and the test specimens are similar in form to those used in tensile testing (see section 27.7). Test-pieces for tensile creep testing may be of either circular or rectangular cross-section but there are no standard sizes. The actual dimensions of the test-pieces used will depend upon the type of creep testing machine used. The basic requirements for a creep testing machine are:

(a) that it must possess means for applying and maintaining a constant tensile load,

(b) there must be a furnace capable of keeping the temperature of the test-piece at the desired value to within very close limits,

(c) there should be means for the accurate measurement of test-piece extension.

Figure 27.15 *String of test-pieces for stress-rupture testing.*

This last requirement is not necessary if the equipment is to be used only for the determination of stress-to-rupture data.

Tensile creep test machines are designed so that the test-piece is mounted vertically and, generally, the axial load is applied to the specimen holder by dead weights and a lever system. It is essential that the temperature of the test-piece be very closely controlled for the duration of a test and that the temperature be uniform along the length of the specimen. The usual arrangement is to have an electric resistance tubular furnace mounted on the frame of the testing machine and movable in the vertical plane. BS 3500 (1987) and ASTM E 139-83 cover the standard conditions for creep testing and specify the very close temperature tolerances required. Temperatures must be maintained to within $\pm 2°C$ for tests at temperatures up to 600°C, to $\pm 2.5°C$ for tests at temperatures between 600°C and 800°C, and to $\pm 3°C$ for temperatures between 800°C and 1000°C.

Accurate measurement of strain is necessary during the course of a creep test, except in the case of creep stress-to-rupture tests, and generally the extensometers used are of the mirror type (see Figure 27.2) capable of measuring extensions of the order 10^{-3} or 10^{-4} mm. Extension rods made from a heat resistant alloy connect the gripping points at the test-piece to the extensometer outside the furnace.

When the time to rupture at some particular combination of temperature and stress is required, and there is no need to monitor the progress of creep strain, it is possible to achieve some economy in the use of creep testing machines and test a number of samples simultaneously within the same machine. A series of test-pieces can be mounted in line as a long string within a furnace and the required load applied (Figure 27.15). As in other forms of creep testing, individual thermocouples are attached to each specimen and the temperature of test must be controlled within the close limits listed above. When one of the test-pieces breaks, the load on all the remaining specimens will be released. Usually the equipment is arranged so that load release accompanying a fracture will open a micro switch and switch off both the furnace and the clock which records the test duration. The broken test-piece can then be removed and the test continued. It is a requirement of BS 3500, Part 1 that the time to rupture be measured to an accuracy of ± 1 per cent.

27.16 Relaxation testing

The principle of relaxation testing is to stress a material by straining the sample to some predetermined value. The strain is then maintained at a constant level and the variation of stress with time is recorded. As in creep testing, close control of temperature is required. A test-piece, with thermocouples attached, is located in a furnace positioned within a rigid straining frame. When the specimen is at the required control temperature, tensile strain is applied by turning a straining screw at one end of the frame. The tensile force developed may be measured by means of a load cell and the changes in load on the sample with time are monitored over the duration of the test. The general requirements for the conduct of relaxation tests are given in BS 3500, Part 6 (1987) and ASTM E 328-86 and D 2991-84.

SUMMARY

There are many different types of test which are used for determining the hardness, strength and toughness of materials. Most of these are tests in which the test-pieces are fractured in the test. Standard conditions are laid down for the preparation of test-pieces and the conduct of tests with the aim of ensuring that the results obtained from small test-pieces are representative of the bulk material. Also, the fact that different laboratories and test houses are conducting tests to the same standards allows the user to have confidence in the test results, irrespective of where the tests were carried out. In the United Kingdom it is the British Standards Institution (BSI) which is the body for compiling and publishing the standardised procedures for testing. The American Society for Testing Materials (ASTM) performs the same function in the USA. Similar national organisations exist in all the developed countries. The International Standards Organisation (ISO), with representatives from all national standards bodies, exists to promote standardisation of procedures across all countries and many ISO standards are now in use.

Tests for determining the hardness of materials are based on resistance to abrasion, and resistance to indentation, a measurement of elastic resilience. The only method for assessing the hardness of ceramics and glasses, non-ductile materials, is Mohs' test which is a measure of the resistance to abrasion. Indentation tests, in which a steel or diamond indentor is forced into the surface under a known load for a specific time, are used for metals and some thermoplastic materials. The most commonly used methods are the Brinell, Vickers diamond and Rockwell tests. The Shore test, in which the rebound of a small falling weight is measured, is used for extremely hard metals and also for elastomers. There are empirical relationships between the hardness and tensile strength of metals and alloys.

The tensile test is a widely used test for measuring the tensile characteristics of metals and polymeric materials. Tensile yield strength, tensile strength and modulus of elasticity are determined in a tensile test, together with the percentage elongation on gauge length or the percentage reduction in area. These latter are measures of the ductility of the material. Thermoplastics are strain-rate sensitive and the rate of strain to be used during testing is specified. The tensile strength of brittle materials cannot be determined in the same way as for metals and thermoplastics and may be measured using either a split cylinder test, which induces tensile stresses in the circumference of a cylinder, or by flexure of a test-piece in a three-point loading arrangement. The fracture strength given by this latter method is termed the modulus of rupture and is approximately double the true tensile strength of the material. Tests to determine compressive, shear and torsional strength have been developed but these are not used widely.

Brittle materials show a major variation in test results because of the presence of pores and cracks within the material. Many test results are required in order to obtain a statistical mean. Statistical analysis permits one to quote a probability of survival for a brittle material subjected to some particular value of stress.

Notch impact testing, in which a test-piece is struck by a fast moving heavy pendulum, and which measures the amount of energy absorbed in fracturing the test-piece, can reveal a tendency to brittle behaviour which is not shown by the standard tensile test. This type of test can also be used to determine the ductile–brittle transition temperature for metals. Tests have been developed also for the determination of the fatigue and creep characteristics of materials.

Questions

27.1 A consistent value of Brinell hardness for a material can be obtained for different size indentors if the ratio F/D^2 is maintained constant. Examine this statement in the light of the experimental data given:

Test	Indenting force (kgf)	Indentor diameter (mm)	Indentation diameter (mm)
1	30	1	0.473
2	60	2	0.698
3	120	2	0.944
4	750	5	2.35
5	1500	5	3.14
6	3000	10	4.75

27.2 Why is it necessary to use different loads when making Brinell or Vickers hardness impressions on two different materials, for example, pure copper and bronze?

Brinell hardness impressions were made on samples of pure copper and a phosphor bronze using a 2 mm diameter ball indentor. The results were

Pure copper – 20 kg load – impression diameter = 0.772 mm
Phosphor bronze – 40 kg load – impression diameter = 0.750 mm

Calculate the Brinell hardness of the materials.

27.3 Vickers and Brinell hardness tests are made on a sample of a steel and the following results were recorded:

Vickers diamond test with 30 kg load – the mean lengths of impression diagonals were: (a) 0.527 mm, (b) 0.481 mm, (c) 0.497 mm

Brinell test using 10 mm ball indentor and 3000 kg load – impression diameters were: (a) 4.01 mm, (b) 4.005 mm, (c) 4.02 mm

Calculate the Vickers diamond and Brinell hardness numbers for the steel and explain any variations in the results.

27.4 Brinell hardness impressions on a sample of annealed copper, using a 5 mm diameter ball indentor, gave the values in the table below:

Load (kg)	125	250	375
Impression diameter (mm)	2.20	2.70	3.10

Determine whether these data satisfy the Meyer relationship and, if so, evaluate the Meyer constants for the material.

27.5 (a) List the main parameters which may be determined in a full tensile test.
   (b) What are the main differences between the standard tensile test for metals and that for plastics?
   (c) A tensile test on a polymeric material, using a test-piece of dimensions 20 mm wide, 3 mm thickness and a gauge length of 50 mm, produced the force–extension curve shown in Figure 27.16.

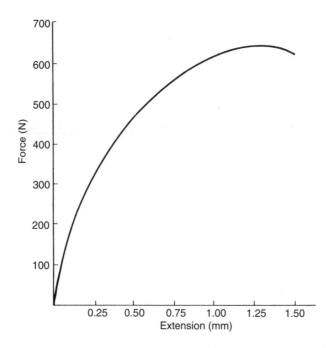

Figure 27.16

Determine: (i) tensile strength, (ii) initial modulus, (iii) secant modulus at 0.2 per cent strain, and (iv) percentage elongation at break.

27.6 The data in the table below were determined in a full tensile test. Evaluate the tensile properties and identify the material. Test-piece dimensions: width = 12.61 mm, thickness = 3.47 mm, gauge length = 50 mm.

Force (N)	Extension (mm)	Force (N)	Extension (mm)	Force (N)	Extension (mm)
25	0.018	125	0.121	225	0.293
50	0.040	150	0.153	250	0.355
75	0.064	175	0.192	275	0.425
100	0.090	200	0.238	300	0.520

Maximum force in test = 1290 N; length between gauge marks at break = 97 mm.

27.7 From the data in the table below plot a force–extension diagram and determine values for: (a) tensile strength, (b) modulus of elasticity, (c) 0.2 per cent proof stress, (d) percentage reduction of area, and (e) the true stress at a nominal strain of 8 per cent.

Force (kN)	39.4	67.5	84.4	90.0	95.6	112.5	123.8	131.1	131.1	123.8
Extension (mm)	0.25	0.40	0.50	0.60	0.75	1.75	3.00	5.00	6.50	8.00

The test-piece original diameter was 11.26 mm, the diameter after fracture was 9.34 mm and the gauge length was 50 mm.

27.8 A metal sample of 10 mm diameter was stressed in tension and the following information obtained. A force of 30 kN gave a total extension of 2.0 mm and a force of 34 kN gave a total extension of 11.0 mm over an original gauge length of 40 mm.
Assuming that the true stress–true strain relationship, $\sigma_t = k\varepsilon_t^n$, applies determine:
(a) the values of the coefficient k and the exponent n,
(b) the total extension at the maximum load,
(c) the maximum load applied, and
(d) the tensile strength of the material.

27.9 Figure 27.17 gives details of a tensile test on a material. From the data determine values for:
(a) E, (b) 0.1 per cent proof stress, (c) tensile strength, and (d) percentage elongation, and identify the type of material tested.

Figure 27.17

27.10 (a) A glass test-piece with the dimensions length = 95 mm, thickness = 3 mm and width = 15 mm is simple supported over a span of 75 mm and a force applied at mid-span. The load required to break the glass is 35 N. Calculate the tensile strength (modulus of rupture) of the glass.
(b) A concrete cylinder of 100 mm diameter and 100 mm length is subjected to a diametral compressive force. Calculate the tensile strength of the concrete if the maximum force at the failure point is 47.5 kN.

27.11 (a) What are the main reasons for using notch impact tests?
(b) What is the major advantage of the Charpy test over the Izod test?
(c) A notch impact test sample of 10 mm square cross-section with a standard milled V-notch 2 mm deep absorbs 115 J of energy in fracture. How should this test result be expressed?

27.12 (a) What is meant by the term *fatigue limit*?
(b) State the principle of the Wöhler fatigue test.
(c) What are the main advantages and disadvantages of rotating beam fatigue tests?

27.13 (a) What are the main reasons for using stress-rupture testing?

(b) How does stress-rupture testing differ from full tensile creep tests?

27.14 A component which will be required to operate for long periods at a constant high temperature is designed with an initial length of 150 mm. The maximum increase in length that can be tolerated in service is 2 mm. What will be the maximum value of creep strain rate at the operating temperature which can be permitted if the required design life of the component is 15,000 hours?

Non-destructive testing

Why learn about non-destructive testing?

There is a need to assess the quality of part-finished and finished components and structures. If a destructive test is used then only a small sample can be tested and, if 1 per cent of production is tested and found to be faulty, does it mean that the remaining 99 per cent is faulty also? Or if the testing of a small sample shows the quality to be acceptable, does this mean that the remainder is equally good? If non-destructive methods of testing are used then, if necessary, 100 per cent of production can be checked, with none of the output being destroyed as part of sample testing. Some materials may contain internal porosity, cracks or other flaws, and some of the non-destructive test methods are capable of detecting such defects. Through effective use of non-destructive testing engineers and users can have greater confidence in the integrity of materials and components.

LEARNING OBJECTIVES

1. List the main methods for the non-destructive testing of materials.

2. Distinguish between surface flaw and internal flaw detection methods.

3. Understand the principles and main applications of liquid penetrant and magnetic particle inspection for the detection of surface flaws.

4. Understand the principles and main applications of eddy current inspection.

5. Describe the principles of sound transmission through materials, the generation of compression, shear, and Rayleigh waves and their applications in inspection.

6. Describe the generation of both X- and γ-radiation, the use of these types of radiation in defect detection, and the hazards associated with their use.

28.1 | Introduction

Defects, such as cracks, porosity and inclusions, which may be potentially damaging may be introduced into materials or components during manufacture, and other defects, such as fatigue cracks, may be generated during service. It is necessary to be able to detect and identify such defects and to ascertain their position and size so that decisions can be taken as to whether specific defects can be tolerated or not. A range of non-destructive test methods is available for the inspection of materials and components and the most widely used techniques are:

liquid penetrant,

magnetic particle inspection,

eddy current systems,

ultrasonics, and

radiography.

Some of the features and applications of the main test methods in use are given in Table 28.1. All these NDT systems co-exist and, depending on the application, may either be used singly or in conjunction with one another. There is some overlap between the various test methods but they are complementary to one another. The fact that, for example, ultrasonic testing can reveal both internal and surface flaws does not necessarily mean that it will be the best method for all inspection applications. Much will depend upon the type of flaw present and the shape and size of the components to be examined.

28.2 | Visual inspection

Often the first stage in the examination of a component is visual inspection. Examination by naked eye will only reveal relatively large defects which break the surface but the effectiveness of visual inspection for external surfaces can be improved considerably through use of a hand lens or stereoscopic microscope. Generally, high magnifications are not necessary for this type of inspection. Optical inspection probes, both rigid and flexible, which can be inserted into cavities, ducts and pipes have been

Table 28.1 *Summary of non-destructive test methods.*

System	Features	Applicability
Visual inspection probes	Detection of defects which break the surface, surface corrosion, etc.	Interior of ducts, pipes and assemblies
Liquid penetrant	Detection of defects which break the surface	Can be used for any metal, many plastics, glass and glazed ceramics
Magnetic particle	Detection of defects which break the surface and sub-surface defects close to the surface	Can only be used for ferromagnetic materials (most steels and irons)
Electrical methods (eddy currents)	Detection of surface defects and some sub-surface defects. Can also be used to measure the thickness of non-conductive coatings, e.g. paint on a metal	Can be used for any metal
Ultrasonic testing	Detection of internal defects but can also detect surface flaws	Can be used for most materials
Radiography	Detection of internal defects, surface defects and to check correctness of assemblies	Can be used for most materials but there are limitations on the maximum material thickness

developed for the inspection of internal surfaces. An optical inspection probe comprises an objective lens system at the working end and a viewing eyepiece at the other end, with a fibre optic coherent image guide linking the two (see Figure 28.1). Illuminating light is conveyed to the working end of the probe through an optical fibre light guide, and both the optical and illumination systems are contained within either a stainless steel tube, for rigid probes, or a flexible plastic or braided metal sheathing in the case of flexible probes. Inspection probes are made in many sizes with, for rigid probes, diameters ranging from about 2 mm up to about 20 mm. The minimum diameter for flexible probes is about 4 mm. Probe lengths may vary considerably also, and the maximum working length for a 2 mm probe is about 150 mm. The maximum permissible working length increases as probe diameter increases and may be up to 5 m for a 20 mm diameter probe.

Inspection probes can be designed to give either direct viewing ahead of the probe end, or to a view at some angle to the line of the probe. It is possible to mount a miniature television camera in place of the normal eyepiece lens system and display an image on a monitor screen (Figure 28.2).

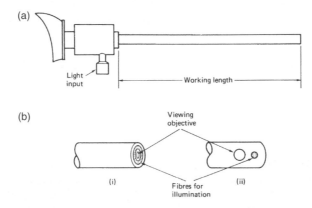

Figure 28.1 *(a) Rigid optical inspection probe. (b) Probe ends (i) for direct viewing, (ii) for angled viewing.*

Figure 28.2 *Optical inspection probe with television camera and monitor.*

28.3 | Liquid penetrant inspection

Liquid penetrant inspection is a technique which can be used to detect defects in a wide range of components, provided that the defect breaks the surface of the material. The principle of the technique is that a liquid is drawn by capillary attraction into the defect and, after subsequent development, any surface-breaking defects may be rendered visible to the human eye. In order to achieve good defect visibility, the penetrating liquid will either be coloured with a bright and persistent dye or else contain a fluorescent compound. In the former type, the dye is generally red and the developed surface can be viewed in natural or artificial light but in the latter case the component must be viewed under ultra-violet light if indications of defects are to be seen.

Liquid penetrant inspection is an important industrial method and it can be used to indicate the presence of defects such as cracks, laminations, laps and zones of surface porosity in a wide variety of components. The method is applicable to almost any component, whether it be large or small, of single or complex configuration, and it is employed for the inspection of wrought and cast products in both ferrous and non-ferrous metals and alloys, ceramics, glassware and some polymer components.

There are five essential steps in the penetrant inspection method. These are:

surface preparation,

application of penetrant,

removal of excess penetrant,

development,

observation and inspection.

The surface of a component must be thoroughly cleaned as it is important that any surface to be examined must be free from oil, water, grease or other contaminants if the successful indication of defects is to be achieved. After surface preparation, liquid penetrant is applied by brush, spray or immersion so as to form a film of penetrant over the component surface. Penetrant will seep into fairly large flaws in a few seconds but it may take up to 30 minutes for the liquid to penetrate into very small defects and tight cracks.

After the penetration stage, it is necessary to remove excess penetrant from the surface of the component. Some penetrants can be washed off the surface with water, while others require the use of specific solvents. The development stage is necessary to reveal clearly the presence of any defect. The developer is usually a very fine chalk powder. This may be applied dry, but more commonly is applied by spraying the surface with chalk dust suspended in a volatile carrier fluid. A thin uniform layer of chalk is deposited on the surface of the component. Penetrant liquid present within defects will be slowly drawn by capillary action into the pores of the chalk. There will be some spread of penetrant within the developer and this will magnify the apparent width of a defect. When dye penetrant is used, the dye colour must be in sharp contrast to the uniform white of the chalk covered surface. The development stage may sometimes be omitted when a fluorescent penetrant is used. If it takes a long time for penetrant to be drawn into a tight crack then it follows that a similar length of time will be needed for liquid to be drawn from the defect by the developer. In consequence, development times of between 10 and 30 minutes are generally required to ensure that all defect indications are visible.

After an optimum developing time has been allowed, the component surface is inspected for indications of penetrant 'bleedback' into the developer. Dye-penetrant inspection is carried out in strong lighting conditions, while fluorescent-penetrant inspection is performed in a suitable screened area using ultra-violet light. The latter technique causes the penetrant to emit visible light, and defects are brilliantly outlined. The five essential operations are shown in Figure 28.3.

The liquid penetrant process is comparatively simple as no electronic systems are involved, and the equipment necessary is cheaper than that required for other non-destructive testing systems. The obvious major limitation of liquid penetrant systems is that they can detect surface-breaking defects only.

The method is not suitable for use with naturally porous materials such as unglazed ceramics but is used to detect glazing defects. Some thermoplastics may be affected by the penetrant fluid, an organic solvent.

The range of applications of liquid penetrant testing is extremely wide and varied. The system is used in the aerospace, automotive and general manufacturing industries for the quality control of production and by users during regular maintenance and safety checks. Typical components which are checked by this system are turbine rotor discs and blades, pistons, cast cylinder heads, wheels, forged components and welded assemblies. Figure 28.4 shows the inspection of the inlet assembly of an

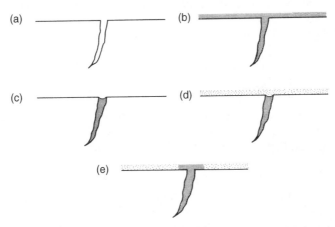

Figure 28.3 *Stages in penetrant testing: (a) material surface clean and grease-free; (b) penetrant absorbed into defect; (c) excess penetrant removed, but liquid remains in defect; (d) developer applied to surface; (e) penetrant absorbed into developer giving indication of defect.*

Figure 28.4 *Inspection of inlet assembly of aircraft gas turbine using ultra-violet light after application of fluorescent penetrant. (Courtesy of Magnaflux Ltd)*

aircraft engine. The operator is looking for surface cracks under ultra-violet light, and after the assembly has been processed with fluorescent penetrant.

28.4 | Magnetic particle inspection

Magnetic particle inspection is a sensitive method of locating surface and some sub-surface defects in ferromagnetic components. When a ferromagnetic component is magnetised, magnetic discontinuities that lie in a direction approximately perpendicular to the field direction will result in the formation of a strong 'leakage field'. This leakage field is present at and above the surface of the magnetised component, and its presence can be detected using finely divided magnetic particles. The application of dry particles or wet particles in a liquid carrier, over the surface of the component, results in a collection of magnetic particles at a discontinuity. This 'magnetic bridge' indicates the location, size and shape of the discontinuity.

Current passing through any straight conductor such as a wire or bar creates a circular magnetic field around the conductor. When the conductor is a ferromagnetic material, the current induces a magnetic field within the conductor as well as within the surrounding space. Hence, a component magnetised in this manner is given a circular magnetisation, as shown in Figure 28.5(a). An electric current can also be used to create a longitudinal magnetic field in components. When current is passed through a coil of one or more turns surrounding a component, a longitudinal magnetic field is generated within the work-piece as shown in Figure 28.5(b).

The effectiveness of defect indication will depend on the orientation of the flaw to the induced magnetic field and will be greatest when the defect is perpendicular to the field. This is shown schematically in Figure 28.6. In this figure, defect A will give a strong indication. Defects B and C are sub-surface defects. An indication will be obtained in respect of defect B as it is normal to the magnetic field but defect C will not be indicated. In the case of defect C, it is too deep within the material to give an indication but, even if such a defect were close to the surface, its alignment with the magnetic field would render detection unlikely.

Generally, in order to indicate the presence of all flaws a component will need to be magnetised more than once. For components of relatively simple shape this is achieved by, firstly, inducing circular magnetisation to indicate longitudinal defects. The component is then demagnetised before magnetising for a second time within a coil to induce longitudinal magnetisation which will enable transverse defects to be located.

Figure 28.5 *(a) Current passes through work-piece inducing circular magnetisation; (b) Longitudinal magnetisation induced by placing work-piece within a coil.*

Figure 28.6 *Magnetic flaw detection. Detectable surface leakage fields produced by defects A and B. Defect C is likely to remain undetected.*

Figure 28.7 *Magnetic test bench with a crankshaft in position. (Courtesy of Magnaflux Ltd)*

A typical magnetic test bench is shown in Figure 28.7. The crankshaft to be tested is clamped between two electrical contacts and a current can be passed through the work to induce circular magnetisation. When the large coil visible in the illustration is energised, longitudinal magnetisation will be induced in the work-piece.

Magnetisation of a component could be accomplished using permanent magnets but generally magnetic fields are induced by passing a heavy current through the component, by placing a coil around or close to the component under test, or by making the component part of a magnetic circuit, for example, by means of a hand yoke. The actual method used will depend on the size, shape and complexity of the parts to be inspected and, for components to be inspected *in situ*, the accessibility to such components.

The magnetic particles which are used for inspection may be made from any ferromagnetic material of low remanence and they are usually finely divided powders of either metal oxides or metals. The particles are classified as *dry* or *wet*. Dry particles are carried in air or gas suspension while wet particles are carried in liquid suspension. Very high sensitivities are possible with wet particle inspection, particularly when a fluorescent chemical is adsorbed on the particles and inspection is made under ultra-violet light. In the best conditions, it is possible to detect cracks with a width of as little as 10^{-3} mm.

Magnetic particle inspection is a sensitive means of detecting very fine surface flaws. It is also possible to obtain indications from some discontinuities that do not break through the surface provided that they are close to the surface. The major limitations of the technique are that it is only suitable for ferromagnetic materials, that for the best results the induced magnetic field should be normal to any defect and thus two or more magnetising sequences will be necessary, and that a demagnetising procedure will need to be carried out for many components after inspection. The sensitivity of magnetic particle inspection is very good generally, but this will be reduced if the surface of the component is covered by a film of paint or other non-magnetic layer.

The principal industrial uses of magnetic particle inspection are for in-process inspection, final inspection and the inspection of components as part of planned maintenance and overhaul schedules. The methods are well suited to the inspection of castings and forgings and components such as crankshafts, connecting rods, flywheels, crane hooks, axles and shafts.

28.5 | Electrical test methods

The basic principle underlying the electrical test methods is that electrical eddy currents and/or magnetic effects are induced within the material or component under test and, from an assessment of the effects, deductions can be made about the nature and condition of the test-piece. These techniques are highly versatile and, with the appropriate equipment and test method, can be used to detect surface and sub-surface defects within components, to determine the thickness of surface coatings, to provide information about structural features such as crystal grain size and heat treatment condition, and also to measure physical properties including electrical conductivity, magnetic permeability and physical hardness.

If a coil carrying an alternating current is placed in proximity to a conductive material, secondary or eddy currents will be induced within the material. The induced currents will produce a magnetic field surrounding the coil. This interaction between fields causes a back e.m.f. in the coil and, hence, a change in the coil impedance value. If a material is uniform in composition and dimensions, the impedance value of a search coil placed close to the surface should be the same at all points on the surface, apart from some variation observed close to the edges of the sample. If the material contains a discontinuity, the distribution of eddy currents, and their magnitude, will be altered in its vicinity and there will be a consequent reduction in the magnetic field associated with the eddy currents, so altering the coil impedance value.

Eddy currents flow in closed loops within a material and both the magnitude and the timing or phase of the currents will depend on a number of factors. These factors include the magnitude of the magnetic field surrounding the primary coil, the electrical and magnetic properties of the material and the presence or otherwise of discontinuities or dimensional changes within the material. Several types of search coil are used, two common types being the flat or pancake type coil which is suitable for the examination of flat surfaces and the solenoid-type coil which can be used in conjunction with solid or tubular cylindrical parts. For tubes, a solenoid-type coil may be placed around the tube or inserted into the bore.

If a component contains a crack or other discontinuity, the flow pattern of eddy currents will be altered and this will cause a change in the magnetic field and, hence, a change in coil impedance. A schematic representation of the effect of a discontinuity on eddy current pattern is shown in Figure 28.8. The impedance of a coil can be determined by measuring the voltage across it. In eddy current test equipment, changes in coil impedance can be indicated on a meter or a chart recorder or displayed on the screen of a cathode ray tube. At component edges, eddy current flow is distorted, because the eddy currents are unable to flow beyond this limiting barrier. The magnitude of this edge effect is usually very large, and hence, inspection is inadvisable close to edges. In general, it is recommended to limit inspection to an approach of 3 mm from a component edge.

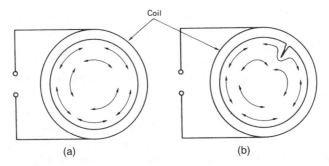

Figure 28.8 *Cross-section of a bar within a solenoid-type coil showing eddy current pattern: (a) defect-free section – uniform eddy currents; (b) eddy current pattern distorted by the presence of a defect.*

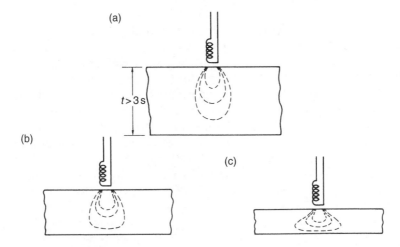

Figure 28.9 *Schematic representation of distortion of eddy current distribution in thin sections.*

Eddy currents are not distributed uniformly throughout a part under inspection. They are most dense at the component surface, immediately beneath a coil, and become progressively less dense with increasing distance from the surface. At some distance below the surface of large components, eddy current flow is negligible. This phenomenon is commonly termed '*skin effect*'. When the thickness of a test-piece is small, the distribution pattern of the eddy currents will become distorted and the extent of such distortion will vary with the thickness of the material. This effect is shown schematically in Figure 28.9. It follows then, that for materials of thin section a change in thickness will alter the impedance value of a test coil. An eddy current system, therefore, when calibrated against known standards, can be used successfully for the accurate measurement of the thickness of thin materials. Thickness measurements can also be made using ultrasonic techniques but, in this case, the degree of accuracy possibly diminishes when the material is very thin. The reverse is true for thickness measurements made using eddy current techniques and so the two different methods become complementary to each other.

The inspection frequencies used in electrical techniques range from 20 Hz to 10 MHz. Inspection of non-magnetic materials is performed at frequencies within the range 1 kHz–5 MHz, while frequencies lower than 1 kHz are often employed with magnetic materials. The actual frequency used in any specific test is often a compromise. Sensitivity for the detection of small surface flaws is greatest at high frequencies but detection of sub-surface flaws requires low frequencies. Inspection of ferromagnetic materials demands very low frequencies because of the relatively low penetration depth of eddy currents in these materials.

The electrical eddy current system is a highly versatile system which can be used to detect not only cracks, but several other conditions, including corrosion. The test equipment may be small and portable and, with a suitable selection of test probes, can be used in many situations. However, the correct interpretation of signal indications does require considerable skill and experience on the part of the operator. The eddy current equipment which is used for the quality control inspection of the products of many material manufacturing processes is often completely automatic and is highly sophisticated. Production of tubing, bar stock and wire is checked in this way by many manufacturers. The range of applications in which eddy current inspection is used successfully is almost infinite. Some typical uses are the routine in-service inspection of many components in aircraft, including undercarriage wheels, routine inspection of railway track and accurate determination of thickness, including wall-thinning due to corrosion and measurement of the thickness of surface coatings on metals.

28.6 Ultrasonic inspection

Ultrasonic techniques are very widely used for the detection of internal defects in materials, but they can also be used for the detection of small surface cracks. Ultrasonics are used for the quality control inspection of part processed material, such as rolled slabs, as well as for the inspection of finished components. The techniques are also in regular use for the in-service testing of parts and assemblies.

Sound waves are elastic waves which can be transmitted through both fluid and solid media. The audible range of frequency is from about 20 Hz to about 20 kHz but it is possible to produce elastic waves of the same nature as sound at frequencies up to 500 MHz. Elastic waves with frequencies higher than the audio range are described as ultrasonic. The waves used for the non-destructive inspection of materials are usually within the frequency range 0.5 MHz to 20 MHz. Sound waves in fluids are of the longitudinal compression type in which particle displacement is in the direction of wave propagation but shear waves, with particle displacement normal to the direction of wave travel, and elastic surface waves, or Rayleigh waves, can also be transmitted in solids.

Ultrasound is generated by piezoelectric transducers. Certain crystalline materials show the piezo-electric effect, namely, the crystal will dilate or strain if a voltage is applied across the crystal faces. Conversely, an electrical field will be created in such a crystal if it is subjected to a mechanical strain, and the voltage produced will be proportional to the amount of strain. The original piezoelectric material used was natural quartz. Quartz is still used to some extent but other materials including barium titanate, lead metaniobate and lead zirconate are used widely. When an alternating voltage is applied across the thickness of a disc of piezoelectric material, the disc will contract and expand and in so doing will generate a compression wave normal to the disc in the surrounding medium. A transducer for sound generation will also detect sound. An ultrasonic wave incident on a crystal will cause it to vibrate, producing an alternating current across the crystal faces. In some ultrasonic testing techniques, two transducers are used, one to transmit the beam and the other acting as the receiver, but in very many cases only one transducer is necessary. This acts as both transmitter and receiver. Ultrasound is transmitted as a series of pulses of extremely short duration and during the time interval between transmissions the crystal can detect reflected signals.

When a beam of longitudinal compression sound waves reaches a boundary between two media, a proportion of the incident waves will be reflected at the interface and a proportion will be transmitted across the interface. At an air/metal interface, reflection of sound waves will be almost 100 per cent. A fluid, oil or water, is needed as a couplant between an ultrasound transmitter and metal to permit some of the sound energy to be transmitted across the interface into the metal. For normal incidence waves, the transmission across the interface will be of the compression wave type. When the incident beam is at some angle other than normal that portion of the beam which is transmitted across the interface will be refracted. However, there may be two refracted beams transmitted into the metal, because part of the transmitted energy is converted into the shear wave mode. One refracted beam will be of the compression type while the other will be a shear wave, as shown in Figure 28.10.

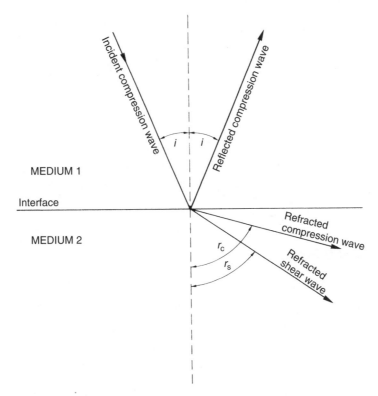

Figure 28.10 *Reflection and refraction at an interface.*

Much ultrasonic inspection uses normal incidence waves but, in many instances, waves at some angle other than normal are required. The presence of two types of wave with differing velocities within the material would give confusing results and so the angle of incidence is adjusted to be greater than the critical angle for compression wave refraction so that only a shear wave is transmitted into the material. This critical angle for total reflection of an incident compression wave at a perspex/steel interface is 27.5°.

If it is required to generate a surface, or Rayleigh, wave, the angle of incidence should be adjusted to a second critical angle to produce a Rayleigh wave at a refracted angle of 90°. The value of this critical angle for a perspex/steel interface is 57°.

An ultrasonic beam transmitted through a metal will be totally reflected at the far surface of the material, a metal/air interface. It will also be wholly or partially reflected by any internal surface, namely cracks or laminations, porosity and non-metallic inclusions, subject to the limitation that the size of the object is not less than one wavelength. From this it follows that the sensitivity and defect resolution will increase as the frequency of the beam is increased. The wavelengths of sound at frequencies of 1 MHz and 5 MHz are 6.19 mm and 1.24 mm respectively in aluminium and 5.81 mm and 1.16 mm respectively in steel. At an inspection frequency of 1 MHz defects in an aluminium component with a size smaller than 6.2 mm would not be detectable but would be observed if higher frequencies were employed.

Some metallic materials can only be inspected satisfactorily with a relatively low-frequency sound beam because the use of high frequencies could cause a mass of reflections from a normal structural constituent which would mask the indications from defects. This situation arises in the inspection of grey cast iron components where the flakes or nodules of graphite in the iron structure may have a size of several millimetres.

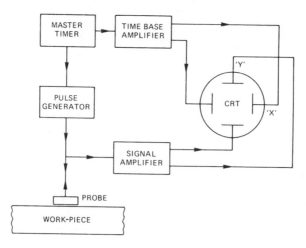

Figure 28.11 *Block diagram for an ultrasonic flaw detector.*

Figure 28.12 *'A' scan display: (a) reflections obtained from defect and backwall; (b) representation of 'A' scan screen display.*

In most ultrasonic test equipment, the signals are displayed on the screen of a cathode ray oscilloscope. The basic block diagram for a typical flaw detector is shown in Figure 28.11.

The information obtained during an ultrasonic test can be displayed in several ways. The most commonly used system is the *'A' scan display* (see Figure 28.12). A blip appears on the CRT screen at the left-hand side corresponding to the initial pulse and further blips appear on the time base corresponding to any signal echoes received. The height of the echo is generally proportional to the size of the reflecting surface but it is affected by the distance travelled by the signal and attenuation effects within the material. The linear position of the echo is proportional to the distance of the reflecting surface from the probe, assuming a linear timebase. This is the normal type of display for hand-probe inspection techniques.

The presence of a defect within a material may be found using ultrasonics with either a reflection or a transmission technique. The most widely used inspection technique is *normal probe reflection,* as illustrated in Figure 28.12. The pulse is wholly or partially reflected by any defect in the material and received by the single probe, which combines as transmitter and receiver.

There are certain testing situations in which it is not possible to place a normal probe at right angles to a defect and the only reasonable solution is offered by angle probes. A good example of this technique is in the inspection of butt welds in parallel sided plate. The transmitter and receiver probes are arranged as in Figure 28.13(a).

If there is any defect in the weld zone, this will cause a reduction in the received signal strength. Distance AB is known as the skip distance and for the complete scanning of a weld the probes should

Figure 28.13 *Angle probe transmission method: (a) probe positions and skip distance; (b) scanning method for complete inspection of butt weld.*

Figure 28.14 *Crack detection using a surface wave probe.*

be moved over the plate surface as shown in Figure 28.13(b). In practice, both probes would be mounted in a jig so that they are always at the correct separation distance.

A Rayleigh, or surface, wave can be used for the detection of surface cracks (see Figure 28.14). The presence of a surface defect will reflect the surface wave to give an echo signal in the usual way. Surface waves will follow the surface contours and so the method is suitable for shaped components such as turbine blades.

Figures 28.15 and 28.16 show the compact and portable nature of modern ultrasonic test equipment.

Ultrasonic test methods are suitable for the detection, identification and size assessment of a wide variety of both surface and sub-surface defects in materials, provided that there is access to one surface. Using hand held probes, many types of component can be tested, including *in-situ* testing. This latter capability makes the method particularly attractive for the routine inspection of aircraft and road and rail vehicles in the search for incipient fatigue cracks. There are automated systems also which are highly suitable for the routine inspection of production items at both an intermediate stage and the final stage of manufacture.

28.7 | Principles of radiography

Very short wavelength electromagnetic radiation, namely X- or γ-rays, will penetrate through solid media but will be partially absorbed by the medium. The amount of absorption which will occur will depend upon the density and thickness of the material that the radiation is passing through and also upon the characteristics of the radiation. The radiation which passes through the material can be detected and recorded on either film or sensitised paper, viewed on a fluorescent screen or detected and monitored by electronic sensing equipment. Strictly speaking, the term *radiography* implies a process in which an image is produced on film.

Figure 28.15 *Inspection of fan blades. (Courtesy of Krautkramer UK Ltd)*

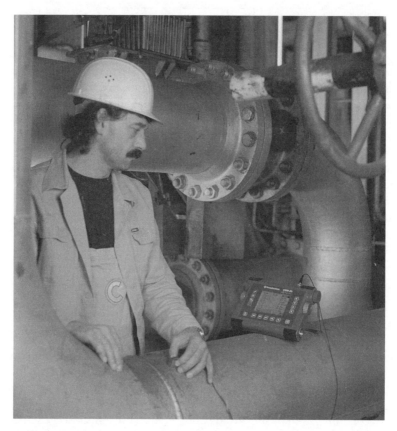

Figure 28.16 *Ultrasonic inspection of weld in stainless steel pipework using a shear wave probe. (Courtesy of Krautkramer UK Ltd)*

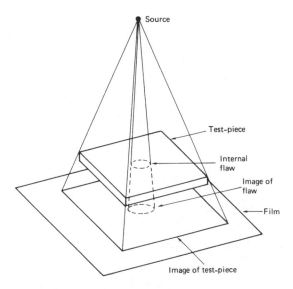

Figure 28.17 *Schematic diagram showing a radiographic system.*

The basic principle of radiographic inspection is that the object to be examined is placed in the path of a beam of radiation from an X- or γ-ray source. A recording medium, usually film, is placed close to the object being examined but on the opposite side from the beam source (as shown in Figure 28.17).

X- or γ-radiation cannot be focused as visible light can be focused and, in many instances, the radiation will come from the source as a conical beam. Some of the radiation will be absorbed by the object but some will travel through the object and impinge on the film producing a latent image. If the object contains a flaw which has a different absorptive power from that of the object material the amount of radiation emerging from the object directly beneath the flaw will differ from that emerging from adjacent flaw-free regions. When the film has been developed, there will be an area of different image density which corresponds to the flaw in the material. Thus the flaw will be seen as a shadow within the developed radiograph. This shadow may be of lesser or greater density than the surrounding image depending on the nature of the defect and its relative absorptive characteristics.

X- and γ-rays are indistinguishable from one another. The only difference between them is the manner of their formation. X-rays are formed by bombarding a metal target material with a stream of high-velocity electrons within an X-ray tube. γ-rays, on the other hand, are emitted as part of the decay process of radioactive substances.

The major components of an X-ray tube are a cathode to emit electrons and an anode target, both being contained within an evacuated tube or envelope. The general arrangement is shown in Figure 28.18. Electrons, liberated from the cathode by thermionic emission, are accelerated towards the target anode by a large potential difference, the tube voltage. X-ray tube voltages generally range from 50 kV to 1 MV. The anode target is usually tungsten, embedded in a water cooled copper block. The envelope of an industrial X-ray tube is usually of ceramic/metal construction and contains a window opposite the anode to permit X-radiation to emerge from the tube. A 3 to 4 mm thickness of beryllium is generally used as a window material.

There are three important variables in X-ray tubes. These are the filament current, the tube voltage and the tube current. A change in the filament current will alter the temperature of the filament which will change the rate of thermionic emission of electrons. An increase in the tube voltage, the potential difference between cathode and anode, will increase the energy of the electron beam and, hence, will increase the energy and penetrating power of the X-ray beam which is produced. The third variable, the tube current, is the magnitude of the electron flow between cathode and anode and is directly

Figure 28.18 *Schematic view of an X-ray tube.*

Table 28.2 *Penetration ability of X-rays.*

Tube voltage (kV)	Photon energy for lowest wavelength radiation (MeV)	Penetration ability (mm of steel)
150	0.15	up to 25
250	0.25	up to 70
400	0.4	up to 100
1000	1.0	5 to 140

related to the filament temperature. The tube current is usually referred to as the milliamperage of the tube. The intensity of the X-ray beam produced by the tube is approximately proportional to the tube milliamperage. The upper practical limit for the voltage of an X-ray tube is about 1000 kV and this will give an X-ray spectrum in which the shortest wavelength radiation will have photon energies of about 1 MeV. Radiation of this high frequency will penetrate approximately 140 mm of steel. The penetrating ability of X-rays is given in Table 28.2.

γ-radiation is emitted during the decay of radioactive nuclei. Unlike the broad 'white' spectrum of radiation obtained from an X-ray tube, a γ-ray emitter will give one or more discrete radiation wavelengths, each one with its own characteristic photon energy. Radium, a naturally occurring radioactive element, has been used as a source of γ-radiation for radiography but it is far more usual to use radioisotopes produced in a nuclear reactor.

The specific isotopes which are generally used as γ-ray sources for radiography are caesium-137, cobalt-60, iridium-192 and thulium-170. (The numbers are the atomic mass number of the radioactive nuclei.) There is a continuous reduction in the intensity of radiation emitted from a γ-ray source as more and more unstable nuclei decay. The rate of decay decreases exponentially with time according to

$$I_t = I_0 e^{-kt}$$

where I_0 is the initial intensity, I_t is the intensity of radiation at time t and k is a constant for a particular disintegrating atomic species. An important characteristic of each particular radioisotope is its *half-life* period. This is the time taken for the intensity of the emitted radiation to fall to one-half of its original value. After two half-life time intervals, the intensity will fall to $\frac{1}{4}$ of the original value, after three half-life time intervals, the intensity will fall to $\frac{1}{8}$ of the original value, and so on. Another characteristic of a γ-ray source is the source strength. The source strength is the number of atomic disintegrations per second and is measured in curies (one curie is 3.7×10^{10} disintegrations per second). The source strength decreases exponentially with time, and the source strength at any given time can be determined using the expression

$$S_t = S_0 e^{-kt}$$

The radiation intensity, usually measured in roentgens per hour at one metre (rhm) is given by source strength (in curies) × radiation output (rhm per curie). (One roentgen is the amount of radiation that will produce ions carrying one electrostatic unit of energy in 1.293 mg of air.) The value of radiation output is a constant for any particular isotope. Another term used in connection with γ-ray sources is *specific activity*. This characteristic, expressed in curies per gramme, is a measure of the degree of concentration of the source.

The characteristics of the commonly used γ-sources for radiography are given in Table 28.3. Commercial radioactive sources are usually metallic in nature, but may be chemical salts or gases adsorbed on carbon. The source is encapsulated in a thin protective covering. This may be a thin sheath of stainless steel or aluminium. Containing the radioactive material in a capsule of this type prevents spillage or leakage of the material and reduces the possibility of accidental mishandling. The encapsulated source is housed in a lead-lined steel container.

Table 28.3 *Characteristics of γ-ray sources.*

Source isotope	Half-life period	Photon energy (MeV)	Radiation output (rhm/curie)	Effective penetrating power (mm of steel)
Caesium-137	33 years	0.66	0.39	75
Cobalt-60	5.3 years	1.17, 1.33	1.35	225
Iridium-192	74 days	12 rays from 0.13 to 0.61	0.55	75
Thulium-170	128 days	0.084	0.0045	12 (aluminium, not steel)

The obtaining of satisfactory radiographs requires a very high degree of skill and expertise on the part of the radiographer as there are very many factors which affect the formation of an X- or γ-ray image. These include the composition, density and dimensions of the object, the type and intensity of the radiation, and the characteristics of the radiographic film. The correct exposure for a particular application may be determined by a process of trial and error or by using an aid such as an exposure chart which relates to a specific grade of film. Often screens are used to improve contrast and reduce exposure times. A lead screen, which is a very thin film of lead bonded to thin card, is placed in contact with both sides of the radiographic film and serves two functions. It absorbs low-energy scattered radiation and interacts with incident high-energy radiation with a resulting emission of electrons. The emitted electrons activate the film emulsion giving improved developed film densities and enhanced contrast. Another type of intensifying screen is the fluorescent screen. Small crystals of calcium tungstate on a thin card base will fluoresce, emitting visible light, when subjected to radiation. They can intensify a radiographic image by a factor of up to 100, thus reducing exposure times, but they do not have a filtering effect on scattered radiation. A third type of screen is the fluoro-metallic screen, comprising both lead and fluorescent crystals, thus combining the advantages of both the former types.

Radiography usually involves exposing radiographic film and then inspection of the film after it has been developed, but inspection may be carried out in real-time. In the system known as *fluoroscopy* the X-ray image is not recorded on radiographic film. The radiation, after passing through the object under examination, strikes a fluorescent screen. The brightness of the fluorescence is greatest in those parts of the screen receiving the most radiation. Fluoroscopy is a useful inspection system for the routine examination of mass-produced items. Figure 28.19 shows such a system for the examination of die cast

(a)

(b)

Figure 28.19 *Fluoroscopy inspection system: (a) wheels entering X-ray cabinet; (b) remote viewing room and control panel.*

automotive wheels. Figure 28.19(a) shows wheels on a conveyor entering the shielded X-ray cabinet, while Figure 28.19(b) shows the images being viewed on a remote television monitor.

Devices termed *Image Quality Indicators* or *penetrameters* are used as a means of assessing the sensitivity of radiographic inspection. Several designs of image quality indicators have been devised by the various standards organisations and they are generally made in the form of steps or wires of varying thickness in the same or a similar material to that being inspected. The relevant British Standard (BS 3971) covers both step-hole and wire IQI types, and these are shown in Figure 28.20. A straight, six-stage step-hole IQI is shown in Figure 28.20(a) while Figure 28.20(b) shows the wire type IQI.

Figure 28.20 *Penetrameters to BS 3971: (a) step-hole; (b) wire type.*

Identification markers must be put on to indicate the type of material and the thickness range of the IQI. In Figure 28.20(a), the coding 8 AL 13 indicates that the IQI is in aluminium with the thinnest step being No. 8 (0.630 mm) and the thickest step being No. 13 (2.00 mm) while in Figure 28.20(b) the coding 9 CU 15 means that there are seven copper wires ranging in diameter from wire No. 9 (0.200 mm diam.) to wire No. 15 (0.80 mm diam.). With the aid of penetrameters, the image quality, or sensitivity, can be expressed as a percentage. The sensitivity is the thickness of the thinnest wire or step or hole visible in the developed radiograph expressed as a percentage of the thickness of the test-piece. In the USA the ASTM standards E 142-86, E 747-90 and E 1025-89 cover the use of IQIs and penetrameters.

The placement of penetrameters is important. They should be placed on the source side of the test-piece and at the edge of the area, namely, in the outer zone of the radiation beam with the thinnest step or wire being outermost.

EXAMPLE 28.1

A radiograph is taken of a bronze casting with a maximum thickness, in the direction of the radiation, of 25 mm. A 9 CU 15 wire IQI is used and the thinnest wire clearly visible in the developed radiograph has a thickness of 0.320 mm. Determine the sensitivity of the radiograph.

SOLUTION

The sensitivity is the thickness of the thinnest wire visible, expressed as a percentage of the casting thickness, and is $(0.320/25) \times 100 = 1.3$ per cent.

A radiograph is valueless unless the developed image can be sensibly interpreted, and correct interpretation needs a person who possesses a considerable amount of knowledge, skill and experience. The interpreter, therefore, needs to have a thorough knowledge of the principles of radiography and to be fully aware of the capabilities and the limitations of the techniques and equipment. In addition, the interpreter should have knowledge of the components to be inspected and the variables in the manufacturing processes which may give rise to defects. For example, in the inspection of castings, it would be beneficial if the interpreter is aware of the way in which defects such as gas porosity, shrinkage and cold shuts can occur and the most likely areas in the particular casting where they may be found. The radiographic interpreter is looking for changes in image density in the radiograph. Density changes may be caused by one of three factors, namely a change in the thickness of the test-piece, including visible surface indentations or protuberances, internal flaws within the component, and density changes which may be induced by faulty processing, mishandling or bad film storage conditions and it is important that the interpreter can assess the nature and cause of each density difference observed.

The conditions in which radiographs are viewed, therefore, are highly important, and the films should be correctly illuminated by means of a purpose-built light source which will give good illumination without glare or dazzle. The radiograph should be viewed in a darkened room so that there will be no light reflections from the surface of the film and the image is seen solely by means of light transmitted through the film. Viewing in poor conditions will cause rapid onset of eye fatigue and so it is also important that the interpreter is in a comfortable position and has no undue distractions. Ultimately, the efficiency of flaw detection is determined by the skill and experience of the interpreter and a highly experienced radiograph interpreter may locate defect indications which could be missed by a less experienced person.

X- and γ-radiation can cause damage to body tissue and blood, but any damage caused is not immediately apparent. The effects of any small doses of radiation received over a period of time is cumulative and so all workers who may be exposed to even small quantities of radiation should have a periodic blood count and medical examination. Strict regulations cover the use of X- and γ-rays and the quantity of radiation to which workers may be exposed. The extent of radiation which may be received by classified workers in the field of radiography must be monitored and this is best achieved by recording the dosage received on a radiation monitoring film (film badge) or by using a pocket ionisation chamber.

Radiography is capable of detecting any feature in a component or structure provided that there are sufficient differences in thickness or density within the test-piece. Large differences are more readily detected than small differences. The main types of defect which can be distinguished are porosity and other voids and inclusions, where the density of the inclusion differs from that of the basis material. Generally speaking, the best results will be obtained when the defect has an appreciable thickness in a direction parallel to the radiation beam. Plane defects such as cracks are not always detectable and the ability to locate a crack will depend upon its orientation to the beam. The sensitivity possible in radiography depends upon many factors but generally if a feature causes a change in absorption of 2 per cent or more, compared with the surrounding material, then it will be detectable.

Radiography and ultrasonics are the two methods which are generally used for the successful detection of internal flaws that are located well below the surface, but neither method is restricted to the detection of this type of defect. The methods are complementary to one another in that radiography tends to be more effective when flaws are non-planar in type whereas ultrasonics tends to be more effective when the defects are planar. Radiographic inspection techniques are frequently used for the checking of welds and castings and in many instances radiography is specified for inspection of these components. This is the case for weldments and thick-wall castings which form part of high-pressure systems. Radiography can also be used to inspect assemblies to check the condition and proper placement of components. One application for which radiography is very well suited is the inspection of electrical and electronic component assemblies to detect cracks, broken wires, missing or misplaced components and unsoldered connections. Radiography can be used to inspect most types of solid material but there could be problems with very high or very low density materials. Non-metallic and metallic materials, both ferrous and non-ferrous, can be radiographed and there is a fairly wide range of material thicknesses that can be inspected. The sensitivities of the radiography processes are affected by a number of factors including the type and geometry of the material and the type of flaw.

Most commercial X-ray sets emit a narrow cone of radiation from a small window but panoramic units are available. These possess a circumferential window with a 360° emission of radiation. Figure 28.21 shows such an X-ray unit positioned within a large vessel for the radiography of a circumferential weld. A continuous strip of radiographic film is positioned around the weld on the outside of the vessel and held in place by magnetic clamps. In this way the complete weld can be radiographed in one exposure.

Figure 28.21   *Radiography of a circumferential weld in a large vessel using a panoramic X-ray set.*

28.8 | Acoustic emission inspection

When strain energy is released as a consequence of structural changes taking place within a material, some of the energy is emitted as high-frequency waves with frequencies in the range from 50 kHz to 10 MHz. Phase transformations, plastic yielding and crack growth generate acoustic signals which may be detected and analysed. Basically, there are two types of acoustic emission from materials, continuous and an intermittent or 'burst' type. Continuous emission is usually of low amplitude and is associated with dislocation movement and plastic deformation, while burst type emissions are of high amplitude and short duration which result from the development and growth of cracks.

Acoustic emission inspection can offer several advantages over other forms of non-destructive inspection. For example, it can assess the dynamic response of a flaw to an imposed stress. There is a marked increase in the intensity of emission when a crack approaches critical size and, thus, a warning is given of instability and catastrophic failure. It is also possible to detect growing cracks of about 2×10^{-4} mm in length. This is a much smaller size than is detectable by other techniques.

Acoustic emissions are sensed by a transducer close-coupled to the component and the signal from the transducer is amplified, filtered and processed to give an audio and/or video output. In some applications, for example, inspection of pressure vessels, it is necessary to detect both the nature of and the precise location of the source of acoustic signals. In such cases, several transducers are used, spaced over the surface of the vessel, and computer-assisted monitoring of time of arrival of signals to the various transducer locations permits analysis of both the source type and its location.

A wide variety of materials can be inspected, including metals, ceramics, polymers, timber and composites. Although the emission sources from each material type may differ, the characteristic signals received can be correlated with the integrity of the material.

28.9 | Vibration testing

There is a natural frequency of vibration for any object. This natural frequency is dependent on the density and the elastic constants of the material. A cracked bell will have a different natural frequency (that is, emit a different sound) from a sound bell of the same dimensions and made from the same material. In this case the presence of a crack has changed the dimensional factors determining the natural frequency. This principle of stimulating vibrations and listening to the sound produced has been used as a non-destructive test technique, for example, in the checking of railway rolling stock wheels by tapping them with a hammer.

Nowadays, vibration testing is used in fundamental research as an extremely accurate method for the determination of the elastic constants of materials. For bars of uniform cross-section the relationships between the frequency of vibration and the constants of the material are as follows:

For bar of circular or square cross-section, clamped centrally, and stimulated into longitudinal vibration

$$f = \frac{n}{2l}\sqrt{\frac{E}{\rho}}$$

where f is the frequency, n is an integer, l is the length of the bar, E is Young's modulus and ρ is the density of the material.

For a bar of circular cross-section, clamped centrally, and stimulated into torsional vibration

$$f = \frac{n}{2l}\sqrt{\frac{G}{\rho}}$$

where G is the modulus of rigidity.

Poisson's ratio, ν, is a function of a ratio of the longitudinal and torsional vibration frequencies for the same bar:

$$\frac{f_L}{f_T} = \sqrt{[2(\nu + 1)]}$$

It is possible to determine the elastic constants of a material to an accuracy of up to one part in a hundred thousand in a vibration test.

The damping capacity of a material can also be determined using a vibration test technique. The damping capacity of a material is its ability to attenuate vibrations. The damping capacity, ζ, is given by

$$\zeta = \frac{1}{n}$$

where n is the number of vibrations of a test bar in free attenuation from an amplitude A to an amplitude of A/e, that is $0.3684\,A$ (e is the exponential function).

The damping capacity of a material is highly dependent on its structure, and minute changes in structure are reflected in changes in the damping characteristics. This feature is used in much fundamental research in materials.

SUMMARY
••

A range of methods has been developed for the non-destructive inspection of materials and components. Some of the methods can detect surface flaws, including microcracks too small to be seen with the naked eye, whereas other methods are capable of detecting internal defects. These methods should be seen as being complementary to one another, rather than in competition. Non-destructive testing is used at many stages in the life of a product. It is used to check the quality and integrity of part processed stock, for example, rolled aluminium alloy plate for use in aircraft wing panels before expensive milling operations commence, as well as for the inspection of finished components. Non-destructive testing is used also to monitor components and structures during their working life, for example to check the rate of growth of fatigue cracks or to measure the loss of material due to corrosion.

The simplest, and often effective, method of non-destructive examination is visual inspection. A useful aid for visual examination is the optical inspection probe or endoscope. Such probes can be inserted into pipes, ducts and cavities and permit close examination of internal surfaces for surface breaking defects and evidence of corrosion.

Liquid penetrant inspection is suitable for the detection of fine cracks which break the surface of a material and which would be extremely difficult to observe with the naked eye. It is a qualitative technique and further examination using another principle must be used if the depth of located cracks is to be determined. The method can be used with any non-absorbent or porous material.

Magnetic particle inspection can be used with ferromagnetic materials only but it is a reliable method for the detection of fine cracks which break the surface and, possibly, internal defects if they lie close to the surface. Again this method is qualitative only.

Electrical or eddy current inspection is a versatile technique which can be used with any metallic material. It is capable of detecting both surface and sub-surface defects and, when correctly calibrated, can identify defect type, location and magnitude. Electrical test methods can also be used to measure wall thickness, thickness of surface coatings and provide information about structural features such as grain size and heat treatment condition.

In ultrasonic inspection pulses of high-frequency sound energy are transmitted through the material. The sound waves, which may be of the compressive, shear or Rayleigh type, will be wholly or partially reflected by any internal surface such as cracks, porosity and inclusions. Ultrasonic inspection can detect both surface and internal flaws but it is used mainly for internal defects and both the location and size of defects can be determined.

Radiography, using either X- or γ-radiation, is the method most suited for the detection of internal defects and is most effective when any defect has a significant thickness in the direction of the radiation. Generally, a defect will not be visible in a radiograph unless its thickness is at least 1–2 per cent of the sectional thickness of the material. A variant of radiography, fluoroscopy, can be used to inspect components in real-time. X- and γ-radiation can cause damage to body tissue and stringent precautions must be taken to protect personnel involved with radiography.

Questions

28.1 (a) List and describe the five stages in the examination of components using a liquid penetrant inspection process.

(b) For what materials can liquid penetrant inspection be used and what types of flaws or defects can be detected?

28.2 (a) Forged steel automotive crankshafts are to be inspected for the occurrence of hairline cracks or other flaws which could act as starting points for fatigue, and magnetic particle inspection is suggested as a suitable method. State, giving reasons, whether the inspection should take place before or after the forgings are machined.

(b) What procedures should be followed to ensure that all surface defects, whether longitudinal or transverse, be detected?

28.3 (a) Explain, briefly, the principles involved in defect detection using the electrical eddy current system.

(b) How does changing the test frequency affect the sensitivity of the method for:
  (i) location of surface defects, and
  (ii) location of sub-surface defects?

28.4 (a) Describe how the following types of ultrasonic energy can be generated and propagated through solid materials: (i) compression waves, (ii) shear waves and (iii) Rayleigh waves.

(b) What type of ultrasonic inspection technique would be best suited for detection of the following types of defect:
  (i) small planar defects in aluminium plate of 20 mm thickness,
  (ii) surface breaking cracks in a steel shaft of 35 mm diameter, when access to only a portion of the shaft surface is possible, and
  (iii) possible cracks and/or porosity in a butt weld between two 25 mm thick steel plates?

28.5 The velocity of compression waves in aluminium is 6190 m/s. Determine the wavelength of ultra-sound waves in aluminium at a frequency of 1.5 MHz. What would be the minimum frequency which could be used if it was a requirement that all defects of size 3 mm and upwards be detected in samples of aluminium using ultrasonic inspection?

28.6 Explain how the sensitivity of radiographic inspection can be assessed.

In a radiograph of an aluminium component with a maximum sectional thickness in the line of the radiation of 28 mm, the thinnest portion of a step-hole IQI which is clearly visible has a thickness of 0.500 mm. Determine the sensitivity of the radiograph.

28.7 The dimensions of a metal bar of circular cross-section are measured accurately and are: length – 100.12 mm, diameter – 10.06 mm, mass – 62.58 g. The bar is then clamped midway along its length and stimulated into longitudinal vibration. The first two resonant frequencies are found to be 25.75 kHz and 51.5 kHz. Determine the value of Young's modulus for the material.

CHAPTER 29

Macro- and micro-examination

Why learn about macro- and micro-examination?

As discussed in many of the earlier chapters, the properties of materials are structure sensitive and, for example, for metals the control of microstructure by alloying, mechanical processing and heat treatment enables us to put the material into the optimum condition for a specific engineering application. Microscopy is a most useful tool and the study of microstructures by scientists and technologists has enabled us to obtain a much fuller understanding of the relationships between structure and properties than would have been possible otherwise.

In this short chapter a brief outline is given of the various macro- and micro-examination techniques used.

LEARNING OBJECTIVES

1. Distinguish between the aims of macro- and micro-examination.
2. Describe the preparation of materials for macro- and micro-examination.
3. Distinguish between transmission and scanning electron microscopy.

29.1 Macro-examination

Macro-examination, examination with the naked eye or at a low magnification, is a useful technique for the inspection of fracture surfaces, and for determining some of the characteristics of metal structures.

Much useful information can be obtained from visual examination of a fracture surface with the naked eye. The difference between a tough fibrous fracture and a brittle cleavage fracture can be readily observed. A fatigue fracture also possesses a characteristic appearance (refer to Chapter 25 and Figure 25.2). It may also be possible to detect the presence of slag inclusions and porosity on a fracture surface, and such defects may have been the points of initiation of the failure. This type of surface detail will be revealed with greater clarity if the surface is further examined with the aid of a low-power (up to ×50 magnification) stereo microscope.

Table 29.1 *Some etchants for macro-examination.*

Composition	Method of application	Uses
Hydrochloric acid 140 ml Sulphuric acid 3 ml Water 50 ml	Immersion in solution for 15–30 minutes at 90°C	A deep etch for steels
Copper ammonium chloride 9 g Water 91 ml	Immersion in solution for 0.5–4 hours	To reveal dendritic structures in steels
Ferric chloride 25 g Hydrochloric acid 25 ml Water 100 ml	Immersion	For copper and its alloys
Ammonium persulphate 10 g Water 100 ml Ammonium hydroxide 50 ml	Immersion	For copper and its alloys
Hydrofluoric acid 20 ml Water 80 ml	Swab surface of sample with etchant	For aluminium and its alloys

In order to obtain information about the structure of a metal it is necessary to section the material. The cut surface is then ground flat and etched with a chemical reagent. A high degree of surface finish on the specimen is not essential for macro-examination, and grinding can be finished with a grade 0 emery paper. During the etching treatment the surface layer of metal is removed and there is a preferential attack on certain constituents and inclusions. There are very many macro-etching agents and the choice of agent will be dependent on the nature of the metal and the type of feature that it is desired to reveal. Details of some macro-etchants are given in Table 29.1. Macro-examination after etching will reveal such detail as defects, segregation effects, and fibre structure in wrought metals.

Sulphur printing is a macro-examination technique that is suitable for the examination of plain carbon steels. Steels contain small sulphide inclusions, and the distribution of these inclusions within the steel is a guide to the distribution of all non-metallic inclusions. The sample of steel to be examined is sectioned and the cut surface is ground flat. A piece of bromide photographic paper is soaked in a 3 per cent solution of sulphuric acid for two minutes and it is then carefully placed, with the emulsion side downward, on the prepared steel surface. The sulphuric acid reacts with sulphide inclusions forming hydrogen sulphide gas. This gas reacts with the silver bromide in the emulsion forming a brown deposit of silver sulphide. The bromide paper should be in contact with the steel surface for about three minutes. The bromide paper is then removed from the steel, washed in water, and the print is 'fixed' by immersion in a 'hypo' solution for a few minutes. Dark-brown areas on the print indicate areas containing sulphide inclusions in the steel section.

29.2 Micro-examination of metals

Microscopic examination of metals is used to reveal fine details of structure. Metals may be examined at magnifications of up to ×3000 with the aid of an optical microscope. In order to examine the structure of a metal at high magnification it is necessary that the metal specimen be carefully prepared and, because metals are opaque to light, incident illumination must be used. Oblique incident illumination is perfectly satisfactory for macro-examination, but normal incident illumination must be used for examination at magnifications of ×50 and greater. The construction of a metallurgical microscope differs from that of a biological microscope (using transmitted light) to cater for this. The metallurgical microscope possesses a built-in light source. A partially reflecting mirror is situated in the microscope tube, and this will reflect the illuminating light through the objective lens on to the specimen surface (Figure 29.1). Alternatively, a small reflecting prism may be used in the microscope tube.

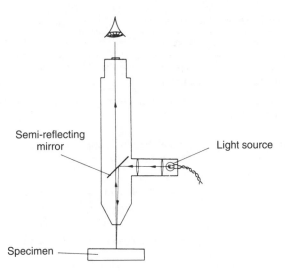

Figure 29.1 *Principle of metallurgical microscope for normal incident illumination.*

A metal specimen for micro-examination is prepared by cutting a small, but representative, sample from a metal component, followed by grinding and polishing a surface of the specimen to a mirror finish. This is achieved by a series of successive hand-grinding operations using progressively finer grades of paper. The specimen should be rotated through 90°C at each change of paper. The final grinding should be on grade 500 or 600 silicon carbide paper. Paraffin should be used as a lubricant for the grinding of very soft metals. After grinding is complete the very fine surface scratches are removed by polishing the surface to a mirror finish. The polishing powders that may be used are alumina, jeweller's rouge, or magnesia. These polishing powders are suitable for the polishing of steels and many other materials, and are used in water suspension in conjunction with a Selvyt or broad-cloth polishing cloth. The polishing cloth is mounted on a flat rotating disc and the specimen is held against this under light pressure. The proprietary metal polishes 'Brasso' and 'Silvo' are highly suitable for polishing copper and copper alloy specimens. Pastes containing very fine diamond dust (particle sizes ranging from $\frac{1}{2}$ to 3 microns) are also widely used for metal polishing. Alternatively, the finely ground surface may be polished electrolytically. This entails anodic solution of the ground surface in a suitable electrolyte.

The flat polished surface of the specimen should be examined under the microscope. The highly polished metal surface will appear bright and will show no structure, as it acts as a mirror to the normal incident light, but the following features, if present, can be observed:

(a) cracks and porosity,

(b) non-metallic inclusions (slag inclusions, and graphite in cast irons),

(c) hard constituents, which stand out in relief from the matrix.

In order to reveal the complete structure of the specimen it is necessary to etch the polished surface in a dilute chemical reagent. The choice of etchant will depend on the nature of the material and the type of feature being investigated. Details of some micro-etchants are given in Table 29.2. The action of the etching agent will be selective and there will be a preferential attack at crystal grain boundaries and at constituent boundaries. Examination of the etched specimen under the microscope will now shown the grain boundaries revealed (Figure 29.2). The effect of an etching agent is not usually a general solution of the metal surface. Often the attack produces a faceted surface, and the orientation of the surface facets may vary from grain to grain. In this case the amount of reflected light recaptured by the objective lens of the microscope will vary from one crystal grain to another. This will lead to some crystals appearing to be light in colour and other crystals appearing dark, even though all the crystals

Table 29.2 *Some etchants for micro-examination.*

Etchant	Uses
Nital: 2 ml nitric acid in 100 ml industrial alcohol	Excellent general etchant for irons and steels, other than stainless steels
Ammonium persulphate 10 g Ammonium hydroxide 20 ml Water 80 ml	Good general etchant for copper and its alloys
Ferric chloride 10 g Hydrochloric acid 30 ml Water 100 ml	Good etchant for copper and its alloys; provides more contrast than ammoniacal ammonium persulphate solution
Hydrofluoric acid 5 ml Nitric acid 1 ml Water 100 ml	Etchant for aluminium alloys; surface of the specimen should be swabbed with cotton wool soaked in etchant
Sodium hydroxide 1 g Water 100 ml	General etchant for aluminium alloys

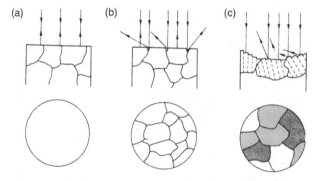

Figure 29.2 *(a) Polished metal surface. No detail revealed. (b) Etched surface revealing grain boundaries. (c) Etched surface revealing grain boundaries, but with grain shading.*

may be identical in composition and type. The photomicrographs of copper and α bronze (Figures 15.5 and 15.12) show this effect. In each of these alloys there is only one phase present, even though there are many shades present in the micrographs.

29.3 Micro-examination of non-metals

Unlike metals, most non-metallic materials are transparent in thin section. For micro-examination, ceramics and polymer materials are prepared as very thin sections mounted on a glass slide and illuminated for viewing by transmitted light. The preparation method for a ceramic material for microscopic examination involves, first, cutting a small thin sample using an abrasive cutting wheel, smoothing one surface to give a fine surface finish and cementing this sample, smooth surface downwards, onto a glass slide. The cement generally used is Canada balsam. After mounting, the exposed surface of the ceramic specimen is then ground down by means of carborundum, emery, or diamond abrasives to a thickness of about 30 μm. Finally, the prepared surface is covered with a thin glass strip, again cemented with Canada balsam.

While considerable information can be gained from viewing a thin slice of material using ordinary transmitted light, further information, leading to positive identification of mineral constituents, will be obtained if the specimens are viewed using polarised light. Polarising microscopes are fitted with two Nicol prisms, one, termed the *polariser*, mounted below the specimen stage and the other, the *analyser*, mounted in the microscope tube. A Nicol prism is a rhomb-shaped prism composed of two pieces of clear calcite bonded with Canada balsam and when incident ordinary light is transmitted into one end of the prism, a ray of plane polarised light emerges from the other end. Many materials show the property of *birefringence*, meaning that they can divide a ray of polarised light into two rays. There are two refractive indices, one for each ray, and the two rays possess different velocities in the material. In the micro-examination of a transparent material in polarised light, the incident ray is divided into two in passing through a birefringent material. After passing through the analysing Nicol the two rays emerge polarised in the same plane but out of phase with one another (as they possess differing velocities) and the resulting interference produces an interference colour. This effect is of diagnostic value. In addition to its use in the examination of ceramics and minerals, polarised light is used in microscopic studies to determine the fibre orientation in composite materials.

29.4 Transmission electron microscopy

The wavelength of visible light is of the order of 5×10^{-5} m and this means that it is impossible to resolve objects of smaller size than about 1 micron using an optical microscope. For magnifications in excess of about ×2000 it is necessary to use the electron microscope. Electrons possess wave characteristics and the wavelength of electrons in a 100 kV electron microscope is about 3.5×10^{-12} m. In an electron microscope a beam of electrons is produced by an electron 'gun'. The electron beam passes through an evacuated tube and is focused on to the specimen by means of magnetic lenses. The specimen is inserted into the vacuum chamber via an air-lock. The electrons pass through the specimen and through the magnetic objective and projector ('eye-piece') lens systems of the microscope to be projected on a fluorescent screen. Magnifications of up to ×100,000 may be obtained with electron microscopes, and objects of sizes of the order of 10^{-9} m may be resolved. Because the electron microscope uses a transmission technique and metal specimens readily absorb electrons, metal samples can be examined directly only if in the form of very thin foils (foil thicknesses of the order of 10^{-6} m). These thin foils of metals have to be prepared very carefully. This is usually done by thinning a small piece of the metal by chemical or electrolytic solution. Transmission electron microscopy of metal foils has been used as a research technique since the mid-1950s. Before that date metal samples were not viewed directly and replica techniques were used, A replica of an etched metal surface can be obtained by depositing a thin film of a plastic on the surface. The replica can then be stripped from the metal surface and viewed in an electron microscope. Both replica and thin foil techniques are used today.

29.5 Scanning electron microscopy

Another type of electron microscope that is used as a research tool is the scanning electron microscope. This type of microscope can accommodate a fairly large specimen and is used for examining the surface topography of a material, for example a fracture surface. A very fine electron beam scans the surface of the specimen, in a similar manner to the scanning of a television picture. The electron beam is scattered from the specimen surface and scattered electrons are picked up by a collector. The signal produced from the collected electrons is used to modulate the scanning beam of a cathode ray tube and this produces a picture of the surface area of the specimen under examination. One major advantage of the scanning electron microscope is that it possesses a large depth of focus. Magnifications of up to ×40,000 are possible with this type of instrument.

29.6 | Other analytical techniques

An electron beam will suffer diffraction by the regularly spaced rows or planes of atoms and the diffraction pattern is characteristic of the type of crystal structure. This diffraction effect can be used for the analysis of crystal structures in the same manner as for X-ray diffraction, as described in section 5.8.

Spectrographic analysis

When a material is excited in some way, the energy emission which follows excitation is characteristic of the chemical elements present in the material. Each chemical element gives rise to a unique line spectrum emission. This is used as the basis of a rapid and accurate method of chemical analysis. A line spectrum, produced by spark excitation, is analysed. In one method, the spectrum is recorded on a photographic plate and the image examined. A qualitative analysis is obtained by observing the number of characteristic spectra present. From measurement of the relative intensities of certain spectral lines, a quantitative analysis can be obtained. In a direct reading spectrograph, the complete spectrum is not recorded but photomultiplier tubes are set in positions to receive emissions corresponding to specific lines for various elements, and the quantity of light received is directly related to the amount of each element present, thus enabling a quantitative chemical analysis to be given.

Electron probe microanalyser

A fine electron beam can be directed at a specific area or areas of a test-piece. The beam stimulates an X-ray emission from the excited area, this emission being characteristic of the chemical element or elements in the scanned area. From an analysis of the emission, both a qualitative and quantitative assessment of composition can be obtained.

SUMMARY

Microscopy covers a widely used set of techniques for examining the structures of materials. Examination at magnifications of up to ×50 is termed macro-examination and is useful for examining fracture surfaces and, on metal specimens with ground and etched surfaces, looking at the fibre structure or the distribution of impurities.

The micro-examination of metals is for the purpose of revealing fine details of structure, for example, crystal grain size and shape, the presence and distribution of second phase particles, and so on. Also, it can yield information on the type of heat treatment given to the material. Because metals are opaque to light, metallurgical microscopes are designed to give normal incident illumination to the surface under examination. Ceramics and other non-metallic materials for micro-examination are carefully prepared as very thin sections, mounted on a glass slide, and viewed using transmitted light. Polarised light may be used also to aid identification of substances.

Electron microscopes may be either of the transmission type or the scanning type. Transmission microscopy can be used for high-resolution examination of microstructures and, at the higher magnifications, it is possible also to observe dislocations. Scanning microscopy is used for the examination of surfaces at a high magnification.

CHAPTER 30

Materials selection

Why learn about materials selection?

There is a vast range of materials available today for use by engineers and designers. It is important that the choice of a material to meet a particular design requirement be made most carefully. An incorrect or inappropriate choice could result in premature failure of a component or structure, possibly with disastrous consequences. Or the wrong choice of material may mean that the costs of maintenance during the life of the component or structure will be much higher than might otherwise have been the case. It is necessary, therefore, that engineers not only possess knowledge about the properties of materials but also that they are fully aware of all the factors that should be taken into consideration when selecting a material to meet a particular requirement. Even so, how can an informed choice be made from such a wide range of materials? It is obvious that there must be a systematic approach to the problem and it is vital that students appreciate this and possess some knowledge of the procedures and protocols associated with the selection process.

LEARNING OBJECTIVES

1. List the parameters to be considered when selecting materials to meet design specifications.

2. Describe a systematic approach to the selection problem.

3. Understand that there is not always a unique solution to the problem and that optimisation is necessary.

4. Appreciate that design and materials selection must consider the complete product life-cycle, including end-of-use disposal and recycling.

30.1 | Introduction

At some stage in the process of converting a design idea into 'hardware', decisions must be taken on the choice of material and the manufacturing route. These decisions should be taken at the earliest possible moment as there are a series of complex interactions between the three elements of materials, manufacturing and design, as illustrated in Figure 30.1.

Figure 30.1 *Interrelationships in material selection.*

Let us look at some simple examples of these interactions. The rough size and shape of a component design will have an influence on both the process route and the choice of material. If die-casting or injection moulding are considered as the best methods to produce the desired shape then it follows that the choice of possible materials is restricted to the lower melting point metals and melt processable thermoplastics.

The type of material considered will also influence design decisions. Details of design will differ between, for example, metals and polymers. Also, specific characteristics of a material may permit differing approaches to certain design features. An example of this is the low stiffness and good 'spring-back' of plastics which permits design for 'clip-fit' assembly. Similar components made from metals would require the use of spot welds, rivets or screws for assembly.

The properties of a material also exert major influences on methods of manufacture. The material properties may preclude the use of some process methods or make processing difficult and costly. The use of extremely hard metals could give rise to the necessity for expensive machining processes to obtain the required dimensional tolerances, although, as an alternative, precision casting using the investment process may be feasible. Conversely, the method of manufacture will affect the properties of the material. Wrought metal products are often stronger than castings made from metals or alloys of similar composition.

In the selection of a suitable material to satisfy a particular design and product requirement, it is necessary to look at many aspects to ensure that the component or assembly can be manufactured within the resources available, that the completed product will function satisfactorily throughout its design life, and that all this can be achieved at an acceptable cost. Also, these days, it is necessary to consider the problems associated with the disposal and recycling of products when they have reached the end of their useful life. In the EU and some other countries legislation is being enacted which will make it the product manufacturer's responsibility to deal with the decommissioning and recycling of some products at end of use.

30.2 | Parameters to be considered

The parameters which need to be considered when selecting a material to fit a design specification include:

Material properties – mechanical   elastic moduli and stiffness
yield and maximum strengths
fatigue strength
creep strength
fracture toughness
hardness

	ductility
	abrasion resistance
physical	density
	T_m and T_g values
	electrical conductivity
	magnetic properties
	thermal conductivity
	thermal expansion
	thermal stability
chemical	resistance to chemicals and solvents
	corrosion resistance
	oxidation resistance
	weathering resistance
Manufacturing characteristics –	castability
	formability
	machinability
Cost and availability –	material cost
	manufacturing cost
	availability
	price stability

With reference to material properties, it is often very useful, when comparing the relative merits of a range of materials, to look at *specific properties*, that is, the ratio of the particular property to the density of the material. Figures 30.2 and 30.3 are bar charts showing the ranges of specific modulus and specific tensile strength respectively for materials. It will be seen from these figures that the specific modulus, that is, specific stiffness, is very similar for all the main engineering metals, and very much less than it is for carbon fibre composites and a number of ceramics. Similarly, the specific strengths of

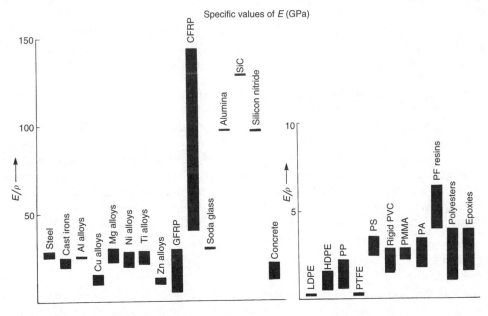

Figure 30.2 *Specific values of E (GPa/g cm^{-3}).*

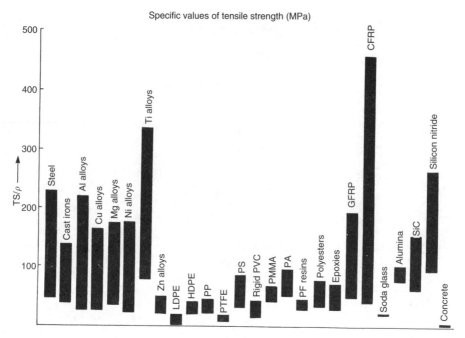

Figure 30.3 *Specific tensile strength (MPa/g cm⁻³).*

titanium and carbon fibre composites are considerably higher than those of most other materials. Weight saving is of prime importance in aerospace applications and, because of the extremely high specific strength and modulus offered by carbon fibre composites, these materials were investigated extensively for use as fan blade material for turbofan aero engines. Unfortunately, the lower abrasion resistance and fracture toughness, in comparison with titanium, meant that they could not be used for this application. Solid titanium alloy blades were used instead to satisfy this design requirement but the latest development, to reduce weight and increase stiffness, is to manufacture composite construction titanium fan blades, composed of a titanium honeycomb clad with titanium sheet.

Similar effects can be seen with respect to other properties, for example, the electrical conductivity of pure copper is some 60 per cent higher than that of aluminium, but a conducting wire of aluminium made to have the same electrical resistance per unit length as a copper wire, although of larger cross-sectional area, is approximately one-half of the weight of the copper, hence the use of aluminium conductors for overhead electricity transmission cables.

Material cost is another important parameter that may be viewed in more than one way. It is customary to see the cost of materials quoted per unit mass. This may give a misleading picture as, frequently, the size of a component is determined by other factors and a cost comparison between materials on the basis of cost per unit volume would be more appropriate. The relative positions of materials in a league table of costs change when the criterion is changed from £ per unit mass to £ per unit volume, as shown in Table 1.4. Yet another method may be used to good effect when attempting to compare the relative costs of competing materials, that is to establish tables on the basis of cost per unit of property.

30.3 | Costs

Cost and quality are all-important. For a product to be marketable, it must be supplied at an *acceptable cost* and be of an *acceptable quality*. The full cost of any product to the customer is made up of a number of items, as shown in Figure 30.4.

Figure 30.4 *Product cost build-up.*

It is difficult to generalise, but in many manufacturing industries the cost of materials accounts for about 50 per cent of the total works cost of the finished product. From this, it follows that the use of a cheaper raw material will have a significant effect on the final selling price. While this is generally true, it is not always so. In some cases, the high cost of a more expensive material may be offset by the material having better processing characteristics. For example, maraging steels are more expensive than high strength low alloy steels but the former materials may be machined to final dimensions and tolerances when in a soft condition, with final high strength and hardness being developed by a low temperature ageing heat treatment. The low alloy steel, on the other hand, may require to be machined to final dimensions after all hardening heat treatments have been completed. If the component is of complex geometry, the use of costly machining processes on a very hard material may make the total cost of the low alloy steel product higher than that of one in maraging steel.

Labour, general manufacturing costs and overheads account for about 50 per cent of total works cost and anything which can reduce this aspect of costs will be worthwhile. Major savings could accrue if the final component shape can be produced with the minimum number of process operations. The production of 'near-net' shapes by a single direct process, such as die-casting, injection moulding, compaction and sintering from powder, etc., should lead to a lower works cost than use of a processing route involving many intermediate stages and costly machining processes.

Some examples of intelligent design, materials selection and processing route are given as Case Studies in section 30.6.

30.4 ▍ Recycling

In earlier centuries mankind was profligate in the use of many of the earth's natural resources and, also, made little or no attempt to lessen the impact on the environment of many industrial activities. Evidence of this profligacy and the apparent lack of concern for the environment can be seen in the vast spoil tips associated with some mineral mining activities, and in the increased levels of atmospheric, water and land pollution found in many areas. Today, some mineral resources are scarce and known reserves may be completely used up within a relatively short time scale. The earth's resources are finite and there is a need both for conservation and for lessening the impact of many industrial activities on our environment. Engineers and designers have a major responsibility in this respect.

Another vital and costly resource is energy and considerable quantities of energy are consumed in the processing of many materials. Consider a metal such as aluminium. Energy is used in each of the many stages in the process route from ore to final product, for example aluminium sheet. Firstly the ore is mined from the earth and then concentrated to separate the bauxite mineral from spoil. In the next stage the bauxite is processed into pure alumina and this is dissolved in molten cryolite mineral to give the molten electrolyte which is electrolytically smelted to give aluminium. Energy is consumed in all these processes, especially in the electrolytic reduction of alumina to aluminium metal. The pure aluminium ingots produced have to be remelted and alloying additions made before casting into large ingots for rolling. These alloy ingots are then reheated to the correct temperature for hot rolling. Energy is needed to power the hot and cold rolling mills and to operate the furnaces for annealing cold rolled sheet metal. It is possible to calculate the energy content of materials, that is the amount of energy required to produce unit quantities of materials from their raw ingredients and approximate values of energy content for a range of engineering materials are given in Table 30.1. Ever increasing energy costs and the decline in reserves of some minerals of economic importance make it more important than ever that, wherever possible, materials are recycled. The energy needed to process secondary aluminium, that is aluminium metal produced from recycled scrap, is comparatively small, being less than one-fifth of the energy used to produce aluminium from primary sources.

Table 30.1 *Approximate energy content of some materials (GJ/tonne).*

A Materials derived from primary sources			
Titanium bar	560	Zinc (castings)	70
Magnesium extruded bar	425	Mild steel (rod)	60
Aluminium (ingot)	280	Cast iron (castings)	50
Aluminium (sheet)	300	Glass	20
PVC (rod or tube)	180	Reinforced concrete	12
Nylon 66 (rod)	180	Cement	8
Polyethylene (sheet)	110	Brick	4
Stainless steel (sheet)	110	Timber	2
Copper (pipe)	100	Gravel	0.1
B Additional energy content of metals from secondary sources			
Aluminium	45	Mild steel	20
Copper	30	Cast iron	17

It is evident from the above that materials should be recycled wherever possible as this not only helps to conserve valuable resources but also helps to reduce energy consumption. It is not only metals that can be recycled. Many plastics materials, some rubbers and glass can be recycled also. However, the recycled material is not necessarily of the same high quality as the original and there may be some restrictions on the uses to which recycled materials may be put. Some aspects of the recycling of materials are considered in the following paragraphs.

Metals

It is possible to reclaim and recycle most metals and alloys although it is not always feasible to do so. The first condition is that scrap metals for recycling should be carefully segregated and sorted before reprocessing. If metals and alloys can be fully sorted by alloy type then the scrap will have high value and can be used for reprocessing into the same grade material. If, however, the scrap is mixed its value is reduced and it may only be possible to use it for the manufacture of lower-grade materials. In the case of steels, for example, scrap material can be introduced, with blast furnace iron, into the initial steel making process in which impurities are removed by oxidation. This is fine for plain carbon steel scrap but if alloy steels have not been segregated from the scrap then valuable alloying elements will be lost during reprocessing. Alloy steel scrap should be remelted and used for new alloy steel manufacture.

Similarly, scrap aluminium should be separated by alloy type, if possible, and reused in the manufacture of the appropriate alloy type. Some aluminium alloy specifications for low-grade applications have relatively wide tolerance limits for the amounts of some impurity elements and unsegregated scrap or material of dubious quality may be used in the production of these alloys. During World War

II a large furnace was built in Britain to take and melt down the remains of crashed German and allied aircraft, after removal of engines, radio and some other equipment. The resultant material was used for low-grade applications only but, nevertheless, it was a useful source of secondary metal. Used drinks cans form a major source of aluminium scrap for recycling. A problem is that the can bodies and can lids are made from two different aluminium alloys. The bodies are formed from an Al–Mn alloy while the can lids are made from an Al–Mg alloy. There are several methods for dealing with this type of scrap. The two alloy types may be separated using a thermomechanical process from where they can be reprocessed separately, or the complete cans can be remelted. The molten alloy can then either be treated with chlorine gas to remove magnesium or used to make a lower-grade alloy which can tolerate either manganese or magnesium as an impurity element.

Similarly for copper, its alloys, and most other metals, the scrap material can be reprocessed to give a valuable source of secondary metal.

Polymeric materials

Some polymeric materials can be recycled while others cannot. Thermoplastics soften when heated and become more rigid when cooled but there is invariably some degradation and loss of properties with each heating and cooling cycle. The polymers which are most easily recycled are polyethylene (both LDPE and HDPE), polypropylene, polystyrene and polyethylene teraphthalate (PET). As stated in Chapter 17 polymeric moulding materials contain fillers, dyes and other additives and their presence may sometimes render recycling difficult. As with metals, the polymer materials to be reclaimed and recycled must be sorted by type before reprocessing. It is also useful to sort according to colour otherwise the recycled material could end up with an unusual colour with little customer appeal. If the material is not colour sorted then additional colorants can be added to give the reprocessed material a dark brown or black appearance. The fact that there is some degradation of the polymer during recycling means that the reprocessed material often cannot be used for the same applications as new material. As examples of this, major applications of HDPE are in the manufacture of bottles and containers for water, milk and foodstuffs. Recycled HDPE is not used in these applications but for making waste bins, flower pots, toys and traffic cones. Similarly much LDPE is used in the form of blown film for freezer bags and other food packaging whereas recycled LDPE is used for refuse sacks, bin liners and other general-purpose packaging.

The recycling of thermosets is difficult as these materials possess network structures created by non-reversible chemical reactions during the first moulding. Some used thermosets are ground to powder which can then be used as a filler material in the preparation of thermoset moulding compounds using new ingredients. Similarly, most elastomer materials are vulcanised during manufacture and the cross-linked structures of these materials makes reprocessing difficult. A major source of scrap rubber is from vehicle tyres and the high tonnage of this waste produced annually causes major problems. At one time some of this was used as a fuel in some furnaces but the flue gases have a high sulphur content. Other reclamation methods are to grind the old tyres into small particles. These have then been utilised in a number of ways. One is to mix with an adhesive and use the mixture to make mouldings for non-demanding applications such as rubber mats for car interiors and vehicle mud flaps. Another use has been to mix with bitumen for the manufacture of surfacing materials. The inclusion of reclaimed rubber in asphalt road surfaces gives a hard wearing surface with reduced traffic noise levels, while an increased rubber content in bitumen can provide an impact absorbing asphalt surface useful for sports venues. Unfortunately, at present, these applications can only use a very small percentage of the used tyres generated each year.

Polymer-based composites, such as GFRP and CFRP, are virtually impossible to recycle because of the extreme difficulty in separating the reinforcing material from the matrix.

Unfortunately much of the polymer material waste generated cannot be reprocessed and most of it is not biodegradable. The only disposal methods available are in land-fill or incineration. Even then many polymers cannot be incinerated in safety as toxins and other pollutants may be generated by combustion.

Glass

Glass may be recycled. Much glass is used in the manufacture of bottles for foodstuffs and alcoholic drinks, thus generating a large quantity of waste glass annually. As the raw materials for glass manufacture (silica sand, soda and lime) are comparatively cheap the market value of used glass is quite low. For this reason much used glass ends up in land-fill sites. For recycling to be economically feasible the glass must be sorted according to both type and colour. Most of the glass which is recycled is from bottles and other glass containers. Generally these are made in three colours, clear, brown and green, and the most economic system for waste glass collection is to place 'bottle banks' at strategic sites in towns or hypermarkets where the public can deposit their waste bottles, doing the colour sorting by dumping the bottles into the appropriate container.

30.5 | A rationale for selection

There are some instances where the choice of a material is so severely restricted by service requirements that there is either no choice or only an extremely limited choice of possible materials. For example, if it is considered essential to the design that a component possess an electrical resistivity of less than $2.0 \times 10^{-8}\,\Omega\,\mathrm{m}$, then the only choices are silver, high-conductivity copper, or a few copper alloys. However, perhaps the design requirement, by quoting a maximum resistivity value, is too restrictive. On a weight basis, aluminium is a more efficient electrical conductor than copper, and is used for many high conductivity applications. It is another factor, not conductivity, which prevents aluminium replacing copper for all electrical applications, namely, the fact that copper can be soft soldered with ease, permitting connections to be made with greater ease than is possible with aluminium.

In most cases, the problem is not simple as there may be very many materials which could be considered as possible contenders to meet a design specification. In many cases, there may be conflicting requirements, for example, high strength coupled with high ductility, and compromises have to be made. Frequently, there will be no perfect solution, but several alternative viable solutions can be achieved. The solution eventually arrived at will be the result of an optimisation process.

How can we come to a decision? In order to take an informed decision two things are necessary:

(a) Reliable property data on materials.

(b) Full appreciation of the service conditions that the component or assembly is expected to operate under: working stresses, possibility of overloads, environmental conditions leading to weathering or corrosion, etc.

In the case of most well-established metals and alloys, there is much reliable property data available. Also, major initiatives have taken place in recent years to standardise many commercial alloy compositions and specifications internationally. In the case of new alloy compositions, much testing and property evaluation has to take place before the material can enter service. Some of this testing, such as creep and corrosion tests, is very long-term. The situation with regard to polymer and ceramic materials is less clear. There is far less standardisation. The amounts and exact nature of fillers and additives in, say, a polypropylene moulding compound may vary significantly from one major manufacturer to another and the user must rely on property data supplied by the specific manufacturer, or else embark on a lengthy and costly programme of testing for himself. In some cases, a new material may be used extensively on the basis of its attractive properties, as determined by short-term tests with subsequent problems developing as a result of long-term ageing and weathering problems. An example of this is high-alumina cement. This is a high-strength cement with a very rapid hardening rate. When first developed, it was widely used in the construction industry for structural concrete. Unfortunately, in the long-term, a structural change occurs in one of the cement constituents leading to loss of strength and an increase in porosity which gives a greater susceptibility to weathering. A number of structural failures ensued and the use of high alumina cement in structural concrete is now greatly restricted.

There is a vast quantity of property data available for materials, much of it quantitative but some of a qualitative nature. In recent years there have been numerous developments in the production of materials property databases and materials selection computer programs. Some of these are available commercially, either as subscription services or for outright purchase.

Professor Michael Ashby of Cambridge University has developed an interesting series of materials property charts which can be used as an aid to selection. In these charts the range of certain properties, say strength against density, for all materials is plotted using logarithmic scales. The property range for a group of related materials, for example engineering ceramics, is shown as an envelope within which are smaller property range envelopes for specific ceramics. By using logarithmic scales covering five or six decades all material classes can be included on the same chart. Any property combination can be plotted in this manner, including modulus against density, fracture toughness against strength, specific strength against specific modulus, and so on. Reference to these charts enable one to see at a glance which groups of materials will possess the minimum acceptable property values for some application, therefore worthy of closer examination, and which groups can be eliminated.

The second aspect, a thorough analysis of the service conditions which a component or assembly has to operate under, and the conditions which may cause premature failure, requires the exercise of considerable skill and experience by the design and materials engineers. As stated earlier, the selection of a suitable material to fit a requirement often involves compromise, with a 'trade-off' between properties. As the selection process is one of optimisation, it is of extreme importance for the engineer to be able to determine which is the most important property to satisfy a particular design – is stiffness more important than fatigue strength, or is fracture toughness more critical than yield strength?

A systematic approach to the problem of materials selection is necessary and the following is cited as one possible approach.

Stage 1

Analyse fully the product specification and determine the minimum acceptable values for all the relevant material properties.

Stage 2

Make first selection by eliminating all materials which do not possess all the minimum criteria.

Stage 3

Assess the degree of relative importance of the various required properties from essential through to desirable, and for each property place the potential materials in ranking order.

Stage 4

Evaluate the materials and process costs for each material.

Stage 5

On the basis of the decisions made in Stage 3 and the data from Stage 4, optimise to determine the materials which give the best overall combination of properties for the least cost.

It is hoped that the foregoing paragraphs give the reader some awareness of the complexities of materials selection problems and highlight the necessity for considering the possible materials and processing routes from the commencement of any product design study. It is only in this way that effective and efficient utilisation of materials will be achieved.

30.6 Case Studies

I have selected a few Case Studies to illustrate the following concepts:

(a) how, by making progressive design changes involving materials and processes, costs can be reduced;

(b) how, by changing the material, a simpler processing route can be achieved;

(c) how, by changing the material, significant weight reductions can be achieved which can then have implications for energy saving;

(d) how, by changing the processing method, there can be a more efficient utilisation of material;

(e) how advanced composite materials can be used to good effect in the construction and aerospace industries;

(f) how new materials are developed to meet new and exacting requirements.

Case Study 1 – Progressive design changes of a pulley

The design of a simple part may change progressively over a period of time even though the basic function of the part remains the same. An example of this occurs in the case of a pulley used as part of the drive mechanism in a domestic washing machine. Initially the pulley was made as a sand casting in iron. Although the material and the sand casting process are relatively inexpensive the castings required a considerable amount of machining which increased the final cost of the component. The quantity of production needed was high enough to justify die casting and the part was redesigned as an aluminium alloy die casting. The material was more expensive but little machining was required and the component cost was reduced by about 33 per cent. The next design change was to make the pulley wheel as a two-piece die casting in a zinc alloy. Although the two parts had to be assembled there are no machining operations required and the cost of the component was reduced further by about 30 per cent. Finally, it was redesigned yet again to be manufactured as an injection moulding in acetal. This gave a pulley with a unit cost approximately 70 per cent less than the original cast iron part.

It is worth noting at this stage that sometimes a company may wish to make a design change but cannot because of other constraints. For example, a company may have invested a large amount of capital in, say, a die-casting facility and will be reluctant to change material or process if it means that facility becomes redundant.

Case Study 2 – Simplification of design

It is often possible to replace a complex assembly with a single component. Figure 30.5 shows a combined gear/cam which is a component in a business machine. The original component was an assembly of five individual parts, a gear and a cam, both blanked out from low carbon steel sheet, a central boss and two rivets. The replacement component was designed in PA 6.6 and made as a one-piece injection moulding. The moulding was designed to have the same envelope shape and dimensions as the original so that no design changes in other parts of the machine were needed. This change resulted in a reduction of the unit cost of the component of almost 80 per cent.

Figure 30.5 *Combined gear/cam for an office machine showing the original component made from five metal parts and the replacement component, a single injection moulding in PA 6.6.*

Case Study 3 – Containers for drinks

The market for drinks is enormous and a wide range of beverages including water, juices, carbonated and alcoholic drinks are sold in sealed containers in a variety of sizes from 250 ml capacity upwards. The main types of container used are glass bottles, blow moulded bottles in HDPE for water and juices, or in PET for carbonated drinks, and aluminium cans for juices, carbonated drinks and beer.

Let us compare the three types of material, glass, polymer and aluminium, as containers for liquids. The costs of containers of, say 330 ml capacity, are very similar for each of the three materials with glass being slightly less expensive than PET or aluminium, and all three materials are recyclable. The biggest difference is in the weight of the containers. The weight of a 330 ml can or PET bottle is approximately 15 g whereas a 330 ml glass bottle weighs approximately 185 g. This difference may not be a problem if you are purchasing one or two bottles but it is highly significant if you are concerned with bulk handling and transport of bottled and canned drinks. For example, 3000 filled aluminium cans or PET bottles weigh just over 1 tonne but the same quantity of full glass bottles has 50 per cent greater weight at more than 1.5 tonnes. This means that more energy is used in handling and transporting the same quantity of liquid because of the massive increase in container weight. Because of this the use of aluminium cans and plastic, either HDPE or PET, has superseded glass for most beverages. Glass is still the main container material for wines and spirits although some wines and spirits are bottled in PET containers. Beers may be sealed in cans or in glass bottles. PET is very slightly permeable to oxygen and any slight ingress of oxygen would adversely affect the quality of beer held in PET bottles. However, this situation could be about to change. Recently there have been developments with the aim of producing blow moulded PET bottles with a barrier coating which is totally impermeable to oxygen. Two different techniques have been used experimentally. In one the barrier material is co-extruded around the PET parison for blow moulding while the other technique being developed is to apply a coating to the moulded bottles. The plastic beer bottle may become a reality very shortly.

Case Study 4 – Advantages of powder processing

As stated earlier the unit cost of a finished component is the sum of the costs of material and processing and must take in the costs of reworking process waste. Consider the timing chain sprocket for a car

Figure 30.6   *Timing chain sprocket made by a powder metallurgy process.*

Figure 30.7   *Components made by powder processing. (Courtesy of Schunk GmbH)*

engine shown in Figure 30.6. This component is made from a low alloy steel, is of 125 mm diameter and has a mass of 290 g.

Formerly this component was manufactured by hot stamping a blank followed by machining operations. Forming the component in this manner meant that the finished part was about 75 per cent of the metal input, the remainder returning as process scrap. The forming process was changed and the component was made by compaction and sintering an alloy steel powder mix. The component is produced to close tolerances and with a good surface finish so that no machining operations are required.

There is virtually no wastage of powder so, although metal powder, as a raw material, is more expensive per unit mass than steel forging blanks the better utilisation of material and the simplification of processing gives a lower unit cost for the finished component.

Other examples of components produced by powder processing are shown in Figure 30.7.

Case Study 5 – Metal injection moulding of watch parts

Metal injection moulding (MIM) is a development of the powder metallurgy process. The metal powder is mixed with a plastic binder to render it suitable for injection moulding. After the moulded parts have been made, the binder is removed carefully and the porous powder metal moulding sintered. During sintering densification and consolidation of the moulding occur to give a finished part of almost theoretical density. There is a large volume shrinkage during sintering but if the original powder/plastic mix was uniform then the shrinkage will be uniform and predictable with no warping or other distortion.

The watch cases illustrated in Figure 30.8(a) were formerly made by hot forming from austenitic stainless steel sheet. The Swiss company ETA SA Fabriques d'Ebauches investigated the use of metal injection moulding for making these cases for the range of SWATCH watches with the aim of making cost savings. The injection moulded parts, which are of the same basic design as hot formed cases, are produced to give a stainless steel product which, after sintering, has a density of 7950 kg/m³. Generally the tolerances on the cases are maintained at ± 0.02 mm. After sintering the cases are polished to give a mirror finish. The change of process from hot forming to injection moulding yielded significant cost savings.

Similarly, another company, HOPTEC of France, has used metal injection moulding to manufacture the wristwatch bracelet clasp components (shown in Figure 30.8(b)) from stainless steel powder. Previously these small parts were made either by machining or by blanking and forming from stainless steel sheet. The injection moulded stainless steel segments are sintered to give a dense product with a density of 7860 kg/m³ and to very close dimensional tolerances, ± 0.05 mm generally. The two parts have overall dimensions of $40 \times 10 \times 2$ mm and $40 \times 13 \times 2$ mm and weigh 2.56 g and 3.72 g respectively. The fact that the clasp components are sintered to a high density enables them to be given a highly polished finish, a quality required by the watch manufacturer. The two parts are assembled and

(a) (b)

Figure 30.8 *(a) MIM watch cases in stainless steel. (Courtesy of EPMA and ETA SA Fabriques d'Ebauches, Switzerland). (b) MIM watch bracelet clasp components in stainless steel. (Courtesy of EPMA and HOPTEC, France)*

the final operation is bending to fit the wearer's wrist. The change to metal injection moulding from traditional processing methods gave a cost saving.

Case Study 6 – The Chrysler Composite Concept Vehicle

When a concept car is launched by a major car manufacturer usually it is the embodiment of a new philosophy in vehicle design which gives an indication of future trends in the automotive industry. This is the case with the Chrysler Composite Concept Vehicle (CCV) (Figure 30.9), a concept which could revolutionise car production. The body shell of the Chrysler CCV is made completely of injection mouldings of an impact modified, reinforced polyethylene teraphthalate resin (PET). Why has this not been done before? Firstly, to produce the large body parts required for car manufacture requires the use of injection moulding machines considerably larger than those in current use and, secondly, it requires a mouldable polymer which is sufficiently strong and tough and with good dimensional stability. The impact modified PET was developed by Ticona, in conjunction with Chrysler.

The structural body shell of the vehicle (Figure 30.10) comprises four large injection mouldings which are directly bonded to the chassis. Six large injection moulded panels are used to complete the body. The completed body has a weight which is approximately one-half the weight of a steel body for a vehicle of similar dimensions and this will lead to extremely low fuel consumption. Manufacturing costs will be greatly reduced and it is predicted that if the vehicle goes into quantity production the manufacturing costs could be reduced by up to 80 per cent. It is estimated that the assembly time for the CCV with the PET body would be about 6.5 hours. This compares with an assembly time of about 19 hours for a conventional vehicle of similar size. Additionally, the total area required for vehicle assembly would be very much less than the area needed for conventionally built vehicles. Painting and drying facilities would be rendered unnecessary as the required colour is compounded into the PET moulding compound.

Figure 30.9 *The Chrysler Composite Concept Vehicle. (Courtesy of Ticona)*

Case Study 7 – The Aberfeldy bridge

The world's first major footbridge constructed in fibre reinforced composite material was constructed across the river Tay at Aberfeldy in Scotland in 1992. This is a suspension bridge with a deck of 120 m in length and a central span between the towers of 63 m (see Figure 30.11). Why was a glass fibre

Figure 30.10   *CCV body shell comprising four PET injection mouldings. (Courtesy of Ticona)*

Figure 30.11   *The Aberfeldy footbridge. (Courtesy of Maunsell Structural Plastics)*

reinforced plastic chosen as the construction material for this structure? The client, the Aberfeldy Golf Club, wished to extend its course and was granted planning permission to build 9 holes on the opposite bank of the river Tay on condition that it also provided a footbridge across the river. The fact that the bridge was to be built in a rural site with limited access for heavy construction equipment and that the client wanted a low-cost solution suggested that the design team investigate the use of advanced reinforced structural plastics.

The bridge, which was designed by Maunsell Structural Plastics, has two A-shaped suspension pylons 18 m in height with Kevlar suspension cables supporting the bridge deck. The suspension cables are sheathed for protection within an LDPE coating. Both the pylons and the bridge deck were constructed from standard cellular GFRP structural sections with cross-sectional dimensions of 600 mm × 80 mm and these individual sections joined to one another by specially designed connectors and adhesive. Because of their light weight the two 18 m lengths and the cross-piece for each suspension pylon could be delivered to site by a single lorry and, after assembly, the A-frame was raised into the vertical using guy ropes and a single 5 tonne capacity winch, thus eliminating the need for a large crane on site. The erection of the bridge on site was completed by a small team in a period of ten weeks and it is estimated that the bridge will not require any maintenance or repair work until more than 20 years after erection.

Case Study 8 – EADS Eurofighter (Typhoon)

In aircraft construction any saving which can be made in structural weight will yield bonuses in other directions. In military aircraft the saving of weight in the structure, provided of course that structural strength is not compromised, will result in improved performance, or an increase in the amount of fuel which can be carried, thus increasing operational range, or an increase in the weight of stores that can be carried. From its inception the international design team for the Eurofighter aircraft, now named the Typhoon (Figure 30.12), decided to use a large quantity of low-density structural composites in the aircraft. The EADS (European Aeronautical Defence and Space) consortium have produced a Mach 2.0 aircraft of great agility with a high thrust-to-weight ratio, low wing loading and capable of high rates of turn over the whole of the speed range. This has been achieved by using a considerable amount of

Figure 30.12   *EADS Typhoon aircraft. (Courtesy of BAe Systems)*

Figure 30.13 *Materials utilisation in Typhoon (Eurofighter). (Sketch based on information provided by BAe Systems)*

carbon fibre and glass fibre composite materials in the structure. Yet the airframe is extremely strong and has been designed to tolerate accelerations of up to $+9g$ and $-3g$ and to have a minimum service life of 25 years or 6000 flying hours.

The utilisation of structural composites in the Typhoon aircraft is as follows:

Fuselage. The main frame structure of the fuselage is largely composed of high-strength aluminium alloys but the bulk of the fuselage skin panels are of carbon fibre composite, with the nose radome cone being of glass fibre reinforced plastic.

Wing. The wing spars, ribs and skin panels are all constructed in carbon fibre composites, with the exception of the outboard flaperons (titanium), the wing leading edges and the wing tip housings for defensive aids (both in aluminium).

Tailfin. The vertical tailfin and rudder are manufactured largely from carbon fibre composites but aluminium alloys are used for the leading and trailing edges and the fin tip is made from a glass fibre reinforced plastic. All in all some 70 per cent of the total surface area of the aircraft is panelled in carbon fibre composites, with a further 12 per cent of the skin being of glass fibre composites. This is shown in Figure 30.13.

Case Study 9 – Airbus 380

The saving of structural weight in a civil aircraft will have an impact on the operational profitability of the airline flying the aircraft. The saving of 1 kg of structural weight could mean a gain of about £100,000 to the operator over the 30 year service life of a large civil aircraft. The international European consortium, Airbus Industrie, which was formed in 1971, has been a pioneer in the use of composite materials in aircraft structures. In the beginning, carbon fibre composites were used for applications such as cabin flooring, parts of seating, and so on, but later, as knowledge of, and confidence in these materials increased, so they came to be used in secondary structures such as control surfaces. By 1985 the designers started to use carbon fibre composites for some primary structures, specifically tailplane and tailfin, on some members of the Airbus range. In the 1990s tailplane and fin structures were fabricated from carbon fibre composites for most of the aircraft in the Airbus range. Today the keel beam for the large A340-600 is made of a carbon fibre composite.

Figure 30.14 *Airbus 380. (Courtesy of Airbus UK)*

Approx. 40%
of wing ribs

CFRP upper deck
floor beams

Fin box, rudder,
tailplane box & elevators
in monolithic CFRP

Upper fuselage
panels in GLARE

GLARE upper
fuselage panels

CFRP pressure bulkhead

Thermoplastic fixed
wing leading edge

CFRP centre wing box

Figure 30.15 *Distribution of composite materials in Airbus A380. (Sketch based on information provided by Airbus UK)*

The newest design in the Airbus family is the A380-800 (Figure 30.14). This very large two-deck aircraft, which is scheduled to enter airline service in 2006, will embrace the latest technologies and is designed to have a greater utilisation of composite materials in its structure than earlier Airbus aircraft. Currently the design envisages use of carbon fibre reinforced plastics in the following sectors of the structure: the centre wing box, a number of the wing ribs (approximately 40 per cent of the total), the tailplane and elevators, the tailfin and rudder, the rear cabin pressure bulkhead and the upper deck floor beams. The fixed wing leading edge will be in a thermoplastic material and, in addition, it is planned to make the upper fuselage panels from another composite material, known as GLARE, which comprises glass fibre prepreg tape sandwiched between layers of aluminium alloy. The distribution of composites in the structure of the A380 is shown in Figure 30.15.

Case Study 10 – Space re-entry vehicles (the Space Shuttle)

When a body travels through air at a high velocity the friction between the body and the air molecules will produce a kinetic heating effect which causes a rise in the temperature of the body. This heating effect can be highly significant even in the rarefied atmospheres at high altitudes. For example, the normal flight operating conditions for the Concorde supersonic airliner are Mach 2.0 at 18,300 m (60,000 ft). Under these conditions the kinetic heating will raise the temperature of parts of the aircraft skin by approximately 175 °C. As the outside air temperature at 18,300 m is about −57°C, this means that the skin temperature of the aircraft is about 118 °C.

This kinetic heating effect is a major problem for the re-entry of vehicles into earth's atmosphere from space. The thermal protection of early re-entry vehicles, such as the Mercury and Apollo capsules of the National Aeronautics and Space Administration of the USA (NASA), took the form of thick ablative heat shields which were largely destroyed during the period of re-entry. The heat shield material was based on thermoset polymers which, under the great heat generated during re-entry, degraded and carbonised and then suffered ablation, or vaporisation, of the carbon. The very large latent heat of vaporisation of carbon allows for extremely rapid dissipation of the large amount of heat energy generated by friction during re-entry and gives thermal protection to the capsule.

During the 1970s NASA designed an orbiter, now known as the Space Shuttle, which while launched by a large rocket would re-enter earth's atmosphere and land on a runway in the same manner as an aircraft. The Space Shuttle was designed as a reusable space vehicle which would have an operational life of 100 space missions. The main airframe of the craft was constructed largely of aluminium alloys and the maximum temperature that any part of this airframe should be subjected to, even for short periods, was 175 °C. For such a space vehicle to be completely reusable it was necessary to develop thermally protective materials which could be used to fabricate an outer skin to cover and protect the main airframe from the great heat generated during re-entry. Although the design and materials development work was conducted some 30 years ago, it is still valid to discuss it today to show how the application of well-known principles can be applied for the solution of advanced technological problems.

The temperature build-up during re-entry is not the same for all parts of the space vehicle with the highest temperatures occurring at the nose, and at the wing and fin leading edges. During re-entry the temperature at the nose of the shuttle can reach almost 1500 °C and that at the leading edges of wings and fin can reach about 1260 °C. Most of the undersurface of the fuselage and wings reach temperatures of 1000 °C and above but the temperature build-up on the upper surfaces is much less and the temperatures of parts of the upper fuselage and upper wing surfaces do not reach much above 300 °C. This is illustrated in Figure 30.16. Because of the widely differing requirements for different parts of the structure a number of different materials were developed.

In order to function successfully and give the necessary thermal protection to the Space Shuttle airframe the materials have to possess properties that meet a set of extremely exacting conditions. These are:

(a) be capable of withstanding a wide range of temperatures from −110 °C to various upper limits ranging from 300 °C to 1500 °C,

(b) be capable of withstanding severe thermal shock and very large temperature gradients,

(c) possess sufficient strength to withstand the mechanical stresses developed during launch and the thermal stresses developed during re-entry,

(d) have the necessary thermal properties to keep the maximum airframe surface temperature below 175 °C.

(e) be of low density,

(f) have a smooth outer surface for good aerodynamics,

(g) have the ability to be fixed successfully to an aluminium airframe structure, and

(h) possess a life expectancy of at least 100 space missions.

The most demanding application, in terms of the maximum temperature reached, is for the nose and leading edges of the shuttle, where temperatures of up to 1500–1600 °C are reached. The material developed to meet this application was a carbon–carbon composite, that is a composite composed of pyrolitic carbon reinforced with carbon fibre. This is a low density material with high strength and a very low coefficient of thermal expansion. It is resistant to thermal shock and because of the low

1000°C

500–650°C

450°C

650°C

1465°C

320°C

650°C   1260°C

Under surfaces 980°C–1260°C

Figure 30.16   *Sketch of Space Shuttle indicating maximum temperatures during re-entry at various points on the vehicle.*

expansion coefficient the level of thermal stresses developed is kept to a minimum. After fabrication to the required shapes the composite surface is coated with a thin layer of silicon carbide to protect the material from surface oxidation.

Much of the surface of the Space Shuttle, about 70 per cent, is covered with ceramic tiles for thermal protection. Generally, a ceramic material has a high melting point, retains its strength at temperatures close to its melting point, and possesses a low value of thermal conductivity. The tiles are used for those areas where the surface temperatures reach values from just over 300 °C to 1260 °C during re-entry. The size of the tiles range between 150 mm × 150 mm to about 200 mm × 200 mm and the thicknesses vary between 5 mm and 90 mm. More than 24,000 tiles are used to cover the relevant sections of each Space Shuttle but each tile is accurately machined to a precise and individual shape to mate with the contours of the shuttle airframe. What of the material? Silica, SiO_2, has a melting temperature of 1723 °C and a low thermal conductivity and forms the basis material of the tiles. The best thermal insulation medium is a vacuum and in the design, development and manufacture of the ceramic tile material this principle was used, in that the ceramic tiles contain more than 90 per cent of void space. How is this achieved? The material that was developed to meet the exacting requirements is composed of short, fine silica fibres which are bonded together by sintering at 1370 °C to give a very low density material with a very high void space. Several grades of tile material have been developed but the main type used is sintered to give a material with the very low density of 140 kg/m³. Thermal conduction through a tile with such a large void space is extremely low, such that if one surface of the tile is heated to 1200 °C the opposite surface will remain cool enough to touch. For those areas of the shuttle surface where a higher strength material is needed, namely around doors and access panels, tiles composed of a mixture of silica fibre and aluminium borosilicate fibre are used. This material has the somewhat higher density of 190 kg/m³. The surface of the tiles is coated with a glass to give smooth surface finish.

Two main categories of protective tiles are used to coat the shuttle airframe and these are referred to as low-temperature reusable surface insulation (LRSI) and high-temperature reusable surface insulation (HRSI) respectively. The difference between these two types is in the glass used to coat the tile surface. LRSI tiles are coated with a borosilicate glass and they are white in colour. The maximum temperature that these tiles should be subjected to is 650 °C. HRSI tiles are coated with a borosilicate glass and added silicon tetraboride (SiB_4). The surface of these tiles is black and they can withstand surface temperatures of up to 1260 °C. LRSI tiles are used to cover the upper wing and tail surfaces and major portions of the upper fuselage. The white tile surface reflects the sun's rays and this helps to keep the shuttle relatively cool when in orbit. HRSI tiles are used to cover the lower wing and underfuselage surfaces, portions of the front upper fuselage and tailfin leading and trailing edges. Being black they have high thermal emissivity and can radiate a large proportion of the heat generated during re-entry away from the shuttle.

The thickness of the thermal protection layer required will depend on the maximum surface temperature developed and so thinner tiles can be used to cover those areas where the re-entry temperature is, say, 650 °C than would be required in those areas where the maximum re-entry temperature is, say, 1200 °C.

It was stated above that about 70 per cent of the Space Shuttle is covered with ceramic tiles. What about the remaining 30 per cent? Parts of the upper surfaces of wing and fuselage do not reach such high temperatures with maximum temperatures during re-entry being in the range 300–600 °C. These parts of the airframe structure are clad with a felt reusable surface insulation material (FRSI). Two types are used in the shuttle construction. One type is a nylon felt blanket with a top surface coating of a silicone elastomer to give the desired thermal properties. This material can operate up to a maximum temperature of 400 °C. The other type, capable of operating up to maximum temperatures of 800 °C, is made from a silica fibre felt sandwiched between a layer of glass fabric on one side and a layer of woven silica fibre on the other with the three layers stitched together with silica fibre thread. When installing this type of felt the silica fibre fabric must be outermost.

All of these insulating materials, LRSI tiles, HRSI tiles and the two types of FRSI materials are bonded to the aluminium airframe structure using a silicone adhesive which cures at room temperature and which will withstand repeated heating to temperatures not exceeding 290 °C. A layer of a nylon felt is interposed between the aluminium airframe and the ceramic tiles, the purpose of this being to largely isolate the brittle ceramic tiles from major mechanical and thermal strains which may develop in the airframe.

The first Space Shuttle, named *Columbia*, was launched successfully in 1981. Four other shuttles were built subsequently but one, *Challenger*, was destroyed in a tragic accident during launch. The four remaining Space Shuttles have performed many highly successful space missions during the last two decades and are scheduled for many more missions in the future. The thermal protection system has worked extremely well, apart from some problems in the early missions with the fracture and dislodgement of some of the ceramic tiles. This problem was soon rectified and the result is a space re-entry vehicle which is a tribute to the skills of the materials and aerospace design teams.

SUMMARY

In this concluding chapter we have examined the various parameters which must be considered when selecting both a material and a manufacturing process to fit a particular design requirement. These parameters are the material properties, the manufacturing characteristics, and cost and availability, including the use of secondary or recycled materials where appropriate.

The engineer and designer must consider total design, including disposal and recycling at the end of the service life of the product.

The selection process is essentially one of optimisation and a strategy for selection is proposed.

Finally, a number of Case Studies are included to illustrate the question of materials selection.

Physical properties of some pure metals

Metal	Symbol	Melting point (°C)	Density (kg/m³ × 10⁻³)	Crystal structure*	Elastic constants (GPa)		Poisson's ratio	Transformation temperature (°C)
					E	G		
Aluminium	Al	660	2.7	f.c.c.	70.5	27.0	0.34	
Antimony	Sb	630	6.67	r.				
Beryllium	Be	1280	1.85	c.p.h.	313	160	0.28	
Bismuth	Bi	271	9.8	r.	32	12.5	0.33	
Chromium	Cr	1888	7.1	b.c.c.	238	88.5	0.3	
Cobalt	Co	1492	8.7	αc.p.h.	203	75	0.31	
				βf.c.c.				430† $\alpha \to \beta$
Copper	Cu	1083	8.9	f.c.c.	122.5	45.6	0.35	
Gold	Au	1063	19.3	f.c.c.	80	28.3	0.42	
Iron	Fe	1535	7.87	αb.c.c.	215	84.8	0.29	
				γf.c.c.				908 $\alpha \to \gamma$
				δb.c.c.				1388 $\gamma \to \delta$
Lead	Pb	327	11.3	f.c.c.	16.5	5.6	0.44	
Magnesium	Mg	649	1.74	c.p.h.	44	17.6	0.28	
Manganese	Mn	1244	7.4	αc.cub.	200	77.6	0.24	
				βc.cub.				700† $\alpha \to \beta$
				γf.c.t.				1100† $\beta \to \gamma$
				δb.c.c.				1140† $\gamma \to \delta$
Molybdenum	Mo	2620	10.2	b.c.c.	338	120	0.3	

Metal	Symbol	Melting point (°C)	Density (kg/m³ × 10⁻³)	Crystal structure*	Elastic constants (GPa)		Poisson's ratio	Transformation temperature (°C)
					E	G		
Nickel	Ni	1453	8.9	f.c.c.	208	78.7	0.31	
Niobium	Nb	2420	8.57	b.c.c.	104	36.7	0.39	
Platinum	Pt	1769	21.65	f.c.c.	173	61.2	0.39	
Silicon	Si	1412	2.34	d.				
Silver	Ag	961	10.5	f.c.c.	79	29	0.37	
Tin	Sn	232	7.3	αd.	40.8	13.6	0.33	
				βb.c.t.				18 $\alpha \to \beta$
Titanium	Ti	1660	4.51	αc.p.h.	106.4	40	0.36	
				βb.c.c.				880† $\alpha \to \beta$
Tungsten	W	3380	19.3	b.c.c.	393	152	0.30	
Uranium	U	1130	19.05	αorth.				
				βt.				668 $\alpha \to \beta$
				γb.c.c.				774 $\beta \to \gamma$
Vanadium	V	1920	6.15	b.c.c.	127	46.8	0.37	
Zinc	Zn	419	7.14	c.p.h.	92	37.3	0.25	
Zirconium	Zr	1860	6.4	αc.p.h.	95.5	36.4	0.33	
				βb.c.c.				

*Key: b.c.c. – body-centred cubic
    b.c.t. – body-centred tetragonal
    c.cub. – complex cubic
    c.p.h. – close-packed hexagonal
    d. – diamond
    f.c.c. – face-centred cubic
    f.c.t. – face-centred tetragonal
    orth. – orthorhombic
    r. – rhombohedral
    t. – tetragonal

† Approximate value.

Formulae and structures of common polymers

Hydrocarbon polymers

PE (polyethylene)

$$\left[\begin{array}{c} H \\ | \\ C \\ | \\ H \end{array} \begin{array}{c} H \\ | \\ C \\ | \\ H \end{array} \right]_n$$

PP (polypropylene)

$$\left[\begin{array}{c} H \\ | \\ C \\ | \\ H \end{array} \begin{array}{c} H \\ | \\ C \\ | \\ CH_3 \end{array} \right]_n$$

PS (polystyrene)

$$\left[\begin{array}{c} \bigcirc \\ | \\ C \\ | \\ H \end{array} \begin{array}{c} H \\ | \\ C \\ | \\ H \end{array} \right]_n \qquad \bigcirc \quad -C_6H_5$$

Carbon chain polymers

(a) Chlorocarbon and fluorocarbon polymers

PVC (polyvinyl chloride)

$$\left[\begin{array}{c} Cl \\ | \\ C \\ | \\ H \end{array} \begin{array}{c} H \\ | \\ C \\ | \\ H \end{array} \right]_n$$

CPVC (chlorinated polyvinyl chloride)

$$\left[\begin{array}{c} Cl \;\; Cl \\ | \;\;\;\; | \\ C-C \\ | \;\;\;\; | \\ H \;\; H \end{array}\right]_n$$

PVDC (polyvinylidene chloride)

$$\left[\begin{array}{c} Cl \;\; H \\ | \;\;\;\; | \\ C-C \\ | \;\;\;\; | \\ Cl \;\; H \end{array}\right]_n$$

PTFE (polytetrafluoroethylene)

$$\left[\begin{array}{c} F \;\; F \\ | \;\;\; | \\ C-C \\ | \;\;\; | \\ F \;\; F \end{array}\right]_n$$

PVDF (polyvinylidene fluoride)

$$\left[\begin{array}{c} F \;\; H \\ | \;\;\; | \\ C-C \\ | \;\;\; | \\ F \;\; H \end{array}\right]_n$$

(b) Acrylic polymers

PMMA (polymethyl methacrylate)

$$\left[\begin{array}{c} CH_3 \\ | \\ CH_2-C \cdots \\ | \\ C-O-CH_3 \\ \| \\ O \end{array}\right]_n$$

PAN (polyacrylonitrile)

$$\left[\begin{array}{c} CN \;\; H \\ | \;\;\;\; | \\ C-C \\ | \;\;\;\; | \\ H \;\; H \end{array}\right]_n$$

Heterochain polymers

POM (polyoxymethylene-acetal)

$$\left[\begin{array}{c} H \\ | \\ C-O \\ | \\ H \end{array}\right]_n$$

PA (polyamide)

$$\left[\begin{array}{c} H \;\;\;\;\; O \;\; H \;\;\;\;\; O \\ | \;\;\;\;\;\; \| \;\;\; | \;\;\;\;\;\; \| \\ N-R-C-N-R'-C \end{array}\right]_n \quad R= \left[\begin{array}{c} H \\ | \\ C \\ | \\ H \end{array}\right]_x \quad R'= \left[\begin{array}{c} H \\ | \\ C \\ | \\ H \end{array}\right]_y$$

PC (polycarbonate)

$$\left[\begin{array}{c} CH_3 O \\ -\bigcirc-C-\bigcirc-O-C-O- \\ CH_3 \end{array}\right]_n$$

PET (polyethylene teraphthalate)

PBT (polybutylene teraphthalate)

PEEK (polyether ether ketone)

PPO (polyphenylene oxide)

PPS (polyphenylene sulphide)

PSU (polysulphone)

PES (polyether sulphone)

PAE (polyaryl ether)

PI (polyimide)

PEI (polyetherimide)

PAI (polyamideimide)

R class elastomers

NR (natural rubber – polyisoprene)

BR (polybutadiene)

CR (polychloroprene)

IIR (butyl rubber – polyisobutylene)
(copolymerised with a little polyisoprene)

Thermosets

PF (phenol–formaldehyde)

UF (urea–formaldehyde)

MF (melamine–formaldehyde)

EP (epoxides)

$$\left[O-\!\!\left\langle\!\bigcirc\!\right\rangle\!\!-\!\!\overset{\overset{\displaystyle CH_3}{|}}{\underset{\underset{\displaystyle CH_3}{|}}{C}}\!\!-\!\!\left\langle\!\bigcirc\!\right\rangle\!\!-O-\!\!\overset{\overset{\displaystyle H}{|}}{\underset{\underset{\displaystyle H}{|}}{C}}-\!\!\overset{\overset{\displaystyle OH}{|}}{\underset{\underset{\displaystyle H}{|}}{C}}-\!\!\overset{\overset{\displaystyle H}{|}}{\underset{\underset{\displaystyle H}{|}}{C}}\right]_n$$

PUR (polyurethane)

$$\left[\overset{\overset{\displaystyle O}{\|}}{C}-N-R-\overset{\overset{\displaystyle H}{|}}{N}-\overset{\overset{\displaystyle O}{\|}}{C}-O-R' \right]_n$$

Glossary of terms

Chapter 2

Atomic mass number. The sum of the protons and neutrons in an atomic nucleus.

Atomic mass unit (a.m.u.). One-twelfth of the mass of an atom of carbon-12 equal to 1.6598×10^{-27} kg.

Atomic number. The number of protons in an atomic nucleus.

Avogadro's number, N_0. The number of molecules in a mole (atoms in the case of a monatomic element). $N_0 = 6.023 \times 10^{23}$/mol.

Binding energy. This relates to the strength of a bond and is the energy needed to completely separate two bonded atoms.

Co-ordinate bond. A primary bond based on electron sharing in which some of the shared electrons arc provided by an ionic donor atom.

Covalent bond. A primary bond based on the sharing of electrons between atoms.

Electron energy states. The energy levels, or states, as defined by quantum numbers, occupied by the electrons surrounding an atomic nucleus.

Gramme molecule. The amount of a molecular substance with a mass in grammes equal to its molecular mass number.

Ground state. The state of an atom when all its electrons occupy the lowest permitted energy states.

Half-life period. The time required for the activity of a radioactive substance to fall to half of its original value.

Hydrogen bond. When hydrogen is covalently bonded to an electronegative atom, as in H_2O, the molecule becomes highly polarised with hydrogen being positive relative to the oxygen. A relatively strong electrostatic attraction occurs between these polarised molecules.

Ionic bond. A primary bond based on electron transfer between atoms creating ions, with strong electrostatic attraction between positive and negative ions.

Isotopes. Atomic nuclei of an element with differing atomic mass numbers.

Kilogramme molecule. The amount of a molecular substance with a mass in kilogrammes equal to its molecular mass number.

Kilomole. SI unit of molar mass equal to one kilogramme molecule.

Mass defect. The difference between the measured mass of an atom and the sum of the masses of the protons, neutrons and electrons that make up the atom.

Metallic bond A primary bond existing in metals in which the valence electrons are shared between all atoms giving an array of metal ions permeated by an electron 'gas'.

Mixed bonds. Situation in which a substance contains more than one type of interatomic bonding.

Mole. SI unit of molar mass equal to one gramme molecule.

Pauli exclusion principle. This states that no two electrons within an atom can possess the same set of quantum number values.

Periodic table. The arrangement of the chemical elements in a table to highlight groups of elements with similar properties.

Primary bonds. The ionic, covalent and metallic types of strong interatomic bonds.

Quantum numbers. The numbers used to describe the energy states of electrons in atoms. Four quantum numbers are necessary to fully define the state of an electron.

Radioactivity. The spontaneous disintegration of unstable nuclei with emission of α- or β-particles and γ-radiation.

Secondary bonds. Relatively weak intermolecular bonds based on electrostatic attraction between dipoles. Includes hydrogen and van der Waal's bonds.

Valence electrons. The outer shell electrons of an atom.

van der Waal's bonds. Relatively weak intermolecular bonds based on electrostatic attraction between dipoles.

Chapter 3

Co-ordination number. The number of other atoms or ions with which a particular atom or ion is in contact.

Elastomer. A material which has a very low elastic modulus and shows very great extensibility.

Electrolyte. A substance in which transfer of an electric charge (conduction) is by the movement of ions.

Glass state. The state in which the atoms or molecules of a solid are not arranged in a fully symmetrical pattern.

Glass transition temperature. Refer to Chapter 6.

Modulus of elasticity, E. A material constant which is the ratio of direct stress to the strain produced (for fully recoverable elastic deformations).

Vitreous state. See *Glass state*.

Young's modulus. See *Modulus of elasticity*.

Chapter 4

Addition polymerisation. The process in which small molecules react together to form large chain molecules.

Branching. Reaction to create long side chains in a linear polymer.

Condensation polymerisation. Polymerisation reactions in which a small molecular substance, often water, is formed as a by-product.

Cross-linking. Reaction to create cross-linkages between linear polymer molecules.

Degree of polymerisation. The average number of repeating units in a linear polymer.

Elastomer. A polymer material of great flexibility and with a very low elastic modulus.

Initiator. A reactant used to provide chain termination in addition polymerisation.

Linear polymer. A polymer composed of long chain molecules.

Monomer. A comparatively simple compound which can be polymerised either on its own, or in conjunction with other monomers.

Network polymer. A polymer with a very large three-dimensional molecular structure.

Non-linear polymer. A polymer in which molecular chains have major side branches, or the molecular chains have links between them, or in which the molecular structure is a three-dimensional network.

Polymer. A material composed of extremely large molecules which is synthesised from one or more relatively simple compounds.

Thermoplastic. A polymer material which softens on heating but regains its rigidity on cooling.

Thermoset. A polymer material which, on first heating, becomes plastic and can be moulded but which then cures at temperature into a hard, rigid structure.

Chapter 5

Amorphous. Literally meaning 'without form', this term is used to describe non-crystalline structures.

Bragg's law. The relationship between X-ray wavelength, inter-planar spacing and diffraction angle for monochromatic radiation incident on a crystal.

Co-ordination number. The number of atoms with which any one atom is in contact in a crystal.

Crystalline. The state in which the constituent atoms or molecules of the substance are arranged in a regular, repetitive and symmetrical pattern.

Glass. A solid non-crystalline substance, which often shows short range order.

Hexagonal packing. The closest possible packing of uniform spheres in a plane.

Interplanar spacing. The separation distance between adjacent, parallel planes of atoms.

Interstitial sites. The spaces existing between uniform spheres in a crystal system. Atoms of another element of very small diameter may be able to fit into such sites.

Lattice parameters. The constants, namely edge lengths and angles between axes, for a unit cell of a crystal.

Miller indices. A system, based on three reference axes, used to identify specific planes and directions within crystals.

Miller–Bravais system. A system with three co-planar reference axes and a fourth axis, mutually perpendicular to the other three, used to define the hexagonal crystal system. Four Miller–Bravais indices are required to identify specific planes and directions.

Octahedral site. The interstitial site which is the space between six atoms which occupy the corners of an octahedral figure.

Packing fraction. The proportion of total space occupied by atoms in a crystal.

Short range order. An arrangement wherein atoms are in a regular arrangement with their neighbours over a short distance but there is no long-range pattern of symmetry.

Space lattice. The regular, repetitive pattern of atoms in a crystal.

Stacking sequence. The sequence in which planes of atoms are packed to generate a three-dimensional crystal.

Tetrahedral site. The interstitial site which is the space between four atoms that occupy the corners of a regular tetrahedron.

Unit cell. The smallest grouping of atoms showing the symmetry of the crystal type.

Chapter 6

Crystalline polymer. A polymer in which some of the molecular chains form into symmetrical crystalline arrays. Generally, the crystalline areas do not account for more than 50 per cent of the total mass.

Crystallite. A small crystalline area within a polymer. (See also *Spherulite*.)

Fictive temperature. See *Glass transition temperature*.

Glass. A solid material which is not crystalline. A glass may show short range order but not full long-range symmetry.

Glass former. An oxide that can form a strong non-crystalline network structure (glass) easily when cooled from the liquid state.

Glass modifier. An oxide which, when added to a glass former, can cause the network to begin to break up.

Glass transition temperature. The temperature below which a supercooled liquid becomes a solid glass. In thermoplastics this is also known as the *glass–rubber transition* and above which the material is a viscoelastic solid.

Network former. See *Glass former*.

Network modifier. See *Glass modifier*.

Spherulite. Roughly spherical shaped crystalline areas observed in many thermoplastics.

Vitreous silica. The glassy non-crystalline form of silica.

Chapter 7

Bulk modulus of elasticity. With the standard symbol K, this is the ratio of stress to volume strain when a material is subject to equilateral tension or compressional stress.

Direct strain. The dimensional change in a material caused by a direct stress and symbolised ε (see also *Strain*).

Direct stress. A tensile or compressive stress and symbolised σ (see also *Stress*).

Elastic limit. The level of stress beyond which the behaviour of the material ceases to be elastic.

Elasticity. In true elasticity the strain produced is directly proportional to the stress causing it and the strain is fully recovered when the stress is removed.

Modulus of elasticity. With the standard symbol E, this is a constant for a material and is the ratio of direct stress to direct strain for stresses within the elastic limit. It has the same dimensions as stress. An alternative name is *Young's modulus*.

Modulus of rigidity. With the standard symbol G, and also known as the *Shear modulus*, this is the ratio of shear stress to shear strain for stresses within the elastic limit, and has the same dimensions as stress.

Plastic strain. Non-recoverable strain produced when a material is subjected to a stress equal to or greater than its yield stress.

Poisson's ratio. The ratio of lateral to axial strain for a material subject to a direct stress.

Shear modulus. See *Modulus of rigidity*.

Shear strain. The dimensional change, or twist, caused by a shear stress. It is symbolised γ, where $\gamma = \tan \phi$, where ϕ is the angle of twist.

Shear stress. A stress which imparts a twist to the material.

Strain. The dimensional change in a material caused by a stress. For a direct stress it is the ratio of the change in length to the original length and, as a ratio, has no dimensions.

Stress. The balancing force set up within a material in response to an externally acting force. The stress within a material is defined as the force exerted per unit area and its units are Pa (N/m^2) or, more usually, MPa (N/mm^2).

Thermal stress. A stress established in a material when it is heated or cooled and is subject to dimensional constraint.

Yield stress. The level of stress necessary to initiate plastic deformation in a material.

Young's modulus. See *Modulus of elasticity*.

Chapter 8

Age hardening. A form of dispersion hardening in which finely dispersed zones of coherent precipitate form within the structure as a result of diffusion.

Annealing. A heat treatment to bring about recrystallisation and softening of strain hardened metals.

Annealing twins. Twinned crystals which form in annealed f.c.c. metals. (See also *Twinned crystals*.)

Arrhenius' law. An exponential law relating the rate of a process to temperature.

Burger's vector. A parameter of a dislocation which gives both the magnitude and direction of slip.

Coherent precipitate. Precipitate particles which remain fully coherent with the crystal structure of the parent lattice, but which cause major strain within the lattice. Formed during age or precipitation hardening processes.

Cold working. Plastic deformation of a metal at temperatures below the recrystallisation temperature causing strain hardening.

Cottrell atmosphere. Concentrations of solute atoms close to dislocations which tend to lock the dislocations in position.

Critical resolved shear stress. The value of resolved shear stress on a slip system necessary for slip to commence.

Crystal boundaries. The disordered zones, usually several atoms in thickness, separating adjacent crystals. As these zones do not contain regular slip planes they tend to be stronger than the crystals.

Diffusion. The migration of solute atoms through a material.

Dislocation. A line defect in the otherwise regular structure of a crystalline material. Dislocations are described as edge-type, screw type, or mixed, in which the dislocation has both an edge and a screw component.

Dislocation climb. The movement of a dislocation normal to its slip plane brought about by diffusion of vacancies to the dislocation.

Dispersion hardening. The strengthening of a metal by either a very fine dispersion of a particulate phase or of a coherent precipitate.

Elastic limit. The level of applied stress at which the strain changes from elastic to plastic.

Frank–Read source. A dislocation line pinned at both ends which, when stressed, can move and generate additional dislocations.

Grain boundaries. See *Crystal boundaries.*

Grain growth. The migration of crystal boundaries causing some crystals to increase in size at the expense of others. This occurs at temperatures in excess of the recrystallisation temperature.

Hot working. Plastic deformation of a metal at temperatures above the recrystallisation temperature allowing almost instantaneous recrystallisation to occur, thus not causing strain hardening.

Interstitial defect. A stranger atom fitting into an interstice within a crystal lattice. Generally, interstitial atoms are much smaller than the parent metal atoms.

Non-coherent precipitate. Second phase particles whose structures are not coherent with the structure of the parent material.

Petch relationship. A mathematical expression relating the tensile yield strength of a metal to its grain size.

Plastic strain. Non-recoverable or permanent strain caused by the action of a stress beyond the elastic limit.

Point defects. Types of defect found in crystal structures (see *Interstitial defect, Substitutional defect* and *Vacancy*).

Recrystallisation. The heat treatment process in which severely distorted crystal grains in a cold worked metal reform as unstrained crystals.

Schmid's law. The law stating that slip in a metal crystal commences when the shear stress in a slip system reaches a critical value.

Slip. The movement of dislocations across a slip plane under the action of a shear stress, causing plastic deformation.

Slip planes. The crystal planes on which slip occurs. Generally these are the planes of greatest atomic density.

Slip system. The combination of a slip plane and a slip direction within a crystal.

Solution hardening. The strengthening of a metal caused by atoms of a second element occupying either interstitial or substitutional positions within the crystal lattice.

Strain hardening. The strengthening of a metal caused by plastic deformation.

Stress relief. The release of locked-in elastic strains in a cold worked metal brought about by heat treatment at a relatively low temperature.

Substitutional defect. A stranger atom fitting into a normal lattice position within a crystal lattice. Generally, substitutional atoms are of roughly similar size to the parent atoms.

Superplasticity. The ability of some metallic materials to be deformed to an extremely large extent by low deformation forces.

Twinned crystals. Crystals divided by a plane, the twin plane, such that the crystal lattices on either side of the twin plane have a specific geometrical relationship to one another.

Twinning. The process of creating twinned crystals. Some metals may plastically deform by twin formation as opposed to slip. This may happen if the slip planes are not aligned favourably with respect to the axis of stress.

Vacancy. A type of point defect in which a lattice point which should contain an atom has no atom present.

Whiskers. Very small single crystals which are virtually defect free.

Work hardening. See *Strain hardening.*

Yield stress. The stress at which plastic yielding takes place.

Chapter 9

Creep. The continued slow straining with time of a material under constant stress.

Maxwell model. A model for viscoelasticity comprising a Hookean spring and a dashpot in series.

Maxwell–Kelvin model A model for viscoelasticity comprising a Maxwell model arranged in series with a *Voigt–Kelvin model*.

Recovery. The slow recovery of strain with time after a stress has been reduced or removed.

Relaxation. The reduction of stress with time when a material is held in a state of constant strain.

Viscoelasticity. The condition in which a material possesses some elasticity but also shows some of the characteristics of a highly viscous fluid. Material parameters such as modulus and strength are functions of time, temperature and rate of strain.

Viscosity. The constant, η, which relates the rate of strain (viscous flow) to the magnitude of the shear stress causing flow. Its dimensions are Pa s.

Voigt–Kelvin model. A model for viscoelasticity comprising a Hookean spring and a dashpot in parallel.

Chapter 10

Brittle behaviour. Failure with little or no prior plastic deformation.

Critical strain energy release rate. See *Toughness*.

Critical stress intensity factor. See *Fracture toughness*.

Ductile fracture. See *Tough fracture*.

Ductile–brittle transition temperature. The temperature at which the *Notch impact toughness* value for a b.c.c. or h.c.p. metal changes through $200 \, kJ/m^2$, being tough at higher temperatures and brittle at lower temperatures.

Fracture toughness. The resistance of a material to fast crack propagation. This is an intrinsic property of a material, symbolised K_c, and with units of MPa $m^{1/2}$. It is also known as *Critical stress intensity factor*.

Notch impact toughness. The parameter measured in an impact test, such as the Izod or Charpy test. This parameter should not be confused with *Fracture toughness*.

Stress concentration. Certain features, including internal or surface defects, and sharp sectional changes, will act as points of local stress concentration.

Tough fracture. Fracture preceded by plastic deformation. Considerable energy is absorbed in causing this type of fracture.

Toughness. An intrinsic property of a material, symbolised G_c, and with units of J/m^2. Also known as *Critical strain energy release rate*.

Chapter 11

Allotropy. The ability of an element to exist in more than one crystalline form.

Alloy. A mixture of two or more metals, a mixture of a metal and a non-metal, or a mixture of two or more ceramic substances.

Dendrite. The tree-like three-dimensional shape formed during the solidification of a phase in an alloy system.

Equilibrium diagram. See *Phase diagram*.

Eutectic mixture. The fine-grained mixture of two solid phases formed by solidification of the alloy of eutectic composition.

Eutectic point. The point in a phase diagram which marks the lowest melting point alloy composition in a system and at which the liquid alloy solidifies simultaneously into two solid phases.

Eutectoid. Similar to an eutectic but a phase change which takes place entirely within the solid state and in which, on cooling, a single-phase solid solution transforms into two other solid phases.

Intermediate phase. Similar to an *Intermetallic compound* but where the phase exists over a narrow range of compositions rather that at one specific composition.

Intermetallic compound. A compound with a specific composition formed between metallic elements, for example $CuAl_2$. Similar compounds can occur also in ceramic alloy systems.

Liquidus. The line in a phase diagram, above which the alloy is in the fully liquid state.

Peritectic. A reaction within an alloy system in which, during cooling, a solid alloy phase and a liquid phase react together at a constant temperature to form a second solid phase.

Phase. A portion of matter that is homogeneous.

Phase diagram. A map of a system which shows the phases which should exist under equilibrium conditions for any particular combination of temperature and composition.

Solid solution. A single phase in which the atoms of two or more substances take up positions in a single space lattice.

Solidus. The line in a phase diagram, below which the alloy is in the completely solid state.

Solvus. A line in a phase diagram denoting the maximum solubility limit of one substance in another.

Chapter 12

Activation energy. The amount of energy necessary to initiate a process and quoted as either J/molecule or MJ/kmol.

Age hardening. See *Precipitation hardening*.

Arrhenius' rate law. The exponential law relating the process rate (s^{-1}) to temperature (K).

Carburising. The process in which additional carbon is caused to enter the surface layers of a steel to bring about case hardening.

Critical cooling velocity. The specific rate of cooling of an alloy which, if exceeded, will bring about a *martensitic transformation*.

Decarburising. The loss of carbon from the surface layers of a steel caused by heating in an oxidising environment.

Diffusion. The movement of atoms through a space lattice through the mechanism of atom/vacancy jumps.

Diffusionless transformation. See *Martensitic transformation*.

Fick's laws. Laws postulated by Fick giving mathematical expressions for the diffusion process. The First law covers steady-state diffusion and the Second law covers the dynamic condition in which both the rate of diffusion and the concentration gradient change with time.

Martensite. A supersaturated solid solution formed by very rapid cooling which brings about a diffusionless transformation. In steels martensite is extremely hard and brittle.

Martensitic transformation. A diffusionless transformation which can be brought about in some alloy systems by very rapid cooling from a high temperature. During such a transformation the movements of individual atoms will be small, less than one atomic spacing.

Overageing. The loss of hardness which occurs if a precipitation hardened material is overheated causing non-coherent precipitate particles to separate out and releasing strain in the crystal lattice.

Precipitation hardening. A form of dispersion hardening in which finely dispersed zones of coherent precipitate form within the structure as a result of diffusion.

Solution heat treatment. The heating of a normally two-phase alloy in a single-phase region followed by rapid cooling which prevents the separation of the second phase and preserves a single-phase supersaturated solid solution at ordinary temperatures.

Thermally activated process. A process in which the rate increases exponentially with temperature (K).

T–T–T diagram. A Time–Temperature–Transformation diagram is one which shows the relationships between temperature, rate of cooling and the type of transformation structure for an alloy.

Chapter 13

Band structure. In aggregates of atoms the discrete electron energy states of a single atom broaden into bands.

Coercivity. The extent of the reversed magnetising force required to reduce the induced magnetisation in a ferromagnetic material to zero.

Conduction band. The energy band containing excited mobile electrons for electrical conduction.

Conductivity. The parameter of a material relating to the ease with which an electric current can pass through the material, measured in $\Omega^{-1}\,m^{-1}$.

Curie temperature. The temperature above which a ferromagnetic material loses its ferromagnetism, or above which ferroelectric behaviour is lost.

Diamagnetism. When a material is weakly repelled by a magnet.

Dielectric constant. See *Permittivity*.

Domains. The sub-structure in a ferromagnetic material. All the atomic magnetic dipoles are aligned parallel to one another within one domain.

Energy gap. The forbidden gap between the valence and conduction bands. This gap is small in semiconductors but large in insulating materials.

Extrinsic semiconductor. A semiconductor containing a small addition of impurities or dopants to modify the charge carrying properties.

Ferrites. Complex oxide ceramics with a spinel structure which show ferromagnetic behaviour.

Ferroelectric material. A ceramic material that polarises in an electric field and which remains polarised when the field is removed.

Ferromagnetism. The property of a material to be strongly attracted by a magnet.

Hysteresis loop. This, or *B–H loop*, is the enclosed loop formed by plotting the induced magnetisation, *B*, against the magnetising force, *H*, for a complete magnetisation cycle. The enclosed area is a measure of the energy used to realign magnetic domains in a complete cycle, this energy being dissipated as heat.

Impurity semiconductor. See *Extrinsic semiconductor*.

Insulating materials. Ceramic or polymeric materials with measured resistivities of $10^6\,\Omega\,m$ or greater.

Intrinsic semiconductor. Covalently bonded materials whose conductivities are due to the electron energy distribution within the material. The conductivity increases with rise in temperature.

Magnetic annealing. Slow cooling of a ferromagnetic material from the annealing temperature within a strong magnetic field.

Magnetic saturation. The maximum level of magnetisation that can be induced within a ferromagnetic material.

Magnetically hard. A magnetically hard material is one which has high *Remanence* and high *Coercivity*.

Magnetically soft. A magnetically soft material is one which has low *Remanence* and low *Coercivity*.

Paramagnetism. When a material is weakly attracted to a magnet.

Permittivity. The ability of a material to polarise and store an electrical charge. The relative permittivity, which is the permittivity of the material relative to that of a vacuum, is also termed the *Dielectric constant*.

Piezoelectric material. A material that generates an electrical field when physically strained, and which strains when subject to an electrical field.

Remanence. The level of magnetisation remaining in a material when the external magnetising field has reduced to zero.

Resistivity. The parameter of a material relating to the difficulty with which an electric charge can move through a material, measured in $\Omega\,m$.

Transducer. A *piezoelectric* device that converts electrical fields into mechanical vibrations, and vice versa.

Valence band. The energy band containing the valence electrons. This band may only be partially filled in some materials.

Chapter 14

Birefringence. An effect in some non-metallic crystals in which the velocity of light, and hence the refractive index, in the crystal differs in different crystal orientations. Light incident upon such a crystal is split into two beams.

Cathodoluminescence. Excitation of a luminescent material by an electron beam.

Colour centres. Impurity ions in a material which may produce selective absorption of certain frequencies, causing the appearance of a colour.

Damping capacity. This is the ability of a material to attenuate vibrations.

Diffraction. Diffraction is the divergence of a beam of monochromatic electromagnetic radiation after passing through a narrow slit.

Electroluminescence. Excitation of a luminescent material by the effects of an electric field.

Fibre optics. Telecommunications systems in which signal transmission is by means of coherent light pulses sent through optical fibres.

Fluorescence. A luminescent emission of light which occurs only for the duration of the excitation.

Heat capacity. The amount of energy required to raise the temperature of one kilomole of substance through one degree K. (See also *Specific heat capacity*.)

Laser. Light Amplification by Stimulated Emission of Radiation. A device for generating a pulse or continuous wave of monochromatic coherent light.

Luminescence. The general term for the phenomenon of light emission from a material as a result of excitation.

Optical fibre. Drawn glass fibres of very high transparency used in fibre optic systems.

Phonon. The quantum of elastic wave energy produced by the vibration of atoms.

Phosphor. A material which will phosphoresce when excited.

Phosphorescence. A luminescent emission of light which persists for a period of time after excitation has ceased.

Photoconduction. An effect in which incident photon energy induces valence electrons to cross the energy gap in semiconductors, thus increasing the electrical conductivity.

Photoelastic effect. A stress-induced *Birefringence* which occurs in some glassy polymers. The effect can be used to study stress distribution in models.

Photoluminescence. Excitation of a luminescent material by incident photons of ultra-violet or higher-frequency radiation.

Planck's quantum relationship. This relates the energy of a photon, E, to the frequency of the radiation, ν. $E = h\nu$, where h is Planck's constant.

Refractive index. The refractive index of a transparent material is the ratio of the velocity of light in a vacuum to the velocity of light in the material.

Specific heat or *Specific heat capacity*. The amount of energy required to raise the temperature of one kilogramme of a substance through one K.

Thermal conductivity. The measure of the conduction of heat energy through materials. Its units are watts per metre per Kelvin (W/m K).

Thermal expansion coefficient. The constant for a material which relates the dimensional change which occurs during heating or cooling to the temperature change.

Thermal shock resistance. Failure through thermal shock is the failure of a material when heated or cooled rapidly. The resistance to thermal shock is proportional to thermal conductivity and nominal fracture stress, and inversely proportional to the coefficient of thermal expansion and modulus of elasticity.

Chapter 15

Age hardening. The increase in hardness and strength which occurs naturally at ordinary temperatures in some alloys following a solution heat treatment.

Aluminium bronze. An alloy containing, principally, copper and aluminium.

Anodising. An electrolytic process for increasing the thickness of the oxide film on the surface of aluminium and its alloys.

Brass. An alloy containing, principally, copper and zinc.

Bronze. An alloy containing, principally, copper and tin.

Cartridge brass. An α brass containing 30 per cent of zinc.

Cathode copper. Relatively pure copper produced by electrolytic extraction.

Cupronickel. An alloy containing, principally, copper and nickel.

Gunmetal. A bronze containing some added zinc.

High tensile brass. A brass with an $\alpha + \beta$ or β structure containing manganese and other elements to give improved strength. A former name was *Manganese bronze*.

Inconel. A nickel/chromium/iron alloy with excellent corrosion resistance.

Manganese bronze. See *High tensile brass.*

Monel. A highly corrosion resistant nickel alloy containing copper, together with small additions of iron and manganese.

Muntz metal. An $\alpha + \beta$ brass containing 40 per cent of zinc.

Nickel silver. Ternary single-phase alloys containing copper, nickel and zinc.

Nimonic. A series of nickel-based alloys developed for prolonged high temperature service.

Overageing. The softening of a precipitation hardening alloy which will occur when the precipitation heat treatment is conducted either at too high a temperature or for too long a time.

Pewter. Tin alloyed with some antimony and copper.

Precipitation heat treatment. A heat treatment which is carried out at a comparatively low temperature on an alloy which has been solution heat treated to increase the strength and hardness of the material.

Season cracking. A form of stress corrosion that may occur in cold worked α brasses.

Solution heat treatment. Heating an alloy to a high temperature to ensure that second-phase particles are taken into solution, followed by rapid cooling to produce a metastable supersaturated solid solution at ordinary temperatures.

Superalloys. The term used to describe the nickel-base and cobalt-base groups of alloys developed for prolonged high temperature service under stress.

White metal. The term used for tin-based and lead-based bearing metals.

Chapter 16

Alloy steels. Steels that either contain more than 0.5 per cent Si, more than 1.5 per cent Mn, or contain any other element present as a deliberate alloying addition.

Austenite. The face-centred cubic, or γ, form of iron, and f.c.c. ferrous solid solutions.

Bainite. The term given to the decomposition products of austenite when either it transforms isothermally or at a rate of cooling intermediate between the rapid rate required for martensite formation and the slower rate at which pearlite is formed.

Blackheart iron. A type of *Malleable iron* produced by prolonged heating of a white iron casting.

Blast furnace iron. Impure iron, the product of iron ore smelting in a blast furnace.

Carbon equivalent. The effective carbon content of a steel or cast iron, taking account of the presence of other elements.

$$\text{For steels CE} = \text{C} + \frac{\text{Mn}}{6} + \frac{\text{Cr} + \text{Mo} + \text{V}}{5} + \frac{\text{Ni} + \text{Cu}}{15}$$

$$\text{For cast irons CE} = \text{C} + \frac{\text{Si} + \text{P}}{3}$$

Carbonitriding. A process for infusing carbon and nitrogen into the surface of a low carbon steel to increase the surface hardness.

Carburising. A process for infusing additional carbon into the surface of a low carbon steel to increase the surface hardness.

Case hardening. Any one of a number of processes for increasing the surface hardness of a low carbon steel.

Cast iron. Selected grades of pig iron, remelted and used for making castings.

C–C–T diagram. Continuous cooling-transformation diagram showing the influence of cooling rates on the transformation of austenite. (See also *T–T–T diagram.*)

Cementite. The carbide of iron, Fe_3C.

Critical cooling velocity. The rate of cooling which, if exceeded, causes austenite to transform into martensite.

Critical temperatures. The temperatures at which structural changes occur in steels.

Cyaniding. A process for infusing carbon and nitrogen into the surface of a low carbon steel to increase the surface hardness.

Ferrite. The body-centred cubic, or α and δ, forms of iron, and b.c.c. ferrous solid solutions.

Flame hardening. A heat treatment to surface harden a medium carbon steel.

Full anneal. Heating to 30 °C above the A_3 temperature followed by slow furnace cooling to put hypoeutectoid steels into the softest possible condition.

Grey cast iron. A cast iron in which some or all of the carbon is in the form of graphite.

Hadfield's steel. A highly abrasion resistant high carbon steel containing 13 per cent Mn which has an austenitic structure.

Hardenability. A measure of the depth of hardening which can be achieved when a steel is rapidly quenched.

Hardening. A heat treatment to form martensite in steels involving rapid quenching from above the A_3 temperature, for hypoeutectoid steels, or from just above the A_1 temperature, for hypereutectoid steels.

High alloy steels. Steels with a large alloying content, usually more than 5 per cent, which have structures, and require heat treatments, which differ from those of plain carbon steels.

High carbon steels. Also known as *tool steels*, contain more than 0.6 per cent C. Such steels are used in the hardened and tempered condition.

High speed steels. High alloy steels containing tungsten and other carbide forming elements, used for the manufacture of metal cutting tools.

Hypereutectoid steel. A composition with more carbon than the eutectoid composition.

Hypoeutectoid steel. A composition with less carbon than the eutectoid composition.

Ihrigising. See *Siliconising*.

Induction hardening. A treatment to surface harden a medium carbon steel.

Jominy test. A test for assessing the *Hardenability* of a steel.

Lanz–Perlit iron. A type of high strength cast iron.

Low alloy steels. Steels with up to 3 or 4 per cent of one or more alloying elements and having similar microstructures to, and requiring similar heat treatments to, plain carbon steels.

Low carbon steels. Steels containing not more than 0.3 per cent C. They can be cold worked but will not respond to a hardening heat treatment.

Malleable iron. See *Blackheart iron* and *Whiteheart iron*.

Maraging steels. A range of very high strength, high alloy steels rich in nickel and cobalt, in which maximum strength is developed by a precipitation heat treatment.

Martensite. The very hard and brittle constituent formed when a steel is very rapidly cooled from the austenitic state. It is a ferrite, highly supersaturated with dissolved carbon, and referred to as α'.

Medium carbon steels. Steels containing between 0.3 and 0.6 per cent C, used in the hardened and tempered condition.

Meehanite. A type of high strength cast iron.

Mild steel. See *Low carbon steels*.

Mottled iron. A cast iron in which the structure is part *grey* and part *white*.

Nitriding. A treatment to infuse nitrogen into the surface of a steel to give a major hardening effect.

Nodular cast irons. Grey cast irons of high fracture toughness in which the graphite is present in the form of roughly spheroidal nodules. Also known as *SG* or *Spheroidal graphite cast iron*.

Non-alloy steels. See *Plain carbon steels*.

Normalising. A softening treatment for hypoeutectoid steels involving heating at 30 °C above the A_3 temperature, followed by cooling in still air.

Pearlite. The eutectoid mixture of ferrite and cementite formed when austenite transforms during cooling. It consists of alternate layers, or lamellae, of the two constituents.

Pig iron. Small ingots cast from blast furnace iron.

Plain carbon steels. Steels with up to 1.5 per cent C, not more than 0.5 per cent Si, not more than 1.5 per cent Mn and only traces of other elements.

Process anneal. See *Sub-critical anneal.*

Ruling section. The maximum diameter of a steel bar which can be hardened throughout by rapid quenching.

S-curve. See *T–T–T diagram.*

Siliconising. A treatment (also known as *Ihrigising*) to create a silicon-rich surface layer on a low carbon steel to increase surface hardness and corrosion resistance.

Sorbite. A name given to a structure produced when a martensitic structure is tempered and, more properly, termed *tempered martensite.*

Spheroidal graphite (SG) cast iron. See *Nodular cast irons.*

Spheroidise anneal. A softening treatment for hypereutectoid steels which involves heating to just below the A_1 temperature.

Stainless steels. A range of steels containing more than 12 per cent Cr, and often nickel, which are highly resistant to corrosion by virtue of a continuous film of chromic oxide which develops on the steel surface.

Steel. A generic term covering a wide range of iron alloys containing up to 1.5 per cent C, together with other elements.

Sub-critical anneal. An annealing process for cold worked mild steel involving heating at a temperature below the *critical temperatures.*

Tempering. The heat treatment given to a hardened steel to increase its toughness, but at the expense of some hardness.

Tool steels. See *High carbon steels* and *High speed steels.*

Troostite. A name given to a structure produced when a martensitic structure is tempered and, more properly, termed *tempered martensite.*

T–T–T diagram. Time–temperature–transformation diagram showing the influence of cooling rates and the transformation of austenite into non-equilibrium phases.

White cast iron. A cast iron in which all of the carbon is present in combined form as Fe_3C.

Whiteheart iron. A type of *malleable iron* produced by prolonged heating of a white iron casting in an oxidising environment.

Chapter 17

Additive. Any one of a range of materials added to a pure polymer to make a polymer moulding compound.

Antioxidant. An additive in a polymer moulding compound to inhibit deterioration through interaction with oxygen.

Antiozonant. An additive in a polymer moulding compound to inhibit deterioration through interaction with ozone.

Cold setting plastic. A polymer material which will set, or cure, into a fully rigid form at ambient temperature after the active ingredients have been mixed together. The curing rate can be increased by raising the temperature.

Cross-linking. A chemical reaction which causes linkages to form between adjacent molecules in a bulk polymer. Cross-linking increases the hardness and stiffness of the material. (See also *Vulcanisation.*)

Dye. A colouring additive made to a polymer moulding compound. Unlike a *Pigment*, a dye dissolves in the polymer.

Elastomer. A polymer material which has a very low modulus of elasticity and which shows a very large extensibility when stressed, but which returns to its original dimensions when the deforming stress is removed.

Filler. An additive made to a polymer moulding compound either to enhance some property of the material, or to extend and lower the cost of the material.

Pigment. A colouring additive in the form of a fine powder made to a polymer moulding compound. Pigments may be transparent or opaque but they remain as a separate phase in the moulding compound. (See also *Dye*.)

Plasticiser. An additive in a polymer moulding compound which increases the plasticity of the material with a consequent reduction in stiffness and strength but an increase in toughness and impact resistance.

Stabiliser. An additive in a polymer moulding compound which prevents or retards deterioration of the polymer through the action of light or ultra-violet radiation, or the interaction of oxygen or ozone with the polymer.

Thermoplastic. A polymer material which softens when heated and becomes rigid again when cooled.

Thermoset. A polymer material which softens and can be moulded when first heated but which sets, or cures, into a rigid shape at temperature. The cured material cannot be softened afterwards.

Vulcanisation. The reaction between an unsaturated elastomer material and sulphur to cause cross-linkages to form, increasing the hardness and stiffness of the material.

Chapter 18

Annealing point. The temperature at which a glass has a viscosity of 10^{12} Pa s. Internal stresses in the glass are relieved rapidly at this temperature.

China clay. A clay which is almost pure *kaolinite* and which gives a pure white fired product.

Clay. Deposits of complex aluminosilicates, rich in *kaolinite*, used for the manufacture of clay products.

Fictive temperature. See *Glass transition temperature.*

Fireclays. Clays containing mainly *kaolinite* and quartz which, when fired, give refractory materials capable of prolonged service at temperatures of up to 1200 °C.

Glass. A material, usually a complex aluminosilicate, which is amorphous being, in effect, a supercooled liquid.

Glass ceramic. A complex silicate first formed as a glass but converted into a fully crystalline state by a controlled heat treatment.

Glass transition temperature. The temperature, also known as *fictive temperature*, at which viscous flow ceases and taken as the temperature at which the viscosity is 10^{16} Pa s.

Glazing. The process of creating a hard impervious layer on the surface of a fired ceramic during a second firing.

Refractories. Ceramic products which are stable at high temperatures and used in the construction of furnaces and allied equipment.

Strain point. The temperature at which the viscosity of a glass is 10^{13} Pa s. A glass can be cooled rapidly from this temperature without developing internal stresses.

Whiteware. A term covering fine-textured clay ceramics including china, porcelain and stoneware.

Chapter 19

Aramid fibres. High strength fine fibres made from an aromatic polyamide. The most common forms go under the trade name *Kevlar.*

Carbon fibres. Fine fibres of carbon with diameters in the range 6–10 μm possessing high strength and modulus values. There are several different grades available with differing properties.

Cermets. Metal matrix composites containing 80–90 per cent of ceramic particles.

CMC. A ceramic matrix composite material.

e-glass. A non-alkaline borosilicate glass used for the manufacture of glass fibre.

Fibre reinforced material. A composite material in which the matrix is reinforced by fine fibres of a second material. The fibres may be either continuous or discontinuous.

Glass fibres. Fine fibres of glass with diameters in the range 6–10 μm possessing high strengths. There are several types, the two most common being *e-glass* and *s-glass.*

Hard metal. A cermet composed of hard carbides in a matrix of metallic cobalt and used as a cutting tool tip material.

Kevlar. The trade name for a group of aramid fibres.

MMC. A metal matrix composite.

Particulate composite. A composite material in which the strengthening component is in the form of a wide dispersion of discrete particles.

s-glass. A magnesium aluminosilicate glass used for the manufacture of high strength glass fibres.

Sandwich structure. A low density, high stiffness composite structure composed of a lightweight foam or honeycomb core bounded by high strength skins.

Chapter 20

Annealing. A heat treatment given to cold worked material to bring about recrystallisation and softening.

Casting. The shaping of metals by pouring a liquid metal into a prepared mould.

Centrifugal casting. The casting of a metal into a rotating mould. This is suitable for shapes of simple rotational symmetry, for example, pipes.

Chemical vapour deposition (CVD). The deposition of a coating on a substrate from a vapour-phase chemical reaction.

Closed die forging. Forging in which both hammer and anvil are dies with shaped cavities into which the hot metal is caused to flow under pressure.

Cold shut. A casting defect in which two streams of liquid metal have not completely fused together.

Cold working. Plastic deformation carried out at temperatures below the recrystallisation temperature.

Continuous casting. A method of ingot casting in which the liquid metal is rapidly solidified within a shallow water-cooled mould, the solid ingot being withdrawn continuously from the mould.

Cope. The upper half of a sand moulding box.

Core. The shape, made separately, which is inserted into a sand mould to make a hollow casting.

Die casting. The casting of a metal into a metal mould, either under pressure or by gravity alone.

Die forging. See *Closed die forging.*

Drag. The lower half of a sand moulding box.

Drawing. A cold working process in which rod, wire or tubing is pulled through a die with a reducing aperture.

Electroless plating. The deposition of a metal coating on a substrate from a solution of a metallic salt.

Electrophoresis. The transport of electrostatically charged particles in an electric field and used for the deposition of a variety of coatings onto substrates.

Extrusion (of metals). The process in which a preheated ingot is forced to flow, under pressure, through a shaped aperture in a die. Both solid and hollow sections can be formed.

Fibre structure. The pattern, marked by the position of small inclusions and intermetallics, showing the directions of plastic flow in a hot deformed metal component.

Flame spraying. One method of *Thermal spraying* in which a coating material is melted in a flame and the droplets propelled at high velocity onto a substrate.

Forging. The shaping, by plastic deformation, of metal by a series of hammer blows or by slow application of pressure.

Gravity die casting. The casting of a metal into a metal mould under the action of gravity alone.

Greensand. A naturally occurring sand containing some clay and used for making sand moulds.

Hot tear. A casting defect created by the build-up of thermal stresses during cooling within the mould sufficient to cause fracture.

Hot working. Plastic deformation carried out at temperatures above the recrystallisation temperature.

Ingot casting. The casting of liquid metal into regular square, rectangular or circular sections to produce ingots for subsequent working.

Investment casting. The casting of a metal into a mould made by coating a wax pattern with a refractory material and melting out the wax to leave a perfect mould.

Ion implantation. A process for introducing another substance into the surface of a material in which the substance to be added is ionised and the ions accelerated towards the surface by a large potential difference.

Killed steel. A fully deoxidised steel.

Physical vapour deposition (PVD). A group of related processes for applying coatings to materials in which the coating substance is vaporised by intense heat and condenses, forming a coating, on the substrate.

Powder metallurgy. The process of producing components from metal powders by either pressing and sintering or injection moulding and sintering.

Pressure die casting. The casting of a metal under pressure into metal dies or moulds.

Rimmed steel. A steel which is not deoxidised before ingot casting. The cast ingot has a rim of almost pure iron, the impurities being concentrated in the core.

Sand casting. The casting of a liquid metal into a mould made from foundry sand.

Shell moulding. A process for making thin-walled moulds using a sand/thermoset resin mix in hot metal dies. The resin cures, binding the sand grains.

Smith's forging. The shaping of hot metal by hammering the work-piece on an anvil. The smith determines the position and force of each hammer blow.

Thermal spraying. The general name for a range of processes in which a coating material is melted and then projected at high velocity onto a substrate.

Upsetting. A forging operation in which the direction of plastic flow is changed from that of earlier operations, in order to obtain the optimum fibre structure.

Chapter 21

Blow moulding. A process in which a tube, or *parison*, of thermoplastic is extruded into a mould. Air is then blown in to expand the parison into the exact shape of the mould. The process is used for making bottles and other hollow shapes.

Calendering. A process for making thermoplastic sheet in which a hot plastic dough passes through a series of heated rolls, emerging as flat sheet.

Co-extrusion. A process in which two or more thermoplastics are extruded simultaneously from separate extruders to form a laminate.

Compression moulding. A moulding technique for thermosets in which the moulding compound is heated and compressed in a die, and kept in the heated die until curing is complete.

Dough moulding compound. A mixture of uncured resin, chopped reinforcing fibre and inert filler which can be injection moulded, the resin curing in the mould.

Drape forming. A sheet shaping process in which hot plastic sheet is draped and stretched over a male mould shape.

Expanded plastics. Solid foam materials.

Extrusion. A process in which the hot viscous thermoplastic is forced through a shaped die. The hot plastic must be cooled immediately after leaving the die to retain its shape.

Extrusion coating. A process in which a thermoplastic is extruded around a wire core.

Filament winding. Continuous fibre rovings, impregnated with resin, are wound around a former, usually in a helical pattern.

Film blowing. A process in which compressed air is blown into extruded thermoplastic tubing immediately after extrusion to form a thin-walled film bubble.

Injection moulding. A forming process for thermoplastic materials in which the hot viscous fluid is forced under pressure into a shaped mould cavity or die. The accurately shaped part is cooled in the mould before ejection.

Preforms. Formed shapes in fibre, with a dilute resin binding, used for *Resin transfer moulding*.

Prepregs. Pre-impregnated fibre/resin intermediate products which can be later formed into the required shapes and cured.

Pulforming. Similar to *Pultrusion*, but used for shapes with a varying cross-sectional geometry.

Pultrusion. A process in which continuous fibre rovings, impregnated with resin, are pulled through a shaped die and then through a curing oven. Used for solid and hollow sectional shapes.

Reaction injection moulding. A process in which reactants are mixed together immediately before injection into a mould, where they cure rapidly.

Reinforced reaction injection moulding. Similar to reaction injection moulding, but where chopped reinforcing fibre is mixed with one of the reactants.

Resin transfer moulding. A moulding process in which fibre reinforcement is placed in the mould prior to the injection of resin. Curing occurs within the heated mould.

Rotational moulding. A process in which a large mould containing thermoplastic moulding compound is heated and rotated about two axes. Used for large hollow shapes.

Sheet moulding compound. A mixture of uncured resin, chopped reinforcing fibre and inert filler produced in sheet form and which can be shaped by *Vacuum* or *Drape forming*.

Transfer moulding. A moulding process for thermosets in which the moulding compound is heated in an ante-chamber and when plastic is injected into a heated die.

Vacuum forming. A shaping process in which heated thermoplastic sheet is placed over a shaped mould and the intervening space is evacuated, drawing the sheet into the mould shape.

Warp sheet. Prepreg sheet with unidirectional continuous fibre reinforcement.

Chapter 22

Blow moulding. The blowing of a *Parison* of hot glass into a shaped die to produce a hollow shape.

Die pressing (of ceramics). The compaction of a ceramic powder in a metal die.

Firing. The heating of moulded clay articles in a kiln to form a solid ceramic product.

Float-glass process. The process for making distortion-free glass sheet or plate in which molten glass is allowed to flow onto the surface of a bath of molten tin.

Glass blowing. The forming of a hollow shape from a *Parison* of hot glass by means of air pressure.

Glaze. A mixture of glass forming oxides applied to the surface of a fired clay ceramic to give a vitreous, non-porous surface layer after a second firing.

Glost firing. The second firing of a clay ceramic to give a hard, vitreous surface layer.

Hot isostatic pressing. The sintering of a preformed powder compact in an autoclave under high gas pressure.

Hot pressing (of ceramics). The compaction of powder in a heated die so that sintering under pressure occurs, giving a higher density product than would be otherwise obtained.

Hot pressing (of glass). A process analogous to closed-die metal forging in which hot viscous glass is formed into shapes.

Isostatic pressing. The pressing of ceramic powder within a flexible mould which is subject to hydrostatic pressure.

Liquid-phase sintering. Sintering in which one constituent is liquid at the sintering temperature. On cooling, this forms a glass phase at grain boundaries.

Parison. A closed tube of hot glass, the precursor for blow moulding or glass blowing.

Reaction bonding. The conversion of a powder compact into a polycrystalline ceramic by means of a chemical reaction between the powder particles and a gas or liquid.

Reaction sintering. The sintering of a powder compact into a polycrystalline material involving a chemical reaction between two of the components in the powder mix.

Sintering. The heating of a powder compact to a temperature sufficiently high for the material to consolidate into a polycrystalline material. The driving force is the reduction in surface energy as powder particles coalesce.

Slip casting. A process in which a fluid clay slurry is poured into a porous mould.

Tape casting. A process in which a slip is poured onto a continuous moving belt. The tape is dried and pieces can be blanked from the tape for firing.

Chapter 23

Abrasive machining. Material-removal processes involving scratching, grinding or polishing.

Built-up edge. The adherence of some chip material to the cutting face of a tool, effectively increasing the rake angle.

Capstan lathe. See *Turret lathe*.

Carbon tool steels. High carbon non-alloy steels used in the hardened and tempered condition.

Cemented carbides. Tungsten, titanium and other carbides in a matrix of cobalt, compacted and sintered and used as cutting tool materials.

Centre lathe. A machine tool in which the work-piece is rotated and the cutting tool can move both parallel to and normal to the direction of rotation.

Chemical machining. Material removal by controlled chemical attack.

Chip-breaker. A device on a cutting tool to aid discontinuous chip formation.

Clearance angle. The angle between the non-cutting face of a tool and the work-piece surface.

CNC machining. Machining using a computer-controlled NC machine tool.

Coated carbides. Cemented carbides with a thin coating of a hard ceramic.

Continuous chip. The chip formed by machining is continuous, curling into a spiral or helical form.

Cutting fluid. An oil or water/soluble oil emulsion used as a coolant and/or lubricant during machining operations.

Electrochemical machining. A material-removal process based on solution of material within an electrolytic cell.

Electrodischarge machining. A machining process based on the erosive action of an electric spark. Also known as *Spark erosion.*

Electron beam machining. A precision process in which a high energy electron beam melts and vaporises a portion of the work-piece.

Forming. The production of a shape by cutting where the cutting tool has a profile that is the reverse of the shape to be formed.

Generating. Producing a machined surface where the shape is controlled by the movement of the machine tool slides rather than by the shape of the cutting tool.

Grinding. Material removal by abrasion using a grinding wheel or coated abrasive.

High speed steels. Highly alloyed steels containing tungsten, chromium, vanadium, molybdenum and carbon which retain their hardness at relatively high temperatures.

Machineability. The relative ease with which a material can be cut by a machining process.

Machining. The forming of shapes by a material-removal process.

Machining centre. A complex CNC machine tool with one or more machining heads and capable of performing a wide range of cutting operations.

Milling. A cutting process in which the work-piece feeds past a rotating multi-point tool.

NC machine. A numerically controlled machine tool.

Non-traditional machining process. A chipless machining process.

Oblique cutting. Material cutting where the cutting edge of the tool is at some angle other than 90° to the cutting direction.

Orthogonal cutting. Material cutting where the cutting edge of the tool is normal to the cutting direction.

Planing. The production of a flat surface on a work-piece by cutting with a single-point tool.

Rake angle. The angle between the cutting face of a tool and the normal to the work-piece surface.

Shaping. The production of a flat surface on a work-piece by cutting with a single-point tool.

Spark erosion. See *Electrodischarge machining.*

Taylor's law. An empirical law relating cutting tool life to other parameters.

Tool life. The length of time that a cutting tool can be used before tool wear becomes unacceptable.

Turning. A rotational cutting process carried out on a lathe.

Turret lathe. A production lathe with a turret holding up to six tools.

Chapter 24

Arc welding. Fusion welding using an electric arc as the heat source.

Brazing. See *Hard soldering.*

Diffusion joining. Solid-state joining processes in which a prepared assembly is heated *in vacuo* or in an inert atmosphere permitting diffusion of material across the mating surfaces, thus effecting a permanent joint.

Electrogas welding. An automated shielded arc welding process suitable for sectional thicknesses in the range 12–75 mm.

Electron beam welding. A precision fusion welding process using a focused high energy electron beam as the heat source.

Electroslag welding. An automated welding process suitable for sectional thicknesses in the range 50–1000 mm.

Flame cutting. The cutting of steels using a high intensity oxyacetylene flame.

Flash butt welding. A fusion welding process in which an arc is struck between the two surfaces to be welded. When the arc is switched off the two surfaces are brought together under pressure.

Flux. A compound, basically an oxide solvent, used as an aid to soldering, brazing and welding.

Friction welding. A process in which one part is rotated at high speed and then brought into contact, under pressure, with a stationary part. Seizure and welding occurs. Also known as *Inertia welding* and *Spin welding*.

Furnace brazing. The joining of assemblies by brazing in a controlled atmosphere furnace.

Fusion welding. Welding using a molten filler and involving some melting of the parent metal.

Gas welding. Fusion welding using acetylene or propane gas as the heat source.

GMA welding. See *Inert gas shielded arc welding*.

Hard soldering. A joining method for metals using a brass or Cu–Zn–Ag alloy as the filler material.

Heat affected zone. The zone adjacent to a weld in which the microstructure and properties have been modified by the heating associated with the welding process.

Hot gas welding. A fusion welding process for thermoplastics in which the heat source is a jet of hot gas, rather than a flame.

Inert gas shielded arc welding. Arc welding under an inert gas shield using either a tungsten electrode (*TIG welding*) or a consumable metal electrode (*GMA, MAGS* or *MIG* welding).

Inertia welding. See *Friction welding*.

MAGS welding. See *Inert gas shielded arc welding*.

MIG welding. See *Inert gas shielded arc welding*.

Plasma arc welding. A variant on *TIG welding* in which a plasma of extremely hot ionised gas, gives a stable concentrated arc, permitting deep penetration with a narrow heat affected zone.

Pressure welding. Welding processes in which surfaces are brought together under pressure and heated. There is limited melting at the interface.

Projection welding. A variation on *Spot welding* in which several welds are made at the same time.

Seam welding. A variation on *Spot welding* in which a series of overlapping spots gives a continuous seam weld.

Silver soldering. Hard soldering using a Cu–Zn–Ag alloy as the filler metal.

Soft soldering. A joining method for metals using a low melting point filler material.

Solvent welding. A joining process for some thermoplastics in which the mating surfaces are partially dissolved by a solvent, thus allowing them to fuse into one another.

Spin welding. See *Friction welding*.

Spot welding. An electric resistance welding process in which the metals to be joined are under pressure between two electrodes. When a current is passed for a controlled period, there is some melting at the interface and a weld made.

Submerged arc welding. An automated arc welding process in which the arc is struck below the surface of a flux layer.

Thermit welding. A fusion welding process in which the chemical reaction between aluminium and iron oxide is the heat source.

TIG welding. See *Inert gas shielded arc welding*.

Ultrasonic welding. A process suitable for joining some metals and thermoplastics in which the mating surfaces are vibrated at frequencies in the range 10–50 kHz.

Chapter 25

Alternating stress cycle. A stress cycle in which the mean stress is zero.

Beach marks. Also known as *mussel shell markings*, these are marks on the smooth section of a fatigue fracture which indicate stages in the growth of the fatigue crack.

Creep. The continued slow straining of a material subject to a constant stress.

Endurance strength. See *Fatigue limit*.

Fatigue failure. The failure of a material after the application of many stress cycles.

Fatigue limit. Also termed *Endurance strength*, this is the value of maximum stress in an alternating stress cycle below which fatigue failure should not occur. Most steels show a fatigue limit, the value of which is approximately one-half of the tensile strength.

Fluctuating stress cycle. A stress cycle in which neither the minimum stress nor the mean stress is zero.

Gerber equation. An empirical expression relating stress range, mean stress and fatigue limit.

Goodman equation. An empirical expression relating stress range, mean stress and fatigue limit.

High-cycle fatigue. Fatigue due to cyclic stresses, where the maximum stress is below the yield strength. Usually, failure requires some 10^5–10^6 stress cycles.

Low-cycle fatigue. Fatigue due to cyclic stresses where the maximum stress exceeds the yield strength. Generally, the number of cycles to failure is fewer than 1000.

Miner's law. An empirical law used to estimate fatigue life for a component subject to more than one different stress cycles.

Mussel shell marks. See *Beach marks*.

Peening. A treatment which induces residual compressive stresses in the surface layers of a material.

Primary creep. Also known as *Transient creep*. The first stage of creep during which the rate of strain, initially high, reduces to the steady-state value.

Relaxation. A slow reduction of stress with time when a material is subject to constant strain.

Repeating stress cycle. A stress cycle in which the minimum stress is zero.

Secondary creep. The second stage of creep in which the creep rate is constant over a long period.

Soderberg equation. An empirical expression relating stress range, mean stress and fatigue limit.

Steady-state creep. See *Secondary creep*.

Tertiary creep. The final stage of creep in which the creep rate increases, leading rapidly to failure.

Transient creep. See *Primary creep*.

Chapter 26

Cathodic protection. The deliberate creation of a galvanic cell using a sacrificial anode, thus protecting a structure by making it cathodic.

Cavitation attack. Pitting caused on parts moving rapidly through a fluid by a combined corrosion/erosion action when air bubbles collapse producing a compression wave.

Composition cell. A galvanic cell in which two dissimilar metals, or two different constituents in an alloy, are in direct electrical contact and also joined through an electrolyte.

Concentration cell. A galvanic cell formed when a metal is in contact with a continuous electrolyte in which there are varying concentration levels.

Corrosion. The loss of material through electrochemical reactions.

Corrosion fatigue. A corrosive environment accelerates the nucleation and growth of cracks in a part subject to alternating or fluctuating load cycles, reducing fatigue life.

Degradation (of polymers). Any change in a polymeric material which causes some breakdown of its chemical structure.

Electrode potential. See *Standard Electrode Potential*.

Environmental stress cracking. Cracking which occurs in some thermoplastic materials when they are stressed and in contact with certain substances.

Fretting. A form of combined erosion/corrosion which can occur when there is a small relative motion between mating parts.

Galvanic cell. A corrosion cell in which the anode material is dissolved. See also *Composition cell*, *Concentration cell* and *Stress cell*.

Galvanic series. Tables of metals and alloys arranged in order of their tendency to corrode in particular environments.

Impressed voltage. A *Cathodic protection* method in which a direct current is introduced into the structure to be protected, making it cathodic.

Inhibitors. Agents used in closed cooling systems that either control pH values or help to form protective coatings, thus minimising corrosion.

Oxidation. Generally, any reaction involving the loss of electrons from a metal but usually referring to the direct reaction between a metal and oxygen to form an oxide. Oxide scale formation tends to be rapid for many metals at high temperatures.

Passivation. The reduction in corrosion rate that occurs when a product of a corrosion reaction forms an impervious layer over the anode surface.

Pilling–Bedworth ratio. The ratio of the molecular mass number divided by density of the oxide to that for the metal, and it defines the characteristics of the oxide layer on a metal surface.

Pitting corrosion. A highly localised form of corrosion causing the formation of pits.

Polarisation. A reduction in the voltage of a galvanic cell which reduces the rate of corrosion.

Rust. The hydrated oxide of iron, $Fe(OH)_3$.

Standard Electrode Potential. The potential difference between a metal in equilibrium with its ions and a standard hydrogen electrode. Metals may be positive or negative with respect to hydrogen.

Stress cell. A galvanic cell created when a material with an uneven stress distribution is in contact with an electrolyte. The more highly stressed portion will be anodic with respect to other parts.

Stress corrosion. Crack nucleation and growth as a result of the effects of stress combined with a corrosive environment.

Weathering. The deterioration of a material in an outdoor environment, including the effects of air, water, temperature, sunlight and ultra-violet radiation.

Chapter 27

Brinell test. An indentation test for the hardness of metals using a steel ball indentor.

Charpy test. A notch impact test used to determine the toughness of a material.

Extensometer. A device for measuring test-piece extension during a tensile test.

Izod test. A notch impact test used to determine the toughness of a material.

Knoop test. A microhardness test for metals using a diamond indentor.

Meyer index. A work hardening index for metals derived from Brinell hardness tests.

Modulus of elasticity. The ratio of stress to strain for a material in elastic deformation.

Modulus of rupture. This is the flexural strength of a brittle material, as determined in a three-point bend test.

Mohs' scale of hardness. A series of ten standard minerals ranging from soft talc to diamond against which the relative hardness of ceramics and glasses is assessed.

Nominal strain. As measured in a tensile test it is the change in length divided by the original length. (See *True strain.*)

Nominal stress. As measured in a tensile test this is the applied force divided by the original cross-sectional area and does not take into account changes in cross-section during the test. (See *True stress.*)

Offset yield stress. An alternative term for *Proof stress.*

Percentage elongation. The extension in length, expressed as a percentage of the original gauge length, derived from a tensile test to destruction. The gauge length must be specified. It is a measure of ductility.

Percentage reduction of area. The difference between the cross-sectional area of a cylindrical test-piece at the point of fracture and the original cross-sectional area, expressed as a percentage. It is a measure of ductility.

Proof stress. The stress required to cause a specified amount of plastic strain, usually 0.1 or 0.2 per cent as determined in a tensile test. It is the applied load required to cause the specified strain divided by the original cross-sectional area.

Rockwell test. An indentation test for the hardness of metals and thermoplastics. There are several scales, using different indentors and loads, to cover a wide range of materials.

Secant modulus. This is the ratio of stress to strain at the level of stress required to cause a strain of 0.002 (0.2 per cent). This value is quoted for those thermoplastic materials for which it is extremely difficult to determine Young's modulus.

Shore test. A hardness test method in which the rebound height of a falling diamond tipped weight is measured. Used for very hard metals and elastomers.

Tensile strength. The maximum load applied divided by the original cross-sectional area derived from a tensile test to destruction.

Torsionmeter. A device for measuring the angle of twist of a test-piece during a torsion test.

True strain. As measured in a tensile test at any instant, this is the change in length divided by the instantaneous length and is equal to $\ln(L/L_0)$ where L is the instantaneous length and L_0 is the original length.

True stress. This is the applied force divided by the instantaneous cross-sectional area of the test-piece in a tensile test.

Ultimate tensile strength. See *Tensile strength*.

Vickers diamond test. An indentation test for the hardness of metals using a diamond indentor.

Weibull modulus. A measure of the probability of failure of a material. The higher the value of this index, the greater will be the probability of survival of the material when subjected to a certain level of stress.

Wöhler test. A rotating cantilever fatigue test.

Yield stress. The stress required to cause yielding in steels which show a discontinuity in the tensile force–extension curve. It is the applied load to cause yielding divided by the original cross-sectional area. *Proof stress* is sometimes referred to as yield stress.

Young's modulus. See *Modulus of elasticity*.

Chapter 28

Acoustic emission inspection. The analysis of high-frequency sound energy emitted from materials as a result of structural change or the growth of cracks.

Eddy current inspection. A defect detection system using the interactive effects between a high-frequency electrical field and the metal under examination.

Endoscope. An optical inspection probe.

Fluoroscopy. A variation on *Radiography* in which the radiation, after passing through the work-piece, impinges on a fluorescent screen generating an image.

γ-radiation (Gamma radiation). High-frequency electromagnetic radiation emitted during the radioactive decay of a radioisotope.

Half-life period. The time taken for the intensity of emitted radiation from a radioisotope to fall to one-half of its original value.

Image quality indicator. A device used for the determination of the sensitivity of radiography. (Also termed a *Penetrameter*.)

Liquid penetrant inspection. System in which a low viscosity fluid is absorbed into surface defects. Used to locate surface breaking defects.

Magnetic particle inspection. A defect detection system which involves magnetising the work-piece. The presence of leakage fields indicates surface defects.

Penetrameter. See *Image quality indicator*.

Radiography. The exposing of a material to X- or γ-radiation and recording the radiation passing through the work-piece on film. Variations in image density indicate the presence of defects.

Ultrasonic inspection. The transmission of high-frequency sound energy through a material and measuring the reflected echoes received from internal discontinuities.

Ultrasonic probe. A transducer used to generate a high-frequency compression wave, and to convert a received high-frequency wave into an electrical signal.

X-radiation. High-frequency electromagnetic radiation generated in an X-ray tube.

Chapter 29

Etchant. A chemical reagent used to reveal the structure for macro- or micro-examination.

Etching. A controlled chemical attack, using an etchant, to reveal the internal crystal structure of a metallic material.

Macro-examination. The examination of a fracture surface or a prepared surface at magnifications of up to × 50. The examination of prepared surfaces can reveal features such as fibre structure and segregation.

Micro-examination. The examination of prepared surfaces at magnifications of × 50 and greater. Because of their opacity metals need to be viewed under reflected light but transmitted light is used for other materials.

Scanning electron microscopy. A technique for the examination of surface topography at magnifications of up to × 40,000.

Transmission electron microscopy. A technique where the focused electron beam passes through the material under examination and capable of magnifications of between × 2000 and × 100,000. For the examination of metals the object is either a very thin foil or replica of an etched surface.

Answers to numerical questions

Chapter 2

2.1 107.87

2.2 Cu-63 – 34 neutrons, 73%; Cu-65 – 36 neutrons, 37%

2.3 (a) 0.658 kg O_2; (b) 2.477×10^{25}

2.4 5.83×10^{-15}

2.5 (a) $x = 5$, $y = 6$, $z = 2$

(b) Valency of fluorine = 1, valency of magnesium = 2, MgF_2

(c) 62.31 g

2.6 3

2.7 (a) $Z = 13$, $M = 27$; (b) $1s^2$, $2s^2$, $2p^6$, $3s^2$, $3p^1$; (c) 3

2.8 For iron with valency of 2 – $1s^2$, $2s^2$, $2p^6$, $3s^2$, $3p^6$, $4s^2$, $3d^6$

For iron with valency of 3 – $1s^2$, $2s^2$, $2p^6$, $3s^2$, $3p^6$, $4s^2$, $3d^5$, $4p^1$

2.9 (a) $Z = 82$, $M = 208$; (b) $Z = 82$, $M = 214$

2.10 (a) 79.2%; (b) 62.8%; (c) 39.4%

2.11 Approx. 1570 years.

2.12

Nucleus	Symbol	Emission
Uranium	$^{235}_{92}U$	α
Thorium	$^{231}_{90}Th$	β
Protoactinium	$^{231}_{91}Pa$	α
Actinium	$^{227}_{89}Ac$	β
Thorium	$^{227}_{90}Th$	α
Radium	$^{223}_{88}Ra$	α

Nucleus	Symbol	Emission
Radon	$^{219}_{86}\text{Rn}$	α
Polonium	$^{215}_{84}\text{Po}$	β
Astatine	$^{215}_{85}\text{At}$	α
Bismuth	$^{211}_{83}\text{Bi}$	β
Polonium	$^{211}_{84}\text{Po}$	α
Lead	$^{207}_{82}\text{Pb}$	Stable

2.13 (a) ionic; (b) covalent; (c) covalent; (d) metallic

2.14 (a) metallic; (b) covalent; (c) covalent pairs; (d) van der Waal's; (e) metallic; (f) ionic; (g) covalent

2.15 A – van der Waal's, B – metallic, C – ionic

Chapter 4

4.3 42,078

4.4 35

4.5 DP = 3953; av. molecular mass = 340,000

4.6 Styrene/acrylonitrile = 1.455/1

4.7 (a) 3.2% of sites used; (b) 188 kg sulphur/100 kg polyisoprene for full vulcanisation

4.8 $T_m = 324.7\,\text{K}$ at DP = 50; $T_m = 405.2\,\text{K}$ at DP = 500; $T_m = 410.8$ at DP = 1000

Chapter 5

5.1 Plane A $(00\bar{1})$; Plane B (220); Plane C $(\bar{1}1\bar{1})$; Plane D (0001); Plane E $(10\bar{1}0)$

5.2 Direction A $[10\bar{2}]$; Direction B [121]; Direction C [110]; Direction D $[11\bar{2}0]$; Direction E [0001]

5.3 (a)

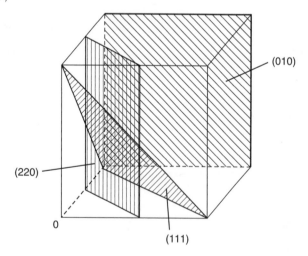

(b) $d_{(220)} = a/\sqrt{8}$; $d_{(010)} = a$; $d_{(111)} = a/\sqrt{3}$

5.5 (a) $9.04 \times 10^3\,\text{kg/m}^3$. (b) This is higher than the actual density. No allowance has been made for the presence of vacancies and other lattice defects

5.6 0.3653 nm

5.7 Reduction in volume of 8.2%

5.8 (a) 0.556 nm; (b) PF = 0.67

5.9 $19.43 \times 10^3\,kg/m^3$

5.11 0.059 nm

5.12 (a) f.c.c.; (b) 0.405 nm; (c) 0.286 nm

5.13 Reflections from the first seven planes, that is up to and including (321)

5.14 0.1277 nm

5.15 (a) b.c.c.; (b) (110), (200), (211), (220), (310), (222) and (321); (c) 0.295 nm; (d) 0.255 nm

5.16 (a) b.c.c.; (b) 0.3177 nm

5.17 Sample A – $a = 0.367$ nm; Sample B – $a = 0.361$ nm. One sample must be impure with solute atoms which strain the lattice, giving a different value of a

5.18 (a) b.c.c.; (b) 0.3159 nm; (c) $19.53 \times 10^3\,kg/m^3$

5.19 (b) 0.125 nm; (c) 5% of added alloy element would increase the lattice parameter by about 0.25%. This difference is too small to be detected using X-ray diffraction. If the addition is greater than 5% a second phase, with its distinctive diffraction pattern, would be present

5.20 Assume a unit cell contains x carbon atoms and $2x$ hydrogen atoms. From data calculate the volume of a unit cell. From the density find the mass of a unit cell. From this x can be evaluated. $x = 3.88$ (approximately 4). The unit cell contains 4 carbon and 8 hydrogen atoms

Chapter 6

6.2

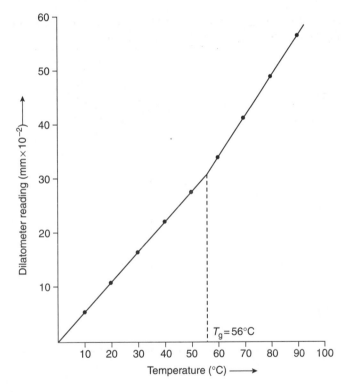

From graph $T_g = 56°C$

6.4 13.38% boric oxide

6.5 Oxygen/silicon ratio = 2.21. Suitable as a glass mix

Chapter 7

7.1 (a) 76.4 MPa; (b) 3.67×10^{-4}; (c) 208.4 GPa

7.2 31.8 MPa (use $2 \times$ c.s.a. as it is double shear)

7.3 (a) 29.85 GPa; (b) 0.316

7.4 (a) 1663 N; (b) 70.6 MPa; 2.5 MPa

7.5 σ_{max} = 846 MPa; E_L = 33 GPa; E_T = 4.9 GPa

7.6 Matrix stress = 0.56 MPa (compressive). Fibre stress = 8.4 MPa (tensile)

7.7 Carbon fibre composite (both meet the minimum strength requirement but the glass fibre composite does not meet the minimum E requirement)

Chapter 8

8.1 (a) 32.14 MPa; (b) 49.24 MPa; (c)43.3 MPa; (d) 17.1 MPa

8.2 50.3 MPa

8.3 (a) 6.24 GPa; (b) 104.7 MPa; (c) theoretical strength assumes a perfect crystal with no dislocations or other defects

8.5 0.297 nm

8.6 12,700 km

8.8 728 vacancies per 10^6 atoms

8.9 16% of atoms are of nickel

8.10 Volume strains – (a) −0.38%; (b) 2.74%; (c) 0.62%

8.11 Radii of interstitial sites – 0.0515 nm in f.c.c. and 0.036 nm in b.c.c. Interstitial solid solution will be much more restricted in b.c.c.

8.13 Plot graph of σ_y against $d^{-0.5}$. From graph K = 21.05 MPa $mm^{0.5}$ and σ_0 = 175 MPa. When d = 0.03 mm, $d^{-0.5}$ = 5.77. From graph σ_y = 297 MPa

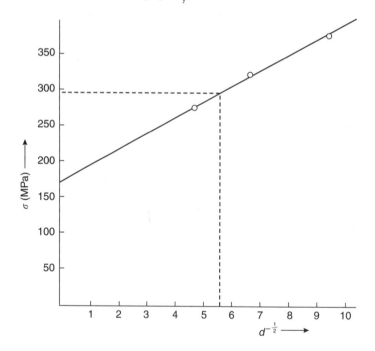

8.14 (a) 46.2 MPa; (b) 31.7 MPa

8.15 (a) 256 grains/mm^2; (b) σ_0 = 55 MPa – estimated strength of a single crystal

Chapter 9

9.2 (a) (i) 0.028; (a) (ii) 0.0083; (b) (i) 0.055; (b) (ii) 0.0099

9.3 (a) 303.3 mm; (b) 304.0 mm; (c) 300.11 mm

9.4 1.25 MPa

9.5 5.07×10^{16} Pa s

Chapter 10

10.2 Approximately 15σ or 2422 MPa

10.3 (a) 25.3 GPa; (b) 9.03 MPa

10.4 3 mm

10.5 Using tan function – 40.06 MPa m$^{\frac{1}{2}}$; using sec function – 40.8 MPa m$^{\frac{1}{2}}$

10.6 160 MPa; 98 MPa

10.7 (a) 367 MPa; (b) 301 MPa

10.8 10.1 mm

10.9 0.28 mm

10.10 σ_f for steel = 981 MPa or $1.13 \times$ design stress; σ_f for aluminium = 312 MPa or $1.06 \times$ design stress; σ_f for titanium = 767 MPa or $1.60 \times$ design stress

10.11 Thickness = 5.6 mm; diameter = 224 m

10.12 $-5°C$.

10.13 $-12°C$.

Chapter 11

11.1 (a) See diagram

(b) Liquidus 830°C
    Solidus 760°C

(c) (i) $L + \gamma$, $L/\gamma = 1/6.4$
    (ii) γ, 100% γ
    (iii) $\gamma + \delta$, $\gamma/\delta = 13.5/1$

(d) 79% of structure is eutectic

11.2 (a)

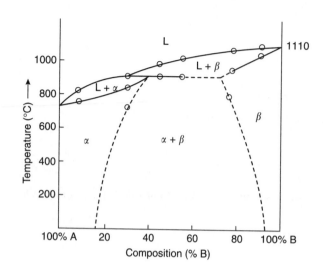

(b) (i) $\alpha + \beta$

(ii) supersaturated β solution

11.3 (a)

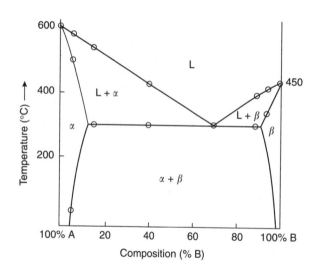

(b) (i) liquid

(ii) liquid $+ \alpha$

(iii) $\alpha + \beta$

11.4 (a) Diagram showing partial solid solubility which may show either a eutectic or peritectic

(b) Diagram showing partial solid solubility with a eutectic

(c) Diagram showing partial solid solubility and at least one intermetallic compound

(d) Diagram showing complete solid solubility

11.5

11.6   (a)

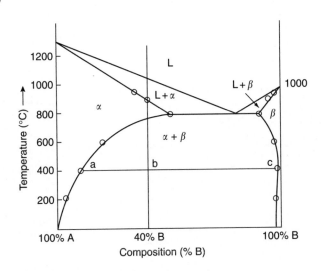

(b) Liquid solidifies into α and at a lower temperature β precipitates from α. 26% β

(c) (i) Solution heat treatment at 800–900°C followed by rapid quenching

(ii) Solution treatment as in (i) and reheated below solvus temperature

11.7 (a) See diagram

(b) (i) $\alpha + \beta$ in ratio 1/1
(ii) $\alpha + \gamma_2$ in ratio 1/1

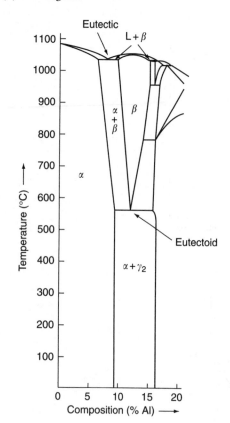

Chapter 12

12.1 Approximately 4
12.2 263.7 MJ/kmol
12.3 15.7 days; 49.86 MJ/kmol
12.4 Approx. 260 days
12.5 52.37 MJ/kmol
12.6 430°C
12.7 (a) 7.79×10^{-19} J/atom; (b) 2.0×10^{-2} mA/mm^2
12.8 22.14 hours, 31.78 MJ/kmol
12.9 (a) 150.6 MJ/kmol; (b) 3.24×10^{-11} m^2 s^{-1}
12.10 231 min at 180°C, 22 min at 220°C
12.11 (a) The Arrhenius law applies; $Q = 175.5$ MJ/kmol, $A = 2.44 \times 10^{-4}$
   (b) D at 500°C $= 3.37 \times 10^{-16}$ m^2 s^{-1}
12.12 7.5×10^{-9} s^{-1}
12.13 (a) 0.335 mm; (b) 1.64 mm
12.14 0.33 mm
12.15 0.685% C
12.16 0.144 mm
12.17 170.5 min (2 hours 50.5 min)
12.18 Carburising time must be increased by 79% at 950°C (giving a cost increase of 43%)

12.19 6.16 hours
12.20 −21°C

Chapter 13

13.1 62.4 Ω
13.2 2.12×10^{-7} Ωm
13.3 5.7×10^{-8} Ωm at −40°C, 1.05×10^{-7} Ωm at 120°C
13.7 $B_{max} = 4300$ T, $B_r = 3900$ T, $H_c = 0.96$ A/m
13.8

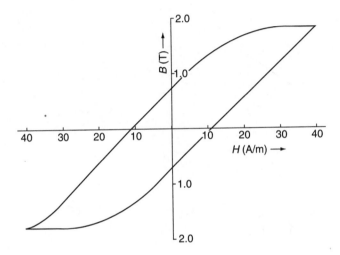

Chapter 14

14.1 3.736×10^{-19} J (2.33 eV)
14.2 Ultra-violet at 4.35×10^{15} Hz
14.3 1.209×10^{19} Hz
14.4 10.5 mm
14.5 3.44°
14.7 1.6%, 0.5588 m diameter
14.8 2.69 GJ/day

Chapter 15

15.4 (a) (i) Liquid + α in the ratio 1/1, (ii) 70% α/30% β'

Chapter 16

16.1 Steel 1 – 0.35% C, Steel 2 – 1.16% C
16.2 (a) $A_1 = 730$°C, $A_3 = 850$°C; (b) 0.3% C
16.3 (a) about 25°C/s; (b) about 2–3°C/min; (c) about 4–5°C/s; (d) about 10 minutes
16.14 1 – 3.93% C, 2 – 4.13% C, 3 – 4.46% C, 4 – 4.68% C, 5 – 3.53% C

Chapter 18

18.4 Apparent porosity 15.5%, total porosity 30%

18.5 Open porosity 35.3%, closed porosity 5.4%

18.6 2.78%

18.7 236°C (509 K)

18.9 20.26 MPa

18.10 160 days

Chapter 19

19.1 (a) 0.907; (b) see diagram

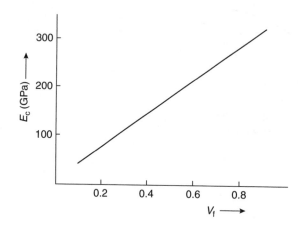

19.2 (a) 1218 MPa; (b) 1812 MPa

19.3 (a) 0.953; (b) 29.66 MPa

19.6 Both the carbon/epoxy and Kevlar/epoxy composites meet the minimum strength and specific modulus requirements

Chapter 20

20.4 (a) £15.52; (b) £11.66; (c) 7237

Chapter 22

22.5 2.83%

22.6 51.47 mm diameter × 4.12 mm thickness

Chapter 23

23.4 107.8 m/min

23.5 High speed steel – tool life reduced by factor of 18, cemented carbide – life reduced by factor of 5, coated carbide – life reduced by factor of about 4.0

Chapter 25

25.2 200 MPa

25.3 (a) From −55 MPa to +505 MPa; (b) from +17.5 MPa to +432.5 MPa; (c) from +46 MPa to +404 MPa

25.4 (a) 2520 MPa; (b) 797 MPa; (c) 252 MPa; (d) 80 MPa; (e) 25 MPa; (f) 8 MPa

25.5 5.3×10^4 cycles

25.6 2.08×10^5 cycles

25.7 35.6 MPa

25.9 492,960 s (137 hours)

25.10 50.6 MPa

25.11 (a) 5.5; (b) 273 MPa

Chapter 26

26.1 Parabolic, 39 g

26.2 (a) 1.59×10^{-4} cm, 1.95×10^{-3} cm, 5.03×10^{-3} cm; (b) 6.3×10^{-3} cm, 7.7×10^{-2} cm, 0.199 cm

26.3 P–B ratio = 1.07, oxide will be protective

Chapter 27

27.1 $F/D^2 = 30$ for tests 1, 3, 4 and 6. H_B values are 160.6, 161.3, 162.8 and 159.1 respectively
In test 2 ($F/D^2 = 15$) H_B is 151.9 and in test 5 ($F/D^2 = 60$) $H_B = 172.2$

27.2 Copper – $H_B = 41.1$, bronze – $H_B = 87.2$

27.3 H_D values – 200.3, 240.4, 225.2 – Av. = 222, H_B values – 227.6, 228.2, 226.4 – Av. = 227.4

27.4 $a = 10$, $n = 3.2$

27.5 (c) (i) 10.7 MPa; (ii) 2.78 GPa; (iii) 1.33 GPa; (iv) 3%

27.6 E (0.2% secant modulus) = 1.28 GPa, tensile strength = 29.5 MPa, % El. = 94%, probably PP or HDPE

27.7 (a) 1330 MPa; (b) 84.8 GPa; (c) 31.1%; (d) 1394 MPa

27.8 (a) $k = 733$ MPa, $n = 0.2$; (b) 8.84 mm; (c) 34.2 kN; (d) 435.4 MPa

27.9 (a) 71.9 GPa; (b) 30.85 MPa; (c) 73.34 MPa; (d) 44%, pure aluminium

27.10 (a) 29.17 MPa; (b) 3.02 MPa

27.11 (c) 14,384 kJ/m^2

27.14 8.89×10^{-7} hr^{-1} (2.47×10^{-10} s^{-1})

Chapter 28

28.5 4.13 mm, 2.06 MHz

28.6 1.8%

28.7 209.07 GPa

Index